Fourier Analysis

PURE AND APPLIED MATHEMATICS

A Wiley-Interscience Series of Texts, Monographs, and Tracts

Founded by RICHARD COURANT
Editors Emeriti: MYRON B. ALLEN III, DAVID A. COX, PETER HILTON, HARRY HOCHSTADT, PETER LAX, JOHN TOLAND

A complete list of the titles in this series appears at the end of this volume.

Fourier Analysis

Eric Stade

A JOHN WILEY & SONS, INC., PUBLICATION

Copyright © 2005 by John Wiley & Sons, Inc. All rights reserved.

Published by John Wiley & Sons, Inc., Hoboken, New Jersey.
Published simultaneously in Canada.

No part of this publication may be reproduced, stored in a retrieval system or transmitted in any form or by any means, electronic, mechanical, photocopying, recording, scanning or otherwise, except as permitted under Section 107 or 108 of the 1976 United States Copyright Act, without either the prior written permission of the Publisher, or authorization through payment of the appropriate per-copy fee to the Copyright Clearance Center, Inc., 222 Rosewood Drive, Danvers, MA 01923, (978) 750-8400, fax (978) 750-4744, or on the web at www.copyright.com. Requests to the Publisher for permission should be addressed to the Permissions Department, John Wiley & Sons, Inc., 111 River Street, Hoboken, NJ 07030, (201) 748-6011, fax (201) 748-6008, e-mail: permreq@wiley.com.

Limit of Liability/Disclaimer of Warranty: While the publisher and author have used their best efforts in preparing this book, they make no representation or warranties with respect to the accuracy or completeness of the contents of this book and specifically disclaim any implied warranties of merchantability or fitness for a particular purpose. No warranty may be created or extended by sales representatives or written sales materials. The advice and strategies contained herein may not be suitable for your situation. You should consult with a professional where appropriate. Neither the publisher nor author shall be liable for any loss of profit or any other commercial damages, including but not limited to special, incidental, consequential, or other damages.

For general information on our other products and services please contact our Customer Care Department within the U.S. at 877-762-2974, outside the U.S. at 317-572-3993 or fax 317-572-4002.

Wiley also publishes its books in a variety of electronic formats. Some content that appears in print, however, may not be available in electronic format.

Library of Congress Cataloging-in-Publication is available.

ISBN 0-471-66984-9

Printed in the United States of America.

10 9 8 7 6 5 4 3 2 1

Contents

	Preface	xv
	Introduction	xix
1	*Fourier Coefficients and Fourier Series*	1
	1.1 *Periodic Functions: Beginning Bits*	1
	1.2 *Fourier Coefficients of 2π-Periodic Functions*	8
	1.3 *More on $P = 2\pi$*	14
	1.4 *Pointwise Convergence of Fourier Series: A Theorem*	25
	1.5 *An Application: Evaluation of Infinite Series*	31
	1.6 *Gibbs' Phenomenon*	36
	1.7 *Uniform Convergence of Fourier Series: A Theorem*	41
	1.8 *Derivatives, Antiderivatives, and Fourier Series*	47
	1.9 *Functions of Other Periods $P > 0$*	55
	1.10 *Amplitude, Phase, and Spectra*	59
	1.11 *Functions on Bounded Intervals: Standard Fourier Series*	65
	1.12 *Other Fourier Series for Functions on Bounded Intervals*	70
2	*Fourier Series and Boundary Value Problems*	79

	2.1	Steady State Temperatures and "the Fourier Method"	79
	2.2	Linear Operators, Homogeneous Equations, and Superposition	88
	2.3	Heat Flow in a Bar I: Neumann and Mixed Boundary Conditions	94
	2.4	Heat Flow in a Bar II: Other Boundary Conditions	100
	2.5	Cylindrical and Polar Coordinates	105
	2.6	Spherical Coordinates	111
	2.7	The Wave Equation I	115
	2.8	The Wave Equation II: Existence and Uniqueness of Solutions	123
	2.9	The Wave Equation III: Fourier Versus d'Alembert	127
	2.10	The Wave Equation IV: Temporally Constant Inhomogeneity	132
	2.11	The Wave Equation V: Temporally Varying Inhomogeneity	136
	2.12	The Wave Equation VI: Drumming Up Some Interest	145
	2.13	Triple Fourier Series	150
3		L^2 Spaces: Optimal Contexts for Fourier Series	155
	3.1	The Mean Square Norm and the Inner Product on $C(\mathbb{T})$	155
	3.2	The Vector Space $L^2(\mathbb{T})$	163
	3.3	More on $L^2(\mathbb{T})$; the Vector Space $L^1(\mathbb{T})$	175
	3.4	Norm Convergence of Fourier Series: A Theorem	179
	3.5	More on Integration	182
	3.6	Orthogonality, Orthonormality, and Fourier Series	187
	3.7	More on the Inner Product	195
	3.8	Orthonormal Bases for Product Domains	202
	3.9	An Application: The Isoperimetric Problem	207
	3.10	What Is $L^2(\mathbb{T})$?	209
4		Sturm-Liouville Problems	217
	4.1	Definitions and Basic Properties	217
	4.2	Some Boundary Value Problems	227
	4.3	Bessel Functions I: Bessel's Equation of Order n	232
	4.4	Bessel Functions II: Fourier-Bessel Series	237
	4.5	Bessel Functions III: Boundary Value Problems	242
	4.6	Orthogonal Polynomials	247
	4.7	More on Legendre Polynomials	252

5 Convolution and the Delta Function: A Splat and a Spike 261
   5.1 Convolution: What Is It? 261
   5.2 Convolution: When Is It Compactly Supported? 265
   5.3 Convolution: When Is It Bounded and Continuous? 269
   5.4 Convolution: When Is It Differentiable? 273
   5.5 Convolution: An Example 278
   5.6 Convolution: When Is It In $L^1(\mathbb{T})$? In $L^2(\mathbb{T})$? 283
   5.7 Approximate Identities and the Dirac Delta "Function" 287

6 Fourier Transforms and Fourier Integrals 297
   6.1 The Fourier Transform on $L^1(\mathbb{R})$: Basics 297
   6.2 More on the Fourier Transform on $L^1(\mathbb{R})$ 303
   6.3 Low-Impact Fourier Transforms (Integration by Differentiation) 311
   6.4 Fourier Inversion on $FL^1(\mathbb{R})$ 316
   6.5 The Fourier Transform and Fourier Inversion on $L^2(\mathbb{R})$ 321
   6.6 Fourier Inversion of Piecewise Smooth, Integrable Functions 327
   6.7 Fourier Cosine and Sine Transforms 333
   6.8 Multivariable Fourier Transforms and Inversion 336
   6.9 Tempered Distributions: A Home for the Delta Spike 343

7 Special Topics and Applications 353
   7.1 Hermite Functions 353
   7.2 Boundary Value Problems 358
   7.3 Multidimensional Fourier Transforms and Wave Equations 365
   7.4 Bandlimited Functions and the Shannon Sampling Theorem 372
   7.5 The Discrete Fourier Transform 379
   7.6 The Fast Fourier Transform, or FFT, algorithm 385
   7.7 Filtering 390
   7.8 Linear Systems; Deconvolution 397
   7.9 Fraunhofer Diffraction and Fourier Optics 403
   7.10 FT-NMR Spectroscopy 412

8 Local Frequency Analysis and Wavelets 421
   8.1 Short-Time, or Windowed, Fourier Transforms 421

8.2	*Finite Windows and the Heisenberg Uncertainty Principle*	*430*
8.3	*Wavelets and Multiresolution Analyses: Basics*	*436*
8.4	*Multiresolution Analyses and Wavelets: A Builder's Guide*	*443*
8.5	*Proof of Theorem* 8.4.1	*457*

Appendix *469*

References *479*

Index *483*

List of Figures

I.1	The functions $\cos 2\pi sx$ (*solid*) *and* $\sin 2\pi sx$ (*dashed*)	xix
I.2	The functions $\cos 2\pi st$ *and* $\sin 2\pi st$	xxii
I.3	The complex exponentials $e^{2\pi isx}$ (*solid*) *and* $e^{-2\pi isx}$ (*dashed*)	xxiv
1.1	The "ah," "eh," and "oh" vowel sounds, respectively, as pronounced by the author. The horizontal axis is in milliseconds	2
1.2	A function $f \in \text{PC}([-4, 4])$	9
1.3	Neither $g(x) = 10\sin(x^{-1})$ (*solid*) *nor* $h(x) = x^{-1/2}$ (*dashed*) *is piecewise continuous on* $(0, b]$ *for* $b > 0$ (*neither* $g(0^+)$ *nor* $h(0^+)$ *exists*)	10
1.4	Top: a 2π-periodic function f. Bottom: f, upon wrapping the x axis around the torus $\mathbb{T}$	10
1.5	$f(x) = e^x$ $(-\pi < x \leq \pi)$, *periodized*	12
1.6	The functions $g_{\pi/4}$ (*solid*) *and* $g_{5\pi/6}$ (*dashed*)	15
1.7	The function h	17
1.8	The function j	20

1.9 Top: x versus $\operatorname{Re} j(x)$. Bottom: x versus $\operatorname{Im} j(x)$ 21

1.10 The sawtooth function σ (solid) and some partial sums S_N^σ (dashed. Shorter dashes connote larger N)   37

1.11 A tube (dotted) of radius $\varepsilon = 0.2$ about the graph of σ (solid) and a partial sum S_N^σ (dashed)   39

1.12 A function $f \in \operatorname{PSC}(\mathbb{T})$ (solid), a narrow tube (dotted) about the graph of f, and the second partial sum (dashed) of the Fourier series for f 45

1.13 A continuous, nowhere differentiable, 2π-periodic function g 46

1.14 The function $k_{2\pi/3}$ 48

1.15 The function v 50

1.16 Top: x versus $\operatorname{Re} f(x)$. Bottom: x versus $\operatorname{Im} f(x)$ 57

1.17 A sinusoid of frequency s, amplitude R, and phase ϕ 61

1.18 Spectra of the author's "ah," "eh," and "oh" vowel sounds, respectively   62

1.19 A function $f \in \operatorname{PSA}(a, b)$ (solid) and its periodic extension f_{per} (dashed and solid combined)   66

1.20 The function f of Example 1.11.2 (solid) and its ℓ-periodic extension f_{per} (dashed and solid combined)   69

1.21 Top: a function $f \in \operatorname{PSA}(0, \ell)$ (solid) and its even 2ℓ-periodic extension f_{even} (dashed and solid combined). Bottom: the same function f (solid) and its odd 2ℓ-periodic extension f_{odd} (dashed and solid combined)   71

1.22 A function $f \in \operatorname{PSA}(0, \ell)$ (solid) and its reflection about $x = \ell$ (dashed). The dashed and solid curves together constitute f_{refl} 74

2.1 A semi-infinite plate   80

2.2 A bar of uniform, circular cross section   95

2.3 The curve $y = (s - 1/s)\sin s - 2\cos s$ 103

2.4 The cylindrical coordinate system   106

2.5 The spherical coordinate system   112

2.6	A string vibrating according to the wave equation (2.149). Solid portions have upward acceleration; dashed portions have downward acceleration	116
2.7	The initial position function f	119
2.8	Snapshots of the vibrating string of Example 2.7.2 (here, time evolves from top to bottom, left to right)	128
2.9	Snapshots of a string vibrating according to (2.149)–(2.151), with initial velocity zero (as before, the chronology is from top to bottom, left to right)	131
2.10	The graph of $T_{n_0}(t)$ for $s = n_0 c/(2\ell)$. (Here $n_0 = c = 1$ and $K = \ell = \pi$)	139
3.1	A few f_N's (shorter dashes connote larger N)	158
3.2	The g_N's converge in norm, but pointwise nowhere (top: g_1; middle: g_2; bottom: g_3)	159
3.3	f_N (solid) and f_M (dashed) for $M > N$	164
3.4	f_2, f_4, and f_8	183
3.5	Geometric interpretation of the formula $x = \sum_{n=1}^{2} \langle x, v_k \rangle v_k$ for $x \in \mathbb{R}^2$ and $\{v_1, v_2\}$ an orthonormal basis for $\mathbb{R}^2$. The vector x is dot-dashed; v_1 and v_2 are solid; the "projections" $\langle x, v_1 \rangle v_1$ and $\langle x, v_2 \rangle v_2$ of x in the directions of v_1 and v_2, respectively, are dashed	191
3.6	A curve Γ satisfying (a)(b)(c) above	208
4.1	The region S	229
4.2	The Bessel function J_5 (solid), and f_5 for f_n as in (4.85) (dashed)	236
5.1	The graph of $\chi_{[-4,0]} * \chi_{[-1,1]}$	267
5.2	The functions $h_{2/5}$, $h_{4/5}$, h_1, $h_{6/5}$, $h_{8/5}$, and $h_{13/5}$ (shorter dashes correspond to larger subscripts)	281
5.3	A discontinuous (top) and a continuous (bottom) compactly supported, piecewise smooth function $f \in L^2(\mathbb{R})$ (solid), and $f * g_{[\varepsilon]}$ (dashed) for g as in (5.107), and for various values of ε. Shorter dashes connote smaller ε	291
5.4	The approximate identity $\{g_{[\varepsilon]} : \varepsilon > 0\}$ generated by the function g of (5.107). Shorter dashes connote smaller ε	293

5.5	δ, in spike form	294
6.1	$F[\chi_{[-1/2,1/2]}(x)]$ *(solid) and* $F[e^{-2\pi|x|}]$ *(dashed)*	301
6.2	The function G	308
6.3	$f \in \mathrm{PS}(\mathbb{R}^+) \cap L^1(\mathbb{R}^+)$ *(solid) and its even extension* f_{even} *(dashed and solid combined)*	334
7.1	A Dirichlet problem on a semidisk	362
7.2	Top: $\widehat{f}$ is supported on $[-2\Omega, -\Omega] \cup [\Omega, 2\Omega]$. Bottom: What happens to the frequency content of f under 2Ω-sampling	374
7.3	$\|\widetilde{g}_n\|$, with $N = 100$ *(empty boxes), and* $\|c_n(g)\|$ *(smaller, filled boxes)*	381
7.4	Local frequency content of an excerpt from the author's composition "Col de Torrent." (Time evolves from top to bottom, left to right.) The horizontal axis is in kilohertz (kHz); the vertical scale is logarithmic	389
7.5	Top: A "true" signal g. Bottom: The "corrupted" signal $g + r$	390
7.6	The graph of $\|\widehat{g+r}\|$, for $g+r$ as in Figure 7.5	392
7.7	The graph of $\|(\widehat{g+r})\Phi_\Omega\|$, for $g+r$ as in Figure 7.5 and a choice of Ω motivated by Figure 7.6	392
7.8	The graph of $F\big[(\widehat{g+r})(s)\Phi_\Omega(s)\big]^-$	393
7.9	Ceci n'est pas une zebra	394
7.10	$\|\widehat{f}\|$, for f the characteristic function of Figure 7.9	394
7.11	$\|\widehat{f}\Phi_P\|$, for a suitable filter Φ_P	395
7.12	It's a horse, of course (the inverse Fourier transform of the function $\widehat{f}\Phi_P$ of Fig. 7.11)	395
7.13	The electric field E for planar light (snapshot at a particular time t)	404
7.14	Object and spectrum planes	406
7.15	Huygen's principle	408

LIST OF FIGURES   xiii

7.16   Top: The real part of a simple damped harmonic oscillator. The resonant frequency s_1 is just the reciprocal of the distance between peaks. Bottom: The real part of a compound damped harmonic oscillator. Who knows what its resonant frequencies are? 413

7.17 $|\hat{g}|$, for g as in (7.166) (and $\operatorname{Re} g$ as depicted in the lower portion of Figure 7.16) 414

7.18   Ethylbenzene 415

7.19   Free induction decay data from a sample of ethylbenzene 415

7.20   Absolute value of the (fast) Fourier transform of the ethylbenzene FID data 416

7.21   A closer look at the C-8 resonance 416

7.22   Possible combinations of spin orientations of the two hydrogen protons at C-7 417

7.23   A closer look at the C-7 resonance 417

7.24   Possible combinations of spin orientations of the three hydrogen protons at C-8 418

7.25   A closer look at the benzene ring resonance 418

8.1 $f\,\overline{r_{(\tau)}}$ (solid) and $(2b)^{-1/2}f$ (dashed and solid combined), for r as in (8.16) 425

8.2   Top: Rectangular (solid), Bartlett (dashed), Hann (dot-dashed), and Gabor (dotted) windows. Bottom: Respective Fourier transforms of these windows 429

8.3   A finite window w centered at $(0,0)$ (solid) and the resulting windowed complex exponential $e_{\sigma,\tau,w}$ (dashed) 432

8.4   The Haar function ψ^H 439

8.5 $\chi_{[0,1)}$ (solid), k_1, k_2, and k_3 (shorter dashes correspond to larger subscripts). Here the k_N's are as in (8.58) 440

8.6   The scaling function φ (top) and wavelet ψ (bottom) of Example 8.4.2 in the case $N=2$ 449

8.7   The scaling function φ (top) and wavelet ψ (bottom) of Example 8.4.3 (with $M=3$) 453

A.1   Real and imaginary parts, complex conjugate, modulus, and argument 472

Preface

This book is aimed primarily at upper level undergraduate and early graduate students in mathematics. I hope and believe, though, that it will also be of value to engineering students (particularly those studying electrical engineering, signal processing, and the like); to students in physics and other sciences; to scientists and technicians who use Fourier analysis "in real life"; and to anyone fostering an appreciation of the beauty and power of Fourier analysis, or a desire to acquire such an appreciation.

The prerequisites are few: A good grasp of calculus in one and several variables will suffice. A number of more advanced concepts are encountered in the course of the text—complex numbers, linear algebra, differential equations, and a good truckload or two of ideas from real analysis. But familiarity with these concepts is not required; in fact, I hope that the reader who has not encountered them previously will find herein a good introduction to them.

I've strived for relatively high degrees of mathematical rigor and completeness. But at the same time, I've tried to place the math in scientific and technological (and historical) context and to infuse due detail into discussions of specific applications. (The detail is, though, for the most part limited to that which relates to Fourier analysis. Real-world applications necessarily entail myriad other considerations; I've avoided reflection on these in order to stay on point, and because I understand them only marginally.)

Cover to cover, the book probably amounts to two semesters; however, a variety of different single-semester tracks may be extracted. For example: a robust course in Fourier series and boundary value problems could be constructed around Chapters 1–4, more or less. Here, "more or less" means the following. Sections 1.5, 1.10, and

3.9 could be skipped; they are not relevant to the solution of boundary value problems. Sections 1.6 and 3.10 are of tangential relevance; they might also be skipped or at most skimmed. Additionally, the proofs in Sections 1.4 and 1.7 and throughout Chapter 3 might also be omitted or treated cursorily, depending on one's focus. Omitting these proofs would likely leave time for some discussion of Fourier *integrals* and boundary value problems; the material for this may be found in Chapters 5 and 6 and Sections 7.1–7.3. (Only the first and last sections of Chapter 5 and the first, second, sixth, seventh, and eighth of Chapter 6 are truly essential here, as long as one is willing to glance back at other sections, upon the occasional reference to results therein.)

Alternatively, a course emphasizing aspects of Fourier analysis relevant to technological and scientific issues (other than the issues arising in the context of boundary value problems) could be based on Chapters 1, 3, and 5–8. (Any of the Sections 1.5, 1.10, 3.9, 6.3, 6.9, and 7.1–7.3 might be omitted.) Or one can teach a very "pure" Fourier theory course, omitting or minimizing discussion of the material in Chapters 2, 4, 7, and 8. Or one can, as I do, mix it up. In my own single-semester course, composed mostly of junior and senior math majors and beginning math graduate students, I focus on Chapter 1 and Sections 3.1–3.7, 5.1–5.4, 5.7, and 6.1–6.6 (from all of this I omit a good many proofs, but include a good many too) and throw in a smattering from the remainder. The smattering varies from year to year. I lean toward Sections 2.1, 2.4, 2.7, 2.9, 4.1, 4.2, and 7.4–7.10, and as much of Chapter 8 as there is time left for; usually, there's not much.

Aditionally, there are excellent arguments to be made for two-semester (undergraduate and graduate) real analysis/Fourier analysis course sequences. These disciplines grew up together; indeed, real analysis was invented (discovered?) expressly to create the proper framework for investigation and formalization of Fourier's assertions. So a course in Fourier analysis constitutes a logical sequel to one in real analysis—and the converse is also true! That is, questions arising out of Fourier analysis *demand*, and therefore *motivate*, many real analysis constructs whose *raisons d'etre* are, sometimes, not otherwise apparent. So, the two semesters of a real analysis/Fourier analysis sequence might reasonably be presented in either order, depending on how one feels about these things, philosophically speaking. I like to think that this text would serve as well at the tail end of a real analysis/Fourier analysis track as it would at the leading end of a Fourier analysis/real analysis track. Either way, Chapters 1, 3, 5, and 6 should be of particular interest. (Perhaps Chapter 8 too. Local frequency analysis and wavelets certainly admit a cornucopia of applications, but the theory, and especially the real analysis, behind them is also of exceptional beauty. Theorem 8.4.1, in particular, encompasses an impressive array of real analysis BIG IDEAS.)

This book could never have come to be without the assistance and support of a number of people. My colleagues Larry Baggett and Dick Holley at the University of Colorado have been endlessly patient with my persistent emails and have done a monumental job of unconfusing me with regard to a number of questions. Edward Burger of Williams College, a better friend even than he is a teacher, has been an unflagging source of constructive criticism and, even more importantly, encouragement. Martin Hairer of HairerSoft has kindly permitted me to use, in the text, output from his excellent "Amadeus II" sound editing software. Joseph Hornak of Magnetic

Resonance Laboratories has generously provided me with the FT-NMR data used in Section 7.10, along with some quite illuminating commentary and explanation regarding the related material.

The crew at John Wiley & Sons has been terrific throughout. I am particularly grateful to Steve Quigley, Laurie Rosatone, Susanne Steitz, and Lisa Van Horn, all of whom had the audacity to believe in this project.

A heartfelt thanks goes to ten years plus of Fourier analysis students, who have greatly inspired and continue to inspire me while also providing free proofreading. Among these students, Marc Lanskey, Tiffany Tasset, and Sonja Wieck have offered especially valuable suggestions regarding the manuscript itself.

Most of all, thanks to my beautiful, exceptional wife Beth and my wonderful, extraordinary boys Jack and Nick. Without you, I am nothing. This book is dedicated to you.

ERIC STADE

Boulder, Colorado

Introduction

Thus there is no function ... which cannot be expressed by a trigonometric series ... [or] definite integral.

—Joseph Fourier [20]

Fourier analysis is the art and science whereby any reasonable function may be realized as a superposition of sinusoids, each of these sinusoids possessing a distinct frequency.

We explain: First, by *superposition* we mean a summation or similar process of amalgamation. We'll elaborate shortly, but for the moment the notion of a sum will be sufficient to capture the general sense of what's going on. Next, by *sinusoid* we mean a linear combination of the two functions $\cos 2\pi s x$ and $\sin 2\pi s x$, where x denotes a real variable and s some nonnegative, real constant. And by *frequency* of such a sinusoid we mean precisely this constant s. See Figure I.1.

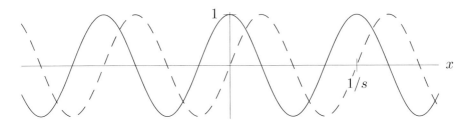

Fig. I.1 The functions $\cos 2\pi s x$ (solid) and $\sin 2\pi s x$ (dashed)

(We'll depict some other sinusoids a bit later: see Figs. I.3 and 1.17.) Note that, in Figure I.1, $s > 0$: A sinusoid of frequency zero is just a constant function, since $\cos 0 = 1$ and $\sin 0 = 0$.

All of this permits the following somewhat more explicit, though still rather rough, description of Fourier analysis: Fourier analysis *is*, roughly, the study of expressions of the form

$$f(x) = \sum_{s \in F_f} \left(A_s(f) \cos 2\pi s x + B_s(f) \sin 2\pi s x \right), \tag{I.1}$$

of conditions (on f) under which such an expression will exist; of the nature of the set F_f, and the coefficients $A_s(f)$ and $B_s(f)$, when it does; of generalizations and abstractions of such expressions; of their applications; and so on.

Fourier analysis is, in fact, often also called "frequency analysis," and the elements

$$A_s(f) \cos 2\pi s x + B_s(f) \sin 2\pi s x \tag{I.2}$$

of a superposition (I.1) the "frequency components" of the given function f. Indeed (I.2) is, for a given s, generally understood as "that part of f having frequency s." Also, the set F_f is frequently (pun intended) called the "frequency domain" of f, and (I.1) itself a "frequency decomposition" (also known as a "sinusoid decomposition") for f.

It turns out that Fourier analysis is of great import and utility, in both "abstract" and "concrete" settings: And why should this be? Why should Fourier analysis be central to so many issues in mathematics and the sciences? The short answer is this: It's because sinusoids do two particularly nice things. First, they *differentiate* in extraordinarily simple, fundamental ways; second, they *cycle* in extraordinarily simple, fundamental ways.

Regarding the first of these nice things, we note that the first derivative—and therefore also any higher derivative—of a sinusoid is another one, of the same frequency. In particular, the frequency component (I.2), let's call it $f_s(x)$, is readily seen to satisfy the basic differential equation

$$f_s''(x) = -(2\pi s)^2 f_s(x). \tag{I.3}$$

In fact, as is generally shown in an introductory differential equations course, *any* function f_s satisfying (I.3), for a given $s > 0$, is a sinusoid of frequency s. (If $s = 0$, then the solutions to (I.3) are the first-degree polynomials $f_0(x) = Ax + B$.)

As a consequence of this, a great variety of differential equations may be solved according to the following strategy. First, it's stipulated that the desired solution f have an expression of the form (I.1), for some set F_f dictated by the specifics of the problem at hand and for as yet unknown coefficients $A_s(f)$ and $B_s(f)$ (possibly dependent on other variables). Next, Fourier analysis and other considerations are applied to the explicit determination of these coefficients, and thus the exact nature of f is uncovered.

It was, in fact, precisely to address the solution of certain differential equations that Joseph Fourier (1768–1830) developed Fourier analysis—hence the name. He was specifically concerned, in his landmark 1807 manuscript *Theory of the Propagation of Heat in Solid Bodies* (and subsequent revisions and expansions thereof, the culmination of these being the 1822 book *The Analytical Theory of Heat* [20]), with the differential equations governing heat conduction. He was able not only to determine the nature of these equations, but also, via his sinusoidal analysis of functions and the *separation of variables* technique—which he also developed, and of which we'll make copious use in coming chapters—to solve them under various sets of assumptions. (Strictly speaking, both frequency decompositions and separation-of-variables arguments predated Fourier, in rudimentary forms. However, he was unquestionably the first to make systematic, fruitful use of either.)

Fourier was aware of the relevance of his ideas to other differential equations, besides those modeling heat flow. In particular he was able, using these ideas, to shed considerable light on the "wave equation," which had previously been studied by Leonard Euler and Daniel Bernoulli, among others. Even so, Fourier himself likely would not have predicted the ubiquity presently enjoyed by his ideas in the theory of differential equations.

Concerning the second particularly nice thing about sinusoids, we recall that they do, indeed, cycle. That is, they repeat themselves at regular intervals; that is, they're *periodic*. See Figure I.1 above. This periodicity is familiar, but its extraordinarily simple, fundamental nature should be emphasized. To this end we consider a point moving, with constant angular velocity, around the perimeter of a circle—such motion is, arguably, the simplest kind of periodic motion imaginable. Let's suppose, to be specific, that the motion is counterclockwise in a circle of radius 1, centered at the origin, and that the point sweeps out $2\pi s$ radians per unit of time t. Then, if this point has coordinates $(1, 0)$ at time zero, it will have coordinates $(\cos 2\pi st, \sin 2\pi st)$ at time t. See Figure I.2.

(We note especially that, because each revolution comprises 2π radians, the point just described has "cycling rate" s. That is, it completes s revolutions per unit t. This justifies our application to s of the term "frequency.")

Thus the simple, fundamental nature of sinusoids, from the perspective of periodicity, is evident. So in light of (I.1) and the discussion surrounding it, we can recognize Fourier analysis as a theory whereby quite general phenomena, whether or not they are themselves periodic or exhibit any obvious overall cyclical behavior, can be understood as *amalgamations* of basic periodic elements of definite frequencies. In particular Fourier analysis, in concert with other ideas and techniques from mathematics, science, and engineering, gives us the ability to investigate, identify, and even alter these elements. And this ability has a large number of practical applications.

Thus we've addressed, however briefly, the what and why of Fourier analysis. But let's turn our attention, even more briefly, to the when and how. Specifically, we now ask: Which functions *are* reasonable enough to be expressible as superpositions of sinusoids, and for those that are, what do these expressions look like and how do we find them?

These are big questions. Their answers, and the very pursuit of those answers, have had a profound impact on civilization as we know it. Our considerations of these answers, and this pursuit, will occupy a large portion of this book. For now, though,

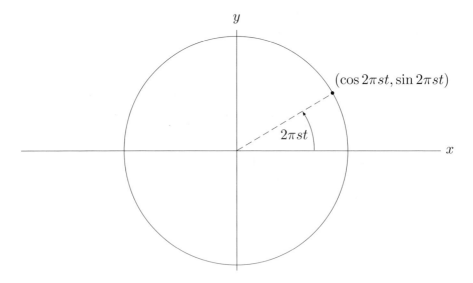

Fig. I.2 The functions $\cos 2\pi st$ and $\sin 2\pi st$

we'll try to say just enough about the how and the when to flesh out our overview.

We do so by addressing some gory details that we've thus far glossed over. Here's an especially gory one: The frequency domain F_f of a given function f is, in fact, usually *infinite*. So generally, for a superposition of sinusoid components, we'll need not a sum *per se*, but some infinite analog thereof. Namely, we'll require either an infinite series, if the set F_f is *discrete* (meaning it comprises isolated points on the real line), as it often is, or a definite integral, if this set is *continuous* (meaning it consists of a continuum of points on the line), as it also often is.

And this gives our discussion a new wrinkle; namely, once infinite processes are thrown into the mix, the issue of *convergence* of these processes must be addressed. That is, if we say a function *has* an expression as a superposition of sinusoids, we really mean this superposition *converges to* the function, but for this to make precise sense, we must be explicit about the meaning of "converges to."

As we'll see in this book, there are various useful notions of convergence of functions. And to each of these corresponds a different notion of reasonable function. That is, to each given sense of convergence will correspond a different set of criteria assuring that a function has a frequency decomposition converging to it.

It should be noted that, at the time of Fourier's work, the concept of convergence of functions — or of convergence at all, for that matter — was not a very well-formed one. Nor was the concept of reasonable function — or of function at all, for that matter. In fact, the first truly systematic studies of functions and convergence grew out of efforts to understand, clarify, elaborate on, generalize, and make rigorous Fourier's decompositions. Such studies then generated distinct mathematical theories of their own; these theories in turn formed the seeds of the vast discipline presently known, in the math world, as "real analysis." So Fourier, in case you were wondering whom to blame, is (indirectly) as responsible as anyone, and more responsible than most, for the development of that discipline.

More specifically, such mathematical quantities, constructs, and results as Riemann sums and integrals; the formal "ε-N" and "ε-δ" definitions of limits; related formal definitions of continuity and differentiability; definitions and theories of pointwise, uniform, and norm convergence of functions; Hilbert spaces; Lebesgue integrals and such theorems regarding them as Fubini's theorem and the Lebesgue dominated convergence theorem; distributions; and so on all arose out of investigations into Fourier's sinusoid decompositions. We'll touch upon each of these results, constructs, and quantities at one place or another in this book, and even so we'll only skim the surface of the vast collection of mathematical ideas developed in response to his work.

Indeed it has been argued, convincingly we believe, that the very standards of precision and rigor that currently prevail in mathematics grew, themselves, out of examination of Fourier's claims.

To summarize the story so far, the idea of a sinusoid decomposition, first espoused (in any generality) by Joseph Fourier in 1807, has wide practical *and* theoretical implications.

But what makes Fourier analysis particularly nifty is the richness of the *interplay* between its theory and its practice. We know of no mathematical discipline where this interplay is deeper, and this includes even disciplines about which we know something.

We'll try, in what follows, to convey not only how the theory makes the practice fly, but also how the practice elucidates and illuminates the theory.

We conclude this section with three remarks. First, considerable gains in mathematical convenience and elegance may be reaped by reformulating (I.1) in terms of the *complex exponentials* $e^{2\pi i s x}$ and $e^{-2\pi i s x}$ (Fig. I.3). Here i denotes the usual imaginary square root of -1, and

$$e^{\pm i\theta} = \cos\theta \pm i\sin\theta \tag{I.4}$$

for $\theta \in \mathbb{R}$ (cf. the Appendix).

The advantages of the complex exponential perspective will manifest themselves amply in the course of things.

Our second remark is that the big picture is actually quite a bit bigger than has been described so far: One may also consider the expression of reasonable functions as superpositions of other, not necessarily sinusoidal, "building blocks." And a bit of investigation reveals not only that decompositions of this sort do exist, but that they're abundant; not only that they're important, but that they're abundantly so. We'll investigate such alternative decompositions a bit later: see especially Chapter 4, Section 7.1, and Chapter 8.

Thirdly, we note that the few details provided to this point concern, strictly speaking, *one-dimensional* Fourier analysis—that is, they apply to functions of a *single* real variable. Certainly *multidimensional* Fourier analysis, wherein functions of several variables $x_1, x_2, \ldots, x_m$ are analyzed (into several-variable analogs of the sinusoid (I.2), or of complex exponentials, or of other sorts of functions), is also of significant interest, and will be given the requisite attention in appropriate contexts. But we take

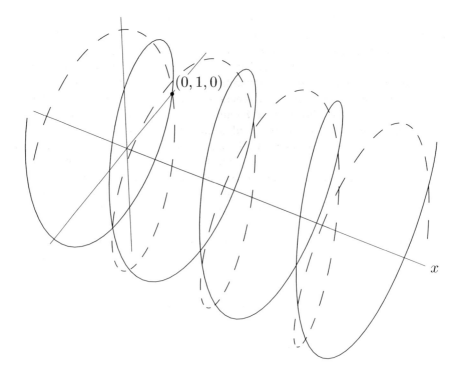

Fig. I.3 The complex exponentials $e^{2\pi i s x}$ (solid) and $e^{-2\pi i s x}$ (dashed)

the point of view here, as we will throughout, that the major issues are best introduced in the one-dimensional setting. From there, the generalization to several dimensions will usually be quite straightforward.

We now proceed to shade in some elements, already outlined, of the big picture. Warning: Our shadings won't always stay within neat, well defined lines. Which is okay because, in fact, Fourier analysis is like fingerpainting: Its various parts and patches are not disjoint or sharply delineated but blend, swirl, and fade gently into each other.

If we're going to fingerpaint, we're necessarily going to get our hands dirty. And there's no point in trying to wash this stuff off: It's Fourier analysis, it's indelible. So ye who enter here, abandon all soap.

Fourier Analysis

1

Fourier Coefficients and Fourier Series

1.1 PERIODIC FUNCTIONS: BEGINNING BITS

As mentioned in the Introduction, sinusoids are periodic, meaning they repeat themselves in a regular fashion. For this reason functions f that are, themselves, periodic are well suited to frequency decompositions. And for the same reason, it's in the periodic setting that the general ideas and many of the specific details of Fourier analysis are most naturally conveyed. Moreover, from that setting the cognitive leap to other, nonperiodic ones will be relatively painless.

So we focus, now, on periodic functions. We begin by defining them precisely.

Definition 1.1.1 Let $P > 0$. A function f of a real variable is P-*periodic* if

$$f(x + P) = f(x) \tag{1.1}$$

for all x in the domain of f.

Geometrically, a P-periodic function is one whose graph looks the same when translated P units horizontally. Of course such a graph will look the same even if such a translation is performed *repeatedly*; so, if f is P-periodic, then it's also nP-periodic for any $n \in \mathbb{Z}^+$. (Throughout this text, $\mathbb{Z}^+$ denotes the set of positive integers, $\mathbb{N}$ the set of nonnegative integers, and $\mathbb{Z}$ the set of all integers.)

A given periodic function therefore has infinitely many periods. The smallest of these, if there is a smallest one, is called the *fundamental period*, or *wavelength*, of the function. Most functions have fundamental periods, though there are exceptions. For example, a constant function is P-periodic for any $P > 0$, while "Dirichlet's function"—see Exercise 1.1.1—has any positive *rational* number P as a period.

2 FOURIER COEFFICIENTS AND FOURIER SERIES

Many "real-life" phenomena can be modeled, at least roughly, by periodic functions. Here are some examples: In each case the specified period P is the fundamental one.

- The distance from the earth to the sun (here $P = 1$ year).

- The phase of the moon (here $P = 28$ days—a "lunar cycle").

- The sound wave—more specifically, the variation in air pressure measured at a given point in space—resulting when the A3 key is struck on a piano (here $P = 1/440$ of a second).

- The sound wave resulting when a vowel sound is pronounced (here P is typically somewhere between $1/100$ and $1/200$ of a second). See Figure 1.1.

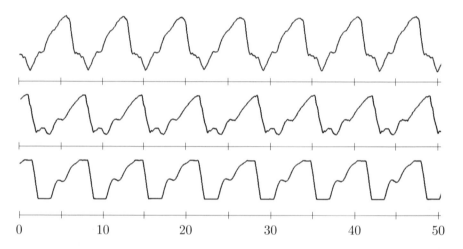

Fig. 1.1 The "ah," "eh," and "oh" vowel sounds, respectively, as pronounced by the author. The horizontal axis is in milliseconds

- The angular displacement of a bob at the end of a swinging pendulum (here P depends on the length, initial position, and initial velocity of the pendulum, but not on the mass of the bob).

- Populations of both predator and prey in certain predator-prey environments (here P depends on many parameters, but will generally be the same for both populations in question).

- An electrocardiogram reading (here P is the reciprocal of the pulse rate).

- The coordinates of a particle moving, at a constant rate, along a closed curve in the plane. (We may quantify this situation by expressing these coordinates as functions of the *distance traveled* along the curve. See Section 3.9. Then P is just the perimeter of the curve.)

- Electrostatic potential in a crystal (actually, crystals are three-dimensional. We'll discuss multidimensional periodicity and sinusoid decompositions later: See Sections 2.12, 2.13, 3.8, and 6.8).
- Most Philip Glass music.

Of course, in real life nothing truly repeats itself *ad infinitum*, or with perfect regularity. Well, okay, except Philip Glass maybe. But remember, while math models real life, math isn't real life, fortunately for real life, and for math. And sometimes a periodic model is appropriate and useful.

Let's now return our attention to frequency analysis. We inquire into the particulars of the decomposition (I.1) when f is P-periodic. And to begin this inquiry, we ask: What should the frequency domain F_f of such an f be?

To answer, we reason as follows: Intuitively a P-periodic function should, assuming it's reasonable enough to admit a decomposition (I.1), decompose into P-periodic sinusoids. And such sinusoids are readily identified. Namely, we have:

Proposition 1.1.1 *Let $P > 0$ be given. The P-periodic sinusoids are precisely those of frequency n/P for some $n \in \mathbb{N}$.*

Proof. We have

$$\cos\frac{2\pi n}{P}(x+P) = \cos\left(\frac{2\pi nx}{P} + 2\pi n\right) = \cos\frac{2\pi nx}{P}, \tag{1.2}$$

and similarly for sine functions. So the sinusoid $A\cos 2\pi sx + B\sin 2\pi sx$ will certainly be P-periodic if $s = n/P$ for some $n \in \mathbb{N}$.

For the converse result, see Exercise 1.1.2. □

To summarize, if f is P-periodic and well-behaved enough to have a decomposition (I.1), then the superposition there should take place over $F_f = \{n/P \colon n \in \mathbb{N}\}$. (Not every element of this latter set need actually contribute to the frequency decomposition of f; some of the coefficients in (I.1) might equal zero.) Or so we've argued in heuristic terms. We've not proved anything to this effect. But we have enough faith in our argument to proclaim the following.

Imprecise conviction 1.1.1 *Suppose f is P-periodic and reasonable. Then there are numbers $a_n(f)$ ($n \in \mathbb{N}$) and $b_n(f)$ ($n \in \mathbb{Z}^+$) such that*

$$f(x) = \frac{a_0(f)}{2} + \sum_{n=1}^{\infty}\left(a_n(f)\cos\frac{2\pi nx}{P} + b_n(f)\sin\frac{2\pi nx}{P}\right). \tag{1.3}$$

Observe that, in passing from the generality of (I.1) to the more specific situation at hand, we've adjusted the notation somewhat. Namely, we've used the less cumbersome symbols $a_n(f)$ and $b_n(f)$ in place of $A_{n/P}(f)$ and $B_{n/P}(f)$, respectively, for

$n \geq 1$. Also, we've replaced the $n = 0$ summand in (I.1), which equals $A_0(f)$, with $a_0(f)/2$. The "2" in the denominator will allow us, in many situations, to discuss all $a_n(f)$'s in the same breath, without having to distinguish between the case $n = 0$ and the case $n > 0$. See especially (1.6) and (1.19).

For various reasons to become clear as we progress, it will be convenient to reformulate Imprecise conviction 1.1.1 in terms of the complex exponentials of (I.4). The following lemma tells us how to do this.

Lemma 1.1.1 *The frequency decomposition of (Imprecise) conviction 1.1.1 is formally (that is, ignoring issues of convergence of the series) equivalent to the decomposition*

$$f(x) = \sum_{n=-\infty}^{\infty} c_n(f) e^{2\pi i n x/P}, \tag{1.4}$$

where

$$c_n(f) = \begin{cases} \frac{1}{2}(a_{-n}(f) + ib_{-n}(f)) & \text{if } n < 0, \\ \frac{1}{2}a_n(f) & \text{if } n = 0, \\ \frac{1}{2}(a_n(f) - ib_n(f)) & \text{if } n > 0, \end{cases} \tag{1.5}$$

or equivalently

$$a_n(f) = c_n(f) + c_{-n}(f) \quad (n \in \mathbb{N}), \tag{1.6}$$
$$b_n(f) = i(c_n(f) - c_{-n}(f)) \quad (n \in \mathbb{Z}^+). \tag{1.7}$$

Proof. We have

$$\cos\theta = \frac{e^{i\theta} + e^{-i\theta}}{2} \quad \text{and} \quad \sin\theta = \frac{e^{i\theta} - e^{-i\theta}}{2i}, \tag{1.8}$$

as may be deduced by solving equations (I.4) for $\cos\theta$ and $\sin\theta$. So, for $n \in \mathbb{Z}^+$,

$$a_n(f)\cos\frac{2\pi n x}{P} + b_n(f)\sin\frac{2\pi n x}{P}$$
$$= a_n(f)\frac{e^{2\pi i n x/P} + e^{-2\pi i n x/P}}{2} + b_n(f)\frac{e^{2\pi i n x/P} - e^{-2\pi i n x/P}}{2i}$$
$$= \left(\frac{a_n(f) + ib_n(f)}{2}\right)e^{-2\pi i n x/P} + \left(\frac{a_n(f) - ib_n(f)}{2}\right)e^{2\pi i n x/P}$$
$$= c_{-n}(f)e^{-2\pi i n x/P} + c_n(f)e^{2\pi i n x/P}. \tag{1.9}$$

(For the second equality, we used the fact that $i = -1/i$; for the third we noted that, by replacing n with $-n$ in the first case of (1.5), we get $c_{-n}(f) = (a_n(f) + ib_n(f))/2$ for $n > 0$.) But then, formally at least,

$$\frac{a_0(f)}{2} + \sum_{n=1}^{\infty} \left(a_n(f) \cos \frac{2\pi n x}{P} + b_n(f) \sin \frac{2\pi n x}{P} \right)$$

$$= c_0(f) + \sum_{n=1}^{\infty} c_{-n}(f) e^{-2\pi i n x/P} + \sum_{n=1}^{\infty} c_n(f) e^{2\pi i n x/P}$$

$$= c_0(f) e^{2\pi i (0) x/P} + \sum_{n=-\infty}^{-1} c_n(f) e^{2\pi i n x/P} + \sum_{n=1}^{\infty} c_n(f) e^{2\pi i n x/P}$$

$$= \sum_{n=-\infty}^{\infty} c_n(f) e^{2\pi i n x/P}. \tag{1.10}$$

(Here, to get the second equality, we replaced n by $-n$ in the first of the sums preceding this equality.)

Finally, the equivalence of (1.5) with (1.6)/(1.7) is a straightforward matter: See Exercise 1.1.3. □

Note in particular, from (1.9), that each frequency component in (1.3), except for the one of frequency zero, is represented by two distinct summands of (1.4). If f is *real-valued*, though, then either of these two summands completely determines the other. See Proposition 1.10.1.

One of the advantages of the complex exponential perspective is that the series in (1.4) is more concise than the one in (1.3). Another advantage is that, should a (sufficiently reasonable) function f have the form (1.4), the coefficients $c_n(f)$ may be expressed in terms of f by a particularly simple, elegant argument. That argument hinges on this lemma.

Lemma 1.1.2 *For $n, k \in \mathbb{Z}$ and $P > 0$, we have*

$$\frac{1}{P} \int_{-P/2}^{P/2} e^{2\pi i n x/P} e^{-2\pi i k x/P} \, dx = \delta_{n,k}, \tag{1.11}$$

where the "Kronecker delta function" $\delta_{n,k}$ is defined by

$$\delta_{n,k} = \begin{cases} 1 & \text{if } n = k, \\ 0 & \text{if not.} \end{cases} \tag{1.12}$$

Proof. If $n \neq k$, then

$$\frac{1}{P} \int_{-P/2}^{P/2} e^{2\pi i n x/P} e^{-2\pi i k x/P} \, dx = \frac{1}{P} \int_{-P/2}^{P/2} e^{2\pi i (n-k) x/P} \, dx \tag{1.13}$$

$$= \frac{e^{2\pi i (n-k) x/P}}{2\pi i (n-k)} \bigg|_{-P/2}^{P/2} = \frac{e^{\pi i (n-k)} - e^{-\pi i (n-k)}}{2\pi i (n-k)}$$

$$= \frac{(-1)^{n-k} - (-1)^{n-k}}{2\pi i (n-k)} = 0. \tag{1.14}$$

The first equality is by Proposition A.0.5(a), the second by antidifferentiating, the third by just plugging in the limits of integration, and the fourth because

$$e^{\pm \pi i m} = (e^{\pm \pi i})^m = (\cos \pm \pi + i \sin \pm \pi)^m = (-1 + i0)^m = (-1)^m \quad (m \in \mathbb{Z}). \tag{1.15}$$

Now the above calculations don't apply when $k = n$ (note the factors of $n - k$ appearing in denominators), but we compute separately that

$$\frac{1}{P} \int_{-P/2}^{P/2} e^{2\pi i n x/P} e^{-2\pi i n x/P} \, dt = \frac{1}{P} \int_{-P/2}^{P/2} 1 \, dx = 1, \tag{1.16}$$

as required. □

We may now, at least formally, compute the coefficients $a_n(f), b_n(f)$, and $c_n(f)$ appearing in the decompositions (1.3) and (1.4), as follows. Let's multiply both sides of (1.4) by $P^{-1} e^{-2\pi i k x/P}$ for some fixed integer k and then integrate over $[-P/2, P/2]$. We get

$$\frac{1}{P} \int_{-P/2}^{P/2} f(x) e^{-2\pi i k x/P} \, dx = \frac{1}{P} \int_{-P/2}^{P/2} \left(\sum_{n=-\infty}^{\infty} c_n(f) e^{2\pi i n x/P} \right) e^{-2\pi i k x/P} \, dx$$

$$= \sum_{n=-\infty}^{\infty} c_n(f) \cdot \frac{1}{P} \int_{-P/2}^{P/2} e^{2\pi i n x/P} e^{-2\pi i k x/P} \, dx$$

$$= \sum_{n=-\infty}^{\infty} c_n(f) \delta_{n,k}, \tag{1.17}$$

the last step by Lemma 1.1.2. The right side equals $c_k(f)$ (all summands vanish except the kth one; this one equals $c_k(f) \delta_{k,k} = c_k(f)$). So (1.17) (with k now replaced by n, for convenience) gives

$$c_n(f) = \frac{1}{P} \int_{-P/2}^{P/2} f(x) e^{-2\pi i n x/P} \, dx \quad (n \in \mathbb{Z}). \tag{1.18}$$

Moreover, we deduce from this and (1.6) that

$$a_n(f) = \frac{1}{P} \int_{-P/2}^{P/2} f(x) e^{-2\pi i n x/P} \, dx + \frac{1}{P} \int_{-P/2}^{P/2} f(x) e^{2\pi i n x/P} \, dx$$

$$= \frac{1}{P} \int_{-P/2}^{P/2} f(x) \left(e^{-2\pi i n x/P} + e^{2\pi i n x/P} \right) dx$$

$$= \frac{2}{P} \int_{-P/2}^{P/2} f(x) \cos \frac{2\pi n x}{P} \, dx \quad (n \in \mathbb{N}) \tag{1.19}$$

(the last step by the first part of (1.8)), and similarly (see Exercise 1.1.4) that

$$b_n(f) = \frac{2}{P} \int_{-P/2}^{P/2} f(x) \sin \frac{2\pi n x}{P} \, dx \quad (n \in \mathbb{Z}^+). \tag{1.20}$$

Again, these arguments are formal, but they're compelling enough to bring us to:

Somewhat less imprecise conviction 1.1.1 *If f is P-periodic and reasonable, then it has decompositions* (1.3) *and* (1.4), *with the coefficients $c_n(f)$, $a_n(f)$, and $b_n(f)$ given by* (1.18)–(1.20).

We remark, in closing, that Lemma 1.1.2 documents the *orthogonality* of the set of functions

$$\{e^{2\pi inx/P} : n \in \mathbb{Z}\}. \tag{1.21}$$

Here we're referring to the following definition: A set of nonzero P-periodic functions is said to be an *orthogonal set* if, given any distinct elements f and g of this set,

$$\int_{-P/2}^{P/2} f(x)\overline{g(x)}\, dx = 0. \tag{1.22}$$

It's also true that

$$\left\{\cos\frac{2\pi nx}{P} x : n \in \mathbb{N}\right\} \cup \left\{\sin\frac{2\pi nx}{P} x : n \in \mathbb{Z}^+\right\} \tag{1.23}$$

is orthogonal. See Exercise 1.1.5.

The notion of orthogonality will recur frequently in the course of our studies. (See Section 3.6 and beyond.)

Exercises

1.1.1 Dirichlet's function $Q\colon \mathbb{R} \to \{0, 1\}$ is defined by

$$Q(x) = \begin{cases} 1 & \text{if } x \text{ is rational}, \\ 0 & \text{if not.} \end{cases}$$

Show that Q is P-periodic for any rational number $P > 0$.

1.1.2 Show that, if $s > 0$, either A or B is nonzero, and the sinusoid $h(x) = A\cos 2\pi sx + B\sin 2\pi sx$ is P-periodic, then necessarily $s = n/P$ for some $n \in \mathbb{Z}^+$. Hint: First suppose $B \neq 0$; use the assumption that $h(P/2) = h(-P/2)$. Next suppose $B = 0$ and $A \neq 0$; use the assumption that $h(0) = h(P)$. (You may take it on faith that, for y real, $\cos y = 1$ if and only if y is an integer multiple of 2π and $\sin y = 0$ if and only if y is an integer multiple of π.)

1.1.3 Derive (1.5) from (1.6)/(1.7), and vice versa.

1.1.4 Derive formula (1.20) from (1.18), (1.8), and (1.7), in the manner used above to deduce (1.19) from (1.18), (1.8), and (1.6).

1.1.5 Show by direct evaluation of the relevant integrals that (1.23) constitutes an orthogonal set of P-periodic functions. To perform the necessary integration, you may want to use the identities of Exercise A.0.16.

1.2 FOURIER COEFFICIENTS OF 2π-PERIODIC FUNCTIONS

If f is P-periodic and we define, for a given $P^* > 0$, the function f^* by

$$f^*(x) = f\left(\frac{P}{P^*}x\right), \tag{1.24}$$

then

$$f^*(x + P^*) = f\left(\frac{P}{P^*}(x + P^*)\right) = f\left(\frac{P}{P^*}x + P\right) = f\left(\frac{P}{P^*}x\right) = f^*(x), \tag{1.25}$$

so f^* is P^*-periodic. Or in geometric terms, a P^*-periodic function is just a P-periodic one rescaled horizontally. So we can develop the theory for general P by focusing on a particular one.

We choose $P = 2\pi$. Then $e^{2\pi i n x/P} = e^{inx}$, and similarly for the corresponding cosine and sine functions, so the notation is particularly compact. Moreover (and not coincidentally), 2π is an especially familiar period: The world's most familiar trig functions $\cos x$ and $\sin x$ have this period.

By Somewhat less imprecise conviction 1.1.1, if f is 2π-periodic and reasonable, then

$$f(x) = \sum_{n=-\infty}^{\infty} c_n(f)e^{inx} = \frac{a_0(f)}{2} + \sum_{n=1}^{\infty}\left(a_n(f)\cos nx + b_n(f)\sin nx\right), \tag{1.26}$$

where

$$c_n(f) = \frac{1}{2\pi}\int_{-\pi}^{\pi} f(x)e^{-inx}\,dx \quad (n \in \mathbb{Z}), \tag{1.27}$$

$$a_n(f) = \frac{1}{\pi}\int_{-\pi}^{\pi} f(x)\cos nx\,dx \quad (n \in \mathbb{N}), \tag{1.28}$$

$$b_n(f) = \frac{1}{\pi}\int_{-\pi}^{\pi} f(x)\sin nx\,dx \quad (n \in \mathbb{Z}^+). \tag{1.29}$$

Now certainly, if we're to proceed any further with this, f will need to be "reasonable" enough that the integrals in (1.27), (1.28), and (1.29) make sense. So we'd like to determine some fairly broad criteria under which they do. We need some definitions.

Definition 1.2.1 (a) Let I be a *bounded* interval, meaning it has finite length. A function f of a real variable is *piecewise continuous* on I if all of the following conditions hold:

(i) The left- and right-hand limits

$$f(x_0^-) = \lim_{x \to x_0^-} f(x) \quad \text{and} \quad f(x_0^+) = \lim_{x \to x_0^+} f(x) \tag{1.30}$$

exist at each $x_0 \in I$ that is not an endpoint of I.

(ii) The one-sided limits $f(a^+)$ and $f(b^-)$, where a is the left endpoint of I and b its right endpoint, both exist.

(iii) There are only finitely many points in I at which f fails to be continuous.

(b) A function f of a real variable is said to be piecewise continuous on an *unbounded* interval J if it's piecewise continuous on every bounded interval $I \subset J$.

(c) For any interval I (bounded or unbounded), we denote by PC(I) the set of all piecewise continuous functions on I.

(Often, when I is understood, we simply say "piecewise continuous" instead of "piecewise continuous on I.")

Figure 1.2 depicts a piecewise continuous function on a closed, bounded interval.

It follows from the definitions that any function *continuous* on such an interval is piecewise continuous there. However, the same may not be said for *open*, or *half-open*, bounded intervals, as Figure 1.3 illustrates.

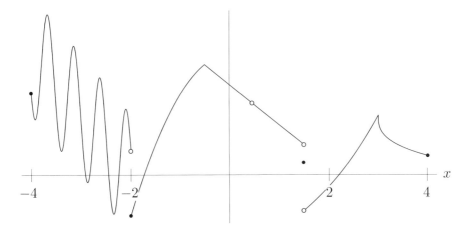

Fig. 1.2 A function $f \in \mathrm{PC}([-4, 4])$

For what follows, it will be useful to "periodize" some of the above notions. To this end, we first need some notation: Namely, we denote by $\mathbb{T}$ the circle $u^2 + v^2 = 1$—also called the "one dimensional torus"—in the uv plane. The reason for introducing $\mathbb{T}$ at this point is this: A 2π-periodic function f can be thought of as situated on $\mathbb{T}$, since if we wrap the x axis around that circle, then the infinitely many iterations of f will collapse into one, yielding a well defined function on $\mathbb{T}$ (Fig. 1.4).

With this in mind, we make:

Definition 1.2.2 We denote by PC($\mathbb{T}$) the set of all 2π-periodic, piecewise continuous functions on $\mathbb{R}$.

10 FOURIER COEFFICIENTS AND FOURIER SERIES

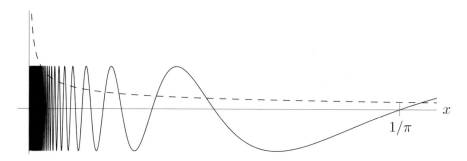

Fig. 1.3 Neither $g(x) = 10\sin(x^{-1})$ (solid) nor $h(x) = x^{-1/2}$ (dashed) is piecewise continuous on $(0, b]$ for $b > 0$ (neither $g(0^+)$ nor $h(0^+)$ exists)

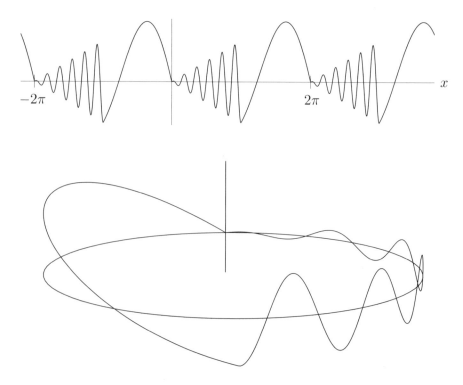

Fig. 1.4 Top: a 2π-periodic function f. Bottom: f, upon wrapping the x axis around the torus $\mathbb{T}$

In other words: PC($\mathbb{T}$) is the set of all 2π-periodic functions having only finitely many discontinuities per bounded interval, and whose left- and right-hand limits exist at every point.

We remark that PC($\mathbb{T}$) (or PC(I), for any interval I) is a *complex vector space*. Recall that this means, essentially, a set of objects that can be multiplied by complex numbers, and added to each other, in such a way that the resulting objects are still in that set. For the formal definition see, for example, [21]. We'll be less than formal, and will consider it sufficient, when checking whether a certain set V constitutes a complex vector space, to check the criteria cited above. These criteria are equivalent to the following condition:

$$\text{If } f_1, f_2 \in V \text{ and } \alpha_1, \alpha_2 \in \mathbb{C}, \text{ then } \alpha_1 f_1 + \alpha_2 f_2 \in V. \tag{1.31}$$

The condition (1.31) is called *closure under linear combinations* ($\alpha_1 f_1 + \alpha_2 f_2$ is called a *linear combination* of f_1 and f_2) and is readily seen to be satisfied when $V = PC(\mathbb{T})$.

We'll often use the shorter term "vector space," or even "space," to refer to a complex vector space.

We also note that, as is well known, a function piecewise continuous on a bounded interval is Riemann integrable there. Moreover, a product of piecewise continuous functions is readily seen to be piecewise continuous; so we can make the following definition.

Definition 1.2.3 If $f \in PC(\mathbb{T})$, then the $c_n(f)$'s, $a_n(f)$'s, and $b_n(f)$'s given by (1.27)–(1.29) are called *Fourier coefficients of f*. The two series in (1.26) are called the *Fourier series* for f—the first of these series is said to be in *cosine-sine* form and the second in *complex exponential* form.

Warning: We still have no concrete results regarding the *convergence*—to f, or to anything for that matter—of the Fourier series for f. But again, as long as $f \in PC(\mathbb{T})$, its Fourier coefficients are defined. We'll say that such an f is *associated to* its Fourier series, and we'll denote such association by "$\sim$" instead of "$=$."

The coefficient $c_0(f) = a_0(f)/2$ has particular significance. Namely, since $e^{i0} = 1$, we have

$$c_0(f) = \frac{1}{2\pi} \int_{-\pi}^{\pi} f(x)\,dx, \tag{1.32}$$

so $c_0(f)$ is equal to the *average value* of f on $[-\pi, \pi]$.

Let's consider an example.

Example 1.2.1 Find the Fourier series for the function $f \in PC(\mathbb{T})$ defined by

$$f(x) = e^x \ (-\pi < x \leq \pi) \tag{1.33}$$

(see Fig. 1.5).

Solution. We'll compute the $c_n(f)$'s first: Since f is itself an exponential function, the integral in (1.27) is particularly easy. We get

$$c_n(f) = \frac{1}{2\pi} \int_{-\pi}^{\pi} f(x) e^{-inx}\,dx = \frac{1}{2\pi} \int_{-\pi}^{\pi} e^x e^{-inx}\,dx = \frac{1}{2\pi} \int_{-\pi}^{\pi} e^{(1-in)x}\,dx$$

$$= \left. \frac{e^{(1-in)x}}{2\pi(1-in)} \right|_{-\pi}^{\pi} = \frac{e^{(1-in)\pi} - e^{-(1-in)\pi}}{2\pi(1-in)} = \frac{(e^{\pi} - e^{-\pi})(-1)^n}{2\pi(1-in)}, \tag{1.34}$$

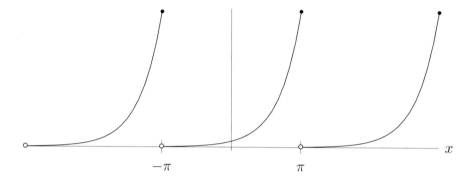

Fig. 1.5 $f(x) = e^x$ $(-\pi < x \leq \pi)$, periodized

the last step by (1.15). So

$$f(x) \sim \sum_{n=-\infty}^{\infty} c_n(f) e^{inx} = \frac{e^\pi - e^{-\pi}}{2\pi} \sum_{n=-\infty}^{\infty} \frac{(-1)^n e^{inx}}{1 - in}. \quad (1.35)$$

We also have, by (1.6) and (1.7) and the fact that $(-1)^n = (-1)^{-n}$,

$$\begin{aligned} a_n(f) &= c_n(f) + c_{-n}(f) \\ &= \frac{e^\pi - e^{-\pi}}{2\pi} (-1)^n \left[\frac{1}{1-in} + \frac{1}{1+in} \right] \\ &= \frac{e^\pi - e^{-\pi}}{2\pi} (-1)^n \left[\frac{1+in+1-in}{(1-in)(1+in)} \right] = \frac{(e^\pi - e^{-\pi})(-1)^n}{\pi(1+n^2)} \end{aligned} \quad (1.36)$$

and

$$\begin{aligned} b_n(f) &= i(c_n(f) - c_{-n}(f)) \\ &= i \frac{e^\pi - e^{-\pi}}{2\pi} (-1)^n \left[\frac{1}{1-in} - \frac{1}{1+in} \right] = -\frac{n(e^\pi - e^{-\pi})(-1)^n}{\pi(1+n^2)}. \end{aligned} \quad (1.37)$$

So

$$\begin{aligned} f(x) &\sim \frac{a_0(f)}{2} + \sum_{n=1}^{\infty} (a_n(f) \cos nx + b_n(f) \sin nx) \\ &= \frac{e^\pi - e^{-\pi}}{2\pi} + \frac{e^\pi - e^{-\pi}}{\pi} \sum_{n=1}^{\infty} \frac{(-1)^n}{1+n^2} (\cos nx - n \sin nx). \end{aligned} \quad (1.38)$$

Exercises

1.2.1 Which of the following functions are continuous on $\mathbb{R}$? Which are piecewise continuous on $\mathbb{R}$? Briefly explain your answers.

    a. $f(x) = x^{1/3}$.

    b. $f(x) = \csc x$.

    c. $f(x) = \begin{cases} 1 & \text{if } x \leq 0, \\ \sin x & \text{if } x > 0. \end{cases}$

    d. $f(x) = \begin{cases} 0 & \text{if } x \leq 0, \\ x \sin(x^{-1}) & \text{if } x > 0. \end{cases}$

    e. $f(x) = \begin{cases} (\sin x)^{1/5} & \text{if } x \leq 0, \\ \cos x & \text{if } x > 0. \end{cases}$

1.2.2 Show that the taking of Fourier coefficients is *linear*. In other words, show $c_n(\alpha_1 f_1 + \alpha_2 f_2) = \alpha_1 c_n(f_1) + \alpha_2 c_n(f_2)$ for all $\alpha_1, \alpha_2 \in \mathbb{C}$, $f_1, f_2 \in \text{PC}(\mathbb{T})$, and $n \in \mathbb{Z}$, and similarly for the $a_n(f)$'s and $b_n(f)$'s.

1.2.3 Use the definition (1.30) of one-sided limits to show carefully that, if $f \in \text{PC}(\mathbb{T})$, then $f(x^-) = f((x + 2k\pi)^-)$ and $f(x^+) = f((x + 2k\pi)^+)$ for all $k \in \mathbb{Z}$ and $x \in \mathbb{R}$.

In each of the Exercises 1.2.4–1.2.14, sketch the graph of, and find both forms of the Fourier series for, the function $f \in \text{PC}(\mathbb{T})$ having the given definition on $(-\pi, \pi]$.

1.2.4
$$f(x) = \begin{cases} e^x & \text{if } -\pi < x \leq 0, \\ e^{x-\pi} & \text{if } 0 < x \leq \pi. \end{cases}$$

1.2.5
$$f(x) = \begin{cases} e^x & \text{if } -\pi < x \leq 0, \\ -e^{x-\pi} & \text{if } 0 < x \leq \pi. \end{cases}$$

1.2.6 $f(x) = \cosh x$ $(-\pi < x \leq \pi)$. (Recall $\cosh x = (e^x + e^{-x})/2$.)

1.2.7 $f(x) = \sinh x$ $(-\pi < x \leq \pi)$. (Recall $\sinh x = (e^x - e^{-x})/2$.)

1.2.8 $f(x) = \sin \sqrt{2}\, x$ $(-\pi < x \leq \pi)$. (Use (1.8).)

1.2.9 $f(x) = e^{-ax}$ $(-\pi < x \leq \pi)$, where $a \in \mathbb{R}$. Sketch for two different choices of a.

1.2.10
$$f(x) = \begin{cases} \dfrac{1}{2a} & \text{if } |x| < a, \\ 0 & \text{elsewhere on } (-\pi, \pi], \end{cases}$$

where $0 < a < \pi$. Sketch for two different choices of a.

1.2.11

$$f(x) = \begin{cases} 0 & \text{if } |x| < a, \\ \cos a - \cos x & \text{elsewhere on } (-\pi, \pi], \end{cases}$$

where $0 < a < \pi$. (Use (1.8).) Sketch for two different choices of a.

1.2.12

$$f(x) = \begin{cases} \cos x & \text{if } |x| < \dfrac{\pi}{2}, \\ \dfrac{\cos 5x}{5} & \text{elsewhere on } (-\pi, \pi]. \end{cases}$$

(Use (1.8).)

1.2.13

$$f(x) = \begin{cases} 0 & \text{if } -\pi < x \leq 0, \\ \sin x & \text{if } 0 < x \leq \pi. \end{cases}$$

(Use (1.8).)

1.2.14 $f(x) = e^{-iax}$ $(-\pi < x \leq \pi)$, where $a \in \mathbb{R}$. Sketch for two different choices of a, one of which should be an *integer*. For integer a, you should obtain *only one* nonzero Fourier coefficient $c_n(f)$! (Which one?) For noninteger a, you should get

$$f(x) \sim \sum_{n=-\infty}^{\infty} \frac{\sin \pi(a+n)}{\pi(a+n)} e^{inx}.$$

1.3 MORE ON $P = 2\pi$

Sometimes useful in the computation of Fourier coefficients is the following observation. As is evident geometrically—see Exercise 1.3.1 for a careful verification—the integral of an element of $PC(\mathbb{T})$ over an interval of length 2π does not depend on where that interval begins. Now the product of 2π-periodic functions is 2π-periodic, so if f is in $PC(\mathbb{T})$, then so is the integrand in (1.27), meaning that formula implies

$$c_n(f) = \int_d^{d+2\pi} f(x) e^{-inx} \, dx \tag{1.39}$$

for such f and for any real number d. Analogous remarks apply to the formulas (1.28) and (1.29) for the $a_n(f)$'s and $b_n(f)$'s.

Example 1.3.1 Find both forms of the Fourier series for the function $g_r \in PC(\mathbb{T})$ defined by

$$g_r(x) = \begin{cases} x & \text{if } -r < x \leq r, \\ \dfrac{r}{\pi - r}(\pi - x) & \text{if } r < x \leq 2\pi - r, \end{cases} \tag{1.40}$$

where $0 < r < \pi$.

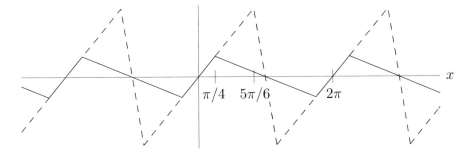

Fig. 1.6 The functions $g_{\pi/4}$ (solid) and $g_{5\pi/6}$ (dashed)

Solution. We sketch g_r for a couple of particular values of r (Fig. 1.6).
We compute the complex exponential coefficients first. We use the formula

$$\int x e^{\alpha x}\, dx = \left(\frac{x}{\alpha} - \frac{1}{\alpha^2}\right) e^{\alpha x} + C \quad (\alpha \neq 0), \tag{1.41}$$

readily obtained through integration by parts. We also use (1.39) with $d = -r$, since g_r is given to us explicitly on $(-r, 2\pi - r]$. We get, for $n \neq 0$,

$$\begin{aligned}
c_n(g_r) &= \frac{1}{2\pi} \int_{-r}^{2\pi - r} g_r(x) e^{-inx}\, dx \\
&= \frac{1}{2\pi} \int_{-r}^{r} x e^{-inx}\, dx + \frac{r}{2\pi(\pi - r)} \int_{r}^{2\pi - r} (\pi - x) e^{-inx}\, dx \\
&= \frac{1}{2\pi} \left(-\frac{x}{in} + \frac{1}{n^2}\right) e^{-inx} \bigg|_{-r}^{r} + \frac{r}{2\pi(\pi - r)} \left(-\frac{\pi - x}{in} - \frac{1}{n^2}\right) e^{-inx} \bigg|_{r}^{2\pi - r} \\
&= \frac{e^{-inr} - e^{inr}}{2(\pi - r) n^2} = \frac{\sin nr}{i(\pi - r) n^2};
\end{aligned} \tag{1.42}$$

the second-to-last step amounts to copious cancellation (using the fact that $e^{-2\pi i n} = 1$ for $n \in \mathbb{Z}$); the last is by (the second part of) (1.8). For the case $n = 0$, we compute separately that

$$c_0(g_r) = \frac{1}{2\pi} \int_{-r}^{r} x\, dx + \frac{r}{2\pi(\pi - r)} \int_{r}^{2\pi - r} (\pi - x)\, dx = 0, \tag{1.43}$$

as is suggested by Figure 1.6. (Recall that $c_0(g_r)$ gives the average value of g_r on $[-\pi, \pi]$.) So

$$g_r(x) \sim \sum_{n=-\infty}^{\infty} c_n(g_r) e^{inx} = \frac{1}{i(\pi - r)} \sum_{n \neq 0} \frac{\sin nr}{n^2} e^{inx}. \tag{1.44}$$

Further, we have

$$a_0(g_r) = 2 c_0(g_r) = 0 \tag{1.45}$$

16 FOURIER COEFFICIENTS AND FOURIER SERIES

and, for $n \geq 1$,

$$a_n(g_r) = c_n(g_r) + c_{-n}(g_r) = \frac{\sin nr}{i(\pi - r)n^2} + \frac{\sin(-n)r}{i(\pi - r)(-n)^2} = 0, \quad (1.46)$$

and

$$b_n(g_r) = i(c_n(g_r) - c_{-n}(g_r)) = i\left(\frac{\sin nr}{i(\pi - r)n^2} + \frac{\sin(-n)r}{i(\pi - r)(-n)^2}\right)$$

$$= \frac{2 \sin nr}{(\pi - r)n^2}. \quad (1.47)$$

So

$$g_r(x) \sim \frac{a_0(g_r)}{2} + \sum_{n=1}^{\infty}(a_n(g_r)\cos nx + b_n(g_r)\sin nx)$$

$$= \frac{2}{\pi - r}\sum_{n=1}^{\infty}\frac{\sin nr}{n^2}\sin nx. \quad (1.48)$$

Note that, in the above example, all of the $a_n(g_r)$'s are zero. This exemplifies how certain geometric properties of a function are reflected in its Fourier coefficients. Specifically, let's recall that a function f is said to be *even* (on its domain) if $f(-x) = f(x)$ for all x (in that domain), and *odd* (on its domain) if $f(-x) = -f(x)$ for all x (in that domain). Let's also recall that geometrically, at least for real-valued functions, evenness means having a graph unchanged by reflection about the vertical axis, while oddness means having a graph unchanged by reflection about the two axes in succession. Regarding complex-valued functions, we note that f is even (respectively, odd) if and only if Re f and Im f are.

The function g_r of Example 1.3.1 is, for any positive $r < \pi$, clearly odd; as it turns out, this is why all the $a_n(g_r)$'s vanish. Indeed, we have:

Proposition 1.3.1 (a) *If $f \in \mathrm{PC}(\mathbb{T})$ is odd, then*

$$a_n(f) = 0 \quad \text{and} \quad b_n(f) = \frac{2}{\pi}\int_0^{\pi} f(x)\sin nx\,dx \quad (1.49)$$

for $n \in \mathbb{N}$ and $n \in \mathbb{Z}^+$, respectively.

(b) *If $f \in \mathrm{PC}(\mathbb{T})$ is even, then*

$$a_n(f) = \frac{2}{\pi}\int_0^{\pi} f(x)\cos nx\,dx \quad \text{and} \quad b_n(f) = 0 \quad (1.50)$$

for $n \in \mathbb{N}$ and $n \in \mathbb{Z}^+$, respectively.

Proof. We consider part (a): Let $f \in \mathrm{PC}(\mathbb{T})$ be odd. Then, for any $n \in \mathbb{Z}$,

$$c_{-n}(f) = \frac{1}{2\pi}\int_{-\pi}^{\pi} f(x)e^{inx}\,dx = \frac{1}{2\pi}\int_{\pi}^{-\pi} f(-u)e^{-inu}\,d(-u)$$

$$= \frac{1}{2\pi}\int_{-\pi}^{\pi} f(-u)e^{-inu}\,du = -\frac{1}{2\pi}\int_{-\pi}^{\pi} f(u)e^{-inu}\,du = -c_n(f). \quad (1.51)$$

(For the second equality we put $u = -x$; for the third we rearranged; for the fourth we used the oddness of f.) So for $n \in \mathbb{N}$ we have, by (1.6),

$$a_n(f) = c_n(f) - c_n(f) = 0, \tag{1.52}$$

while for $n \in \mathbb{Z}^+$ we have, by (1.7),

$$\begin{aligned}
b_n(f) &= 2i\, c_n(f) \\
&= \frac{i}{\pi} \int_{-\pi}^{\pi} f(x) e^{-inx}\, dx = \frac{i}{\pi} \left(\int_0^{\pi} f(x) e^{-inx}\, dx + \int_{-\pi}^0 f(x) e^{-inx}\, dx \right) \\
&= \frac{i}{\pi} \left(\int_0^{\pi} f(x) e^{-inx}\, dx + \int_{\pi}^0 f(-u) e^{inu}\, d(-u) \right) \\
&= \frac{i}{\pi} \left(\int_0^{\pi} f(x) e^{-inx}\, dx - \int_0^{\pi} f(u) e^{inu}\, du \right) \\
&= \frac{i}{\pi} \int_0^{\pi} f(x)(e^{-inx} - e^{inx})\, dx = \frac{2}{\pi} \int_0^{\pi} f(x) \sin nx\, dx. \tag{1.53}
\end{aligned}$$

(The reasoning here is much as in (1.51); we also used, at the end, the identity in (1.8) for $\sin \theta$.)

This proves part (a); we leave the proof of part (b) to Exercise 1.3.2. □

Example 1.3.2 Find both forms of the Fourier series for the function $h \in PC(\mathbb{T})$ defined by

$$h(x) = \begin{cases} 1 + \cos 2x & \text{if } 0 \leq |x| \leq \dfrac{\pi}{2}, \\ -1 - \cos 2x & \text{if } \dfrac{\pi}{2} \leq |x| \leq \pi \end{cases} \tag{1.54}$$

(Fig. 1.7).

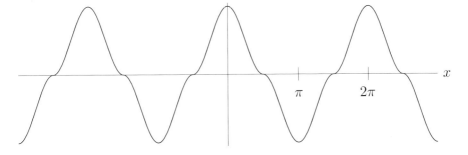

Fig. 1.7 The function h

18   FOURIER COEFFICIENTS AND FOURIER SERIES

Solution. Since h is even, we have $b_n(h) = 0$ for $n \in \mathbb{Z}^+$, by Proposition 1.3.1(b). Also, for $n \in \mathbb{N}$ we have

$$a_n(h) = \frac{2}{\pi} \int_0^\pi h(x) \cos nx \, dx$$

$$= \frac{2}{\pi} \int_0^{\pi/2} (1 + \cos 2x) \cos nx \, dx - \frac{2}{\pi} \int_{\pi/2}^\pi (1 + \cos 2x) \cos nx \, dx. \quad (1.55)$$

Now identities from Exercise A.0.16 may be used to deduce the formula

$$\int \cos kx \cos nx \, dx = \frac{n \sin nx \cos kx - k \sin kx \cos nx}{(n-k)(n+k)} + C \quad (k \neq \pm n). \quad (1.56)$$

So

$$a_n(h) = \frac{2}{\pi n} \sin nx \Big|_0^{\pi/2} + \frac{2n \cos 2x \sin nx - 4 \sin 2x \cos nx}{\pi(n-2)(n+2)} \Big|_0^{\pi/2}$$

$$- \frac{2}{\pi n} \sin nx \Big|_{\pi/2}^\pi - \frac{2n \cos 2x \sin nx - 4 \sin 2x \cos nx}{\pi(n-2)(n+2)} \Big|_{\pi/2}^\pi$$

$$= \frac{4}{\pi n} \sin \frac{n\pi}{2} + \frac{4n \cos \pi \sin \frac{n\pi}{2}}{\pi(n-2)(n+2)} = -\frac{16}{\pi(n-2)n(n+2)} \sin \frac{n\pi}{2} \quad (1.57)$$

for $n \in \mathbb{N}$ not equal to 0 or 2. We compute separately that

$$a_0(h) = \frac{2}{\pi} \int_0^{\pi/2} (1 + \cos 2x) \, dx - \frac{2}{\pi} \int_{\pi/2}^\pi (1 + \cos 2x) \, dx$$

$$= \frac{2}{\pi} \left[x + \frac{\sin 2x}{2} \right]_0^{\pi/2} - \frac{2}{\pi} \left[x + \frac{\sin 2x}{2} \right]_{\pi/2}^\pi = 0 \quad (1.58)$$

and

$$a_2(h) = \frac{2}{\pi} \int_0^{\pi/2} (1 + \cos 2x) \cos 2x \, dx - \frac{2}{\pi} \int_{\pi/2}^\pi (1 + \cos 2x) \cos 2x \, dx$$

$$= \frac{2}{\pi} \int_0^{\pi/2} \left(\cos 2x + \frac{1 + \cos 4x}{2} \right) dx - \frac{2}{\pi} \int_{\pi/2}^\pi \left(\cos 2x + \frac{1 + \cos 4x}{2} \right) dx$$

$$= \frac{2}{\pi} \left[\frac{\sin 2x}{2} + \frac{x}{2} + \frac{\sin 4x}{8} \right]_0^{\pi/2} - \frac{2}{\pi} \left[\frac{\sin 2x}{2} + \frac{x}{2} + \frac{\sin 4x}{8} \right]_{\pi/2}^\pi = 0. \quad (1.59)$$

Since the sine of an integer multiple of π is zero, we have $\sin(n\pi/2) = 0$ for n even, so we may summarize as follows:

$$a_n(h) = \begin{cases} 0 & \text{if } n \text{ is even,} \\ -\dfrac{16}{\pi(n-2)n(n+2)} \sin \dfrac{n\pi}{2} & \text{if } n \text{ is odd.} \end{cases} \quad (1.60)$$

Consequently, from (1.5) we find that

$$c_n(h) = \begin{cases} 0 & \text{if } n \text{ is even,} \\ -\dfrac{8}{\pi(n-2)n(n+2)} \sin\dfrac{n\pi}{2} & \text{if } n \text{ is odd.} \end{cases} \quad (1.61)$$

So we have the Fourier series

$$h(x) \sim -\frac{16}{\pi} \sum_{\text{odd } n \in \mathbb{Z}^+} \frac{\sin(n\pi/2)}{(n-2)n(n+2)} \cos nx$$

$$\sim -\frac{8}{\pi} \sum_{\text{odd } n \in \mathbb{Z}} \frac{\sin(n\pi/2)}{(n-2)n(n+2)} e^{inx}. \quad (1.62)$$

We can clean this up further by putting $n = 2m - 1$ into either sum. We observe: First, as n runs over all odd, positive integers, m runs over *all* positive integers; second, as n runs over all odd integers, m runs over *all* integers; third,

$$\sin\frac{(2m-1)\pi}{2} = \operatorname{Im}\left(e^{i(2m-1)\pi/2}\right) = \operatorname{Im}\left(e^{i\pi m} e^{-i\pi/2}\right)$$

$$= \operatorname{Im}\left((-1)^m(-i)\right) = (-1)^{m+1}. \quad (1.63)$$

(We've used (1.15).) We get

$$h(x) \sim \frac{16}{\pi} \sum_{m=1}^{\infty} \frac{(-1)^m}{(2m-3)(2m-1)(2m+1)} \cos(2m-1)x$$

$$\sim \frac{8}{\pi} \sum_{m=-\infty}^{\infty} \frac{(-1)^m}{(2m-3)(2m-1)(2m+1)} e^{i(2m-1)x}. \quad (1.64)$$

The fact that $a_n(h) = 0$ for all even n, in the above example, is a consequence of the π-*antiperiodicity* of h, meaning the fact (manifest in Fig. 1.7) that $h(x + \pi) = -h(x)$ for all x. It turns out in general that, if $f \in \operatorname{PC}(\mathbb{T})$ is π-antiperiodic, then $c_n(f) = a_n(f) = b_n(f) = 0$ for all even n. Analogously, if $f \in \operatorname{PC}(\mathbb{T})$ is π-periodic, then these coefficients vanish for all odd n. See Exercises 1.2.4 and 1.2.5 for further exemplifications and Exercise 1.3.3 for proofs.

Incidentally, π-antiperiodicity bears no relation to evenness or oddness. The above function h is π-antiperiodic and even, but $g_{\pi/2}$, with g_r as in Example 1.3.1, is π-antiperiodic and odd. DIY[1]: Work out the details. And the function f of Exercise 1.2.5 is π-antiperiodic but neither even nor odd. Similarly, π-periodicity is independent of evenness/oddness.

We consider one more example, which is of interest for various reasons.

---

[1] "Do It Yourself."

Example 1.3.3 Find the complex exponential form of the Fourier series for the function $j \in PC(\mathbb{T})$ defined by

$$j(x) = \frac{e^{ix}}{2 - e^{ix}}. \tag{1.65}$$

Solution. First we sketch j. We can do so in three dimensions—see Figure 1.8—or, as is perhaps more illuminating, we can sketch $\operatorname{Re} j$ and $\operatorname{Im} j$ separately, as in Figure 1.9.

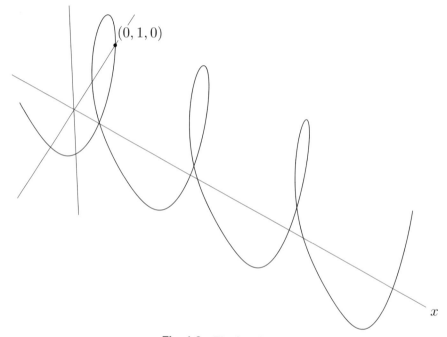

Fig. 1.8 The function j

We have, by the Maclaurin series expansion

$$\frac{z}{1-z} = \sum_{k=1}^{\infty} z^k \tag{1.66}$$

(which is, one shows, valid for z complex as well as real, as long as $|z| < 1$),

$$\begin{aligned}
c_n(j) &= \frac{1}{2\pi} \int_{-\pi}^{\pi} \frac{e^{ix}}{2 - e^{ix}} e^{-inx} \, dx = \frac{1}{2\pi} \int_{-\pi}^{\pi} \frac{e^{ix}/2}{1 - e^{ix}/2} e^{-inx} \, dx \\
&= \frac{1}{2\pi} \int_{-\pi}^{\pi} \sum_{k=1}^{\infty} (e^{ix}/2)^k e^{-inx} \, dx = \frac{1}{2\pi} \sum_{k=1}^{\infty} 2^{-k} \int_{-\pi}^{\pi} e^{i(k-n)x} \, dx \\
&= \begin{cases} 2^{-n} & \text{if } n \geq 1, \\ 0 & \text{if not,} \end{cases}
\end{aligned} \tag{1.67}$$

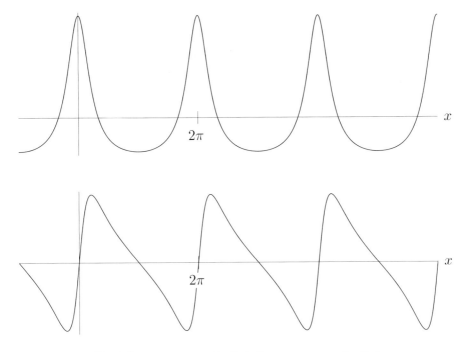

Fig. 1.9 Top: x versus $\operatorname{Re} j(x)$. Bottom: x versus $\operatorname{Im} j(x)$

the last step by Lemma 1.1.2. The application of (1.66) is valid because $|e^{ix}/2| = 1/2 < 1$ for $x \in \mathbb{R}$; the interchange of the sum and the integral may be justified using the convergence of $\sum_{k=1}^{\infty} 2^{-k}$. So

$$j(x) \sim \sum_{n=-\infty}^{\infty} c_n(j) e^{inx} = \sum_{n=1}^{\infty} 2^{-n} e^{inx}. \tag{1.68}$$

Note that the right side of (1.68) in fact *equals* $j(x)$, again by the Maclaurin series (1.66). So Somewhat less imprecise conviction 1.1.1 is accurate at least if, by "reasonable," we mean "equal to j." We'll see in the next section that it's accurate under considerably less stringent reasonability requirements.

The above example illustrates interesting relations between Fourier and Maclaurin series. The general idea is this: Putting $z = \alpha e^{ix}$, or $z = \alpha e^{-ix}$, where α is constant, into a series of the latter type gives one of the former type, under appropriate conditions. We omit further details; most important is a general awareness of the phenomenon.

We conclude with some comparative remarks concerning the functions f, g_r, h, and j of Examples 1.2.1, 1.3.1, 1.3.2, and 1.3.3, respectively. Namely, we observe that j is smoother—that is, of a higher level of continuity and differentiability—than is h, which in turn is smoother than (any of) the g_r's, which in turn are smoother than f. (f is piecewise smooth but discontinuous; each g_r is in PSC($\mathbb{T}$), but no g'_r

22 FOURIER COEFFICIENTS AND FOURIER SERIES

is; h and h' are in PSC($\mathbb{T}$), but h'' is not; all derivatives of j are in PSC($\mathbb{T}$). See Exercise 1.3.18.) We also note that the $c_n(j)$'s decay—that is, approach zero as $n \to \pm\infty$—faster than the $c_n(h)$'s, which decay faster than the $c_n(g_r)$'s (for any $r \in (0, \pi)$), which decay faster than the $c_n(f)$'s. (See Exercise 1.3.19.) A pattern emerges: Seemingly, smoothness of a function is related to rate of decay of its Fourier coefficients.

Here what seems to be the case in fact *is* the case, in a sense that will be fleshed out in Section 1.8. (See Corollary 1.8.1.)

Exercises

1.3.1 Show that, if $g \in$ PC($\mathbb{T}$), then

$$\int_d^{d+2\pi} g(x)\, dx = \int_{-\pi}^{\pi} g(x)\, dx$$

for any $d \in \mathbb{R}$. Hint: Write the integral on the left side as one over $[d, \pi]$ plus one over $[\pi, d + 2\pi]$. Substitute $u = x - 2\pi$ in the second integral; use the periodicity of g to finish.

1.3.2 Prove Proposition 1.3.1(b).

1.3.3 Let f be piecewise continuous on $\mathbb{R}$.

   a. Suppose f is π-antiperiodic, meaning $f(x + \pi) = -f(x)$ for all x. Show that f is 2π-periodic; that $c_n(f) = a_n(f) = b_n(f) = 0$ for all even n; and that $c_n(f) = (1/\pi) \int_0^\pi f(x) e^{-inx}\, dx$ for all odd n. Find analogous formulas for $a_n(f)$ and $b_n(f)$ for n odd. Hint: Write $c_n(f)$ as an integral over $[-\pi, 0]$ plus one over $[0, \pi]$; substitute $u = x + \pi$ in the first integral and add the result to the second one.

   b. Suppose f is π-periodic. Show that f is 2π-periodic; that $c_n(f) = a_n(f) = b_n(f) = 0$ for all odd n; and that $c_n(f) = (1/\pi) \int_0^\pi f(x) e^{-inx}\, dx$ for all even n. Find analogous formulas for $a_n(f)$ and $b_n(f)$ for n even. (Here, by $c_n(f)$, $a_n(f)$, and $b_n(f)$, we mean the Fourier coefficients of f considered as a 2π-periodic function—that is, the Fourier coefficients of (1.27)–(1.29)—not as a π-periodic one.)

   c. According to discussions in Section 1.1, a reasonable π-periodic function should comprise sinusoids of frequencies n/π ($n \in \mathbb{N}$). Explain how the result of part b of this problem reflects those discussions.

1.3.4   **a.** Use (1.41) to show that, for n a nonzero integer,

$$\frac{1}{2\pi} \int_{-\pi}^{\pi} x e^{-inx}\, dx = \frac{i(-1)^n}{n}.$$

   b. Use integration by parts and part a above to show that, for such n,

$$\frac{1}{2\pi} \int_{-\pi}^{\pi} x^2 e^{-inx}\, dx = \frac{2(-1)^n}{n^2}, \qquad \frac{1}{2\pi} \int_{-\pi}^{\pi} x^3 e^{-inx}\, dx = \frac{i(-1)^n \left(\pi^2 n^2 - 6\right)}{n^3}.$$

c. Show that the 2π-periodic function f defined on $(-\pi, \pi]$ by $f(x) = -x^3 + 2\pi x^2 + \pi^2 x$ has Fourier series

$$f(x) \sim \frac{2}{3}\pi^3 + \sum_{n\in\mathbb{Z}-\{0\}} (-1)^n \left[\frac{6i}{n^3} + \frac{4\pi}{n^2}\right] e^{inx}.$$

Also, use formulas (1.6)/(1.7) to find the cosine-sine form of this Fourier series.

1.3.5 Put $\alpha = ib$, where b is a nonzero real number, into (1.41) and equate real and imaginary parts of the result to obtain antiderivatives of $x\cos bx$ and $x\sin bx$.

1.3.6 In each of the Exercises 1.3.7–1.3.15, sketch the graph of, and find both forms of the Fourier series for, the 2π-periodic function f defined by the given formula. Use (1.39), (1.41), Proposition 1.3.1, Exercises 1.3.3 and 1.3.5, and either of the following formulas, when helpful.

$$\int \sin ax \cos bx \, dx = \begin{cases} \dfrac{a\cos ax \cos bx + b\sin ax \sin bx}{b^2 - a^2} + C & \text{if } b \neq a, \\ \dfrac{\cos^2 ax}{2a} + C & \text{if } b = a, \end{cases}$$

$$\int x^2 \cos bx \, dx = \frac{2bx \cos bx + (b^2 x^2 - 2)\sin bx}{b^3} + C \quad (b \neq 0).$$

1.3.7 $f(x) = e^x$ $(d < x \leq d + 2\pi)$, where d is a real number.

1.3.8 $f(x) = x$ $(-\pi < x \leq \pi)$.

1.3.9 $f(x) = x e^{ix/2}$ $(-\pi < x \leq \pi)$. (Sketch Re f and Im f separately.)

1.3.10 $f(x) = \pi \operatorname{sgn} x (\sin x - x) + x^2$ $(-\pi < x \leq \pi)$, where $\operatorname{sgn} x$ is as in (A.12).

1.3.11
$$f(x) = \begin{cases} x & \text{if } -\pi < x \leq 0, \\ ix & \text{if } 0 < x \leq \pi. \end{cases}$$

(Sketch Re f and Im f separately.)

1.3.12 $f(x) = |\sin x|$ $(-\pi < x \leq \pi)$.

1.3.13 $f(x) = |x|$ $(-\pi < x \leq \pi)$.

1.3.14 $f(x) = (\pi - |x|)e^x$ $(-\pi < x \leq \pi)$.

1.3.15
$$f(x) = \begin{cases} px & \text{if } 0 < x \leq L, \\ \dfrac{pL(x - 2\pi)}{L - 2\pi} & \text{if } L < x \leq 2\pi, \end{cases}$$

where $0 < L < 2\pi$ and $p > 0$. (For your sketch, choose convenient particular values of p and L.)

1.3.16 Let $a \in \mathbb{R}$ be a noninteger; let $f \in PC(\mathbb{T})$ be defined by

$$f(x) = (\pi \cos \pi a + ix \sin \pi a)e^{-iax} \quad (-\pi < x \leq \pi).$$

Show that

$$f(x) \sim \frac{1}{\pi} \sum_{n=-\infty}^{\infty} \frac{(-1)^n \sin^2 \pi(a+n)}{(a+n)^2} e^{inx}.$$

It may help to use the fact that $\sin \pi(a+n) = \operatorname{Im} e^{i\pi(a+n)} = \operatorname{Im}((-1)^n e^{i\pi a}) = (-1)^n \sin \pi a$ for $n \in \mathbb{Z}$.

What is the Fourier series for f in the case $a \in \mathbb{Z}$?

1.3.17 (Prelude to part b of the next problem) Suppose that f is continuous at $x = x_0$ and that $\lim_{x \to x_0} f'(x)$ exists. Apply l'Hôpital's rule to the formula

$$f'(x_0) = \lim_{x \to x_0} \frac{f(x) - f(x_0)}{x - x_0}$$

to conclude that f' exists, and is in fact continuous, at $x = x_0$.

1.3.18 Let h be as in Example 1.3.2. Show that:

a. h is continuous on $\mathbb{R}$ (as Fig. 1.7 suggests). Hints: It's clear from the definition of h that it's continuous, and in fact infinitely differentiable, except perhaps at odd multiples of $\pi/2$ and odd multiples of π. So by evenness and periodicity of h, you need only show continuity of h at $x = \pi/2$ and $x = \pi$. That is, show $h(\pi/2) = h((\pi/2)^-) = h((\pi/2)^+)$, and similarly at $x = \pi$. (For the latter, you might want to use the fact that, by Exercise 1.2.3, $h(\pi^+) = h(-\pi^+)$.)

b. The derivative h' is also continuous on $\mathbb{R}$. Again it suffices to consider $x = \pi/2$ and $x = \pi$; you should also use Exercise 1.3.17 above.

c. h'' is discontinuous at $x = \pi/2$.

1.3.19 Let $f, g_r, h,$ and j be as in Examples 1.2.1, 1.3.1, 1.3.2, and 1.3.3, respectively. Show that

$$\lim_{n \to \pm\infty} c_n(f) = \lim_{n \to \pm\infty} n\, c_n(g_r) = \lim_{n \to \pm\infty} n^2\, c_n(h) = \lim_{n \to \pm\infty} n^\ell c_n(j) = 0$$

for any $r \in (0, \pi)$ and $\ell \in \mathbb{Z}^+$, but that none of the limits

$$\lim_{n \to \pm\infty} n\, c_n(f), \quad \lim_{n \to \pm\infty} n^2\, c_n(g_r), \quad \lim_{n \to \pm\infty} n^3\, c_n(h)$$

are zero. What does this say about the relative rates of decay of these Fourier coefficients?

1.4 POINTWISE CONVERGENCE OF FOURIER SERIES: A THEOREM

P. G. L. Dirichlet, in 1829, was the first to meticulously pose and rigorously verify any kind of general result concerning convergence of Fourier series. In this section we present a somewhat simplified variant on his theorem.

We need, first, a fundamental definition and lemma.

Definition 1.4.1 Let $f \in \text{PC}(\mathbb{T})$ and $N \in \mathbb{N}$. *The Nth partial sum S_N^f of the Fourier series for f is defined by*

$$S_N^f(x) = \sum_{n=-N}^{N} c_n(f) e^{inx} = \frac{a_0(f)}{2} + \sum_{n=1}^{N} \bigl(a_n(f) \cos nx + b_n(f) \sin nx\bigr). \tag{1.69}$$

Lemma 1.4.1 (the Riemann-Lebesgue lemma for $\text{PC}(\mathbb{T})$) *If $g \in \text{PC}(\mathbb{T})$, then*

$$\lim_{N \to \infty} a_N(g) = \lim_{N \to \infty} b_N(g) = \lim_{N \to \infty} c_N(g) = \lim_{N \to \infty} c_{-N}(g) = 0. \tag{1.70}$$

Proof. Note that, for $N \in \mathbb{Z}^+$,

$$\sum_{n=-N}^{N} |c_n(g)|^2 - \sum_{n=-(N-1)}^{N-1} |c_n(g)|^2 = |c_N(g)|^2 + |c_{-N}(g)|^2. \tag{1.71}$$

Now if a sequence converges, then the difference between consecutive terms must approach zero as $N \to \infty$. So, if we can show that the sequence whose Nth term is

$$\sum_{n=-N}^{N} |c_n(g)|^2 \tag{1.72}$$

converges, then by (1.71), $c_N(g)$ and $c_{-N}(g)$ will approach zero as $N \to \infty$. But then, by (1.6) and (1.7), so will $a_N(g)$ and $b_N(g)$, and we'll be done.

To demonstrate the required convergence, we first note that, for any functions $g, h \in \text{PC}(\mathbb{T})$,

$$\begin{aligned}
0 &\leq \frac{1}{2\pi} \int_{-\pi}^{\pi} (|g(x)| - |h(x)|)^2 \, dx \\
&= \frac{1}{2\pi} \int_{-\pi}^{\pi} |g(x)|^2 \, dx - 2 \cdot \frac{1}{2\pi} \int_{-\pi}^{\pi} |g(x) h(x)| \, dx + \frac{1}{2\pi} \int_{-\pi}^{\pi} |h(x)|^2 \, dx \\
&\leq \frac{1}{2\pi} \int_{-\pi}^{\pi} |g(x)|^2 \, dx - 2 \left| \frac{1}{2\pi} \int_{-\pi}^{\pi} g(x) \overline{h(x)} \, dx \right| + \frac{1}{2\pi} \int_{-\pi}^{\pi} |h(x)|^2 \, dx. \quad (1.73)
\end{aligned}$$

(To get the last equality, we used the fact that $|ab| = |a\bar{b}|$ and that the integral of an absolute value is greater than or equal to the absolute value of the integral.) Now by

Definition 1.4.1,

$$\frac{1}{2\pi}\int_{-\pi}^{\pi}|S_N^g(x)|^2\,dx = \frac{1}{2\pi}\int_{-\pi}^{\pi}\left(\sum_{n=-N}^{N}c_n(g)e^{inx}\right)\left(\sum_{k=-N}^{N}\overline{c_k(g)}e^{-ikx}\right)dx$$

$$= \sum_{n=-N}^{N}c_n(g)\sum_{k=-N}^{N}\overline{c_k(g)}\cdot\frac{1}{2\pi}\int_{-\pi}^{\pi}e^{inx}e^{-ikx}\,dx$$

$$= \sum_{n=-N}^{N}c_n(g)\sum_{k=-N}^{N}\overline{c_k(g)}\,\delta_{n,k} = \sum_{n=-N}^{N}c_n(g)\overline{c_n(g)}$$

$$= \sum_{n=-N}^{N}|c_n(g)|^2 \qquad (1.74)$$

(we've used Lemma 1.1.2 with $P = 2\pi$) and

$$\frac{1}{2\pi}\int_{-\pi}^{\pi}g(x)\overline{S_N^g(x)}\,dx = \frac{1}{2\pi}\int_{-\pi}^{\pi}g(x)\sum_{n=-N}^{N}\overline{c_n(g)}e^{-inx}\,dx$$

$$= \sum_{n=-N}^{N}\overline{c_n(g)}\cdot\frac{1}{2\pi}\int_{-\pi}^{\pi}g(x)e^{-inx}\,dx$$

$$= \sum_{n=-N}^{N}\overline{c_n(g)}c_n(g) = \sum_{n=-N}^{N}|c_n(g)|^2 \qquad (1.75)$$

as well. So (1.73) with $h = S_N^g$ yields

$$0 \leq \frac{1}{2\pi}\int_{-\pi}^{\pi}|g(x)|^2\,dx - 2\sum_{n=-N}^{N}|c_n(g)|^2 + \sum_{n=-N}^{N}|c_n(g)|^2, \qquad (1.76)$$

or

$$\sum_{n=-N}^{N}|c_n(g)|^2 \leq \frac{1}{2\pi}\int_{-\pi}^{\pi}|g(x)|^2\,dx. \qquad (1.77)$$

Now the *monotone sequence property* of $\mathbb{R}$ says: If $x_N \geq x_{N-1}$ for all $N \in \mathbb{Z}^+$, and if each x_N is less than or equal to a fixed number x, then the sequence x_N has a finite limit L. So (1.77) and the obvious fact that the terms on the left are nondecreasing with N tell us that those terms have a finite limit as $N \to \infty$, which is exactly what we wanted to show. □

Now let us, as is eminently reasonable, phrase the issue of convergence of Fourier series as a question of the approach of S_N^f to f as N gets large. As intimated in the Introduction, there are various plausible notions of "approach" of functions to other functions. In this section we consider perhaps the simplest, which is the "one-point-at-a-time," more formally known as the *pointwise*, notion. In other words, we focus on any particular point x and investigate the behavior of $S_N^f(x)$, which is just a *complex number*, as $N \to \infty$.

We have:

Theorem 1.4.1 *Let $f \in \mathrm{PC}(\mathbb{T})$ and $x \in \mathbb{R}$. If the left- and right-hand limits of the derivative f' both exist at x, then*

$$\lim_{N \to \infty} S_N^f(x) = \frac{1}{2}(f(x^-) + f(x^+)). \tag{1.78}$$

That is, at such a point x, $S_N^f(x)$ converges to the point midway along the "jump" in f at x.

In particular, if x, in addition to satisfying the above conditions, is also a point of continuity of f, then $S_N^f(x)$ converges to $f(x)$.

Proof. Since this is such a big theorem, we should discuss the big idea behind its proof before fleshing out all the details.

The big idea is this: We'll show that, under the stated conditions on f and x,

$$S_N^f(x) - \frac{1}{2}(f(x^-) + f(x^+)) = c_{-(N+1)}(g) - c_N(g) \tag{1.79}$$

for any $N \in \mathbb{N}$, where g is a certain function in $\mathrm{PC}(\mathbb{T})$ that depends on both f and x. By Lemma 1.4.1, $c_{-(N+1)}(g) - c_N(g)$ will go to zero as $N \to \infty$, whence (1.78). The remainder of the theorem will follow from the fact that, if f is continuous at x, then the right side of (1.78) equals $f(x)$.

Now, on to the details. For $f \in \mathrm{PC}(\mathbb{T})$ and $x \in \mathbb{R}$ we compute, using (1.27) (with u as the variable of integration, so as not to confuse this variable with the number x currently under consideration),

$$S_N^f(x) = \sum_{n=-N}^{N} c_n(f) e^{inx} = \sum_{n=-N}^{N} e^{inx} \cdot \frac{1}{2\pi} \int_{-\pi}^{\pi} f(u) e^{-inu} \, du$$

$$= \frac{1}{2\pi} \int_{-\pi}^{\pi} f(u) \left(\sum_{n=-N}^{N} e^{in(x-u)} \right) du = \frac{1}{2\pi} \int_{-\pi}^{\pi} f(u) D_N(x-u) \, du,$$
$$\tag{1.80}$$

where D_N, the "Dirichlet kernel," is defined by

$$D_N(y) = \sum_{n=-N}^{N} e^{iny}. \tag{1.81}$$

Substituting $y = x - u$ gives

$$S_N^f(x) = \frac{1}{2\pi} \int_{x-\pi}^{x+\pi} f(x-y) D_N(y) \, dy = \frac{1}{2\pi} \int_{-\pi}^{\pi} f(x-y) D_N(y) \, dy, \tag{1.82}$$

the last equality because, as observed earlier (see Exercise 1.3.1), the definite integral of a 2π-periodic function over an interval of length 2π does not depend on where that interval starts.

We make some observations regarding D_N, and apply them to (1.82). First, we calculate easily that

$$\frac{1}{2\pi}\int_0^\pi D_N(y)\,dy = \frac{1}{2\pi}\int_{-\pi}^0 D_N(y)\,dy = \frac{1}{2}. \qquad (1.83)$$

(See Exercise 1.4.1.) We may therefore write

$$\frac{1}{2}(f(x^-) + f(x^+)) = f(x^-)\cdot\frac{1}{2\pi}\int_0^\pi D_N(y)\,dy + f(x^+)\cdot\frac{1}{2\pi}\int_{-\pi}^0 D_N(y)\,dy$$

$$= \frac{1}{2\pi}\int_{-\pi}^\pi h(y)D_N(y)\,dy, \qquad (1.84)$$

where

$$h(y) = \begin{cases} f(x^+) & \text{if } -\pi < y < 0, \\ f(x^-) & \text{if } 0 < y \le \pi. \end{cases} \qquad (1.85)$$

Subtracting (1.84) from (1.82), we get

$$S_N^f(x) - \frac{1}{2}(f(x^-) + f(x^+)) = \frac{1}{2\pi}\int_{-\pi}^\pi (f(x-y) - h(y))D_N(y)\,dy. \qquad (1.86)$$

Next we note that, for y not an integer multiple of 2π,

$$D_N(y) = \sum_{n=-N}^N e^{iny} = \sum_{n=-N}^N e^{-iNy}e^{i(N+n)y} = e^{-iNy}\sum_{n=-N}^N (e^{iy})^{N+n}$$

$$= e^{-iNy}\sum_{\ell=0}^{2N}(e^{iy})^\ell = e^{-iNy}\left(\frac{e^{iy(2N+1)}-1}{e^{iy}-1}\right) = \frac{e^{i(N+1)y} - e^{-iNy}}{e^{iy}-1}. \qquad (1.87)$$

To get the fourth equality here, we substituted $\ell = N + n$; to get the fifth, we used the geometric sum formula

$$\sum_{\ell=0}^M r^\ell = \frac{r^{M+1}-1}{r-1} \quad (r\ne 1) \qquad (1.88)$$

(valid here because our restriction on y implies $e^{iy}\ne 1$).

Putting (1.87) into (1.86) gives

$$S_N^f(x) - \frac{1}{2}(f(x^-) + f(x^+)) = \frac{1}{2\pi}\int_{-\pi}^\pi \frac{f(x-y)-h(y)}{e^{iy}-1}(e^{i(N+1)y} - e^{-iNy})\,dy$$

$$= c_{-(N+1)}(g) - c_N(g), \qquad (1.89)$$

where g is the 2π-periodic function defined by

$$g(y) = \frac{f(x-y) - h(y)}{e^{iy} - 1} = \begin{cases} \dfrac{f(x-y) - f(x^+)}{e^{iy} - 1} & \text{if } -\pi < y < 0, \\ \dfrac{f(x-y) - f(x^-)}{e^{iy} - 1} & \text{if } 0 < y \leq \pi. \end{cases} \quad (1.90)$$

We leave g unspecified at $y = 0$—our goal is to show that $g \in \mathrm{PC}(\mathbb{T})$, and toward this end $g(0)$ is immaterial; g needn't even be defined at zero. What is necessary, though, is that g have finite one-sided limits there. We claim that it does, under our hypothesis that $f'(x^-)$ and $f'(x^+)$ both exist.

To prove this claim we note that, by l'Hôpital's rule,

$$\begin{aligned} g(0^+) = \lim_{y \to 0^+} g(y) &= \lim_{y \to 0^+} \frac{f(x-y) - f(x^-)}{e^{iy} - 1} \\ &= \lim_{y \to 0^+} \frac{d(f(x-y) - f(x^-))/dy}{d(e^{iy} - 1)/dy} \\ &= \lim_{y \to 0^+} \frac{-f'(x-y) - 0}{ie^{iy}} = \frac{-f'(x^-)}{i}. \end{aligned} \quad (1.91)$$

The right side exists by assumption, so $g(0^+)$ does too. One shows existence of $g(0^-)$ in much the same way.

But then from (1.90) and the fact that $f \in \mathrm{PC}(\mathbb{T})$, we can conclude that g has at most finitely many discontinuities in $(-\pi, \pi]$, that $g(y^-)$ and $g(y^+)$ exist at each $y \in (-\pi, \pi)$, and that $g(\pi^-)$ and $g(-\pi^+)$ exist too. From this and the 2π-periodicity of g, it follows that $g \in \mathrm{PC}(\mathbb{T})$.

So, by Lemma 1.4.1, the right side of (1.79) approaches zero as $N \to \infty$. But then so does the left side, as claimed, and we're done. □

The above theorem is of such scope, importance, and awe-inspiring beauty, it demands definition of some complex vector spaces that conform especially well to it.

Definition 1.4.2 (a) If f and f' are both piecewise continuous on an interval I (bounded or not), we say f is *piecewise smooth on* I.

We denote by $\mathrm{PS}(I)$ the complex vector space of all piecewise smooth functions on I.

(b) We denote by $\mathrm{PS}(\mathbb{T})$ the complex vector space of all 2π-periodic functions in $\mathrm{PS}(\mathbb{R})$.

(c) Let I be an open interval. If f is piecewise continuous on I, and

$$f(x) = \frac{1}{2}(f(x^-) + f(x^+)) \quad (1.92)$$

for each $x \in I$, then we say f is *averaged* on I.

We denote by PSA(I) the complex vector space of all functions f in PS(I) that are averaged on I.

(d) We denote by PSA($\mathbb{T}$) the complex vector space of all 2π-periodic functions in PSA($\mathbb{R}$) (or equivalently, of all $f \in$ PS($\mathbb{T}$) that are averaged on $\mathbb{R}$).

So a piecewise smooth function f is one such that f and f' are continuous except perhaps at finitely many points on any given bounded interval, and such that the one-sided limits of f and f' exist at every point. An averaged function is one that's piecewise continuous and defined everywhere, and whose value at each point is midway along the jump, if any, in the function at that point.

Theorem 1.4.1 readily implies the following.

Corollary 1.4.1 **(a)** *If $f \in$ PS($\mathbb{T}$), then (1.78) holds for all $x \in \mathbb{R}$. In particular, if such a function f is continuous at x, then*

$$\lim_{N \to \infty} S_N^f(x) = f(x). \tag{1.93}$$

(b) *If $f \in$ PSA($\mathbb{T}$), then (1.93) holds for all $x \in \mathbb{R}$.*

For example, the functions of Examples 1.2.1, 1.3.1, 1.3.2, and 1.3.3 are all in PS($\mathbb{T}$). The function f of Example 1.2.1 is not in PSA($\mathbb{T}$), though, since the value of f at $x = \pi$, or $x = -\pi$, or $x = k\pi$ for any odd integer k, does not equal the average of the limits $f(x^-)$ and $f(x^+)$. See Figure 1.5.

We make one more observation. Corollary 1.4.1(a) says that, if $f \in$ PS($\mathbb{T}$) is continuous everywhere, then S_N^f converges pointwise to f on $\mathbb{R}$, meaning (1.93) holds for all $x \in \mathbb{R}$. For example, the functions g_r, h, and j of Examples 1.3.1, 1.3.2, and 1.3.3, respectively, are all piecewise smooth and continuous, and therefore the Nth partial sum of the Fourier series for any of these functions converges pointwise on $\mathbb{R}$ to that function.

But actually, even nicer things can be said about the convergence of Fourier series of piecewise smooth, continuous functions f. We'll see this in Section 1.7.

Exercises

1.4.1 Use the definition (1.81) of the Dirichlet kernel D_N to prove the formulas (1.83).

1.4.2 Which of the functions of Exercise 1.2.1 are piecewise smooth on $\mathbb{R}$? Which are not? Explain.

1.4.3 **a.** Use Corollary 1.4.1(b) to show that, if $f, g \in$ PSA($\mathbb{T}$) and $c_n(f) = c_n(g)$ for all $n \in \mathbb{Z}$, then $f(x) = g(x)$ for all $x \in \mathbb{R}$.
    b. Explain why the above conclusion need not hold if only $f, g \in$ PS($\mathbb{T}$).
    c. Show that, if $f, g \in$ PS($\mathbb{T}$) and $c_k(f) = c_k(g)$ for all $n \in \mathbb{Z}$, then each bounded interval $[a, b]$ contains at most finitely many points x such that $f(x) \neq g(x)$.

1.5 AN APPLICATION: EVALUATION OF INFINITE SERIES

Evaluation of infinite series is a sport enjoyed by many people. Number theorists, of whom the author is one (as he will admit freely in most social situations), are especially keen on this sport, because of the relevance of these series to the behavior of prime numbers and to other, more or less (or much less) related, phenomena. Others take part because infinite series and their values have applications in chemistry, in physics, and so on. But like all sports, the primary and best reason for participating is just for *fun*. (If anyone knows how to have fun, it's the number theorists, we can assure you.)

One way of evaluating infinite series is via Fourier series. The idea is this: An otherwise inscrutable series can sometimes be expressed in terms of the Fourier series of some appropriate, identifiable function f at some particular point x. Then Theorem 1.4.1, or Corollary 1.4.1, can be applied.

Here's a problem of a kind that a number theorist might enjoy.

Example 1.5.1 Evaluate the alternating sum of reciprocals of products of triples of consecutive odd numbers.

Solution. What does this mean? Well, let's denote an arbitrary odd number by $2m-1$, m being an arbitrary integer. Then the product of that odd number with the preceding and succeeding ones is $(2m-3)(2m-1)(2m+1)$. So the sum in question is

$$\sum_{m=-\infty}^{\infty} \frac{(-1)^m}{(2m-3)(2m-1)(2m+1)} \qquad (1.94)$$

(the $(-1)^m$ is what makes it alternating), which is nice because this series looks a lot like a Fourier series we've seen. It looks like the (complex exponential form of the) Fourier series for the function h of Example 1.3.2.

Indeed, by (1.64) and Corollary 1.4.1(a), and because $h \in \mathrm{PS}(\mathbb{T})$ and h is continuous on $\mathbb{R}$ (see Fig. 1.7), we have

$$\frac{8}{\pi} \sum_{m=-\infty}^{\infty} \frac{(-1)^m}{(2m-3)(2m-1)(2m+1)} e^{i(2m-1)x} = h(x) \qquad (1.95)$$

for any $x \in \mathbb{R}$. To make the left side of this look even more like the series of interest, we let $x = 0$: then (1.95) gives

$$\frac{8}{\pi} \sum_{m=-\infty}^{\infty} \frac{(-1)^m}{(2m-3)(2m-1)(2m+1)} = h(0) = 1 + \cos 0 = 2, \qquad (1.96)$$

or

$$\sum_{m=-\infty}^{\infty} \frac{(-1)^m}{(2m-3)(2m-1)(2m+1)} = \frac{\pi}{4} \qquad (1.97)$$

(which is about 0.785398), and we're done.

32 FOURIER COEFFICIENTS AND FOURIER SERIES

Our next example illustrates, among other things, that care must be taken at points of discontinuity.

Example 1.5.2 Evaluate

$$\sum_{n=1}^{\infty} \frac{1}{1+n^2}. \qquad (1.98)$$

Solution. Again, we seek a function whose Fourier series looks like the infinite series we wish to evaluate. Hey, we've found one: the function f of Example 1.2.1. We have, by (1.38) and Corollary 1.4.1(a),

$$\frac{e^\pi - e^{-\pi}}{2\pi} + \frac{e^\pi - e^{-\pi}}{\pi} \sum_{n=1}^{\infty} \frac{(-1)^n}{1+n^2}(\cos nx - n\sin nx) = \frac{f(x^-) + f(x^+)}{2} \qquad (1.99)$$

for any $x \in \mathbb{R}$. What's a good choice for x? Well, $x = \pi$ is, since not only is $\sin n\pi$ equal to zero, but $\cos n\pi = (-1)^n$ (by equating real parts on either side of (1.15), for example), whence $(-1)^n \cos n\pi = ((-1)^n)^2 = 1$, all of this for $n \in \mathbb{Z}$. So (1.99) gives

$$\frac{e^\pi - e^{-\pi}}{2\pi} + \frac{e^\pi - e^{-\pi}}{\pi} \sum_{n=1}^{\infty} \frac{1}{1+n^2} = \frac{f(\pi^-) + f(\pi^+)}{2}. \qquad (1.100)$$

Now $f(x) = e^x$ on $(-\pi, \pi]$, $f(\pi^-) = e^\pi$, and $f(\pi^+) = f(-\pi^+) = e^{-\pi}$. (We've used Exercise 1.2.3 in the evaluation of $f(\pi^+)$.) So (1.100) reads

$$\frac{e^\pi - e^{-\pi}}{2\pi} + \frac{e^\pi - e^{-\pi}}{\pi} \sum_{n=1}^{\infty} \frac{1}{1+n^2} = \frac{e^\pi + e^{-\pi}}{2}, \qquad (1.101)$$

or

$$\sum_{n=1}^{\infty} \frac{1}{1+n^2} = \frac{\pi}{e^\pi - e^{-\pi}}\left(\frac{e^\pi + e^{-\pi}}{2} - \frac{e^\pi - e^{-\pi}}{2\pi}\right) = \frac{\pi(e^\pi + e^{-\pi})}{2(e^\pi - e^{-\pi})} - \frac{1}{2} \qquad (1.102)$$

(which is approximately equal to 1.07667), and we're done.

The approximate numerical values cited above are of secondary interest. Often, what one really wants to know is the *kind* of number to which a given infinite series converges. Is this number *algebraic*, meaning it solves an equation of the form $p(x) = 0$, where p is a polynomial with integer coefficients? For example, $\frac{4}{5}$ is algebraic, because it solves $5x - 4 = 0$. Similarly, so is any rational number. So is $\sqrt{2}$, since it solves $x^2 - 2 = 0$. So is

$$1 + \sqrt{1 + \sqrt[3]{575}} - \sqrt{2 - \sqrt[3]{575} - \frac{48}{\sqrt{1 + \sqrt[3]{575}}}}, \qquad (1.103)$$

because it solves $x^4 - 4x^3 + 200x - 200 = 0$. (You should check this by hand.[2]) Or is the number in question not algebraic, in which case it's called *transcendental*?

Among the most famous transcendental numbers are π and e. (See [8] for proofs.) So, for example, the series of Example 1.5.1 evaluates to a transcendental number, since a transcendental number times a nonzero rational one is readily seen to be transcendental. (It's not known whether the right side of (1.102) is transcendental.)

Our next example is also of number theoretic interest, and will prepare us for our last, which is a classic.

Example 1.5.3 Evaluate the sum of reciprocals of squares of odd, positive integers.
Solution. The series we're interested in is

$$\sum_{m=1}^{\infty} \frac{1}{(2m-1)^2}. \tag{1.104}$$

Let's consider the function g_r of Example 1.3.1, with $r = x = \pi/2$. By (1.48), Corollary 1.4.1(a), and the fact that each g_r is continuous,

$$\frac{4}{\pi} \sum_{n=1}^{\infty} \frac{\sin^2(n\pi/2)}{n^2} = g_{\pi/2}\left(\frac{\pi}{2}\right) = \frac{\pi}{2}. \tag{1.105}$$

Now $\sin^2(n\pi/2)$ equals zero if n is even and $(\pm 1)^2 = 1$ if it's odd, so we get

$$\frac{4}{\pi} \sum_{\text{odd } n \in \mathbb{Z}^+} \frac{1}{n^2} = \frac{\pi}{2} \tag{1.106}$$

or, putting $n = 2m - 1$,

$$\sum_{m=1}^{\infty} \frac{1}{(2m-1)^2} = \frac{\pi}{2} \cdot \frac{\pi}{4} = \frac{\pi^2}{8}. \tag{1.107}$$

Now for an example that's surely on almost every number theorist's top ten infinite series list.

Example 1.5.4 Evaluate the sum of reciprocals of squares of *all* positive integers.
Solution. Let's give this sum a name:

$$S = \sum_{n=1}^{\infty} \frac{1}{n^2}. \tag{1.108}$$

We express S as a sum of two series, one of which we just evaluated, as follows:

$$S = \sum_{\text{even } n \in \mathbb{Z}^+} \frac{1}{n^2} + \sum_{\text{odd } n \in \mathbb{Z}^+} \frac{1}{n^2} = \sum_{m=1}^{\infty} \frac{1}{(2m)^2} + \sum_{m=1}^{\infty} \frac{1}{(2m-1)^2}. \tag{1.109}$$

---

[2]Just kidding.

Now the series on the far right side equals $\pi^2/8$, by our previous example; moreover,

$$\sum_{m=1}^{\infty} \frac{1}{(2m)^2} = \frac{1}{2^2} \sum_{m=1}^{\infty} \frac{1}{m^2} = \frac{1}{4} S. \tag{1.110}$$

So (1.109) reads

$$S = \frac{1}{4} S + \frac{\pi^2}{8}; \tag{1.111}$$

solving for S gives

$$S = \frac{\pi^2}{6}. \tag{1.112}$$

The series of the above two examples are thus seen to converge to transcendental numbers. (It follows from the definitions that the square of a transcendental number is transcendental.)

The series of the last example is just the value at $s = 2$ of the *Riemann zeta function* $\zeta(s)$, defined by

$$\zeta(s) = \sum_{n=1}^{\infty} \frac{1}{n^s}. \tag{1.113}$$

Here s may be complex but, as one shows, it must have real part larger than one if the series on the right is to converge absolutely. On the other hand, the definition of $\zeta(s)$ may be *extended* in a natural way—this extension is called *the meromorphic continuation of ζ*—to encompass all complex numbers s except for $s = 1$.

The Riemann zeta function was introduced by Bernhard Riemann (hence the name), in 1841 (see [39]). Regarding this function he proved many things, and made several conjectures, some of which are still open. Most notable is the famed *Riemann hypothesis*. This is the conjecture that all solutions to the equation $\zeta(s) = 0$, with the exception of the even, negative integers s (all of which *are* solutions, as is well known), have real part equal to $\frac{1}{2}$. It's fair to say that, now that Fermat's last theorem has been proved, the Riemann hypothesis is foremost among all open questions in number theory.

Another open problem regards the values of ζ at certain special points. The question is this: Is $\zeta(2M + 1)$, for $M \in \mathbb{Z}^+$, algebraic or transcendental? The answer is not known, even for a *single* value of M. It's straightforward enough, using Fourier series arguments, to evaluate $\zeta(s)$ at any *even* positive integer s, much as we just did (in Example 1.5.4) for $s = 2$. See Exercise 1.8.7 for the cases $s = 4$ and $s = 6$, and Exercise 3.7.3 for the cases $s = 8$ and $s = 10$. It turns out that such values of ζ are all transcendental, and in fact are closely related to the so-called *Bernoulli numbers*. But the Fourier series method, like all others so far, for that matter, has failed to handle odd, positive values of s.

It's not even known whether, for all such $s \geq 3$, $\zeta(s)$ is *irrational*. In 1978 Apéry demonstrated the irrationality of $\zeta(3)$; in 2000 Rivoal [40] proved the irrationality of

$\zeta(2M+1)$ for infinitely many $M \in \mathbb{Z}^+$; Zudilin [51] and others have since sharpened Rivoal's results. But the general case remains unsettled.

There's still much more to be said about the zeta function. It plays *the* central role in the original 1896 proofs, by Hadamard and (independently) de la Vallee Poussin, of the "prime number theorem." This celebrated theorem describes how many prime numbers there are, roughly, between 1 and a given positive number x.

And so on. We refer the reader to [3] for more cool facts concerning ζ.

To conclude this section we note, while we're on the subject of Riemann, that is was he who introduced the now standard, ubiquitous Riemann sums and Riemann definite integrals. (Hence the names.) One aspect of Riemann's integration theory that's particularly fascinating from our perspective is that Riemann developed it specifically to address questions regarding convergence of Fourier series. In fact, the paper in which his integrals first appeared is called On the representation of a function by a trigonometric series (see [39]).

Exercises

Evaluate the given infinite series, using the Fourier series for the function from the indicated example or exercise.

1.5.1
$$\sum_{n=1}^{\infty} \frac{(-1)^n}{1+n^2};$$
Example 1.2.1.

1.5.2
$$\sum_{n=1}^{\infty} \frac{\cos(2\pi n/3)}{n^2},$$
Example 1.3.1 and the fact that $\cos 2\theta = 1 - 2\sin^2 \theta$. You may also want to use the result of Example 1.5.4.

1.5.3
$$\sum_{n=1}^{\infty} \frac{\sin na}{n}, \text{ where } 0 < a < \pi;$$
Exercise 1.2.10.

1.5.4
$$\sum_{n=1}^{\infty} \frac{\sin nb}{n}, \text{ where } \pi < b < 2\pi;$$
Exercise 1.2.10 and the fact that $\sin 2\theta = 2\cos\theta \sin\theta$ (with $\theta = nb/2$).

1.5.5
$$\sum_{n=1}^{\infty} \frac{(-1)^n}{4n^2 - 1};$$
Exercise 1.2.14 (with $a = \frac{1}{2}$).

1.5.6
$$\sum_{m=1}^{\infty} \frac{(-1)^m}{(4m^2-1)(4m^2-25)};$$
Exercise 1.2.12.

1.5.7
$$\sum_{n=1}^{\infty} \frac{1}{a^2+n^2}$$
($a \in \mathbb{R}$); Exercise 1.2.9.

1.6 GIBBS' PHENOMENON

> [There is a difference between] *the limit of the graphs*, and ... *the graph of the limit* of the sum. A misunderstanding on this point is a natural consequence of the usage which allows us to omit the word *limit* in certain connections as when we speak of the sum of an infinite series.
>
> —J. W. Gibbs [22]

A. A. Michelson is known especially for his part in the Michelson-Morley experiments on the speed of light, and for his receipt in 1907 of the Nobel Prize, for his design of accurate optical instruments.

Somewhat less well known is the fact that Michelson's talent for invention had an impact on Fourier analysis. It happened like this: Around 1898 Michelson built a "harmonic analyzer," which was a device for *drawing* partial sums S_N^f of Fourier series. And while his was not the first such device—Lord Kelvin had earlier built one—it was, for its time, by far the best, and rendered quite precise drawings. (One of the advantages of Michelson's machine was that it used *springs* where Kelvin's used *strings*, and this allowed for increased precision.)

Michelson generated pictures of S_N^f for a variety of functions f and for N up to about 80. The mathematical physicist J. W. Gibbs (whose *Elementary Principles in Statistical Mechanics* [23] is considered a classic in the field) got his hands, or at least his eyes, on some of these pictures. In a pair of 1899 letters to *Nature* magazine, he commented on some interesting features of these drawings near points of discontinuity of the function being analyzed.

To discuss Gibbs' observations, let's consider, as he did, the "sawtooth function" σ, which is the element of PSA($\mathbb{T}$) defined by

$$\sigma(x) = x \quad (-\pi < x < \pi). \tag{1.114}$$

By Corollary 1.4.1(b), the Fourier series for σ converges pointwise to σ—but, as Gibbs noted, it does so in quite a remarkable manner. See Figure 1.10.

For each value of N, there seem to be points just to the left of $x = \pi$ where S_N^σ *diverges from* σ by a considerable amount—and in an *unexpected* direction! That is, although the precipitous *decline* in S_N^σ, as $x \to \pi^-$, is not surprising, given the

GIBBS' PHENOMENON

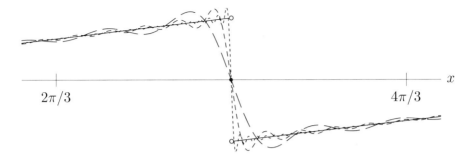

Fig. 1.10 The sawtooth function σ (solid) and some partial sums S_N^σ (dashed. Shorter dashes connote larger N)

downward jump in σ at $x = \pi$, it is quite curious that, before this decline, S_N^σ should *spike upward*! (Similarly curious behavior of this partial sum is exhibited just to the right of $x = \pi$.)

Further, these spikes do not appear to be decreasing in height as N gets larger. And indeed they aren't, as the following proposition testifies.

Proposition 1.6.1 *Let $N \in \mathbb{Z}^+$. If*

$$x_N = \frac{2N-1}{2N+1}\pi, \tag{1.115}$$

then

$$S_N^\sigma(x_N) - \sigma(x_N) > 2\int_0^\pi \frac{\sin \psi}{\psi}\,d\psi - \pi \approx 0.5622. \tag{1.116}$$

Proof. We have

$$S_N^\sigma(x) = \frac{1}{i}\sum_{0<|n|\leq N} \frac{(-1)^{n+1}}{n} e^{inx} = 2\sum_{n=1}^{N} \frac{(-1)^{n+1}}{n} \sin nx. \tag{1.117}$$

(See Exercise 1.3.8.) Differentiating the first equality in (1.117) gives

$$(S_N^\sigma)'(x) = \frac{1}{i}\sum_{0<|n|\leq N} \frac{(-1)^{n+1}}{n} in\, e^{inx} = \sum_{0<|n|\leq N} (-1)^{n+1} e^{inx}$$

$$= -\sum_{0<|n|\leq N} e^{in(x-\pi)}, \tag{1.118}$$

the last step by (1.15). Subtracting both sides of (1.118) from 1 then gives

$$1 - (S_N^\sigma)'(x) = 1 + \sum_{0<|n|\leq N} e^{in(x-\pi)} = \sum_{n=-N}^{N} e^{in(x-\pi)} = D_N(x-\pi), \tag{1.119}$$

where D_N is the Dirichlet kernel of (1.81).

By the fundamental theorem of calculus,

$$\int_{x_N}^{\pi} [1 - (S_N^\sigma)'(x)] \, dx = \pi - S_N^\sigma(\pi) - (x_N - S_N^\sigma(x_N))$$
$$= S_N^\sigma(x_N) - \sigma(x_N) + \pi, \qquad (1.120)$$

since $S_N^\sigma(\pi) = 0$ (for example, by the second equality in (1.117)) and $x_N = \sigma(x_N)$. So if we integrate both sides of (1.119) from x_N to π, we get

$$S_N^\sigma(x_N) - \sigma(x_N) + \pi = \int_{x_N}^{\pi} D_N(x - \pi) \, dx$$
$$= \int_0^{\pi - x_N} D_N(y) \, dy = \int_0^{2\pi/(2N+1)} \frac{\sin(N+1/2)y}{\sin(y/2)} \, dy, \qquad (1.121)$$

the second equality here is by the substitution $y = \pi - x$ and the fact that D_N is even, the third because, by (1.87),

$$D_N(y) = \frac{e^{-iy/2}}{e^{-iy/2}} \cdot \frac{e^{i(N+1)y} - e^{-iNy}}{e^{iy} - 1}$$
$$= \frac{e^{i(N+1/2)y} - e^{-i(N+1/2)y}}{e^{iy/2} - e^{-iy/2}} = \frac{\sin(N+1/2)y}{\sin(y/2)}. \qquad (1.122)$$

Now

$$\frac{1}{\sin x} > \frac{1}{x} \quad \text{for } 0 < x < \pi \qquad (1.123)$$

(see Exercise 1.6.1). So (1.121) gives

$$S_N^\sigma(x_N) - \sigma(x_N) + \pi > \int_0^{2\pi/(2N+1)} \frac{\sin(N+1/2)y}{y/2} \, dy = 2 \int_0^{\pi} \frac{\sin \psi}{\psi} \, d\psi; \qquad (1.124)$$

we got the last equality by substituting $\psi = (N + \frac{1}{2})y$. This gives us the inequality in (1.116); the approximate equality there is obtained by numerical methods. □

Here, then, is what's happening with σ: Pick your favorite small, positive number ε, where "small" means "no larger than the constant to the right of the inequality in (1.116)." Then draw a tube of radius ε about the graph of σ (that is, draw the graphs of $\sigma - \varepsilon$ and $\sigma + \varepsilon$). Then for any particular $x \in \mathbb{R}$ we can, by taking N large enough, ensure that the point $(x, S_N^\sigma(x))$ lies inside this tube. This is what Corollary 1.4.1(b) says. On the other hand, for *any* positive integer N, *some part of the graph of S_N^σ will* "overshoot" the tube near, and in the direction opposite to, the jump in σ at $x = \pi$. This is what Proposition 1.6.1 tells us. (See Fig. 1.11.)

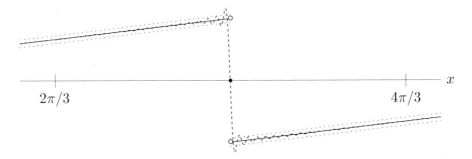

Fig. 1.11 A tube (dotted) of radius $\varepsilon = 0.2$ about the graph of σ (solid) and a partial sum S_N^σ (dashed)

Observe that the jump in σ at $x = \pi$ has height

$$|\sigma(\pi^+) - \sigma(\pi^-)| = |-\pi - \pi| = 2\pi. \tag{1.125}$$

We can then rephrase Proposition 1.6.1 as follows: The amount by which S_N^σ overshoots σ near $x = \pi$ is, for any $N \in \mathbb{Z}^+$, greater than Q times this jump height, where by definition

$$Q = \frac{1}{2\pi}\left[2\int_0^\pi \frac{\sin\psi}{\psi}d\psi - \pi\right]. \tag{1.126}$$

One computes that $Q \approx 0.0894899$. So, in particular, the overshoot is more than 8.9% of the jump height.

The point of this rephrasing is that it exemplifies a general phenomenon. Namely, suppose $f \in \text{PSA}(\mathbb{T})$ is discontinuous at x. It can be shown that, for any N large enough, there is a number x_N such that $S_N^f(x_N)$ will overshoot $f(x_N)$, in a direction opposite the jump in f at x, by an amount larger than $Q|f(x^-) - f(x^+)|$. (The essential idea here is that a given $f \in \text{PSA}(\mathbb{T})$ can be written as a piecewise smooth, *continuous* function plus a linear combination of sawtooth-like functions. See Exercise 1.6.2. The "continuous part" of f will behave nicely, by Theorem 1.7.1 below, but the "sawtooth part" of f will experience Gibbs' phenomenon.) In particular, this overshoot will not approach zero in height as $N \to \infty$.

On the other hand, at least in the case of σ, the overshoot does appear to shrink to zero in *width*, and also to move closer and closer to the point of discontinuity, as N gets larger. (See again Fig. 1.10.) And this is exactly what happens for general $f \in \text{PSA}(\mathbb{T})$.

The behavior just described of partial sums of Fourier series is called "Gibbs' phenomenon." And Gibbs' observations generated a heated discussion on Fourier series and how they converge. To be fair, it should be noted that similar (and in fact somewhat more thorough) observations had actually already been made by the mathematician Wilbraham nearly 60 years earlier. The lack of attention given Wilbraham's remarks may have been due largely to the fact that, at the time he made them, there were no decent harmonic analyzer machines, and therefore no good pictures to back up the words.

Finally, since Gibbs' phenomenon occurs near points of discontinuity, we should be able to avoid this phenomenon by avoiding functions with such points. And indeed we can, in a sense to be discussed in the next section.

Exercises

1.6.1 Prove (1.123). Hint: Consider the behavior of $f(x) = x - \sin x$ on $[0, \pi]$, by looking at f'.

1.6.2 The purpose of this exercise is to show that any $f \in \mathrm{PSA}(\mathbb{T})$ can be written as a function $g \in \mathrm{PSC}(\mathbb{T})$ plus a finite sum of (perhaps shifted and rescaled) sawtooth functions. Here's how:

    a. First consider the case where the discontinuities of $f \in \mathrm{PSA}(\mathbb{T})$ are "like those of σ"; that is, the *only* discontinuity of f on $(-\pi, \pi]$ is at $x = \pi$. Show that, in this case, $f = g + J\sigma$, where $J = (f(\pi^-) - f(\pi^+))/(2\pi)$ and $g \in \mathrm{PSC}(\mathbb{T})$. Hint: define $g = f - J\sigma$. To show $g \in \mathrm{PSC}(\mathbb{T})$, it's enough (explain why) to show $g(\pi^-) = g(\pi^+)$.

    b. Modify your result from part a to suit the case where $f \in \mathrm{PSA}(\mathbb{T})$ still has only one discontinuity on $(-\pi, \pi]$, but now that discontinuity is at an arbitrary point x_0 in this interval.

    c. Generalize your result from part b to suit the case of general $f \in \mathrm{PSA}(\mathbb{T})$; that is, f can now have a finite number of discontinuities on $(-\pi, \pi]$.

1.6.3 Using a calculator or a computer, verify Proposition 1.6.1 numerically for five or six successively larger values of N, at least two of which should be greater than 20.

1.6.4 If $g \in \mathrm{PSA}(\mathbb{T})$ equals 1 on $(0, \pi/2)$, equals -1 on $(\pi/2, \pi)$, and is even and 2π-periodic, then

$$S_N^g(x) = \frac{4}{\pi} \sum_{n=1}^{N} \frac{\sin(\pi n/2)}{n} \cos nx.$$

(This is an easy direct computation; or put $r = \pi/2$ into Example 1.8.1, below.)

    a. Sketch the graph of g.

    b. Using a calculator or a computer, compute $\frac{1}{2}[S_N^g(x_N) - g(x_N)]$, where $x_N = (2N - 3)\pi/(4N + 2)$, for five or six successively larger values of N, at least two of which should be greater than 20. Explain how your results support the general discussion following Figure 1.11.

1.6.5 Using a calculator or a computer, evaluate

$$R_N = \frac{1}{2\pi} + \frac{1}{\pi} \sum_{n=1}^{N} \frac{(-1)^n}{1 + n^2} (\cos nx_N - n \sin nx_N) - \frac{e^{x_N}}{e^{\pi} - e^{-\pi}},$$

with x_N as in (1.115), for five or six successively larger values of N, at least some of which should be greater than 100. What does your result have to do with Gibbs' phenomenon? Explain as completely as you can. Hint: Let f be as in Example 1.2.1. What is $f(x_N)$? What is $S_N^f(x_N)$? What's the jump in f at $x = \pi$?

1.7 UNIFORM CONVERGENCE OF FOURIER SERIES: A THEOREM

By Corollary 1.4.1(b), PSA($\mathbb{T}$) is a pretty good complex vector space from the point of view of convergence of Fourier series. But it has its disadvantages. One of these is this: We can't always assure, given an element f of this space, that $S_N^f - f$ will become "uniformly small" as N gets large. Specifically, if $f \in $ PSA($\mathbb{T}$) has discontinuities then, given ε small enough, there will be no N large enough that the graph of S_N^f sits completely within a tube of radius ε about the graph of f. See Figure 1.11 and, in general, the discussions of the previous section. So $S_N^f - f$, while approaching zero one point at a time, is not doing so in a certain more uniform sense, and we find this vaguely unsettling.

And there are more practical problems with PSA($\mathbb{T}$). Here's a big one: *You can't always differentiate Fourier series of functions there term by term.* For instance, since the sawtooth function $\sigma(x)$, defined in the previous section, equals x on $(-\pi, \pi)$, equation (1.117) and Corollary 1.4.1(b) imply that

$$x = 2 \sum_{n=1}^{\infty} \frac{(-1)^{n+1}}{n} \sin nx \quad (-\pi < x < \pi). \tag{1.127}$$

Formally differentiating term by term gives

$$1 = 2 \sum_{n=1}^{\infty} (-1)^{n+1} \cos nx \quad (-\pi < x < \pi), \tag{1.128}$$

which is *false*: The series on the right doesn't even converge; it's nth term does not approach zero. Indeed, by the Riemann-Lebesgue lemma (Lemma 1.4.1), this series is not the Fourier series for *any* $g \in$ PC($\mathbb{T}$).

We can alleviate such problems as these by introducing a new notion of convergence (whose name we've already intimated) and a new space of 2π-periodic functions whose Fourier series are harmonious with this notion. We begin with the first task, for which we'll need the following.

Definition 1.7.1 Let X be a set; let g be a function from X to $\mathbb{R}$. Then the *supremum*, or "sup," of g on X, denoted

$$\sup_{x \in X} g(x), \tag{1.129}$$

is the least upper bound of the set $\{g(x)\colon x \in X\}$. That is, $\sup_{x \in X} g(x)$ denotes the smallest number c such that

$$g(x) \leq c \text{ for all } x \in X. \tag{1.130}$$

If there is no number c satisfying (1.130), then we write $\sup_{x \in X} g(x) = \infty$.

Some discussion: If the set $\{g(x) \colon x \in X\}$ has a maximum value M, then $\sup_{x \in X} g(x)$ equals M. For example, $\sup_{x \in [0,1]} x(1-x) = 1/4$, which *is* the maximum value of $g(x) = x(1-x)$ on $(0, 1)$. But the point of Definition 1.7.1 is that a supremum may exist where a maximum value doesn't: For instance, $\sup_{x \in (0,2)} x^2 + 1 = 5$, even though $g(x) = x^2 + 1$ never *quite* attains the value 5 on $(0, 2)$. But almost.

We may now present our new, more holistic, notion of convergence.

Definition 1.7.2 We say that a sequence $f_1, f_2, \ldots$ of functions on a set X *converges uniformly to f on X* if

$$\lim_{N \to \infty} \sup_{x \in X} |f_N(x) - f(x)| = 0. \tag{1.131}$$

If X is understood, we simply say that the sequence *converges uniformly to f*.

We will always try to relate new notions of convergence of functions to old ones. For example, we have:

Proposition 1.7.1 *If f_N converges uniformly to f on X, then f_N converges pointwise to f on X.*

Proof. For any particular $x_0 \in X$ we have, by definition of supremum,

$$0 \leq |f_N(x_0) - f(x_0)| \leq \sup_{x \in X} |f_N(x) - f(x)|. \tag{1.132}$$

Suppose f_N converges uniformly to f on X. Then the sup on the right goes to 0 as $N \to \infty$; by the squeeze law, so does $|f_N(x_0) - f(x_0)|$. So $\lim_{N \to \infty} f_N(x_0) = f(x_0)$. This means f_N converges pointwise to f on X (since x_0 was arbitrary), and we're done. □

The converse of the above proposition is false, as Gibbs' phenomenon illustrates nicely. Specifically, for σ the sawtooth function of the previous section, and x_N as in (1.115), we have

$$0 \leq |S_N^\sigma(x_N) - \sigma(x_N)| \leq \sup_{x \in \mathbb{T}} |S_N^\sigma(x) - \sigma(x)|. \tag{1.133}$$

If the sup on the right approached zero as $N \to \infty$, then so would the quantity in the middle, by the squeeze law again. But the latter quantity does not, by (1.116), so the right side does not, so S_N^σ does not converge uniformly to σ on $\mathbb{T}$. (But again, S_N^σ *does* converge pointwise to σ on $\mathbb{T}$.)

Similar remarks apply to any discontinuous function $f \in \text{PSA}(\mathbb{T})$. On the other hand, one *can* show that, if f is an element of this space and I is any closed interval not containing any of the discontinuities of f, then S_N^f converges uniformly to f on

I. We omit this proof, except in the very special case when $I = \mathbb{R}$—that is, when f has *no* discontinuities. This is the case to which we now turn our attention.

We have:

Definition 1.7.3 We denote by PSC($\mathbb{T}$) the complex vector space of all *continuous* functions in PSA($\mathbb{T}$).

It might help to keep in mind the following chain of inclusions:

$$\text{PSC}(\mathbb{T}) \subset \text{PSA}(\mathbb{T}) \subset \text{PS}(\mathbb{T}) \subset \text{PC}(\mathbb{T}). \tag{1.134}$$

Later, we'll tack another space $L^2(\mathbb{T})$ on to the right end of this chain; shortly thereafter, we'll augment the chain still further to the right with the space $L^1(\mathbb{T})$.

For the present, though, our aim is to prove that PSC($\mathbb{T}$) is a very nice place to be, from the point of view of uniform convergence of Fourier series. To this end, we need a couple of lemmas.

Lemma 1.7.1 *If* $f \in \text{PSC}(\mathbb{T})$, *then*

$$c_n(f') = in\, c_n(f), \quad a_n(f') = n\, b_n(f), \quad b_n(f') = -n\, a_n(f) \tag{1.135}$$

for all relevant values of n.

Proof. If $f \in \text{PSC}(\mathbb{T})$, then $f' \in \text{PC}(\mathbb{T})$, so the Fourier coefficients of f' are indeed defined.

To be specific, we have

$$c_n(f') = \frac{1}{2\pi} \int_{-\pi}^{\pi} f'(x) e^{-inx}\, dx. \tag{1.136}$$

We integrate by parts with $u = e^{-inx}$ and $dv = f'(x)\, dx$ to get

$$c_n(f') = \frac{1}{2\pi} f(x) e^{-inx} \Big|_{-\pi}^{\pi} + in \cdot \frac{1}{2\pi} \int_{-\pi}^{\pi} f(x) e^{-inx}\, dx = in\, c_n(f). \tag{1.137}$$

(Since $f(x)e^{-inx}$ is 2π-periodic and continuous, $f(x)e^{-inx}\big|_{-\pi}^{\pi} = 0$.)

The stated formulas for $a_n(f')$ and $b_n(f')$ are left to Exercise 1.7.1. □

Lemma 1.7.2 (Bessel's inequality for PC($\mathbb{T}$)) *If* $g \in \text{PC}(\mathbb{T})$, *then*

$$\sum_{n=-\infty}^{\infty} |c_n(g)|^2 \leq \frac{1}{2\pi} \int_{-\pi}^{\pi} |g(x)|^2\, dx; \tag{1.138}$$

in particular, the sum on the left converges.

Proof. This is just the inequality (1.77) in the limit as $N \to \infty$ (the bounded monotone sequence property of $\mathbb{R}$ tells us that (1.77) continues to hold in this limit). □

(See Exercise 1.7.2 for a cosine/sine form of the above proposition.)

Later (cf. Corollary 3.7.1(b)), we'll see that Bessel's inequality is actually an equality.

We're ready for the main result of this section.

Theorem 1.7.1 *Let $f \in \mathrm{PSC}(\mathbb{T})$. Then S_N^f converges uniformly to f on $\mathbb{T}$.*

Proof. Let $f \in \mathrm{PSC}(\mathbb{T})$; we will first show that

$$\sum_{n=-\infty}^{\infty} |c_n(f)| < \infty. \tag{1.139}$$

To do so, we note that

$$\sum_{n=-\infty}^{\infty} |c_n(f)| = |c_0(f)| + \sum_{n \neq 0} |c_n(f)| = |c_0(f)| + \sum_{n \neq 0} \left| \frac{c_n(f')}{in} \right|, \tag{1.140}$$

the last step by Lemma 1.7.1. Now if $z, w \in \mathbb{C}$, then

$$|z|^2 - 2|zw| + |w|^2 = (|z| - |w|)^2 \geq 0, \tag{1.141}$$

so

$$|zw| \leq \frac{|z|^2 + |w|^2}{2}. \tag{1.142}$$

Applying this with $z_n = c_n(f')$ and $w_n = (in)^{-1}$ for each $n \neq 0$, we deduce from (1.140) that

$$\sum_{n=-\infty}^{\infty} |c_n(f)| \leq |c_0(f)| + \frac{1}{2} \sum_{n \neq 0} |c_n(f')|^2 + \frac{1}{2} \sum_{n \neq 0} \frac{1}{n^2}. \tag{1.143}$$

The first sum on the right converges by Lemma 1.7.2, with $g = f'$, and the second by the p-series test from calculus, with $p = 2$. Thus (1.139) holds, as required.

To demonstrate uniform convergence of S_N^f to f, we note that, if $x \in \mathbb{T}$, then by Corollary 1.4.1(b) and the fact that $\mathrm{PSC}(\mathbb{T}) \subset \mathrm{PSA}(\mathbb{T})$,

$$|S_N^f(x) - f(x)| = \left| \sum_{n=-N}^{N} c_n(f) e^{inx} - \sum_{n=-\infty}^{\infty} c_n(f) e^{inx} \right|$$

$$= \left| \sum_{|n|>N} c_n(f) e^{inx} \right| \leq \sum_{|n|>N} |c_n(f) e^{inx}| = \sum_{|n|>N} |c_n(f)|. \tag{1.144}$$

The inequality follows from the triangle inequality (Proposition A.0.3(c)), extended to infinite series, as is legal provided these series converge absolutely. Now (1.144) says that the sum on the far right side therein is an upper bound for the set $\{|S_N^f(x) - f(x)|: x \in \mathbb{T}\}$. The least upper bound is, by its definition, less than or equal to any upper bound, so (1.144) gives

$$0 \leq \sup_{x \in \mathbb{R}} |S_N^f(x) - f(x)| \leq \sum_{|n|>N} |c_n(f)|. \tag{1.145}$$

As the right side is the tail end of a convergent series—namely, the series in (1.139)—it must approach zero as $N \to \infty$. So by the squeeze law, so does the sup in the the middle, whence our theorem. □

Note that, by (1.139),

$$\sum_{n=-\infty}^{\infty} |c_n(f)e^{inx}| < \infty \tag{1.146}$$

for any $x \in \mathbb{R}$. In other words, Fourier series of functions in PSC($\mathbb{T}$) converge *absolutely*, as well as uniformly, on $\mathbb{R}$.

In geometric terms, Theorem 1.7.1 says this: no matter how narrow a tube we draw about the graph of a given $f \in \text{PSC}(\mathbb{T})$, we can capture S_N^f completely within this tube by choosing N large enough. See Figure 1.12.

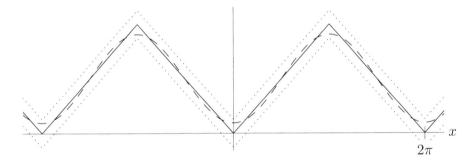

Fig. 1.12 A function $f \in \text{PSC}(\mathbb{T})$ (solid), a narrow tube (dotted) about the graph of f, and the second partial sum (dashed) of the Fourier series for f

We close this section with some general remarks on convergence of Fourier series. Let's, to keep the discussions simple, focus on Corollary 1.4.1(b) and Theorem 1.7.1. Note the parallel between the two: Each identifies a large complex vector space V (equal to PSA($\mathbb{T}$) in the first case and PSC($\mathbb{T}$) in the second) such that, if $f \in V$, then S_N^f converges to f in a certain basic sense (pointwise in the first case; uniformly in the second).

So we could say that PSA($\mathbb{T}$) is *compatible*, from a Fourier series perspective, with pointwise convergence, and PSC($\mathbb{T}$) with uniform convergence. Compatible yes, but not *backward compatible*. In other words, in neither case can the "if" be replaced by

46 FOURIER COEFFICIENTS AND FOURIER SERIES

"if and only if." To see this, we consider the 2π-periodic function g defined by

$$g(x) = \sum_{r=0}^{\infty} \frac{\sin(2^r x)}{2^r}, \qquad (1.147)$$

one of a large class of functions investigated by Weierstrass, ca. 1872. The following may be shown: On the one hand, S_N^g converges uniformly—and hence also pointwise—to g on $\mathbb{T}$; on the other hand, g is *not* in $\mathrm{PSC}(\mathbb{T})$, or even in $\mathrm{PSA}(\mathbb{T})$. In fact, while g is continuous *everywhere*, it's differentiable *nowhere*! See Figure 1.13.

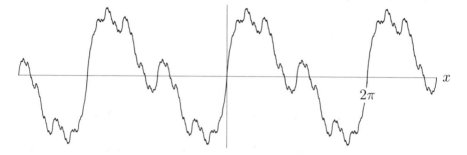

Fig. 1.13 A continuous, nowhere differentiable, 2π-periodic function g

So certainly the hypotheses of both Corollary 1.4.1(b) and Theorem 1.7.1 may be weakened. But how much? One might, on the basis of Weierstrass' example, guess that the hypothesis "f is 2π-periodic and continuous" is sufficient to imply the conclusions of this corollary, or that theorem, or both. Is it?

No! In fact, there's a 2π-periodic, continuous function $h \colon \mathbb{R} \to \mathbb{C}$ and an $x_0 \in \mathbb{R}$ such that $S_N^h(x_0)$ is *unbounded*, meaning for any positive number M there's a positive integer N with $|S_N^h(x_0)| > M$. So certainly $S_N^h(x_0) \not\to h(x_0)$ ($S_N^h(x_0)$ doesn't converge to anything) as $N \to \infty$. The construction of such an h is somewhat, though not terribly, involved; we refer the reader to Chapter X of [30]. We note further that, if one wants to work quite a bit harder, one can obtain a 2π-periodic, continuous $k \colon \mathbb{R} \to \mathbb{C}$ such that $S_N^k(x)$ is unbounded for *infinitely many* $x \in [-\pi, \pi]$! See Chapter VIII in [52].

The moral is this: The complex vector space

$$\{f \colon \mathbb{T} \to \mathbb{C} \colon S_N^f \text{ converges to } f\} \qquad (1.148)$$

is somewhat hard to pin down, at least if, by "converges," one means in either the pointwise or uniform sense. Indeed, for either of these senses, no simple characterization of this space is known.

However, there *is* a complex vector space V, and a meaningful sense of convergence of functions in V, such that V is both forward and backward compatible with this sense, from a Fourier series point of view. In other words, $f \in V$ if and only if S_N^f converges to f in this sense. We're referring here to the vector space $L^2(\mathbb{T})$, and to the notion of *norm* convergence. We'll examine this space and notion in Chapter 3 and will see that they provide, in many ways, the optimal setting for development of Fourier series.

Exercises

1.7.1 Derive the formulas for $a_n(f')$ and $b_n(f')$ of Lemma 1.7.1 in each of the following ways:

    a. From the formula already found for $c_n(f')$, and from (1.6)/(1.7).

    b. Using integration by parts, as in the proof of that lemma, and formulas (1.28)/(1.29).

1.7.2 By writing out the squares of the absolute values (that is, using the fact that $|z|^2 = z\bar{z}$) and applying equations (1.6) and (1.7), show that

$$|a_n(g)|^2 + |b_n(g)|^2 = 2\bigl(|c_n(g)|^2 + |c_{-n}(g)|^2\bigr)$$

for $g \in \text{PC}(\mathbb{T})$ and $n \in \mathbb{Z}^+$. Deduce that, for such g,

$$\frac{1}{4}|a_0(g)|^2 + \frac{1}{2}\sum_{n=1}^{\infty}\bigl(|a_n(g)|^2 + |b_n(g)|^2\bigr) \le \frac{1}{2\pi}\int_{-\pi}^{\pi}|g(x)|^2\,dx.$$

1.7.3 **a.** Is there a function $g \in \text{PC}(\mathbb{T})$ such that $|c_n(g)| > |n|^{-1/3}$ for all $n > 12$? Explain.

    b. Is there a function $h \in \text{PSC}(\mathbb{T})$ such that $|a_n(h)| > |n|^{-14/15}$ for all $n > 6.02 \times 10^{23}$? Explain.

1.8 DERIVATIVES, ANTIDERIVATIVES, AND FOURIER SERIES

Note that Lemma 1.7.1 gives

$$f'(x) \sim \sum_{n=-\infty}^{\infty} in\,c_n(f)e^{inx} \sim \sum_{n=1}^{\infty}\bigl(n\,b_n(f)\cos nx - n\,a_n(f)\sin nx\bigr) \quad (1.149)$$

for $f \in \text{PSC}(\mathbb{T})$. Since

$$in\,c_n(f)e^{inx} = \frac{d}{dx}c_n(f)e^{inx} \quad (1.150)$$

and

$$n\,b_n(f)\cos nx - n\,a_n(f)\sin nx = \frac{d}{dx}\bigl(a_n(f)\cos nx + b_n(f)\sin nx\bigr), \quad (1.151)$$

(1.149) tells us that the Fourier series for f' are obtained simply by termwise differentiation of those for f. Moreover, if f' is nice enough, then $S_N^{f'}$ will converge, to $f'(x)$ or at least to $(f'(x^-) + f'(x^+))/2$, in one sense (like a pointwise one)

48 FOURIER COEFFICIENTS AND FOURIER SERIES

or another (like a uniform one). The upshot is that Fourier series of derivatives of functions in PSC($\mathbb{T}$) behave as we would expect and hope them to. Recall from our discussions in the previous section that the same cannot be said about piecewise smooth, *discontinuous* periodic functions.

Let's consider an application of the above observations.

Example 1.8.1 Find the Fourier series for the function $k_r \in \text{PSA}(\mathbb{T})$ defined by

$$k_r(x) = \begin{cases} 1 & \text{if } -r < x < r, \\ \dfrac{r}{r - \pi} & \text{if } r < x < 2\pi - r, \end{cases} \tag{1.152}$$

where $0 < r < \pi$. (See Fig 1.14.)

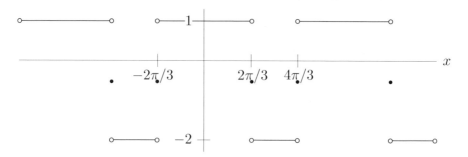

Fig. 1.14 The function $k_{2\pi/3}$

Solution. Note that, if g_r is as in Example 1.3.1, then $k_r(x) = g'_r(x)$ at all points x except those where g_r has sharp corners. (At those points, g'_r is undefined; on the other hand, k_r is defined everywhere, by our assumption that it's periodic and averaged.) And this is enough for k_r and g'_r to have the same Fourier series, since Fourier coefficients are definite integrals, and changing the values of a function at finitely many points on an interval does not change the definite integral of that function there.

Now it's clear from Figure 1.6 that $g_r \in \text{PSC}(\mathbb{T})$, so the Fourier series for g'_r, and thus for k_r, may be obtained by termwise differentiation of the series for g_r. So by (1.44) and (1.48),

$$k_r(x) \sim \frac{1}{\pi - r} \sum_{n \neq 0} \frac{\sin nr}{n} e^{inx} \sim \frac{2}{\pi - r} \sum_{n=1}^{\infty} \frac{\sin nr}{n} \cos nx. \tag{1.153}$$

Since $k_r \in \text{PSA}(\mathbb{T})$ we may, by Corollary 1.4.1(b), replace the above associations with equalities for any $x \in \mathbb{R}$.

Here's another thing to observe about Lemma 1.7.1. Applying it recursively, one gets the following. (Here, as usual, $f^{(m)}$ denotes the mth derivative of f, the 0th derivative meaning the function itself.)

Proposition 1.8.1 *Let* $m \in \mathbb{Z}^+$. *If* $f^{(m-1)} \in \mathrm{PSC}(\mathbb{T})$, *then*

$$c_n(f^{(m)}) = (in)^m c_n(f). \tag{1.154}$$

Writing $n^m = 1/n^{-m}$ on the right side of the above equality and applying the Riemann-Lebesgue lemma (Lemma 1.4.1) to the left, we deduce:

Corollary 1.8.1 *Let* $m \in \mathbb{Z}^+$. *If* $f^{(m-1)} \in \mathrm{PSC}(\mathbb{T})$, *then*

$$\lim_{n \to \pm\infty} \frac{c_n(f)}{n^{-m}} = 0. \tag{1.155}$$

That is, if $f^{(m-1)} \in \mathrm{PSC}(\mathbb{T})$, then $c_n(f)$ decays, meaning it approaches zero as $n \to \pm\infty$, "faster than n^{-m} does."

Now $n^{-(m+1)}$ decays faster than n^{-m}, and moreover a function whose mth derivative is in $\mathrm{PSC}(\mathbb{T})$ is, typically, smoother than one whose $(m-1)$st derivative is. So Corollary 1.8.1 quantifies our observation, made at the end of Section 1.3 (recall also Exercises 1.3.18 and 1.3.19), that increasing the smoothness of f increases the rate of decay of $c_n(f)$. Similar remarks apply to $a_n(f)$'s and $b_n(f)$'s.

This correspondence, between smoothness in the original domain and rate of decay in the frequency domain, is a reflection of the *local/global principle*, which says: Local, or *concentrated*, effects in either domain translate into global, or *spread out*, effects in the other. That Corollary 1.8.1 really does reflect this principle can be seen as follows. Decay of Fourier coefficients *is* a global phenomenon; it's observed as you travel a long, long way (out to ∞) across the frequency domain. On the other hand, smoothness *is* a local property, in the sense that discontinuities, or nondifferentiabilities, can be introduced into a function by altering the values of that function on just a tiny portion of its original domain.

The local/global principle cannot be encapsulated by a single theorem; rather, it's an overriding, somewhat philosophical, perspective with many different specific realizations. We'll encounter more of these realizations as we progress.

But now, back to our program. Let's reexamine some of the above ideas regarding Fourier series and their derivatives from the reverse perspective. That is, let's consider how Fourier series behave with respect to *antidifferentiation*. We have:

Proposition 1.8.2 *Let* $f \in \mathrm{PC}(\mathbb{T})$ *satisfy* $c_0(f) = 0$; *define* $F: \mathbb{R} \to \mathbb{C}$ *by*

$$F(x) = \int_0^x f(y)\, dy. \tag{1.156}$$

Then $F \in \mathrm{PSC}(\mathbb{T})$, *and for* $n \neq 0$,

$$c_n(F) = \frac{c_n(f)}{in}, \quad a_n(F) = -\frac{b_n(f)}{n}, \quad b_n(F) = \frac{a_n(f)}{n}. \tag{1.157}$$

Proof. The piecewise smoothness and continuity of the antiderivative F follow from properties of Riemann integrals. Moreover, for an arbitrary $d \in \mathbb{R}$ we have

$$F(d+2\pi) - F(d) = \int_0^{d+2\pi} f(y)\, dy - \int_0^d f(y)\, dy = \int_d^{d+2\pi} f(y)\, dy = c_0(f), \tag{1.158}$$

the last step by (1.39). But $c_0(f) = 0$ by assumption, so $F(d + 2\pi) - F(d) = 0$ as well. So F is 2π-periodic, and therefore in $\mathrm{PSC}(\mathbb{T})$, as required.

So by Lemma 1.7.1 and the fact that $f = F'$,
$$c_n(f) = in\, c_n(F), \quad a_n(f) = n\, b_n(F), \quad b_n(f) = -n\, a_n(F). \tag{1.159}$$
For $n \neq 0$, these equations are equivalent to (1.157). $\square$

The above proposition says that, for f and F as defined there,
$$F(x) \sim c_0(F) + \sum_{n \neq 0} \frac{c_n(f)}{in} e^{inx}$$
$$\sim \frac{a_0(F)}{2} + \sum_{n=1}^{\infty} \left(-\frac{b_n(f)}{n} \cos nx + \frac{a_n(f)}{n} \sin nx \right). \tag{1.160}$$

In other words, if $f \in \mathrm{PC}(\mathbb{T})$ has constant term zero, then the Fourier series for the antiderivative (1.156) of f is just the termwise antiderivative of the Fourier series for f, with the appropriate constant term $c_0(F) = a_0(F)/2$ thrown in. (This constant term is computed according to the usual integral formulas.) And of course since $F \in \mathrm{PSC}(\mathbb{T})$, its Fourier series converges absolutely, and uniformly to F, on $\mathbb{R}$.

Example 1.8.2 Using Proposition 1.8.2 and the Fourier series for the sawtooth function σ of (1.114), Find the Fourier series for the function $v \in \mathrm{PSC}(\mathbb{T})$ defined by
$$v(x) = x^2 \quad (-\pi < x < \pi). \tag{1.161}$$
(See Fig. 1.15.)

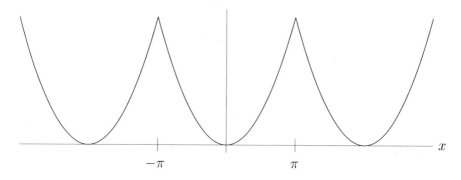

Fig. 1.15 The function v

Solution. By (1.117), we have
$$\sigma(x) \sim 2 \sum_{n=1}^{\infty} \frac{(-1)^{n+1}}{n} \sin nx. \tag{1.162}$$

(The association may be replaced by an equality, by Corollary 1.4.1(b).) In particular, the constant term $c_0(\sigma) = a_0(\sigma)/2$ is zero. So by Proposition 1.8.2 we conclude that, if

$$\Sigma(x) = \int_0^x \sigma(y)\,dy, \tag{1.163}$$

then

$$\Sigma(x) = c_0(\Sigma) + 2\sum_{n=1}^{\infty} \frac{(-1)^n}{n^2} \cos nx. \tag{1.164}$$

(The series on the right in fact converges absolutely, and uniformly to Σ, on $\mathbb{R}$.)

We claim that $\Sigma = v/2$. To prove this we need only, because of the periodicity and continuity of both functions, show that $\Sigma(x) = v(x)/2$ for each $x \in (-\pi, \pi)$. But for such x we have, by (1.163),

$$\Sigma(x) = \int_0^x y\,dy = \frac{x^2}{2} = \frac{v(x)}{2}, \tag{1.165}$$

as required.

So (1.164) and the fact that

$$c_0(\Sigma) = \frac{1}{2\pi}\int_{-\pi}^{\pi} \Sigma(x)\,dx = \frac{1}{2\pi}\int_{-\pi}^{\pi} \frac{x^2}{2}\,dx = \frac{\pi^2}{6} \tag{1.166}$$

imply

$$v(x) = \frac{\pi^2}{3} + 4\sum_{n=1}^{\infty} \frac{(-1)^n}{n^2} \cos nx. \tag{1.167}$$

One may apply (1.167) toward another evaluation of the sum of the squares of the reciprocals of the positive integers: See Exercise 1.8.5.

The condition $c_0(f) = 0$, in Proposition 1.8.2, is crucial; without it F is not 2π-periodic, because of (1.158). But note that, if $f \in PC(\mathbb{T})$ and we define a modified function $g \in PC(\mathbb{T})$ by

$$g(x) = f(x) - c_0(f), \tag{1.168}$$

then g *always* has zeroth Fourier coefficient equal to zero, even when f does not. Proof:

$$c_0(g) = c_0(f - c_0(f)) = c_0(f) - c_0(f) = 0, \tag{1.169}$$

since the zeroth Fourier coefficient of a constant function is equal to that constant. So we can apply Proposition 1.8.2 to g.

Here's an illustration of this idea and some others.

Example 1.8.3 Use the function v of the previous example to evaluate

$$\sum_{m=1}^{\infty} \frac{(-1)^m}{(2m-1)^3}. \tag{1.170}$$

Solution. By (1.167),

$$v(x) - \frac{\pi^2}{3} \sim 4 \sum_{n=1}^{\infty} \frac{(-1)^n}{n^2} \cos nx. \tag{1.171}$$

The function on the left, let's call it w, indeed has zeroth Fourier coefficient equal to zero. So by Proposition 1.8.2, if

$$W(x) = \int_0^x w(y)\,dy = \int_0^x \left(v(y) - \frac{\pi^2}{3}\right) dy, \tag{1.172}$$

then for any x

$$W(x) = c_0(W) + 4\sum_{n=1}^{\infty} \frac{(-1)^n}{n^3} \sin nx. \tag{1.173}$$

(Again, the series on the right converges absolutely, and uniformly to W, on $\mathbb{R}$.)
Now if $-\pi < y < \pi$, then $v(y) = y^2$. So (1.172) gives

$$W(x) = \int_0^x \left(y^2 - \frac{\pi^2}{3}\right) dy = \frac{x^3 - \pi^2 x}{3} \quad (-\pi < x < \pi); \tag{1.174}$$

consequently

$$c_0(W) = \frac{1}{2\pi} \int_{-\pi}^{\pi} \frac{x^3 - \pi^2 x}{3}\,dx = 0. \tag{1.175}$$

So by (1.173),

$$W(x) = 4\sum_{n=1}^{\infty} \frac{(-1)^n}{n^3} \sin nx \quad (x \in \mathbb{R}). \tag{1.176}$$

To finish, we put $x = \pi/2$ into (1.176). Since, by (1.174), $W(\pi/2) = -\pi^3/8$, and since the sine of an integer multiple of π is zero, we get

$$-\frac{\pi^3}{8} = 4 \sum_{\text{odd } n \in \mathbb{Z}^+} \frac{(-1)^n}{n^3} \sin \frac{n\pi}{2} = 4 \sum_{m=1}^{\infty} \frac{(-1)^{2m-1}}{(2m-1)^3} \sin \frac{(2m-1)\pi}{2}$$

$$= 4 \sum_{m=1}^{\infty} \frac{(-1)^m}{(2m-1)^3}, \tag{1.177}$$

the second equality by putting $n = 2m - 1$ and the last by (1.63) (and the fact that $(-1)^{2m-1} = -1$ for any integer m). Solving gives

$$\sum_{m=1}^{\infty} \frac{(-1)^m}{(2m-1)^3} = -\frac{\pi^3}{32}, \tag{1.178}$$

and we're done.

We just evaluated the alternating sum of reciprocals of cubes of odd positive integers. Leave out the adjectives "alternating" and "odd," and the sum we're talking about is

$$\sum_{n=1}^{\infty} \frac{1}{n^3}. \tag{1.179}$$

The latter sum is, in the language of Section 1.5, just $\zeta(3)$. And as we discussed near the end of that section, it would be quite nice to get a handle on $\zeta(3)$.

It sure seems as though this could be accomplished by properly tweaking Example 1.8.3—that is, by choosing a different value of x to plug in, or modifying v or W slightly, or something. But once one actually experiments with such tweaks, it doesn't seem that way at all. Still, it was worth a try.

Exercises

1.8.1 Use (1.149), together with the result of Example 1.3.2, to find the cosine-sine Fourier series for the function $g \in \text{PC}(\mathbb{T})$ defined by

$$g(x) = \begin{cases} \sin 2x & \text{if } 0 \leq |x| \leq \frac{\pi}{2}, \\ -\sin 2x & \text{if } \frac{\pi}{2} \leq |x| \leq \pi. \end{cases}$$

Does this Fourier series converge pointwise to g? Uniformly? Explain.

1.8.2 Find, and discuss convergence of, the cosine-sine Fourier series for the following function $f \in \text{PC}(\mathbb{T})$:

$$f(x) = \begin{cases} \cos 2x & \text{if } 0 \leq |x| \leq \frac{\pi}{2}, \\ -\cos 2x & \text{if } \frac{\pi}{2} \leq |x| \leq \pi. \end{cases}$$

Hint: (1.149) and the previous exercise.

1.8.3 Sketch the graph of the 2π-periodic function g defined on $(-\pi, \pi]$ by $g(x) = e^{ix/2}(2+ix)$. Using (1.149) and Exercise 1.3.9, find the complex exponential Fourier series for g. Also discuss convergence of this series.

1.8.4 Sketch the graph of the 2π-periodic function h defined on $(-\pi, \pi]$ by $h(x) = x(\pi - |x|)$. Using Proposition 1.8.2 and Exercise 1.3.13, find the cosine-sine Fourier series for h. Also discuss convergence of this series.

1.8.5 Use the result of Example 1.8.2 to evaluate (again)
$$\zeta(2) = \sum_{n=1}^{\infty} \frac{1}{n^2}.$$

1.8.6 Start with the result of Example 1.8.3 (see especially (1.174) and (1.176)), and use Proposition 1.8.2 successively to find the cosine-sine Fourier series for each of the following 2π-periodic functions (defined by the given formula on $(-\pi, \pi]$).
  a. $h(x) = x^4 - 2\pi^2 x^2$.
  b. $k(x) = 3x^5 - 10\pi^2 x^3 + 7\pi^4 x$.
  c. $\ell(x) = x^6 - 5\pi^2 x^4 + 7\pi^4 x^2$.

1.8.7 Use parts a and c of the previous exercise to evaluate
$$\zeta(s) = \sum_{n=1}^{\infty} \frac{1}{n^s}$$
for $s = 4$ and $s = 6$.

1.8.8 Let g be the 2π-periodic function defined by
$$g(x) = \begin{cases} -x(2\pi + x) & \text{if } -\pi < x \leq 0; \\ \pi x & \text{if } 0 < x \leq \pi. \end{cases}$$

  a. Sketch the graph of g.
  b. What does
$$\sum_{n=-\infty}^{\infty} (-1)^n n \, c_n(g)$$
converge to? To answer, *don't* actually compute the $c_n(g)$'s; just use (1.149) and an appropriate convergence-of-Fourier-series result.

1.8.9 Let ℓ be the 2π-periodic function defined by $\ell(x) = \sqrt{\pi^2 - x^2}$ for $-\pi < x \leq \pi$.
  a. Sketch the graph of ℓ. Is ℓ in $PC(\mathbb{T})$? In $PS(\mathbb{T})$? Explain.
  b. Compute the Fourier series for ℓ. The Fourier coefficients of ℓ (besides the zeroth one) are not expressible in terms of familiar elementary functions, but you can, and should, write them in terms of the J-Bessel function $J_1(r)$, defined, for $r > 0$, by
$$J_1(r) = \frac{2r}{\pi} \int_0^1 \sqrt{1 - t^2} \cos(rt) \, dt.$$

  c. It may be shown that $J_1(r)$ "looks like $(2/(\pi r))^{1/2} \sin(r - \pi/4)$" as $r \to \pm\infty$; specifically, $\lim_{r \to \infty} J_1(r)/[(2/(\pi r))^{1/2} \sin(r - \pi/4)] = 1$. Assuming this, discuss the function ℓ in question here, and its Fourier series, as they relate to the local/global principle expressed above. Hint: From part b of this problem, you should see that the $c_n(\ell)$'s "decay like a constant times $|n|^{-1/2}$" as $n \to \pm\infty$. How does this decay rate compare to those of the Fourier coefficients of the functions of Exercises 1.3.18 and 1.3.19, say? And how smooth is the present function ℓ compared to those functions?

(In Sections 4.3–4.5, we'll discuss Bessel functions in some detail.)

1.9 FUNCTIONS OF OTHER PERIODS $P > 0$

We write $\mathbb{T}_P$ for the torus, or circle, of circumference P—in particular, $\mathbb{T}$ is short for $\mathbb{T}_{2\pi}$. By a function on $\mathbb{T}_P$ we mean one of period P.

Everything we've said so far, in the last seven sections, about functions on $\mathbb{T}$ generalizes in completely natural, unsurprising ways to $\mathbb{T}_P$. Or to put it another way: Our more or less precise convictions of Section 1.1, regarding P-periodic functions, turn out to be accurate, as exemplified throughout the last seven sections by the case $P = 2\pi$.

To see this, we put $P^* = 2\pi$ into the observations with which we commenced Section 1.2. That is, if f is P-periodic then, defining

$$f^*(x) = f\left(\frac{P}{2\pi}x\right), \tag{1.180}$$

we find that f^* is 2π-periodic. It's easy enough to check that f^* is piecewise continuous on $\mathbb{R}$ if and only if f is, and similarly for piecewise smoothness, for continuity, and for the property of being averaged. So we can deduce all sorts of Fourier series facts concerning "reasonable" P-periodic functions f by applying to f^* results already deduced, and then using a change of variable to get back to where we want to be.

Here's an implementation of such arguments.

Theorem 1.9.1 *Let $P > 0$. For f in the space $\mathrm{PC}(\mathbb{T}_P)$ of piecewise continuous, P-periodic functions, $x \in \mathbb{R}$, and $N \in \mathbb{N}$, define*

$$\begin{aligned}S_N^f(x) &= \sum_{n=-N}^{N} c_n(f) e^{2\pi i n x/P} \\ &= \frac{a_0(f)}{2} + \sum_{n=1}^{N}\left(a_n(f)\cos\frac{2\pi nx}{P} + b_n(f)\sin\frac{2\pi nx}{P}\right),\end{aligned} \tag{1.181}$$

with the $c_n(f)$'s, $a_n(f)$'s, and $b_n(f)$'s given by (1.18), (1.19), and (1.20), respectively. If the limits $f'(x^-)$ and $f'(x^+)$ both exist, then

$$\lim_{N\to\infty} S_N^f(x) = \frac{1}{2}(f(x^-) + f(x^+)). \tag{1.182}$$

Proof. If f and x are as stated and f^* is as in (1.180), then $f^* \in \mathrm{PC}(\mathbb{T})$, and moreover the one-sided limits $(f^*)'((2\pi x/P)^-)$ and $(f^*)'((2\pi x/P)^+)$ both exist.

So by Theorem 1.4.1,

$$\lim_{N\to\infty} S_N^{f^*}\left(\frac{2\pi x}{P}\right) = \frac{1}{2}\left[f^*\left(\left(\frac{2\pi x}{P}\right)^-\right) + f^*\left(\left(\frac{2\pi x}{P}\right)^+\right)\right]. \quad (1.183)$$

But

$$c_n(f) = \frac{1}{P}\int_{-P/2}^{P/2} f(x)e^{-2\pi inx/P}\,dx = \frac{1}{P}\cdot\frac{P}{2\pi}\int_{-\pi}^{\pi} f(u)e^{-2\pi in(Pu/(2\pi))/P}\,du$$

$$= \frac{1}{2\pi}\int_{-\pi}^{\pi} f(u)e^{-inu}\,du = c_n(f^*), \quad (1.184)$$

the second equality by the substitution $u = 2\pi x/P$. Consequently,

$$S_N^f(x) = \sum_{n=-N}^{N} c_n(f)e^{2\pi inx/P} = \sum_{n=-N}^{N} c_n(f^*)e^{in(2\pi x/P)} = S_N^{f^*}\left(\frac{2\pi x}{P}\right). \quad (1.185)$$

Putting this into (1.183) gives

$$\lim_{N\to\infty} S_N^f(x) = \frac{1}{2}\left[f^*\left(\left(\frac{2\pi x}{P}\right)^-\right) + f^*\left(\left(\frac{2\pi x}{P}\right)^+\right)\right] = \frac{1}{2}(f(x^-) + f(x^+)), \quad (1.186)$$

the last step following from the definition (1.180) of f^*. We're done. □

As in the case $P = 2\pi$, we call the $a_n(f)$'s, $b_n(f)$'s, and $c_n(f)$'s the *Fourier coefficients of* f; we note that, in the formulas defining these coefficients, the domain of integration may be replaced by any interval of length P. Also as before, we call S_N^f the *Nth partial sum of the Fourier series for* f; we call the series in (1.3) and (1.4) the *Fourier series for* f.

By direct computation or by the conversion formulas (1.5)–(1.7), we see that (1.184) still holds if c_n is replaced by a_n or by b_n on both sides. In sum, Fourier coefficients are invariant under horizontal rescaling of the function in question.

In addition to Theorem 1.9.1, one also has the completely natural, unsurprising P-periodic generalizations of Corollary 1.4.1, of Gibbs' phenomenon, of Theorem 1.7.1, and for that matter of every other 2π-periodic result obtained above.

Example 1.9.1 Compute, and discuss convergence of, the Fourier series for the 2-periodic, averaged function f defined by

$$f(x) = e^{-\pi ix/3} \quad (-1 < x < 1). \quad (1.187)$$

(See Fig. 1.16.)

FUNCTIONS OF OTHER PERIODS $P > 0$

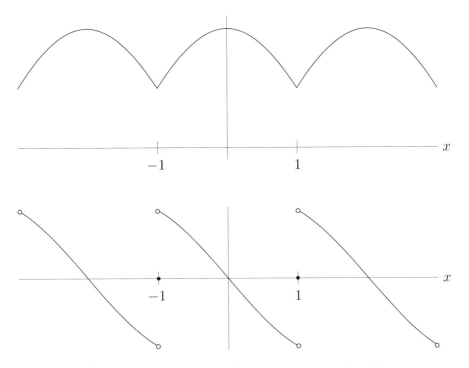

Fig. 1.16 Top: x versus Re $f(x)$. Bottom: x versus Im $f(x)$

Solution. By (1.18),

$$c_n(f) = \frac{1}{2}\int_{-1}^{1} f(x) e^{-2\pi i n x/2}\, dx = \frac{1}{2}\int_{-1}^{1} e^{-\pi i(n+1/3)x}\, dx$$

$$= \left.\frac{e^{-\pi i(n+1/3)x}}{-2\pi i(n+1/3)}\right|_{-1}^{1} = \frac{e^{\pi i(n+1/3)} - e^{-\pi i(n+1/3)}}{2\pi i(n+1/3)} = \frac{3\sqrt{3}(-1)^n}{2\pi(3n+1)}, \tag{1.188}$$

the last step by (1.15), the fact that $(2i)^{-1}(e^{\pi i/3} - e^{-\pi i/3}) = \sin(\pi/3) = \sqrt{3}/2$, and some algebra. So

$$f(x) \sim \frac{3\sqrt{3}}{2\pi} \sum_{n=-\infty}^{\infty} \frac{(-1)^n}{3n+1} e^{\pi i n x}. \tag{1.189}$$

We also have, from (1.19),

$$a_n(f) = \frac{3\sqrt{3}(-1)^n}{2\pi(3n+1)} + \frac{3\sqrt{3}(-1)^{-n}}{2\pi(-3n+1)} = \frac{3\sqrt{3}(-1)^n}{2\pi}\left[\frac{1}{3n+1} + \frac{1}{-3n+1}\right]$$

$$= \frac{3\sqrt{3}(-1)^n}{2\pi}\left[\frac{-3n+1+3n+1}{1-9n^2}\right] = \frac{3\sqrt{3}(-1)^n}{\pi(1-9n^2)}. \tag{1.190}$$

Similarly (DIY: fill in the details),

$$b_n(f) = -\frac{9i\sqrt{3}(-1)^n n}{\pi(1-9n^2)}. \tag{1.191}$$

So

$$f(x) \sim \frac{3\sqrt{3}}{2\pi} + \frac{3\sqrt{3}}{\pi} \sum_{n=1}^{\infty} \frac{(-1)^n}{1-9n^2}[\cos \pi nx - 3i n \sin \pi nx]. \tag{1.192}$$

It may readily be shown that, as the graph suggests, f is piecewise smooth but not continuous. So, by the 2-periodic generalizations of Corollary 1.4.1(b) and of the discussions in Sections 1.6 and 1.7, the Fourier series for f converges pointwise, but not uniformly, to f.

In the above example, we performed P-periodic Fourier analysis by way of explicit formulas from Section 1.1. But sometimes, it's easier to work directly with the function f^* of (1.180), or with the change of variable used to define that function. This will certainly be the case when f^* is a function whose Fourier coefficients we already computed, sometime back when we were considering the case $P = 2\pi$, since computations are particularly easy when they've already been done.

For instance:

Example 1.9.2 Use Example 1.8.2 to show that

$$x^2 = 3 + \frac{36}{\pi^2} \sum_{n=1}^{\infty} \frac{(-1)^n}{n^2} \cos \frac{\pi nx}{3} \quad (-3 < x < 3). \tag{1.193}$$

Solution. In that example, we saw that

$$x^2 = \frac{\pi^2}{3} + 4 \sum_{n=1}^{\infty} \frac{(-1)^n}{n^2} \cos nx \quad (-\pi < x < \pi). \tag{1.194}$$

Let's replace x by $\pi x/3$ throughout:

$$\frac{\pi^2 x^2}{9} = \frac{\pi^2}{3} + 4 \sum_{n=1}^{\infty} \frac{(-1)^n}{n^2} \cos \frac{\pi nx}{3} \quad (-\pi < \frac{\pi x}{3} < \pi). \tag{1.195}$$

Multiplying both sides of the equality by $9/\pi^2$, and each element in the chain of inequalities by $3/\pi$, gives (1.193), and we're done.

The right side of (1.193) converges, for *any* $x \in \mathbb{R}$, to the 6-periodic, averaged function h defined on $(-3, 3)$ by $h(x) = x^2$. Indeed, h is piecewise smooth and continuous (its graph is just that of the function v of Figure 1.15 rescaled horizontally and vertically), so by the 6-periodic analog of Theorem 1.7.1, its Fourier series converges absolutely, and uniformly to h, on $\mathbb{R}$.

Exercises

In Exercises 1.9.1 and 1.9.2 find, both by direct computation *and* by the method of Example 1.9.2 (that is, by invocation of an appropriate result from 2π-periodic setting), the Fourier series (both forms) for the given function.

1.9.1 $f(x) = x$ ($-4 < x \leq 4$), extended 8-periodically. See Exercise 1.3.8 (and also Section 1.6, in particular (1.117)).

1.9.2 $f(x) = e^{\pi x/\ell}$ ($d < x \leq d + 2\ell$, where d is a real number), extended 2ℓ-periodically. See Exercise 1.3.7. Also, if g is 2ℓ-periodic, then the result of Exercise 1.3.1 holds with ℓ in place of π.

In Exercises 1.9.3 and 1.9.4, find the Fourier series (both forms) for the given function, using the method of Example 1.9.2.

1.9.3 $f(x) = -4x^3 + 4x^2 + x$ ($-\frac{1}{2} < x \leq \frac{1}{2}$), extended 1-periodically. See Exercise 1.3.4.

1.9.4 $f(x) = (Q \cos Qa + ix \sin Qa)e^{-iax}$ ($-Q < x \leq Q$), extended $2Q$-periodically. Here $a \in \mathbb{R}$. See Exercise 1.3.16.

1.9.5 Use Theorem 1.9.1 and Example 1.9.1 to evaluate

$$\sum_{n=1}^{\infty} \frac{1}{1 - 9n^2}.$$

1.9.6 Use Example 1.9.2 to show that, for $-3 < x < 3$,

$$3 + \frac{36}{\pi^2} \sum_{n=1}^{\infty} \frac{(-1)^n}{n^2} \cos \frac{\pi n x}{3} = \frac{\pi^2}{3} + 4 \sum_{n=1}^{\infty} \frac{(-1)^n}{n^2} \cos nx.$$

What does either side equal at $x = \pi$? Explain.

1.10 AMPLITUDE, PHASE, AND SPECTRA

If a reasonable periodic function assumes only real values, as will often be the case in "real life," then its Fourier series admits an additional useful formulation, to go along with the cosine-sine and complex exponential ones. That formulation amounts to this proposition.

Proposition 1.10.1 *Let $f \in PC(\mathbb{T}_P)$ be real-valued. Then the nth frequency component (1.9) of f is, for $n \in \mathbb{Z}^+$, equal to*

$$2|c_n(f)| \cos\left(\frac{2\pi n x}{P} + \operatorname{Arg} c_n(f)\right) \qquad (1.196)$$

(where the argument $\operatorname{Arg} c_n(f)$ of $c_n(f)$ is as in Definition A.0.2(e)).

Proof. Suppose f is as stated. Then

$$c_{-n}(f) = \frac{1}{2\pi}\int_{-\pi}^{\pi} f(x)e^{2\pi inx/P}\,dx = \frac{1}{2\pi}\int_{-\pi}^{\pi} \overline{f(x)}\,\overline{e^{-2\pi inx/P}}\,dx$$

$$= \overline{\frac{1}{2\pi}\int_{-\pi}^{\pi} f(x)e^{-2\pi inx/P}\,dx} = \overline{c_n(f)}. \tag{1.197}$$

Note that f being real-valued is crucial here.

Writing $c_n(f)$ in complex exponential form $c_n(f) = |c_n(f)|\,e^{i\,\text{Arg}\,c_n(f)}$ (see (A.34)), we get

$$c_{-n}(f)e^{-2\pi inx/P} + c_n(f)e^{2\pi inx/P}$$

$$= \overline{|c_n(f)|e^{i\,\text{Arg}\,c_n(f)}}\,e^{-2\pi inx/P} + |c_n(f)|e^{i\,\text{Arg}\,c_n(f)}e^{2\pi inx/P}$$

$$= |c_n(f)|\left(e^{-i(\text{Arg}\,c_n(f)+2\pi nx/P)} + e^{i(\text{Arg}\,c_n(f)+2\pi nx/P)}\right)$$

$$= 2|c_n(f)|\cos\left(\frac{2\pi nx}{P} + \text{Arg}\,c_n(f)\right), \tag{1.198}$$

as promised. □

Until further notice, let f be real-valued. Then if $f \in \text{PC}(\mathbb{T}_P)$ we have

$$f(x) \sim c_0(f) + 2\sum_{n=1}^{\infty} |c_n(f)|\cos\left(\frac{2\pi nx}{P} + \text{Arg}\,c_n(f)\right) \tag{1.199}$$

by Proposition 1.10.1. The "$\sim$" is replaceable by "$=$" under the usual additional assumptions on f. The expression on the right side of (1.199) is called the *amplitude-phase* form of the Fourier series for f.

The quantity $2|c_n(f)|$, called the *amplitude* of the frequency component (1.196), is equal to the maximum value of that component (since the cosine function attains a maximum value of 1). Also $\text{Arg}\,c_n(f)$, called the *phase* of this component, measures its position along the horizontal axis. (Here, again, $n \geq 1$. The component $c_0(f)$ is just a constant.) More generally, any sinusoid $R\cos(2\pi sx+\phi)$, where $s>0, R\geq 0$, and $-\pi < \phi \leq \pi$, is said to have amplitude R and phase ϕ.

Figure 1.17 captures these ideas nicely.

Note by (1.19) and (1.20) (or by (1.6), (1.7), and (1.197)) that, for f as in Proposition 1.10.1, the $a_n(f)$'s and $b_n(f)$'s will be real numbers. Then (1.5) and the definition of absolute value give us

$$2|c_n(f)| = \sqrt{a_n(f)^2 + b_n(f)^2}, \tag{1.200}$$

while (1.5) and the definition of argument give us

$$\text{Arg}\,c_n(f) = \begin{cases} \arccos\left(\dfrac{a_n(f)}{\sqrt{a_n(f)^2+b_n(f)^2}}\right) & \text{if } b_n(f) \leq 0, \\ -\arccos\left(\dfrac{a_n(f)}{\sqrt{a_n(f)^2+b_n(f)^2}}\right) & \text{if not}, \end{cases} \tag{1.201}$$

AMPLITUDE, PHASE, AND SPECTRA   61

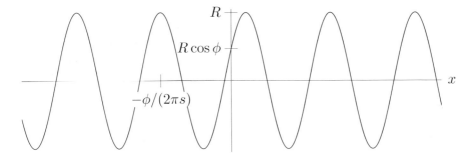

Fig. 1.17 A sinusoid of frequency s, amplitude R, and phase ϕ

all for $n \in \mathbb{Z}^+$. Equation (1.201) is, strictly speaking, for $c_n(f)$ not equal to 0; again, we define $\operatorname{Arg} 0 = 0$.

Example 1.10.1 Find the amplitude-phase form of the Fourier series for the function f of Example 1.2.1.

Solution. By (1.34),

$$2|c_n(f)| = 2\sqrt{c_n(f)\overline{c_n(f)}} = 2\sqrt{\frac{(e^\pi - e^{-\pi})(-1)^n}{2\pi(1-in)} \cdot \frac{(e^\pi - e^{-\pi})(-1)^n}{2\pi(1+in)}}$$

$$= \frac{e^\pi - e^{-\pi}}{\pi\sqrt{1+n^2}}. \tag{1.202}$$

For our phase computations, we note the following. By (1.200) and (1.201),

$$\operatorname{Arg} c_n(f) = \begin{cases} \arccos\left(\dfrac{a_n(f)}{2|c_n(f)|}\right) & \text{if } b_n(f) \le 0, \\ -\arccos\left(\dfrac{a_n(f)}{2|c_n(f)|}\right) & \text{if not.} \end{cases} \tag{1.203}$$

Now for our present function f, $b_n(f)$ is, by (1.37), nonzero for $n \ge 1$ and negative precisely when $(-1)^n$ is positive. So (1.203) gives

$$\operatorname{Arg} c_n(f) = (-1)^n \arccos\left(\frac{a_n(f)}{2|c_n(f)|}\right) = (-1)^n \arccos\left((-1)^n(n^2+1)^{-1/2}\right) \tag{1.204}$$

for $n \in \mathbb{Z}^+$, the last step by (1.36) and (1.202). So

$$f(x) \sim \frac{e^\pi - e^{-\pi}}{2\pi}$$
$$+ \frac{e^\pi - e^{-\pi}}{\pi} \sum_{n=1}^\infty \frac{1}{\sqrt{1+n^2}} \cos\left(nx + (-1)^n \arccos\left((-1)^n(n^2+1)^{-1/2}\right)\right). \tag{1.205}$$

62 FOURIER COEFFICIENTS AND FOURIER SERIES

Yuck: The phase calculations, and results, are a bit messy. As they tend to be. On the other hand, in many situations explicit phase information is not required; the reason for this is as follows. Amplitude is, again, a maximum value, so comparing amplitudes of frequency components is a good way of understanding the relative extents to which the associated frequencies contribute to the makeup of the function in question. If one is concerned with these extents only, then questions of phase are immaterial.

In such situations one may simply consider the graph of amplitude versus frequency; such a graph is called an *amplitude spectrum* or, simply, *spectrum*. Spectra can convey all sorts of interesting information regarding frequency content of the function at hand. For a "real-life" example, let's look at the spectra in Figure 1.18 of the author's vocalizations of Figure 1.1.

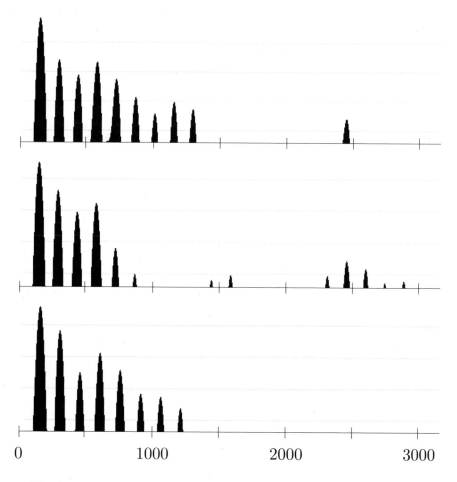

Fig. 1.18 Spectra of the author's "ah," "eh," and "oh" vowel sounds, respectively

The horizontal scale here is in Hertz (Hz), meaning cycles per second. The vertical scale is a logarithmic, or decibel (dB), scale. That is, what we're plotting here, essentially, are the logarithms of the amplitudes of the frequency components.

We glean from these pictures a few interesting things. First, all three sounds have roughly the same "base pitch," or *fundamental frequency*, meaning reciprocal of the fundamental period. This pitch is about 150 Hz.

Of course, it's evident enough from the waveforms in Figure 1.1 that the fundamental periods are nearly the same; we don't really need the spectra to tell us that. But the spectra *do* reveal similarities not immediately apparent in the "time domain," meaning in those original waveforms. Note in particular the rise in amplitude at the "fourth harmonic" (that is, the component whose frequency is four times the fundamental) of each vowel sound.

Figure 1.18 illustrates convincingly, we hope, the applicability of Fourier analysis to voice recognition, speech analysis, and the like, and is strongly indicative of its relevance to the study of many many other related—or quite unrelated—periodic phenomena.

Not to mention (as mentioned in the Introduction) nonperiodic phenomena: Again, *whenever* one has frequency components, one has Fourier analysis. And if the phenomenon in question is real-valued, as it so often is, then amplitude-phase analyses will apply to its frequency components in much the same way as in the periodic situation discussed above.

In fact, now is a good time to discuss, in general terms not limited to the periodic setting, what's called *the problem of phase*. Here's the problem, Part I: While amplitude information is often, as discussed above, sufficient for the study at hand, many other times it is *not*. In particular one *cannot*, generally, reconstruct a function from only the amplitudes of its frequency components; one also needs their phases. And here's the problem, Part II: In many physical contexts, phase is *quite difficult to measure or record*.

Why should this be? The answer varies with the situation. But what's going on can be understood in a broad sense as follows: Think of a wave hitting a seawall. The watermark left on the wall records the amplitude of that wave quite nicely, but says nothing about its phase (meaning the particular times at which the wave crests against the wall). Many means of recording or documenting phenomena, information, frequency components, signals, and so on behave like that seawall.

One solution to the problem of phase, which works beautifully in theory though not always quite as well in practice, is this. Suppose one is able to superimpose the component in question with sinusoids of the same frequency and amplitude as that component, but with known, varying phases. Suppose one is also able to measure the *amplitudes* (alone) of the sinusoids resulting from such superposition. Then one can compute the phase of the original component using the following observation.

Proposition 1.10.2 *For $a, b \in \mathbb{R}$, the sinusoid*

$$\cos(\theta + a) + \cos(\theta + b) \tag{1.206}$$

has amplitude

$$\left| 2\cos\frac{a-b}{2} \right|. \qquad (1.207)$$

In particular, if a is fixed then, as a function of b, the amplitude of (1.206) attains its maximum at $b = a$.

Proof. Using the identity $\cos(X+Y) + \cos(X-Y) = 2\cos X \cos Y$ (see Exercise A.0.16), with $X = \theta + (a+b)/2$ and $Y = (a-b)/2$, we find that

$$\cos(\theta + a) + \cos(\theta + b) = 2\cos\left(\theta + \frac{a+b}{2}\right)\cos\frac{a-b}{2}. \qquad (1.208)$$

The amplitude of the right side, meaning its maximum value as a function of θ, is just (1.207) (since $|\cos(\theta + (a+b)/2)|$ attains a maximum value of 1, whatever a and b are), as required.

Finally, (1.207) itself is, as a function of b, clearly maximized when $b = a$. □

Here's how the proposition applies to the problem of phase. Suppose one has a situation where amplitudes, but not phases, of frequency components may be measured directly. Then presumably, for a function f under consideration—let's take f to be P-periodic again although, as remarked earlier, similar considerations apply in other situations—one may determine the amplitude $2|c_n(f)|$ of the frequency component (1.196). One then constructs, for any $B \in (-\pi, \pi]$, a sinusoidal signal

$$2|c_n(f)|\cos\left(\frac{2\pi nx}{P} + B\right), \qquad (1.209)$$

combines that signal with—that is, adds it to—the component in question, and measures the amplitude of the result. This amplitude varies with B, and by observing this variation directly, one can document the particular value $B = B_0$ that yields the largest amplitude. By Proposition 1.10.2 we have $B_0 = \operatorname{Arg} c_n(f)$, whence we've found $\operatorname{Arg} c_n(f)$, as desired.

A nice application of the above method is to photographic images, whose "problem" is that they appear flat. This is because photographic film does not capture phase information. But such information *can* be collected, by merging the light waves corresponding to the image source with waves of known phase and using the amplitudes of the output to determine the phase of the source, as discussed above. When this phase information is combined with the original, photographic amplitude information, one gets a *hologram*, which looks three-dimensional!

It should be noted, though, that the method just discussed for handling the problem of phase has its own problems. The measurements and combinations and calculations required are often, practically speaking, difficult or impossible to perform. In such situations one needs additional arguments and results, which can be quite discipline specific and quite sophisticated as well.

Sophisticated enough, for example, to merit a Nobel Prize, such as the one awarded to Hauptmann and Karle in 1985 for their work on the problem of phase in the field of X-ray diffraction. Here, one often seeks to understand a molecule's structure by studying the manner in which X rays, incident on the molecule, scatter off of it. It's well known that the observed scatter pattern is, essentially, the amplitude spectrum of the function that models this structure. (This is because of the phenomenon of *Fraunhofer diffraction*, cf. Section 7.9.) To reconstruct that function from its frequency components, one needs, again, not only the amplitudes but also the phases of these components. The work of Hauptmann and Karle provides a method markedly superior to previous ones for determining these phases and, consequently, this structure function.

Exercises

In Exercises 1.10.1–1.10.4, find the amplitude-phase form of the Fourier series from the cited exercise or example.

1.10.1 Example 1.3.1. (Leave your answer in terms of sgn nr, cf. (A.12).)

1.10.2 Exercise 1.3.4.

1.10.3 Example 1.3.2.

1.10.4 Exercise 1.3.14.

1.10.5 Write down the amplitude $2|c_n(f)|$ and phase $\operatorname{Arg} c_n(f)$ of a real-valued $f \in \operatorname{PC}(\mathbb{T})$:

   a. In terms of the $a_n(f)$'s if f is even.
   b. In terms of the $b_n(f)$'s if f is odd.

1.10.6 We might be tempted to define the amplitude R and phase ϕ of a sinusoid $A \cos 2\pi sx + B \sin 2\pi sx$, with A and B possibly *complex* (but s and x still real), by writing

$$A \cos 2\pi sx + B \sin 2\pi sx = R \cos(2\pi sx + \phi)$$

for possibly complex numbers R and ϕ (with the cosine of a complex number being defined by (1.8)). Show, though, that this is not always possible. Hint: Try $A = 1$, $B = i$. Expand out $\cos(2\pi sx + \phi)$ using an identity from Exercise A.0.15 (these remain valid for complex θ_1, θ_2).

1.11 FUNCTIONS ON BOUNDED INTERVALS: STANDARD FOURIER SERIES

Example 1.9.2 takes us in the direction we most want to go for the purposes of the applications in the next chapter. Namely, it takes us toward the notion of *Fourier series for functions on bounded intervals*. And the idea, as illustrated in that example, is simple: To effect Fourier analysis of a function f on such an interval (in the above example, the interval is $(-3, 3)$ and $f(x) = x$ on that interval), simply consider f a "piece" of a periodic function; the frequency decomposition of that periodic function then implies one for f.

Let's, to be definite, assume that $f \in \mathrm{PSA}(a,b)$, where *from now on* we write $\mathrm{PSA}(a,b)$ as shorthand for $\mathrm{PSA}((a,b))$. (Similarly, $\mathrm{PC}(a,b)$ denotes $\mathrm{PC}((a,b))$, and so on.) This situation will suffice for our purposes. We then consider the $(b-a)$-periodic, averaged function, let's denote it f_{per}, that agrees with f on (a,b). We call f_{per} the *periodic extension* of f; the idea, in visual terms, is illustrated by Figure 1.19.

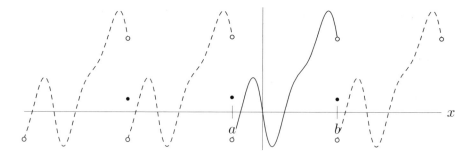

Fig. 1.19 A function $f \in \mathrm{PSA}(a,b)$ (solid) and its periodic extension f_{per} (dashed and solid combined)

We have the following theorem, which gives the desired frequency decomposition for f.

Theorem 1.11.1 *If $f \in \mathrm{PSA}(a,b)$, then, for any $x \in (a,b)$,*

$$f(x) = \sum_{n=-\infty}^{\infty} c_n(f) e^{2\pi i n x/(b-a)}$$
$$= \frac{a_0(f)}{2} + \sum_{n=1}^{\infty} \left(a_n(f) \cos \frac{2\pi n x}{b-a} + b_n(f) \sin \frac{2\pi n x}{b-a} \right), \quad (1.210)$$

where

$$c_n(f) = \frac{1}{b-a} \int_a^b f(x) e^{-2\pi i n x/(b-a)} \, dx \quad (n \in \mathbb{Z}), \quad (1.211)$$

$$a_n(f) = \frac{2}{b-a} \int_a^b f(x) \cos \frac{2\pi n x}{b-a} \, dx \quad (n \in \mathbb{N}), \quad (1.212)$$

$$b_n(f) = \frac{2}{b-a} \int_a^b f(x) \sin \frac{2\pi n x}{b-a} \, dx \quad (n \in \mathbb{Z}^+). \quad (1.213)$$

Proof. Since f is piecewise smooth and averaged, the $(b-a)$-periodic function f_{per} defined above is too. So by the $(b-a)$-periodic generalization of Corollary 1.4.1(b),

$$f_{\text{per}}(x) = \sum_{n=-\infty}^{\infty} c_n(f_{\text{per}}) e^{2\pi i n x/(b-a)} \tag{1.214}$$

for $x \in \mathbb{R}$, where

$$c_n(f_{\text{per}}) = \frac{1}{b-a} \int_{-(b-a)/2}^{(b-a)/2} f_{\text{per}}(x) e^{-2\pi i n x/(b-a)} \, dx \tag{1.215}$$

for $n \in \mathbb{Z}$. Now as we noted just after the proof of Theorem 1.9.1, we may in fact integrate over any interval of length $b-a$. So

$$\begin{aligned} c_n(f_{\text{per}}) &= \frac{1}{b-a} \int_a^b f_{\text{per}}(x) e^{-2\pi i n x/(b-a)} \, dx \\ &= \frac{1}{b-a} \int_a^b f(x) e^{-2\pi i n x/(b-a)} \, dx = c_n(f). \end{aligned} \tag{1.216}$$

Here the next to the last equality is because, by definition of f_{per},

$$f_{\text{per}}(x) = f(x) \quad \text{for } a < x < b; \tag{1.217}$$

the last equality in (1.216) is just the definition (1.211).

Putting (1.217) and (1.216) into (1.214) gives

$$f(x) = \sum_{n=-\infty}^{\infty} c_n(f) e^{2\pi i n x/(b-a)} \quad (a < x < b), \tag{1.218}$$

which is the desired result, at least in complex exponential form. To get the corresponding cosine-sine representation of f on (a, b), we simply apply the conversion formulas (1.6) and (1.7). □

The $a_n(f)$'s, $b_n(f)$'s, and $c_n(f)$'s are called the *Fourier coefficients of f*. The two series in (1.210) are called the *standard Fourier series for f*; the term "standard" is employed to distinguish these series from some variants to be discussed in the next section.

Note that $a_n(f)$, or $b_n(f)$, or $c_n(f)$ can denote one of two things: a Fourier coefficient of a periodic function or a Fourier coefficient of a function on a bounded interval. That's okay: The two things are close enough in spirit that, at least in certain contexts, confusing them is not such a bad thing (which is fortunate for the author, who is easily confused).

On the other hand, care should be exercised on the following point: Since a function defined (only) on a bounded interval (a, b) is *not* periodic, we aren't free to obtain its Fourier coefficients by integrating over whatever interval of length $b-a$ we choose. We must integrate over (a, b).

Example 1.11.1 Show that

$$e^x \sin x = \frac{e^{2\pi} + e^{\pi}}{\pi} \sum_{n=-\infty}^{\infty} \frac{1}{(4n^2 + 4in - 2)} e^{2inx} \quad (\pi < x < 2\pi). \quad (1.219)$$

Solution. Let $g(x) = e^x \sin x$ for $x \in (\pi, 2\pi)$. By (1.211) and the second part of (1.8),

$$c_n(g) = \frac{1}{2\pi - \pi} \int_{\pi}^{2\pi} e^x \left(\frac{e^{ix} - e^{-ix}}{2i} \right) e^{-2\pi inx/(2\pi - \pi)} \, dx$$

$$= \frac{1}{2\pi i} \int_{\pi}^{2\pi} \left(e^{(1+i(1-2n))x} - e^{(1-i(1+2n))x} \right) dx$$

$$= \frac{1}{2\pi i} \left[\frac{e^{(1+i(1-2n))x}}{1+i(1-2n)} - \frac{e^{(1-i(1+2n))x}}{1-i(1+2n)} \right]_{\pi}^{2\pi} = \frac{e^{2\pi} + e^{\pi}}{\pi(4n^2 + 4in - 2)}; \quad (1.220)$$

the last step follows from some simplification, which we omit.
  Theorem 1.11.1 then gives the desired result.

Our next example compares in some interesting ways to the one above, and in some other interesting ways to some examples we'll consider in the next section.

Example 1.11.2 Show that

$$x = \frac{\ell}{2} - \frac{\ell}{\pi} \sum_{n=1}^{\infty} \frac{1}{n} \sin \frac{2\pi nx}{\ell} \quad (0 < x < \ell) \quad (1.221)$$

for any $\ell > 0$.

Solution. If f is the function defined on $(0, \ell)$ by $f(x) = x$, then by (1.211) and the integration-by-parts formula (1.41),

$$c_n(f) = \frac{1}{\ell} \int_0^{\ell} x \, e^{-2\pi inx/\ell} \, dx = \frac{1}{\ell} \left[\left(\frac{\ell x}{-2\pi in} + \frac{\ell^2}{4\pi^2 n^2} \right) e^{-2\pi inx/\ell} \right]_0^{\ell}$$

$$= \left(\frac{\ell}{-2\pi in} + \frac{\ell}{4\pi^2 n^2} \right) - \frac{\ell^2}{4\pi^2 n^2} = -\frac{\ell}{2\pi in} \quad (1.222)$$

for $n \neq 0$, while

$$c_0(f) = \frac{1}{\ell} \int_0^{\ell} x \, dx = \frac{\ell}{2}. \quad (1.223)$$

So from the conversion formulas (1.6) and (1.7), we find that $a_0(f) = \ell$ and that, for $n > 0$,

$$a_n(f) = 0 \quad \text{and} \quad b_n(f) = -\frac{\ell}{\pi n}. \quad (1.224)$$

Theorem 1.11.1 then gives (1.221).

Recall from Theorem 1.7.1 that a function piecewise smooth and continuous "on the torus $\mathbb{T}$," meaning technically a function in $\text{PSC}(\mathbb{T})$, has a Fourier series converging uniformly to it. Certainly the generalization of that theorem to piecewise smooth, continuous, P-periodic functions continues to hold. It's interesting, though, that on a *bounded interval*, being piecewise smooth and continuous does *not* assure a uniformly convergent Fourier series. Indeed, the above example illustrates this: While the function f there is piecewise smooth and continuous on $(0, \ell)$, the series on the right side of (1.221) is *not* uniformly convergent there. The problem may be attributed to the rate of decay of the $b_n(f)$'s: Specifically, $b_n(f)$ is a constant multiple of $1/n$, which is not small enough to make the Fourier series converge uniformly on $(0, \ell)$.

Another, perhaps more illuminating, way to understand what's happening here is to look at the ℓ-periodic extension f_{per} of f (Fig. 1.20).

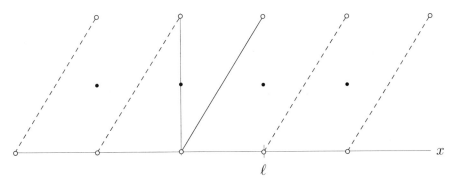

Fig. 1.20 The function f of Example 1.11.2 (solid) and its ℓ-periodic extension f_{per} (dashed and solid combined)

Clearly f_{per}, while piecewise smooth on $\mathbb{R}$, has points of discontinuity there. This implies, as discussed in Sections 1.6 and 1.7 (in the analogous 2π-periodic context), that $S_N^{f_{\text{per}}}$ cannot converge uniformly to f_{per} on $(0, \ell)$.

On the other hand, the series on the right side of (1.219) *does* converge uniformly, on $(\pi, 2\pi)$, to the function on the left. This may similarly be seen in one of two ways: The first is by considering the series explicitly and noting that its coefficients "decay like $1/n^2$," which (one shows) is fast enough to assure the requisite uniform convergence. The second way is by checking that, as a quick sketch makes apparent, the π-periodic extension g_{per} of the function g of Example 1.11.1 is piecewise smooth and *continuous* on $\mathbb{R}$, whence the π-periodic analog of Theorem 1.7.1 applies. (See also Exercise 1.11.4.)

In general, if f is piecewise smooth and continuous on (a, b), then the $(b - a)$-periodic extension f_{per} will be piecewise smooth and continuous on $\mathbb{R}$ if and only if things "match up" at the endpoints, meaning $f(a^+) = f(b^-)$. Thus uniform convergence of Fourier series will occur only in such cases. For similar reasons, term-by-term differentiation of Fourier series will be valid only in such cases.

Exercises

1.11.1 What does

$$\sum_{n=-\infty}^{\infty} c_n(f) e^{2\pi i n a/(b-a)}$$

converge to for $f \in \text{PS}(a,b)$? (Here the $c_n(f)$'s are as in (1.211).) Express your answer in terms of (limits of) values of f on (a,b) only. (You may want to recall Exercise 1.2.3.)

1.11.2 Let $f \in \text{PSA}(0,\pi)$ be defined by $f(x) = e^x \sin x$. Show that f_{per}, the π-periodic extension of f, is a constant multiple of g_{per}, where g is as in Example 1.11.1. (Hint: By periodicity, it suffices to show that f_{per} is a constant times g_{per} on $(\pi, 2\pi)$. But if $x \in (\pi, 2\pi)$, then $x - \pi \in (0, \pi)$, so $f_{\text{per}}(x) = f_{\text{per}}(x - \pi) = f(x - \pi)$.) Use this result and that example to find the complex exponential Fourier series for f without having to perform any additional integration.

1.11.3 Proceeding in a manner similar to that of the previous exercise, use Example 1.9.2 to find the Fourier series for the function $f \in \text{PSA}(3,9)$ defined by $f(x) = x^2 - 12x$. Hint: $x^2 - 12x = (x-6)^2 - 36$.

1.11.4 Note that the Fourier coefficients $c_n(f)$ of the function f of Example 1.11.1 satisfy $\lim_{n \to \infty} n^2 c_n(f) = (e^{2\pi} + e^\pi)/(4\pi)$. In light of this and Corollary 1.8.1 (which holds in period $P > 0$ just as it does in period 2π), determine *without any computation* whether $f'(\pi^+) = f'(2\pi^-)$. Explain. Then compute these limits, to check your answer.

1.12 OTHER FOURIER SERIES FOR FUNCTIONS ON BOUNDED INTERVALS

In the next chapter we'll be especially interested in functions on intervals of the form $(0, \ell)$. One can expand such a function into sinusoids of frequencies n/ℓ, using Theorem 1.11.1 with $a = 0$ and $b = \ell$, as we did in Example 1.11.2. But there are other ways, involving frequencies that are integer multiples of $1/(2\ell)$, that are better suited to our intended applications, and that the following theorem describes.

Theorem 1.12.1 *If $f \in \text{PSA}(0, \ell)$, then for any $x \in (0, \ell)$,*

$$f(x) = \frac{\alpha_0(f)}{2} + \sum_{n=1}^{\infty} \alpha_n(f) \cos \frac{\pi n x}{\ell} = \sum_{n=1}^{\infty} \beta_n(f) \sin \frac{\pi n x}{\ell}, \qquad (1.225)$$

where

$$\alpha_n(f) = \frac{2}{\ell} \int_0^\ell f(x) \cos \frac{\pi n x}{\ell}\, dx \quad (n \in \mathbb{N}), \qquad (1.226)$$

$$\beta_n(f) = \frac{2}{\ell} \int_0^\ell f(x) \sin \frac{\pi n x}{\ell}\, dx \quad (n \in \mathbb{Z}^+). \qquad (1.227)$$

Proof. As in the proof of Theorem 1.11.1, the strategy is to extend f periodically, but now we do so in a different way. Specifically: instead of forming the ℓ-periodic extension f_{per}, we form the *even periodic extension* f_{even} and the *odd periodic extension* f_{odd}. These are the piecewise smooth, averaged functions of period 2ℓ, rather than ℓ, that agree with f on $(0, \ell)$ and are, respectively, even and odd. Geometrically, this is illustrated in Figure 1.21.

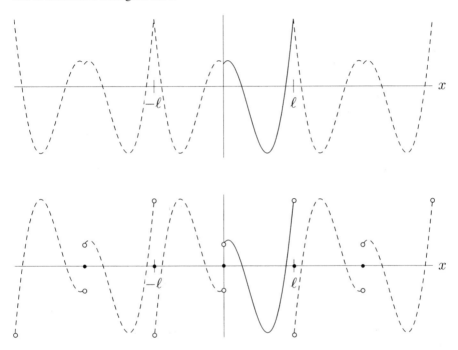

Fig. 1.21 Top: a function $f \in \text{PSA}(0, \ell)$ (solid) and its even 2ℓ-periodic extension f_{even} (dashed and solid combined). Bottom: the same function f (solid) and its odd 2ℓ-periodic extension f_{odd} (dashed and solid combined)

Let's focus on f_{odd} for the moment: The 2ℓ-periodic generalization of Corollary 1.4.1(a) tells us that, for any $x \in \mathbb{R}$,

$$f_{\text{odd}}(x) = \frac{a_0(f_{\text{odd}})}{2} + \sum_{n=1}^{\infty} \left(a_n(f_{\text{odd}}) \cos \frac{\pi n x}{\ell} + b_n(f_{\text{odd}}) \sin \frac{\pi n x}{\ell} \right). \quad (1.228)$$

Also, the 2ℓ-periodic generalization of Proposition 1.3.1(a) tells us that

$$a_n(f_{\text{odd}}) = 0 \quad (1.229)$$

and that

$$b_n(f_{\text{odd}}) = \frac{2}{\ell} \int_0^\ell f_{\text{odd}}(x) \sin \frac{\pi n x}{\ell} \, dx = \frac{2}{\ell} \int_0^\ell f(x) \sin \frac{\pi n x}{\ell} \, dx = \beta_n(f) \quad (1.230)$$

72   FOURIER COEFFICIENTS AND FOURIER SERIES

for all n. (The next to the last equality in (1.230) is because

$$f_{\text{odd}}(x) = f(x) \quad \text{for } 0 < x < \ell, \tag{1.231}$$

and the last is just the definition (1.227) of $\beta_n(f)$.) Putting (1.229), (1.230), and (1.231) into (1.228) yields

$$f(x) = \sum_{n=1}^{\infty} \beta_n(f) \sin \frac{\pi n x}{\ell} \quad (0 < x < \ell), \tag{1.232}$$

which is the second of our desired results. The first follows by similar arguments, involving f_{even} and the 2ℓ-periodic generalization of Proposition 1.3.1(b). □

We call the two series in (1.225) the *Fourier cosine* and *Fourier sine series for f*, respectively. Warning: Neither of these series is the same as the standard Fourier series for f given by Theorem 1.11.1. The following example illustrates the distinction, among other things.

Example 1.12.1 Find the Fourier cosine and sine series representations of the function f on $(0, \ell)$ defined by $f(x) = x$.

Solution. For the cosine series, we use (1.226): we get

$$\alpha_n(f) = \frac{2}{\ell} \int_0^\ell x \cos \frac{\pi n x}{\ell} \, dx = \frac{2}{\ell} \left[\frac{\ell^2 \cos(\pi n x/\ell)}{\pi^2 n^2} + \frac{\ell x \sin(\pi n x/\ell)}{\pi n} \right]_0^\ell$$

$$= \frac{2}{\ell} \left[\frac{\ell^2 (\cos \pi n - \cos 0)}{\pi^2 n^2} + \frac{\ell(\ell \sin \pi n - 0 \sin 0)}{\pi n} \right] = \frac{2\ell((-1)^n - 1)}{\pi^2 n^2} \tag{1.233}$$

for $n \neq 0$ (the employed integration-by-parts formula comes from Exercise 1.3.5); also

$$\alpha_0(f) = \frac{2}{\ell} \int_0^\ell x \, dx = \ell. \tag{1.234}$$

So by Theorem 1.12.1,

$$x = \frac{\ell}{2} + \frac{2\ell}{\pi^2} \sum_{n=1}^{\infty} \frac{(-1)^n - 1}{n^2} \cos \frac{\pi n x}{\ell}$$

$$= \frac{\ell}{2} - \frac{4\ell}{\pi^2} \sum_{m=1}^{\infty} \frac{1}{(2m-1)^2} \cos \frac{\pi(2m-1)x}{\ell} \quad (0 < x < \ell). \tag{1.235}$$

Similarly,

$$\beta_n(f) = \frac{2}{\ell} \int_0^\ell x \sin \frac{\pi n x}{\ell} \, dx = \frac{2}{\ell} \left[\frac{\ell^2 \sin(\pi n x/\ell)}{\pi^2 n^2} - \frac{\ell x \cos(\pi n x/\ell)}{\pi n} \right]_0^\ell$$

$$= \frac{2}{\ell} \left[\frac{\ell^2 (\sin \pi n - \sin 0)}{\pi^2 n^2} - \frac{\ell(\ell \cos \pi n - 0 \cos 0)}{\pi n} \right] = \frac{2\ell(-1)^{n+1}}{\pi n} \tag{1.236}$$

(again, we used Exercise 1.3.5 to integrate by parts). So by Theorem 1.12.1 again,

$$x = \frac{2\ell}{\pi} \sum_{n=1}^{\infty} \frac{(-1)^{n+1}}{n} \sin \frac{\pi n x}{\ell} \quad (0 < x < \ell). \tag{1.237}$$

Observe that the $\alpha_n(f)$'s in the above example decay like $1/n^2$, while the $\beta_n(f)$'s do so only like $1/n$. Consequently, as one checks, the Fourier cosine series converges uniformly to f on $(0, \ell)$, while the Fourier sine series does not. What's happening here is that f_{even} is continuous, but f_{odd} is not (DIY: check this). So the 2ℓ-periodic generalization of Theorem 1.7.1 applies to the former periodic extension but not to the latter.

Similar things occur in general: Namely, suppose f is piecewise smooth and continuous on $(0, \ell)$. Then f_{even} is piecewise smooth and continuous on $\mathbb{R}$; f_{odd}, while certainly piecewise smooth on $\mathbb{R}$, is continuous there if and only if $f(0^+) = f(\ell^-) = 0$. The reader should draw some pictures and reflect on the definitions to see this.

Another thing that (1.235) and (1.237), and for that matter (1.221), illustrate is this: Frequency decompositions of functions on bounded intervals (a, b) are *not unique*. That is, there are many distinct ways to decompose such functions into frequency components. This is because there are many possible *periodic extensions* of such functions.

To push this point even further, let's observe that, already, we've seen many other ways of expressing the function $f: (0, \ell) \to \mathbb{R}$ given by $f(x) = x$. For example, suppose ℓ is less than π, and let's choose a positive number r with $\ell < r < \pi$. The function g_r of Example 1.3.1 satisfies $g_r(x) = x = f(x)$ on $(0, r)$, and therefore on $(0, \ell)$, so by (1.48) and Corollary 1.4.1(b),

$$f(x) = \frac{2}{\pi - r} \sum_{n=1}^{\infty} \frac{\sin nr}{n^2} \sin nx \quad (0 < x < \ell). \tag{1.238}$$

Of course this gives *infinitely* many distinct decompositions of f, because of the infinitude of r's between ℓ and π.

Let's agree that, should we speak of THE frequency domain F_f or THE frequency components of a given piecewise smooth, averaged function f on (a, b), we mean the ones implicit in Theorem 1.11.1. That is, we take this domain to be the set

$$\{n/(b-a): n \in \mathbb{N}\}, \tag{1.239}$$

and we take these components to be the summands $a_0(f)/2 = c_0(f)$ and

$$a_n(f) \cos \frac{2\pi n x}{b-a} + b_n(f) \sin \frac{2\pi n x}{b-a} = c_n(f) e^{2\pi i n x/(b-a)} + c_{-n}(f) e^{-2\pi i n x/(b-a)} \tag{1.240}$$

of (1.210). Again, these are the frequency domain and frequency components, respectively, of f_{per}, which is unique among all $(b-a)$-periodic extensions of f. And f_{per} itself has only *one* possible frequency decomposition: See Corollary 3.6.1.

None of which is to say we should *ignore* other possible periodic extensions, of period P *larger* than the length of the domain of f, such as those of Theorem 1.12.1. Or such as the one that's documented by the following theorem, and that will be useful in the next chapter (see Example 2.3.2).

Theorem 1.12.2 *Under the same hypotheses as in Theorem 1.12.1,*

$$f(x) = \sum_{m=1}^{\infty} \chi_m(f) \sin \frac{\pi(2m-1)x}{2\ell} \quad (0 < x < \ell), \qquad (1.241)$$

where

$$\chi_m(f) = \frac{2}{\ell} \int_0^\ell f(x) \sin \frac{\pi(2m-1)x}{2\ell} \, dx \quad (m \in \mathbb{Z}^+). \qquad (1.242)$$

Proof. We let f_{refl} denote the function in $\text{PSA}(0, 2\ell)$ formed by joining f to its reflection about the line $x = \ell$ (Fig. 1.22). That is,

$$f_{\text{refl}}(x) = \begin{cases} f(x) & \text{if } 0 < x < \ell, \\ f(\ell^-) & \text{if } x = \ell, \\ f(2\ell - x) & \text{if } \ell < x < 2\ell. \end{cases} \qquad (1.243)$$

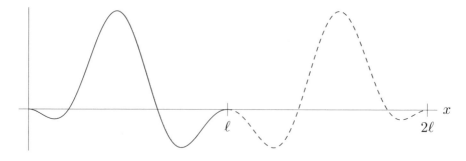

Fig. 1.22 A function $f \in \text{PSA}(0, \ell)$ (solid) and its reflection about $x = \ell$ (dashed). The dashed and solid curves together constitute f_{refl}

Theorem 1.12.1 then gives a Fourier sine series for f_{refl}:

$$f_{\text{refl}}(x) = \sum_{n=1}^{\infty} \beta_n(f_{\text{refl}}) \sin \frac{\pi n x}{2\ell} \quad (0 < x < 2\ell), \qquad (1.244)$$

where

$$\begin{aligned}
\beta_n(f_{\text{refl}}) &= \frac{2}{2\ell} \int_0^{2\ell} f_{\text{refl}}(x) \sin \frac{\pi n x}{2\ell} \, dx \\
&= \frac{1}{\ell} \left[\int_0^\ell f(x) \sin \frac{\pi n x}{2\ell} \, dx + \int_\ell^{2\ell} f(2\ell - x) \sin \frac{\pi n x}{2\ell} \, dx \right] \\
&= \frac{1}{\ell} \left[\int_0^\ell f(x) \sin \frac{\pi n x}{2\ell} \, dx + \int_0^\ell f(u) \sin \frac{\pi n(2\ell - u)}{2\ell} \, du \right], \qquad (1.245)
\end{aligned}$$

the second step by the definition of f_refl and the last by the substitution $u = 2\ell - x$. Now

$$\sin\frac{\pi n(2\ell - u)}{2\ell} = \sin\left(\pi n - \frac{\pi n u}{2\ell}\right) = (-1)^{n+1}\sin\frac{\pi n u}{2\ell}, \qquad (1.246)$$

so (1.245) gives

$$\beta_n(f_\text{refl}) = \frac{1 + (-1)^{n+1}}{\ell}\int_0^\ell f(x)\sin\frac{\pi n x}{2\ell}\,dx. \qquad (1.247)$$

The right side equals 0 if n is even and equals $\chi_m(f)$, as defined by (1.242), if n is an odd integer $2m - 1$. Then (1.244) and the fact that $f = f_\text{refl}$ on $(0, \ell)$ yield our desired result. □

The series (1.241) is called the *Fourier halfsine series* for f; the $\chi_m(f)$'s are called the *Fourier halfsine coefficients* of f.

We take one final look, for now, at the function $f(x) = x$ $(0 < x < \ell)$.

Example 1.12.2 Show that, for $\ell > 0$,

$$x = \frac{8\ell}{\pi^2}\sum_{m=1}^\infty \frac{(-1)^{m+1}}{(2m-1)^2}\sin\frac{\pi(2m-1)x}{2\ell} \qquad (0 < x < \ell). \qquad (1.248)$$

Solution. As usual we let $f(x) = x$ on $(0, \ell)$; then by (1.242) and integration by parts,

$$\begin{aligned}
\chi_m(f) &= \frac{2}{\ell}\int_0^\ell x\sin\frac{\pi(2m-1)x}{2\ell}\,dx \\
&= \frac{2}{\ell}\left[\frac{4\ell^2\sin\frac{\pi(2m-1)x}{2\ell}}{\pi^2(2m-1)^2} - \frac{2\ell x\cos\frac{\pi(2m-1)x}{2\ell}}{\pi(2m-1)}\right]_0^\ell \\
&= \frac{2}{\ell}\left[\frac{4\ell^2(\sin\pi(m-\frac{1}{2}) - \sin 0)}{\pi^2(2m-1)^2} - \frac{2\ell(\ell\cos\pi(m-\frac{1}{2}) - 0\cos 0)}{\pi(2m-1)}\right] \\
&= \frac{8\ell(-1)^{m+1}}{\pi^2(2m-1)^2}, \qquad (1.249)
\end{aligned}$$

the last step by (1.63) and the similarly derived fact that $\cos\pi(m-\frac{1}{2}) = 0$ for $m \in \mathbb{Z}$. Theorem 1.12.2 then gives us the desired result (1.248).

Note the $(2m-1)^2$ in the denominator of $\chi_m(f)$; the latter therefore decays like $1/m^2$, which is sufficient to assure absolute, uniform convergence, on $(0, \ell)$, of the series in (1.248). What's going on here is this: The 4ℓ-periodic, odd extension of f_refl is, in this case, continuous (as you should check), so the 4ℓ-periodic generalization of Theorem 1.7.1 applies to this extension.

Exercises

1.12.1 Find the Fourier cosine, sine, and halfsine series on $(0, \ell)$ for the function $f(x) = \cos(\pi x/\ell)$.

1.12.2 **a.** Differentiate the series in (1.235) and (1.248) term by term to show that

$$1 = \frac{4}{\pi} \sum_{m=1}^{\infty} \frac{1}{2m-1} \sin \frac{\pi(2m-1)x}{\ell}$$

$$= \frac{4}{\pi} \sum_{m=1}^{\infty} \frac{(-1)^{m+1}}{2m-1} \cos \frac{\pi(2m-1)x}{2\ell} \quad (0 < x < \ell).$$

Explain carefully why this differentiation is valid. Hint: Consider the discussions, following Examples 1.12.1 and 1.12.2, concerning smoothness of f_{even} and f_{refl} for $f(x) = x$ on $(0, \ell)$. Also note that Lemma 1.7.1 generalizes to arbitrary periods $P > 0$.

  b. What is the Fourier cosine series for the constant function 1 on $(0, \ell)$? Why don't you arrive at this result if you differentiate the series in (1.237) term by term?

1.12.3 Show that

$$x(\ell - x) = \frac{8\ell^2}{\pi^3} \sum_{m=1}^{\infty} \frac{1}{(2m-1)^3} \sin \frac{\pi(2m-1)x}{\ell} \quad (0 < x < \ell).$$

Hint: antidifferentiate (1.235). Why does this work here? (See Proposition 1.8.2.)

1.12.4 **a.** Calculate directly the Fourier halfsine series

$$1 = \frac{4}{\pi} \sum_{m=1}^{\infty} \frac{1}{2m-1} \sin \frac{\pi(2m-1)x}{2\ell} \quad (0 < x < \ell).$$

  b. By antidifferentiating, show that

$$x = \ell - \frac{8\ell}{\pi^2} \sum_{m=1}^{\infty} \frac{1}{(2m-1)^2} \cos \frac{\pi(2m-1)x}{2\ell} \quad (0 < x < \ell).$$

(Consider Proposition 1.8.2. In particular, explain the contant term here.)

  c. By antidifferentiating, show that

$$x(2\ell - x) = \frac{32\ell^2}{\pi^3} \sum_{m=1}^{\infty} \frac{1}{(2m-1)^3} \sin \frac{\pi(2m-1)x}{2\ell} \quad (0 < x < \ell).$$

(Again, be careful with constant terms. See the comments following Example 1.8.2.)

1.12.5 Note that, if you simply replace ℓ by 2ℓ in the first equality in Exercise 1.12.2a, you get the equality in Exercise 1.12.4a. (There's an analogous correspondence between Exercises 1.12.3 and 1.12.4c). On the other hand, replacing ℓ by 2ℓ in the Fourier sine series identity (1.237) for x *does not* give the Fourier halfsine series formula (1.248) for x. Why do you think this is happening? Hint: Compare the constant function 1 on $(0, 2\ell)$ to the function f_{refl}, where f is the constant function 1 on $(0, \ell)$. Now repeat with x instead of 1.

2
Fourier Series and Boundary Value Problems

When a differential equation, pertaining throughout some specified domain, is paired with conditions to be met on the boundary of that domain, one has what's called a *boundary value problem*. Solution of such a problem means determination of the function or functions that satisfy both the differential equation and the boundary conditions.

It was Fourier himself who recognized the relevance of sinusoid decompositions to the study of boundary value problems. In fact, he developed his theory of the former for the express purpose of solving a variety of the latter. In this chapter we look at a number of physical situations, studied by Fourier and others, that may be modeled by boundary value problems, and demonstrate the extraordinary utility of Fourier series in the solution of these problems.

2.1 STEADY STATE TEMPERATURES AND "THE FOURIER METHOD"

In his landmark, revolutionary 1807 work *Theory of the Propagation of Heat in Solid Bodies*, Joseph Fourier was concerned with the theory of the propagation of heat in solid bodies. In this section we consider a particular boundary value problem associated with this theory. This problem is one of the first Fourier considered, and leads naturally to the notion of a Fourier series.

We begin, as he did, with a long, thin rectangular plate, infinite in one direction. Imagining the width of the plate to be zero, we can take the subset

$$\{(x, y) \in \mathbb{R}^2 : 0 \leq x \leq \pi,\ y \geq 0\} \tag{2.1}$$

of the xy plane as a model for it (Fig. 2.1).

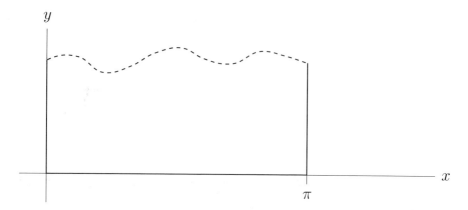

Fig. 2.1 A semi-infinite plate

(Actually, Fourier put his plate between $x = -\pi/2$ and $x = \pi/2$. Our choice of horizontal positioning simplifies things somewhat. See Exercise 2.1.4 below for a look at what his setup yields.)

A prescribed heat source is applied to the lower edge of the plate, and the long vertical edges are held at constant temperature equal to zero (in the chosen temperature scale. "Zero" here does not necessarily, and generally will not, denote *absolute zero*). We assume there is no heat flow across the faces of the plate.

We are interested in steady state, meaning time-independent, temperature distributions $u(x, y)$ through this plate. (We may imagine that such a distribution is what results from switching on and leaving on the heat source, and allowing enough time to elapse that the system reaches an equilibrium state.) Fourier showed that such a distribution satisfies the so-called "Laplace equation"

$$u_{xx} + u_{yy} = 0 \quad (0 < x < \pi,\ y > 0), \tag{2.2}$$

which had been encountered by Laplace (hence the name) and others in different contexts. (Here and throughout, we use standard shorthand for partial derivatives: We write u_x for $\partial u/\partial x$, u_{xy} for $\partial^2 u/(\partial y \partial x)$, u_{xx} for $\partial^2 u/\partial x^2$, and so on.)

Fourier further argued that the temperature should drop off to zero far from the heat source—in other words,

$$\lim_{y \to \infty} u(x, y) = 0 \quad (0 < x < \pi). \tag{2.3}$$

The fixed temperatures along the three edges of the plate amount to "boundary conditions"

$$u(0, y) = 0, \quad u(\pi, y) = 0 \quad (y > 0) \tag{2.4}$$

and

$$u(x, 0) = f(x) \quad (0 < x < \pi), \tag{2.5}$$

where f is some known function. Note that (2.3) can itself be understood as a boundary condition on the "edge at infinity" of the plate.

Incidentally, boundary conditions like those of (2.3), (2.4), and (2.5), specifying behavior of the function u itself along the boundary, are called *Dirichlet conditions*. Alternatively, one may stipulate that the normal derivative of u have prescribed behavior at the boundary: Such stipulations are called *Neumann conditions*.

Fourier now sought a solution to equations (2.2)–(2.5). His approach relied on what is presently known as the *superposition principle*, which we'll formalize later in this chapter. (See Section 2.2.) What it says in the present context is this: Given any one of the equations (2.2)–(2.4), and any number of functions satisfying that equation, any finite linear combination of those functions will do the same. This is fairly evident and readily checked. But note that it's true *only* of these three equations, and *not* equation (2.5). Indeed, if $u_1(x,0) = u_2(x,0) = f(x)$, then $c_1 u_1(x,0) + c_2 u_2(x,0) = (c_1 + c_2)f(x)$, which does not equal $f(x)$ throughout $(0, \pi)$ unless either f is identically zero there or $c_1 + c_2 = 1$.

So Fourier adopted the following strategy: He focused first on (2.2)–(2.4) alone, and sought a large collection of functions u solving these three equations simultaneously. He then sought to satisfy (2.5) through some *particular* linear combination of these solutions. And to reiterate an important point: Such a linear combination would, by the superposition principle, still satisfy equations (2.2)–(2.4).

Fourier's approach to those three equations was rather bold: He considered them "one variable at a time" or, in more modern parlance, he *separated variables*. That is, he sought *product* solutions

$$u(x,y) = X(x)Y(y). \qquad (2.6)$$

This approach, as it turns out, is of such dramatic impact and utility that we should separate it out (pun intended), and study it in and of itself.

Example 2.1.1 Find nonzero solutions of the form (2.6) to equations (2.2)–(2.4).

Solution. Putting (2.6) into (2.2) yields

$$X''(x)Y(y) + X(x)Y''(y) = 0 \quad (0 < x < \pi,\ y > 0). \qquad (2.7)$$

Here's the *trick* to solving this differential equation: We divide through by $u(x,y) = X(x)Y(y)$ to get

$$\frac{X''(x)}{X(x)} + \frac{Y''(y)}{Y(y)} = 0 \quad (0 < x < \pi,\ y > 0). \qquad (2.8)$$

(Strictly speaking, we should take care about dividing by zero. But it's not really a problem, as we will explain very shortly.) That is, $Y''/Y = -X''/X$. This latter equation will certainly hold if both sides equal the same *constant*. Let's call this constant λ: We then find that, to solve (2.7), it suffices to solve the two *ordinary differential equations*

$$X''(x) + \lambda X(x) = 0 \quad (0 < x < \pi), \quad Y''(y) - \lambda Y(y) = 0 \quad (y > 0). \qquad (2.9)$$

(The reader should check that equations (2.9) really do imply (2.7), whether or not $X(x)$ or $Y(y)$ is ever zero on the domain in question. Thus the potential division by zero in (2.8) in fact is *not* a problem. Nor will it be in similar situations; so we will not worry about it in such situations.) Note that we could just as well have set things up so that the "$+\lambda$" goes with the equation in Y; in other words we could have, at the outset, put X'' equal to λX and Y'' equal to $-\lambda Y$. It's just a matter of renaming things: we've named them as we have so that the "$+\lambda$" appears with the equation whose boundary conditions are "linear and homogeneous." (The meaning of this will be explained in the next section.) This is the traditional way of doing things, and we will abide by it.

At any rate, equations (2.9) are readily solved. Specifically, the most general solution to this pair of equations is given by

$$X(x) = \gamma + \delta x \text{ and } Y(y) = \sigma + \tau y \text{ if } \lambda = 0, \qquad (2.10)$$

$$X(x) = \gamma \cos\sqrt{\lambda}\,x + \delta \sin\sqrt{\lambda}\,x \text{ and } Y(y) = \sigma e^{\sqrt{\lambda}\,y} + \tau e^{-\sqrt{\lambda}\,y} \text{ if } \lambda \neq 0, \qquad (2.11)$$

where $\gamma, \delta, \sigma, \tau$ are arbitrary complex numbers. We'll see shortly that, if the boundary conditions (2.4) are to be met, then in fact λ must be real and positive. In the meantime, we note that the given solutions to (2.9) make sense and are valid for arbitrary complex λ, provided we define $\sqrt{\lambda}$ as in Exercise A.0.4 (or A.0.13) and the cosine and sine of a complex number by equations (1.8).

Let's now try, using additional information from the problem at hand, to narrow down the possibilities for $\gamma, \delta, \sigma, \tau$, and λ. We do so by looking at some of our *boundary conditions*. Specifically, if we plug (2.6) into the conditions (2.4), we get

$$X(0)Y(y) = 0, \quad X(\pi)Y(y) = 0 \quad (y > 0). \qquad (2.12)$$

We certainly don't want $Y(y)$ to be zero for all such y, because this would make our solution (2.6) identically zero, which we certainly don't want. But then (2.12) is equivalent to

$$X(0) = 0, \quad X(\pi) = 0. \qquad (2.13)$$

We consider first what this tells us about the case $\lambda = 0$. To this end, we observe that the function $X(x) = \gamma + \delta x$ of (2.10) satisfies $X(0) = X(\pi) = 0$ if and only if $\gamma = \delta = 0$. So in fact, the case $\lambda = 0$ yields no nontrivial solutions. We therefore assume $\lambda \neq 0$, and focus on the equations (2.11).

Applying the above condition $X(0) = 0$ to the formula for X in (2.11) gives $0 = \gamma \cos 0 + \delta \sin 0$, so $\gamma = 0$, so

$$X(x) = \delta \sin\sqrt{\lambda}\,x. \qquad (2.14)$$

We assume $\delta \neq 0$, so that X is not identically zero. But then, applying to (2.14) the condition $X(\pi) = 0$, we get

$$\sin \pi\sqrt{\lambda} = 0. \qquad (2.15)$$

This last equation is true if and only if $\sqrt{\lambda}$ is an integer. (see Exercise 2.1.1a.)
And again, we're presently in the case $\lambda \neq 0$. From (2.14), then, we conclude that

$$X(x) = \delta \sin nx, \qquad (2.16)$$

where n is any nonzero integer. Then from the second equation in (2.11), we get

$$Y(y) = \sigma e^{ny} + \tau e^{-ny}. \qquad (2.17)$$

We thus have the separated solution

$$u(x,y) = X(x)Y(y) = \delta \sin nx (\sigma e^{-ny} + \tau e^{ny}) = \sin nx (\chi e^{-ny} + \omega e^{ny}) \qquad (2.18)$$

to (2.2) and (2.4), for n any nonzero integer and χ and ω arbitrary complex numbers.

Note that the set of solutions $u(x,y)$ corresponding to a given integer n is the same as that corresponding to $-n$. So we get just as many nonzero solutions if we consider only $n \in \mathbb{Z}^+$, which we do. But if $n > 0$, then (2.18) will not approach zero as $y \to \infty$—that is, it will not satisfy the boundary condition (2.3)—unless $\omega = 0$. So (ignoring the arbitrary constant χ) we then have, for each positive integer n, a "separated" solution

$$u(x,y) = u_n(x,y) = e^{-ny} \sin nx \qquad (2.19)$$

to (2.2)–(2.4).

Fourier was the first to make systematic use of the separation-of-variables technique illustrated above. As we'll soon see, this technique is now omnipresent in the study of boundary value problems.

But for now, let's return to the heat flow problem at hand. We have the infinitely many solutions (2.19) to (2.2)–(2.4). So, as already discussed, this heat flow problem amounts to that of putting those solutions together into a linear combination

$$u(x,y) = \sum_{n=1}^{\infty} D_n u_n(x,y) = \sum_{n=1}^{\infty} D_n e^{-ny} \sin nx \qquad (2.20)$$

also satisfying (2.5). But the latter problem is, at least formally, straightforward: Putting $y = 0$ into (2.20) gives

$$u(x,0) = \sum_{n=1}^{\infty} D_n \sin nx, \qquad (2.21)$$

so if we want to make (2.5) true, we'll need

$$f(x) = \sum_{n=1}^{\infty} D_n \sin nx \quad (0 < x < \pi). \qquad (2.22)$$

Check it out: (2.22) *is* a sinusoid decomposition for f!

To recap the story so far, solution of the temperature distribution problem (2.2)–(2.5) requires only the determination of coefficients D_n satisfying (2.22). Indeed, given such D_n's, (2.20) provides precisely this solution. Which means we're done, because we *know*, by virtue of Theorem 1.12.1 with $\ell = \pi$, which D_n's make (2.22) true—at least under appropriate conditions on f. We therefore have the following.

Proposition 2.1.1 *If* $f \in \text{PSA}(0, \pi)$, *then the steady state temperature boundary value problem* (2.2)–(2.5) *has solution* $u(x, y)$ *given by* (2.20), *where*

$$D_n = \beta_n(f) = \frac{2}{\pi} \int_0^\pi f(x) \sin nx \, dx \quad (n \in \mathbb{Z}^+). \tag{2.23}$$

For instance, if $f(x) = x$ on $(0, \pi)$, then we obtain, by virtue of Example 1.12.1 with $\ell = \pi$, the solution

$$u(x, y) = 2 \sum_{n=1}^\infty \frac{(-1)^{n+1}}{n} e^{-ny} \sin nx \tag{2.24}$$

to (2.2)–(2.5).

Actually, our arguments in this section fall somewhat short of *proving* Proposition 2.1.1 rigorously. Indeed, while finite linear combinations of solutions to (2.2)–(2.4) are, as previously observed, themselves solutions, it's not so obvious that this should continue to hold for *infinite* linear combinations, of which, after all, the series (2.20) is an instance. That such a series should really solve (2.2)–(2.4) requires that it converge sufficiently well—well enough that it may be twice differentiated, term by term, in both x and y; well enough that it still approaches zero as $y \to \infty$. We have not verified that these convergence conditions are, in fact, met.

But it turns out that they are, under the stated conditions on f. Why? Because for such f, the D_n's go to zero with n, by the Riemann-Lebesgue lemma (Lemma 1.4.1)—or, more exactly, its analog for $\text{PC}(0, \pi)$. This, together with the rapid decay toward zero of e^{-ny}, as a function of $n \in \mathbb{Z}^+$ (for any fixed $y > 0$), assure that (2.20) converges as nicely as is required, and in fact even more nicely than that, on the interior of the plate. We omit the details; for these see Section 94 of [7].

Here's another crucial question: Could we not have solved the boundary value problem (2.2)–(2.5) in a much simpler way? Indeed, doesn't the function u defined by

$$u(x, y) = \begin{cases} 0 & \text{if } 0 \leq x \leq \pi \text{ and } y > 0, \\ f(x) & \text{if } 0 < x < \pi \text{ and } y = 0 \end{cases} \tag{2.25}$$

solve this problem?

The answer is: Sure it does, but it's not *really* what we have in mind. What we *really* have in mind is a solution that behaves better than does (2.25) at the boundary. What we *really* have in mind is a solution that behaves *near* the boundary in the same way that it does *at* the boundary. After all, boundary problems are, usually, physically motivated, and in physical situations one would expect such "continuous" behavior at the boundary to take place.

Let's make this more precise. Let's agree that, underlying the *explicit* conditions (2.4) and (2.5), we have the *implicit* continuity conditions

$$u(0^+, y) = 0, \quad u(\pi^-, y) = 0 \quad (y > 0) \tag{2.26}$$

and

$$u(x, 0^+) = f(x) \quad (0 < x < \pi). \tag{2.27}$$

Certainly (2.25) does not satisfy the last of these conditions, unless f itself is identically zero, in which case the entire boundary value problem is quite uninteresting. It's perhaps less evident that the solution of Proposition 2.1.1 *does* satisfy all of the above continuity conditions, but it does, if f is as stated; for the kinds of arguments necessary to prove this, see Sections 88 and 89 of [7].

More generally, let's adhere to the time-honored convention that, in a boundary value problem, any specification of the *values* of a function at the boundary is understood as a specification of the appropriate one-sided *limits* of that function. So a Dirichlet condition $u(a, y) = g_1(y)$, for example, tacitly implies the "continuity condition" $u(a^+, y) = g_1(y)$ for all relevant values of y. Similarly, $u(b, y) = g_2(y)$ implies $u(b^-, y) = g_2(y)$; $u(x, c) = f_1(x)$ implies $u(x, c^+) = f_1(x)$; and $u(x, d) = f_2(x)$ implies $u(x, d^-) = f_2(x)$—again for relevant values of the variables in question. Analogous conventions apply to boundary value problems involving Neumann conditions, and those involving any number of variables.

The "Fourier method," by which we mean the strategy we employed in this section, and will spell out more carefully in the next section, does, generally, yield solutions continuous at the boundary in the above sense. At least, it does so under appropriate conditions on the functions (such as the function f of (2.5)) prescribing the behavior at the boundary.

We conclude this section by inquiring into the *uniqueness* of the solution we've obtained. How do we know this solution is the only such? We have not demonstrated this, but it does turn out to be so. One way to see this, in the case of the present and of many other boundary value problems, is to consider the (or a) physical situation that the problem embodies, and to convince oneself that such a situation can pan out in only one way. Another, more rigorous and perhaps more satisfying, way is to appeal to a *uniqueness theorem*. Many such exist in the theory of differential equations and boundary value problems. A good source for several of these is Chapter 10 of [7].

We will not supply here the uniqueness theorem relevant to our steady state temperature problem. But we will demonstrate uniqueness in the case of the *wave equation*—such demonstration being, in that case, particularly pleasant—when we get there (in Section 2.8). Moreover, we'll take that particular demonstration as sufficient tonic to ease any queasiness we might feel over the uniqueness question in general.

Exercises

2.1.1 a. Show that, for θ *complex* and $\sin\theta$ defined by (1.8), we have $\sin\theta = 0$ if and only if θ is an integer multiple of π. Hint: Use Exercise A.0.11. You may take on faith the same facts regarding the cosine and sine of a real variable as were assumed in that exercise.

b. Show that, for θ *complex* and $\cos\theta$ defined by (1.8), we have $\cos\theta = 1$ if and only if θ is an integer multiple of 2π. Hint: Same as in part a of this exercise. Also, $(e^{i\theta} + e^{-i\theta})/2 - 1 = (e^{i\theta/2} - e^{-i\theta/2})^2/2$.

2.1.2 Using separation of variables (see Example 2.1.1), the previous exercise, and the already cited solutions (2.10)/(2.11) to (2.9), find, for each integer n in an appropriate range, separated solutions (that is, solutions that are products of single-variable functions) to the given differential equation plus boundary conditions.

a.
$$u_{xx} + u_{yy} = 0 \quad (0 < x < \pi,\ y > 0),$$
$$u(x, y) \text{ is bounded as } y \to \infty \quad (0 < x < \pi),$$
$$u_x(0, y) = 0, \quad u_x(\pi, y) = 0 \quad (y > 0).$$

(This problem corresponds to a situation like that of Example 2.1.1, but with *insulated* vertical edges instead of edges of fixed temperature zero.) ANSWER: $u(x,y) = \alpha e^{-ny}\cos nx$ for $\alpha \in \mathbb{C}$ and $n \in \mathbb{N}$. (Why not $n < 0$ as well?)

b.
$$c^2 y_{xx} = y_{tt} \quad (0 < x < \pi,\ t > 0),$$
$$y(0, t) = 0, \quad y(\pi, t) = 0 \quad (t > 0),$$

where $c > 0$. (This problem represents wave motion; see Section 2.7.) ANSWER: $y(x, t) = \sin nx\,(\gamma\cos nct + \delta\sin nct)$ for $\gamma, \delta \in \mathbb{C}$ and $n \in \mathbb{Z}^+$. (Why not $n \leq 0$ as well?)

c.
$$c^2 y_{xx} = y_{tt} + a^2 y \quad (0 < x < \pi,\ t > 0),$$
$$y(0, t) = 0, \quad y(\pi, t) = 0 \quad (t > 0),$$

where $c, a > 0$. (This problem represents wave motion in an elastic medium.) Your answer should involve cosines and/or sines of nx, as well as cosines and/or sines of $\sqrt{n^2 c^2 + a^2}\, t$, for $n \in \mathbb{Z}^+$.

2.1.3 Separate variables to reduce each of the following partial differential equations to a system of ordinary differential equations. You don't have to solve any differential equations.

a. $yu_{xx} + xu_{yy} = 0$. ANSWER: $X''(x) + \lambda x X(x) = 0;\ Y''(y) - \lambda y Y(y) = 0$.

b. $u_x + u_{xy} + u_{yy} = 0$. Hint: Try $X'(x) + \lambda X(x) = 0$.

c. $u_{xx} + u_{yy} + u_{zz} = 0$. ANSWER: $X''(x) + \lambda X(x) = 0,\ Y''(y) + \mu Y(y) = 0,\ Z''(z) - (\lambda + \mu)Z(z) = 0$.

d. $3y(u_x + u_{xx}) + 4x(u + u_{yy}) + 5xyu_{zz} = 0$.

e. $u_{xyz} + u_{xxy} + u_z = 0$.

2.1.4 Here we "shift" the above boundary value problem (2.2)–(2.5), to create a situation more like the one Fourier considered.

a. Let g be piecewise smooth and averaged on $(-\pi/2, \pi/2)$; let $f(x) = g(x-\pi/2)$. Show that $u(x, y)$ solves (2.2)–(2.5) if and only if $v(x, y) = u(x+\pi/2, y)$ solves the following boundary value problem:

$$v_{xx} + v_{yy} = 0 \quad (-\pi/2 < x < \pi/2,\ y > 0),$$
$$\lim_{y \to \infty} v(x, y) = 0 \quad (-\pi/2 < x < \pi/2),$$
$$v(-\pi/2, y) = 0, \quad v(\pi/2, y) = 0 \quad (y > 0),$$
$$v(x, 0) = g(x) \quad (-\pi/2 < x < \pi/2).$$

b. Use Proposition 2.1.1 to show that the boundary value problem in part a above has formal solution

$$v(x, y) = \sum_{m=1}^{\infty} \left(F_m e^{-(2m-1)y} \cos(2m-1)x + G_m e^{-2my} \sin 2mx \right),$$

where

$$F_m = \frac{2}{\pi} \int_{-\pi/2}^{\pi/2} g(x) \cos(2m-1)x\, dx, \quad G_m = \frac{2}{\pi} \int_{-\pi/2}^{\pi/2} g(x) \sin 2mx\, dx$$

for $m \in \mathbb{Z}^+$. Hint: Use a trig identity, or the fact that $\sin \theta = \operatorname{Im} e^{i\theta}$ for $\theta \in \mathbb{R}$, to show that

$$\sin n\left(x + \frac{\pi}{2}\right) = \begin{cases} (-1)^{m+1} \cos(2m-1)x & \text{if } n \text{ is odd}, n = 2m-1, \\ (-1)^m \sin 2mx & \text{if } n \text{ is even}, n = 2m. \end{cases}$$

c. Fourier restricted his attention mostly to *even* initial temperature functions g. How does the above solution simplify for such g?

2.1.5 a. Solve the boundary value problem of Exercise 2.1.4 when $g(x) = 1$ for all $x \in (-\pi/2, \pi/2)$.

b. Using the Maclaurin series

$$\arctan z = \sum_{m=1}^{\infty} \frac{(-1)^{m+1}}{2m-1} z^{2m-1}$$

(valid for complex as well as real z, provided $|z| < 1$—this series may be used to *define* arctan z for such z), express your answer from part a of this exercise in terms of $\arctan e^{-(y+ix)}$ and $\arctan e^{-(y-ix)}$.

c. Assuming the identity

$$\arctan p + \arctan q = \arctan \frac{p+q}{1-pq}$$

(to see that this holds at least for p, q in a suitable range, write $p = \tan \theta_1$ and $q = \tan \theta_2$, and then take tangents of both sides, to get a familiar trig identity), express your answer from part b of this exercise in terms of $\arctan(\cos x / \sinh y)$.

d. Recalling that $d \arctan u / du = 1/(1 + u^2)$ (and that $d \cos x / dx = -\sin x$ and $d \sinh y / dy = \cosh y$), show directly that your answer from part c of this exercise satisfies the boundary value problem of part a.

2.2 LINEAR OPERATORS, HOMOGENEOUS EQUATIONS, AND SUPERPOSITION

Here's a general summary of the techniques employed in the previous section to solve the boundary value problem (2.2)–(2.5): (1) we first produced the relatively elementary solutions (2.19) to (2.2)–(2.4), via *separation of variables*. (2) From these solutions we constructed, through infinite linear combination, more complex ones (2.20). (3) To the latter we then applied results from the theory of Fourier series to obtain a solution also conforming to (2.5).

And there you have it: These are the three main elements of the "Fourier method" for solving boundary value problems. In this section, we wish to look more closely at step 2. Specifically we ask: Under what general conditions will a differential equation or boundary condition behave like equations (2.2)–(2.4), in that any linear combination of solutions is itself a solution? The answer lies in the notions of *linearity* and *homogeneity*.

Definition 2.2.1 (a) A *linear operator* L on a complex vector space V is a rule that associates, to any element u of V, an element Lu of another complex vector space W, in such a way that "L of a linear combination is the linear combination of the L's." That is, L is a linear operator on V if

$$L(\alpha_1 u_1 + \alpha_2 u_2) = \alpha_1 L u_1 + \alpha_2 L u_2 \tag{2.28}$$

for any $u_1, u_2 \in V$ and $\alpha_1, \alpha_2 \in \mathbb{C}$.

(b) A differential equation on a specified domain, or a boundary condition pertaining on the boundary of that domain, is called *linear* if it may be written in the form

$$Lu = f, \tag{2.29}$$

where L is a linear operator (on some space V of functions on the domain in question) and f some prescribed function (on that domain).

(c) The equation (2.29) is called *homogeneous* if f is identically zero.

Some examples are in order:

LINEAR OPERATORS, HOMOGENEOUS EQUATIONS, AND SUPERPOSITION

Example 2.2.1 (a) Let c be a fixed element of $\mathbb{C}$. Then

$$L_c z = cz \tag{2.30}$$

defines a linear operator L_c from the complex vector space $\mathbb{C}$ into itself.

(b) Let $V = \mathrm{PC}(a, b)$. Then

$$Iu = \int_a^b u(x)\, dx \tag{2.31}$$

defines a linear operator I from V into $\mathbb{C}$. More generally, if v is a fixed function in V, then

$$I_v u = \int_a^b u(x)v(x)\, dx \tag{2.32}$$

defines a linear operator $I_v : V \to \mathbb{C}$.

(c) The differential operator D defined by

$$Du = xu_y + y^2 (u_{xz})^3, \tag{2.33}$$

where V is a complex vector space of three-variable functions $u(x, y, z)$ such that the relevant partial derivatives exist, is not linear on V. (For example, $D(5xz) \neq 5D(xz)$.) Its nonlinearity is because of the cubing of u_{xz}. Similarly,

$$Eu = x^3 \cos u + e^y u_x \tag{2.34}$$

defines a nonlinear (because of the $\cos u$ factor) differential operator. On the other hand:

(d) An operator of the form

$$Lu = Fu + \sum_{k=1}^{m} G_k u_{x_k} + \sum_{k,n=1}^{m} H_{kn} u_{x_k x_n} \tag{2.35}$$

is called a *linear second-order differential operator*. Here u is assumed to belong to some complex vector space V of functions of the real variables $x_1, x_2, \ldots, x_m$. Also F, the G_k's, and the H_{kn}'s are specified elements of V, and at least one H_{kn} is not the zero function. The boundary value problems with which we'll be concerned in the coming sections will involve differential operators of this kind.

We'll be rather nonspecific for now about V, although we'll have a few things to say about "where u lives" in the context of some of the particular boundary value problems we consider. And we will *always* assume that elements u of V have equal mixed second-order partial derivatives, meaning $u_{x_k x_n} = u_{x_n x_k}$ for all k and n. This will obtain if all such derivatives are continuous, for example.

Note that, because differentiation is itself a linear operator, as is multiplication by a fixed function, a linear second-order differential operator is indeed a linear operator, in the sense of part (a) of this definition.

For instance, the "Laplace operator"

$$Lu = u_{xx} + u_{yy} \tag{2.36}$$

is a linear, second-order differential operator.

(e) The Laplace equation (2.2) is linear and homogeneous; it has the form $Lu = 0$, where L is the Laplace operator just described.

(f) Let V be a complex vector space of functions $u(x, y)$ of two real variables. If $x_0 \in \mathbb{R}$, then the operator P_{x_0} defined by $(P_{x_0} u)(y) = u(x_0, y)$ (that is, P_{x_0} takes u to the function of y whose values are given by $u(x_0, y)$) is linear. It's called an *evaluation operator*. Similarly there is, for each $y_0 \in \mathbb{R}$, a linear evaluation operator Q_{y_0} given by $(Q_{y_0} u)(x) = u(x, y_0)$. Analogous evaluation operators may be defined on complex vector spaces of functions of any number of variables.

Many boundary conditions are of the form $Bu = f$, where B is an evaluation operator: For example, the condition (2.5) may be written in this way, with B equal to the evaluation operator Q_0 as described above. Many other boundary conditions are *compositions* of linear differential and evaluation operators: For instance, a Neumann condition $u_x(\pi, y) = g(y)$, say, may be written $Bu = g$, where $B = P_\pi D$, with P_π as above and $D = \partial/\partial x$. (Juxtaposition of operators denotes composition: $(P_\pi D)u$ by definition equals $P_\pi(Du)$.)

(g) The boundary conditions given by equations (2.3) and (2.4) are linear and homogeneous; the condition (2.5) is linear and *inhomogeneous* (not homogeneous).

Parts (e) and (g) of the above example, together with the following proposition, explain why our earlier approach to (2.2)–(2.4) worked. The proposition also suggests more general situations in which such an approach might be indicated.

Proposition 2.2.1 (the superposition principle I) *Let L be a linear operator on a complex vector space V. If $L(u_k) = 0$ for $1 \leq k \leq N$, then*

$$L\left(\sum_{k=1}^{N} \alpha_k u_k\right) = 0 \tag{2.37}$$

for any complex numbers $\alpha_1, \alpha_2, \ldots, \alpha_N$.

Proof. By an induction argument, the linearity property (2.28) generalizes as follows:

$$L\left(\sum_{k=1}^{N} \alpha_k u_k\right) = \sum_{k=1}^{N} \alpha_k L u_k \tag{2.38}$$

for V any complex vector space, L any linear operator on V, $\alpha_1, \alpha_2, \ldots, \alpha_N \in \mathbb{C}$, and $u_1, u_2, \ldots, u_N \in V$.

But then, under the stipulated conditions of the present proposition,

$$L\left(\sum_{k=1}^{N} \alpha_k u_k\right) = \sum_{k=1}^{N} \alpha_k L u_k = \sum_{k=1}^{N} \alpha_k \cdot 0 = 0, \qquad (2.39)$$

as desired. □

(Part II of the superposition principle appears in Section 2.10.)

An infinite linear combination is a limit of finite ones, so one can construct quite complex solutions to *linear, homogeneous* differential equations and boundary conditions by forming infinite linear combinations of simpler solutions, provided the relevant limiting processes behave appropriately, as they often do under mild conditions. (Thus, for example, we obtained solutions of the form (2.20) to (2.2)–(2.4), from the more elementary, separated solutions u_n of (2.19).) One may then seek, out of all such linear combinations, a particular one that satisfies any linear, *inhomogeneous* boundary conditions present. In the previous section, it was the theory of Fourier series that told us how to make this choice, and thereby allowed us to solve (2.2)–(2.5) completely.

We make a crucial point: It's only because our separated solutions $X(x)Y(y)$ to (2.2)–(2.4) were *sinusoidal* in x that (2.20), when subjected to the inhomogeneous boundary condition (2.5), took the form of a *Fourier series*. Other kinds of $X(x)$'s might result in decidedly different and unfamiliar infinite series of functions of x. Which begs (at least) three questions. First, what general features of our boundary value problem made the $X(x)$'s what they were? Second, was this a fluke, or is there some reason to believe that many other such problems, when approached as we did this one (using "the Fourier method": separation of variables, superposition, and then imposition of inhomogeneous boundary conditions), will also give rise to Fourier series? And third, if such a problem, thus treated, gives rise to some *other* kind of series—if it yields a series like (2.20), but with some other, not necessarily trigonometric, functions of x in place of the $\sin nx$'s—then is there some analog of Fourier theory that will allow us to determine the coefficients of that series?

The third question is essentially the subject of Chapter 4. But let's, for now, say a few words about the first two. The answer to the first question is this: Our separation-of-variables argument in Example 2.1.1 led us to the pursuit of pairs (λ, X) (with $\lambda \in \mathbb{C}$ and X a function from $[0, \pi]$ to $\mathbb{C}$) such that

$$X''(x) + \lambda X(x) = 0 \quad (0 < x < \pi). \qquad (2.40)$$

And as we saw, any such function X is, under appropriate boundary conditions, a sinusoid.

Which brings us back to the second question above. To which the answer is: No, it was not a fluke, and yes, many boundary problems of physical interest give rise, in a similar manner, to Fourier series. Why? Because many such problems lead, when subject to the separation-of-variables technique, to the differential equation in (2.40).

Which perhaps does not really answer the question; it does not tell us *why* the differential equation in (2.40) should arise so often in "real life." Why does it? Because of the phenomenon of *spatial symmetry*.

We explain. A linear operator is called *spatially symmetric* if the following is true: Translating or rotating your spatial coordinates and then applying the operator gives the same result as *first* applying the operator and *then* making this change of coordinates. (That is, "been there and done that" is the same as "done that and been there.") Many physical processes are spatially symmetric by nature, and may be modeled by spatially symmetric linear, second-order differential operators. But: *An m-variable linear differential operator L, of order less than or equal to 2, is spatially symmetric if and only if it's a linear combination of a constant and the operator ∇_m^2 defined by*

$$\nabla_m^2 u = \sum_{k=1}^{m} u_{x_k x_k}. \tag{2.41}$$

(We omit the proof.) The relevance of this fact to the differential equation in (2.40) is clear: Indeed, that differential equation reads $\nabla_1^2 X + \lambda X = 0$. Further, (2.2) itself is the equation $\nabla_2^2 u = 0$.

We call ∇_m^2 the *Laplace operator*, or *Laplacian*, of dimension m. A boundary value problem combining the equation $\nabla_m^2 = 0$ with Dirichlet boundary conditions is called a *Dirichlet problem*; if instead all boundary conditions are of Neumann type, one has a *Neumann problem*.

Recall that it was a separation-of-variables argument, applied to (2.2), that resulted in (2.40). So in sum, we have the following answer to our second question above, regarding the frequency of occurrence of sinusoids in the solution of boundary value problems: (1) For reasons of spatial symmetry, many boundary value problems involve the Laplacian. (2) Applying to such a problem the method of separation of variables often leads to the differential equation in (2.40), whose solutions *are*, under suitable boundary conditions, sinusoids.

The boundary value problem given by (2.40) and (2.13) is an example of what we'll call an *eigenvalue problem*, by which we mean the following.

Definition 2.2.2 An *eigenvalue problem* on a domain R is a problem comprising the following elements.

(a) An equation of the form

$$LX(x) + \lambda X(x) = 0 \quad (x \text{ in the interior of R}), \tag{2.42}$$

where L is a specified linear operator on some complex vector space V of functions on R, λ is an unspecified complex number, and X is an unspecified function.

(b) (Perhaps, some) boundary conditions on X.

If (λ, X), with $\lambda \in \mathbb{C}$ and X a *nonzero* function in V, solves this eigenvalue problem, then λ is called an *eigenvalue* of the problem; X is called an *eigenfunction* (or *eigenvector*) of the problem; and we say X *corresponds to* λ (and vice versa).

We note that certain other sources would, in the context just described, call $-\lambda$ rather than λ an eigenvalue. However, the above definition is more consistent with the language of *Sturm-Liouville* theory, which subsumes much of the material in this chapter, and will be studied *per se* in Chapter 4.

We also remark that broader (and narrower!) definitions of eigenvalue problem exist in the literature. But ours will suffice for our purposes.

We saw in the course of Example 2.1.1 that the eigenvalue problem of (2.40) and (2.13) has, for each $n \in \mathbb{Z}^+$, an eigenvalue $\lambda = n^2$ and corresponding eigenfunction $X(x) = \sin\sqrt{\lambda}\,x = \sin nx$. (Of course, a nonzero constant multiple of an eigenfunction is also an eigenfunction.) Every boundary value problem to be considered in this text gives rise to an eigenvalue problem. In fact, it turns out that each of the latter is of a particular type, called *Sturm-Liouville* type, to be described and discussed more thoroughly in Chapter 4.

We turn, for the remainder of this chapter, to further examples of boundary value problems, whose solutions will entail techniques and ideas outlined above, and some extensions of those ideas and techniques.

Exercises

2.2.1 Show that the operator L is linear on V, in the sense of Definition 2.2.1(a), if and only if

$$L(u_1 + u_2) = Lu_1 + Lu_2, \quad L(\alpha u_1) = \alpha Lu_1$$

for all $u_1, u_2 \in V$ and $\alpha \in \mathbb{C}$.

2.2.2 **a.** Suppose L and M are linear operators, with L taking the vector space V into some vector space W and M taking W into some vector space U. Show that the composition ML is itself a linear operator on V.

 b. Show that composition of linear operators is *not* commutative: LMu does not always equal MLu, even if L and M both take V into itself. Hint: Let V be the space of all functions on $\mathbb{R}$ having continuous derivatives of all orders; consider differentiation versus multiplication by a fixed element of V.

2.2.3 Let D denote the differentiation operator on the space V of all continuously differentiable functions u on $(-1, 1)$ such that $u(0) = 0$. Find an inverse for D, meaning an operator E such that $EDu = DEu = u$ for all u in V. Hint: the fundamental theorem of calculus.

2.2.4 Let P be the operator $Pu(x) = x^{-1/2} u'(\sqrt{x})$, defined on the space V of continuously differentiable functions u on $(0, \infty)$ such that $u(1) = 0$. Show that $Qu(x) = \frac{1}{2}\int_1^{x^2} u(t)\,dt$ defines an inverse to P: $QPu = PQu$ for all u in V.

2.2.5 For any suitable vector space V of 2π-periodic functions, let S be the linear operator defined by $Sf = (c_n(f))_{n=-\infty}^{\infty}$ (the right side denotes the infinite sequence of Fourier coefficients of f). (See Exercise 1.2.2 for linearity of this operator.) Using results from Sections 1.7 and 1.8:

a. Show that, if $V = \text{PC}(\mathbb{T})$, then S takes V into the vector space l^2 of infinite sequences $(r_n)_{n=-\infty}^{\infty}$ such that $\sum_{n=-\infty}^{\infty} |r_n|^2 < \infty$.

b. Show that, if $V = \text{PSC}(\mathbb{T})$, then S takes V into the vector space l^1 of infinite sequences $(s_n)_{n=-\infty}^{\infty}$ such that $\sum_{n=-\infty}^{\infty} |s_n| < \infty$.

c. Find an operator T on l^2 and an operator O on $\text{PSC}(\mathbb{T})$ such that $TSf = SOf$ for all $f \in \text{PSC}(\mathbb{T})$.

d. Find an $x_0 \in \mathbb{R}$ such that, if S is as above and Σ is the operator on l^1 defined by

$$\Sigma(s_n)_{n=-\infty}^{\infty} = \sum_{n=-\infty}^{\infty} s_n,$$

then ΣS, as an operator on $\text{PSC}(\mathbb{T})$, equals the evaluation operator P_{x_0} defined by $P_{x_0} f = f(x_0)$.

2.2.6 Let $Lu = u' - e^{-u}$, defined on the vector space of differentiable functions u on $(0, \infty)$, say.

a. Show that both $u(x) = u_1(x) = \ln(x+1)$ and $u(x) = u_2(x) = \ln(x+2)$ satisfy $Lu = 0$.

b. Show that $u(x) = u_1(x) + u_2(x)$ does not.

c. Does this contradict Proposition 2.2.1? Explain.

2.2.7 Show that the Laplace equation $\nabla_m^2 u = 0$ has solution:

a. $u(x, y) = \ln(x^2 + y^2)$ (if $(x, y) \neq (0, 0)$) in dimension $m = 2$;

b. $u(x, y, z) = 1/\sqrt{x^2 + y^2 + z^2}$ (if $(x, y, z) \neq (0, 0, 0)$) in dimension $m = 3$.

2.3 HEAT FLOW IN A BAR I: NEUMANN AND MIXED BOUNDARY CONDITIONS

We now investigate some time-dependent heat conduction problems. Throughout we will use the fact, first shown by Fourier, that heat flow in m dimensions is governed by the *heat equation*

$$k \nabla_m^2 u = u_t, \tag{2.43}$$

with ∇_m^2 the Laplacian of (2.41) and $k > 0$ a constant called the *thermal diffusivity*. Note that the heat equation is linear and homogeneous; it may be written $Lu = 0$, where L is defined by $Lu = k \nabla_m^2 u - u_t$.

We begin by investigating heat flow in a bar of uniform cross section—the shape of this cross section is immaterial; we'll take it to be circular, to fix ideas—and finite length, let's call it ℓ. Let's place this bar along the x axis from $x = 0$ to $x = \ell$ (Fig. 2.2).

We assume that there is no heat source within the bar, and further that the lateral surface of the bar, meaning all of its surface except perhaps the ends at $x = 0$ and

HEAT FLOW IN A BAR I: NEUMANN AND MIXED BOUNDARY CONDITIONS

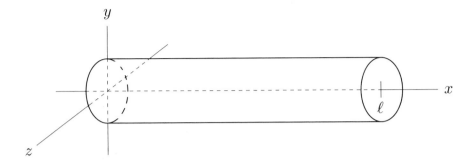

Fig. 2.2 A bar of uniform, circular cross section

$x = \ell$, is insulated, which is to say there is no heat flowing across it. We'll discuss what happens at the ends shortly.

Suppose this bar is subject, at a time that we'll call $t = 0$, to an initial temperature distribution $f(x)$, independent of the other two perpendicular spatial coordinates. This means the resulting temperature distribution u will depend only on x and on t: so $u = u(x, t)$. So we have the one-dimensional heat equation

$$k\, u_{xx} = u_t \quad (0 < x < \ell,\ t > 0). \tag{2.44}$$

Various kinds of conditions may apply at the ends of the bar; in this section we consider two different sets of conditions there. The first is that these ends be insulated: This means we have the Neumann conditions

$$u_x(0, t) = 0, \quad u_x(\ell, t) = 0 \quad (t > 0). \tag{2.45}$$

The second set of conditions is that one end—the left one, say—be held at temperature zero, and the other insulated. This means there is a Dirichlet condition at one end and a Neumann condition at the other:

$$u(0, t) = 0, \quad u_x(\ell, t) = 0 \quad (t > 0). \tag{2.46}$$

(One might also impose Dirichlet conditions at both ends. If, for example, one holds both ends at temperature zero, then the situation and the solution are similar in many respects—though certainly not all—to those of (2.2)–(2.5). See Exercise 2.3.3.)

In any case, the imposition of initial temperatures $f(x)$ amounts to the linear, inhomogeneous Dirichlet condition

$$u(x, 0) = f(x) \quad (0 < x < \ell) \tag{2.47}$$

"at the temporal boundary $t = 0$." As is standard, we will often refer to temporal boundary conditions as "initial conditions."

Let's solve these boundary value problems.

Example 2.3.1 Solve the Neumann heat conduction problem (2.44), (2.45), (2.47).

Solution. As we did in Example 2.1.1, we apply separation of variables to the determination of nonzero solutions to our homogeneous equations — in the present case, these are equations (2.44) and (2.45). That is, we look for solutions

$$u(x,t) = X(x)T(t) \tag{2.48}$$

to these equations, where neither X nor T is identically zero.

Putting this separated form for u into the heat equation (2.44) gives

$$k X''(x)T(t) = X(x)T'(t) \quad (0 < x < \ell,\; t > 0). \tag{2.49}$$

Dividing through by $u = XT$ gives $k X''/X = T'/T$ on the indicated domain; this will certainly be the case if both sides equals the same constant on that domain. Let's say $X''/X = -\lambda$; thus we are led to consideration of the ordinary differential equations

$$X''(x) + \lambda X(x) = 0 \quad (0 < x < \ell), \quad T'(t) + \lambda k\, T(t) = 0 \quad (t > 0). \tag{2.50}$$

The most general simultaneous solution to these is given by

$$X(x) = \gamma + \delta x \text{ and } T(t) = \sigma \text{ if } \lambda = 0, \tag{2.51}$$

$$X(x) = \gamma \cos \sqrt{\lambda}\, x + \delta \sin \sqrt{\lambda}\, x \text{ and } T(t) = \sigma e^{-\lambda k t} \text{ if } \lambda \neq 0, \tag{2.52}$$

where γ, δ, and σ are arbitrary complex numbers. (The solution for X we already discussed in the course of Example 2.1.1. The solution for T is straightforward; see Exercise 2.3.1.)

Next, the boundary conditions (2.45), when applied to the separated form (2.48), give

$$X'(0)T(t) = 0, \quad X'(\ell)T(t) = 0 \quad (t > 0). \tag{2.53}$$

Since T is not to be identically zero, we conclude

$$X'(0) = 0, \quad X'(\ell) = 0. \tag{2.54}$$

We apply these latter conditions first to the case $\lambda = 0$. Here, we observe that a nonzero linear function $X(x) = \gamma + \delta x$ satisfies both of these conditions only if $\delta = 0$. So the eigenvalue $\lambda = 0$ yields, by (2.51), the separated solutions

$$u_0(x,t) = X(x)T(t) = \frac{D_0}{2}, \tag{2.55}$$

where D_0 is an arbitrary constant. (The factor of 2 in the denominator will be convenient for later application of Fourier series arguments.)

We turn now to the consideration of nonzero eigenvalues, and to the corresponding equations (2.52). Into the first of these equations, we put the above condition $X'(0) = 0$; we get $\delta \sqrt{\lambda} = 0$, whence $\delta = 0$, since again we're now in the case $\lambda \neq 0$. So by the first equation in (2.52),

$$X(x) = \gamma \cos \sqrt{\lambda}\, x. \tag{2.56}$$

Plugging into this the condition $X'(\ell) = 0$ gives $-\gamma\sqrt{\lambda}\sin\sqrt{\lambda}\,\ell = 0$ so that, ignoring the trivial case $\gamma = 0$, we find $\sqrt{\lambda}\sin\sqrt{\lambda}\,\ell = 0$. Since again $\lambda \neq 0$, this happens if and only if

$$\sqrt{\lambda} = \frac{\pi n}{\ell} \tag{2.57}$$

for some nonzero integer n. So we get

$$X(x) = \gamma \cos\frac{\pi n x}{\ell}; \tag{2.58}$$

moreover, again by (2.52),

$$T(t) = \sigma e^{-(\pi n/\ell)^2 kt} = \sigma e^{-\pi^2 n^2 kt/\ell^2}. \tag{2.59}$$

In sum we have, for any nonzero integer n and $D_n \in \mathbb{C}$, a solution

$$u_n(x,t) = X(x)T(t) = D_n e^{-\pi^2 n^2 kt/\ell^2} \cos\frac{\pi n x}{\ell} \tag{2.60}$$

to (2.44) and (2.45). And, since the set of solutions corresponding to the integer $-n$ is the same as that corresponding to n, it suffices to consider $n \in \mathbb{Z}^+$.

Since all equations in (2.44) and (2.45) are linear and homogeneous, we may apply the superposition principle (Proposition 2.2.1) to our separated solutions $u_0(x,t)$ and $u_n(x,t)$ ($n \geq 1$) to these equations, to obtain the solution

$$u(x,t) = \sum_{n=0}^{\infty} u_n(x,t) = \frac{D_0}{2} + \sum_{n=1}^{\infty} D_n e^{-\pi^2 n^2 kt/\ell^2} \cos\frac{\pi n x}{\ell} \tag{2.61}$$

to them. Imposing the linear, inhomogeneous boundary condition (2.47) then gives

$$f(x) = \frac{D_0}{2} + \sum_{n=1}^{\infty} D_n \cos\frac{\pi n x}{\ell} \quad (0 < x < \ell), \tag{2.62}$$

and Theorem 1.12.1 tells us how to find D_n's to make this last equation true. Specifically, we have the following.

Proposition 2.3.1 *If $f \in \mathrm{PSA}(0,\ell)$, then the boundary value problem (2.44), (2.45), (2.47) has solution $u(x,y)$ given by (2.61), where*

$$D_n = \alpha_n(f) = \frac{2}{\ell}\int_0^\ell f(x)\cos\frac{\pi n x}{\ell}\,dx \quad (n \in \mathbb{N}). \tag{2.63}$$

As a concrete instance we let $f(x) = x$ on $(0,\ell)$; then we obtain, by Example 1.12.1, the solution

$$\begin{aligned}u(x,t) &= \frac{\ell}{2} + \frac{2\ell}{\pi^2}\sum_{n=1}^{\infty}\frac{(-1)^n - 1}{n^2}e^{-\pi^2 n^2 kt/\ell^2}\cos\frac{\pi n x}{\ell} \\ &= \frac{\ell}{2} - \frac{4\ell}{\pi^2}\sum_{m=1}^{\infty}\frac{1}{(2m-1)^2}e^{-\pi^2(2m-1)^2 kt/\ell^2}\cos\frac{\pi(2m-1)x}{\ell}\end{aligned} \tag{2.64}$$

to our problem.

As in the case of Proposition 2.1.1, the proof of Proposition 2.3.1 requires, strictly speaking, more than we've supplied: One needs to show that the proposed solution converges suitably. But it does, under the given conditions on f. The salient point here is that one has, in the mth summand of the solution just supplied, the term

$$e^{-\pi^2(2m-1)^2 kt/\ell^2}, \qquad (2.65)$$

which for fixed $t > 0$ is of exponential decay in m and therefore forces everything to converge very nicely. See Section 89 of [7] for rigorous verifications.

We now turn to:

Example 2.3.2 Solve the mixed boundary condition heat conduction problem (2.44), (2.46), (2.47).

Solution. The differential equation here is the same as in the previous example, so it has the same separated solutions (2.51), (2.52). But the boundary conditions are different. Specifically, the mixed conditions (2.46), when applied to the separated form (2.48) for u, give

$$X(0)T(t) = 0, \quad X'(\ell)T(t) = 0 \quad (t > 0). \qquad (2.66)$$

We assume that T is not identically zero, so we conclude

$$X(0) = 0, \quad X'(\ell) = 0. \qquad (2.67)$$

Let's, as before, look first at the case $\lambda = 0$. Putting the conditions (2.67) into the first equation in (2.51) yields $\gamma = \delta = 0$, so this case gives only the trivial solution, meaning zero is not an eigenvalue.

What about $\lambda \neq 0$? To answer, we substitute the above condition $X(0) = 0$ into the equation for X in (2.52); we get $\gamma = 0$, so that

$$X(x) = \delta \sin \sqrt{\lambda}\, x. \qquad (2.68)$$

Plugging into this the condition $X'(\ell) = 0$ gives $\delta\sqrt{\lambda} \cos \sqrt{\lambda}\, \ell = 0$; ignoring the trivial case $\delta = 0$, we find $\cos \sqrt{\lambda}\, \ell = 0$, or

$$\sqrt{\lambda} = \frac{\pi(2n-1)}{2\ell} \qquad (2.69)$$

for some integer n. (That these values of $\sqrt{\lambda}$ are the *only* complex numbers satisfying $\cos \sqrt{\lambda}\, \ell = 0$ follows from arguments similar to those of Exercise 2.1.1, providing one takes it on faith that no other *real* numbers solve this equation.) Combining this with (2.68) and with the equation for T in (2.52), we conclude: For $n \in \mathbb{Z}$ and $D_n \in \mathbb{C}$,

$$u_n(x,t) = D_n e^{-\pi^2(2n-1)^2 kt/(4\ell^2)} \sin \frac{\pi(2n-1)x}{2\ell} \qquad (2.70)$$

solves (2.44), (2.46). Actually, it suffices to consider $n \geq 1$, since $2(1-n) - 1 = -(2n-1)$, so the integer $1-n$ gives the same set of solutions as does n.

Since the heat equation (2.44) and boundary conditions (2.46) are linear and homogeneous, the superposition principle (Proposition 2.2.1) leads us to solutions of the form

$$u(x,t) = \sum_{n=1}^{\infty} u_n(x,t) = \sum_{n=1}^{\infty} D_n e^{-\pi^2 (2n-1)^2 kt/(4\ell^2)} \sin \frac{\pi(2n-1)x}{2\ell}. \quad (2.71)$$

Imposing the linear, inhomogeneous boundary condition (2.47) then gives

$$f(x) = \sum_{n=1}^{\infty} D_n \sin \frac{\pi(2n-1)x}{2\ell}, \quad (2.72)$$

and Theorem 1.12.2 tells us how to find D_n's to make this last equation true. So we have the following.

Proposition 2.3.2 *If* $f \in \mathrm{PSA}(0, \ell)$, *then the mixed boundary value problem* (2.44), (2.46), (2.47) *has solution* $u(x, y)$ *given by* (2.71), *where*

$$D_n = \chi_n(f) = \frac{2}{\ell} \int_0^\ell f(x) \sin \frac{\pi(2n-1)x}{2\ell} \, dx \quad (n \in \mathbb{Z}^+). \quad (2.73)$$

If, for instance, $f(x) = x$ on $(0, \ell)$, then Example 1.12.2 gives the solution

$$u(x,t) = \frac{8\ell}{\pi^2} \sum_{n=1}^{\infty} \frac{(-1)^{n+1}}{(2n-1)^2} e^{-\pi^2 (2n-1)^2 kt/(4\ell^2)} \sin \frac{\pi(2n-1)x}{2\ell} \quad (2.74)$$

to our mixed boundary condition problem.

Of course as always, an infinite linear combination can only be an actual solution if it converges sufficiently well. But in the situation of the above proposition, as in previous ones, it does, as long as f is piecewise smooth and averaged, as stipulated.

Exercises

2.3.1 Solve the equation for T in (2.50) using elementary, one-variable separation techniques. That is, write $dT/dt + \lambda kT = 0$; get all T's on one side and all t's on the other; antidifferentiate; solve for T. (Check your results against (2.51)/(2.52).)

2.3.2 **a.** Show that the Neumann heat conduction problem of Example 2.3.1, with initial temperature distribution $f(x) = x$, has constant temperature equal to $\ell/2$ on the cross section of the bar that's equidistant from both ends; that is, show $u(\ell/2, t) = \ell/2$ for all $t > 0$. (Recall (2.64).)

b. Show that, for any $x \in (0, \ell)$, $\lim_{t \to \infty} u(x, t) = \ell/2$. (You may take it on faith that the limit of the sum in (2.64) equals the sum of the limits.)

c. Note that $\ell/2 = \ell^{-1} \int_0^\ell x \, dx$ is the *average* value of the initial distribution $f(x) = x$ on $(0, \ell)$. In light of this, explain what's happening to the temperature distribution $u(x,t)$ in question as time evolves, and why this makes physical sense.

2.3.3 Find the heat flow in a bar with homogeneous Dirichlet conditions at the ends of the bar. That is, solve (formally) the boundary value problem consisting of the heat equation (2.44), the boundary conditions

$$u(0,t) = 0, \quad u(\ell,t) = 0 \quad (t > 0),$$

and the initial condition (2.47). Also find an explicit series formula for this heat flow in the case $f(x) = x$. (See (1.237).)

2.3.4 Solve the heat problem of the previous exercise in the case where the initial temperature distribution is uniform: $f(x) = \tau_0$ for some constant τ_0. (You may wish to use Exercise 1.12.4.) Show that, in this case, there is no heat flow across the cross section of the bar that's equidistant from both ends; that is, show $u_x(\ell/2, t) = 0$ for all $t > 0$. (You may take it on faith that it's OK to differentiate term by term.) Why does this make sense physically?

2.4 HEAT FLOW IN A BAR II: OTHER BOUNDARY CONDITIONS

Here we look at two additional examples of heat flow in the bar of the previous section. The second of these examples will not be completed until Chapter 4, but our present consideration of it will, among other things, help motivate some of the investigations in which we partake between now and then.

The following problem arises if, for example, one imagines that the bar of Figure 2.2 is bent into a circle and the ends are joined together.

Example 2.4.1 Solve the boundary value problem consisting of the heat equation (2.44), the initial condition (2.47), and the so-called *periodic* boundary conditions

$$u(0,t) = u(\ell,t), \quad u_x(0,t) = u_x(\ell,t) \quad (t > 0). \tag{2.75}$$

Solution. Again we have the separated solutions (2.51), (2.52). Now the above periodic boundary conditions, when applied to the separated form (2.48) of u, give

$$X(0)T(t) = X(\ell)T(t), \quad X'(0)T(t) = X'(\ell)T(t). \tag{2.76}$$

Since T is assumed not identically zero, we conclude

$$X(0) = X(\ell), \quad X'(0) = X'(\ell). \tag{2.77}$$

Regarding the case $\lambda = 0$, we note that the condition $X(0) = X(\ell)$, applied to the equation for X in (2.51), yields $\gamma = \gamma + \delta\ell$; the condition $X'(0) = X'(\ell)$, applied similarly, yields $\delta = \delta$. So it's necessary and sufficient that $\delta = 0$. So the eigenvalue $\lambda = 0$ gives, by (2.51), the constant solutions

$$u(x,t) = \frac{D_0}{2} \qquad (2.78)$$

to (2.44) and (2.75).

Now let $\lambda \neq 0$. The periodic condition $X(0) = X(\ell)$, applied to the formula for X in (2.52), gives

$$\gamma = \gamma \cos \sqrt{\lambda}\,\ell + \delta \sin \sqrt{\lambda}\,\ell; \qquad (2.79)$$

the condition $X'(0) = X'(\ell)$, applied to the same, gives

$$\sqrt{\lambda}\,\delta = \sqrt{\lambda}(-\gamma \sin \sqrt{\lambda}\,\ell + \delta \cos \sqrt{\lambda}\,\ell). \qquad (2.80)$$

If we multiply the first of these equations by γ, the second by $\delta/\sqrt{\lambda}$, and add, we get

$$(\gamma^2 + \delta^2) = (\gamma^2 + \delta^2) \cos \sqrt{\lambda}\,\ell. \qquad (2.81)$$

We assume $\gamma^2 + \delta^2 \neq 0$, since we seek a nonzero solution X; we conclude that $\cos \sqrt{\lambda}\,\ell = 1$, which happens if and only if $\sqrt{\lambda} = 2\pi n/\ell$ for some integer n, nonzero because λ is.

Then (2.52) gives, for each such n, the separated solutions

$$u(x,t) = X(x)T(t) = \left(\gamma \cos \frac{2\pi n x}{\ell} + \delta \sin \frac{2\pi n x}{\ell}\right) \sigma e^{-4\pi^2 n^2 kt/\ell^2} \qquad (2.82)$$

to (2.44) and (2.75). These equations are linear and homogeneous (the boundary conditions (2.75) may be written $L_1 u = 0$ and $L_2 u = 0$, respectively, where $(L_1 u)(t) = u(0,t) - u(\ell,t)$ and $(L_2 u)(t) = u_x(0,t) - u_x(\ell,t)$), so the superposition principle points us toward the solutions

$$u(x,t) = \frac{D_0}{2} + \sum_{n=1}^{\infty}\left(D_n \cos \frac{2\pi n x}{\ell} + E_n \sin \frac{2\pi n x}{\ell}\right) e^{-4\pi^2 n^2 kt/\ell^2}. \qquad (2.83)$$

The linear, inhomogeneous boundary condition (2.47) then amounts to the equation

$$f(x) = \frac{D_0}{2} + \sum_{n=1}^{\infty}\left(D_n \cos \frac{2\pi n x}{\ell} + E_n \sin \frac{2\pi n x}{\ell}\right) \quad (0 < x < \ell), \qquad (2.84)$$

and we may then find the D_n's and E_n's using Theorem 1.11.1, with $a = 0$ and $b = \ell$. The upshot is:

Proposition 2.4.1 *If $f \in \mathrm{PSA}(0,\ell)$, then the periodic boundary value problem (2.44), (2.47), (2.75) has solution $u(x,t)$ given by (2.83), where*

$$D_n = a_n(f) = \frac{2}{\ell}\int_0^\ell f(x) \cos \frac{2\pi n x}{\ell}\, dx \quad (n \in \mathbb{N}), \qquad (2.85)$$

$$E_n = b_n(f) = \frac{2}{\ell}\int_0^\ell f(x) \sin \frac{2\pi n x}{\ell}\, dx \quad (n \in \mathbb{Z}^+). \qquad (2.86)$$

To illustrate let's take, as we always do, $f(x) = x$ on $(0, \ell)$. then Example 1.11.2 gives the solution

$$u(x,t) = \frac{\ell}{2} - \frac{\ell}{\pi} \sum_{n=1}^{\infty} \frac{1}{n} e^{-4\pi^2 n^2 kt/\ell^2} \sin \frac{2\pi nx}{\ell} \qquad (2.87)$$

to our boundary value problem.

Yet again, the condition that f be piecewise smooth and averaged really is sufficient, though we haven't proved this, to assure the conclusion of the above proposition.

Let's now consider a situation where the ends of our bar are neither joined together, nor insulated, nor held at constant temperature; rather, heat is permitted to flow across them according to its natural disposition. *Newton's law of cooling* tells us that, in this case—assuming, as we do, that the medium surrounding the bar has constant temperature zero—we have the boundary conditions

$$u_x(0,t) = hu(0,t), \qquad u_x(\ell,t) = -hu(\ell,t) \quad (t > 0), \qquad (2.88)$$

where $h > 0$ is the ratio of the *surface conductance* of the bar to its *thermal conductivity*.

Example 2.4.2 Solve the boundary value problem (2.44), (2.47), (2.88).

Solution. We have, one more time, the separated solutions (2.51), (2.52) to the heat equation (2.44). Moreover, the above boundary conditions (2.88), applied to the usual separated form $u = XT$ for u, yield

$$X'(0)T(t) = hX(0)T(t), \qquad X'(\ell)T(t) = -hX(\ell)T(t) \quad (t > 0), \qquad (2.89)$$

so that

$$X'(0) = hX(0), \qquad X'(\ell) = -hX(\ell). \qquad (2.90)$$

Again, we first consider the case $\lambda = 0$. If we put the condition $X'(0) = hX(0)$ into the formula for X in (2.51), we get $\delta = h\gamma$, so that $X(x) = \gamma(1 + hx)$. But if, into this result, we apply the condition $X'(\ell) = -hX(\ell)$, we get $h\gamma = -h\gamma(1 + h\ell)$, or $h\gamma(2 + h\ell) = 0$. Since h and ℓ are positive, this can happen only if $\gamma = 0$. But the latter makes X identically zero; so $\lambda = 0$ is not an eigenvalue of our eigenvalue problem for X. We assume, then, that $\lambda \neq 0$.

Putting the condition $X'(0) = hX(0)$ into the equation for X in (2.52) gives $\sqrt{\lambda}\delta = h\gamma$. But then that equation reads

$$X(x) = \gamma\left(\cos\sqrt{\lambda}\,x + \frac{h}{\sqrt{\lambda}} \sin\sqrt{\lambda}\,x\right). \qquad (2.91)$$

Into this we put the condition $X'(\ell) = -hX(\ell)$; we get (ignoring the case $\gamma = 0$)

$$-\sqrt{\lambda}\sin\sqrt{\lambda}\,\ell + h\cos\sqrt{\lambda}\,\ell = -h\cos\sqrt{\lambda}\,\ell - \frac{h^2}{\sqrt{\lambda}} \sin\sqrt{\lambda}\,\ell, \qquad (2.92)$$

or

$$\left(\frac{\sqrt{\lambda}}{h} - \frac{h}{\sqrt{\lambda}}\right) \sin \sqrt{\lambda}\,\ell - 2\cos \sqrt{\lambda}\,\ell = 0. \tag{2.93}$$

Then the values of λ we seek are just the ones solving this last equation.

There's no nice closed formula for solutions λ to (2.93). But we can understand them geometrically, at least if we take it on faith that they are *real and positive*, as will in fact be demonstrated in Section 4.2. Then to say that λ is such a solution is precisely to say that the curve

$$y = \left(\frac{s}{h} - \frac{h}{s}\right) \sin s\ell - 2\cos s\ell \tag{2.94}$$

in the sy plane intersects the s axis at $s = \sqrt{\lambda} > 0$.

This curve is illustrated in Figure 2.3; there we've chosen $h = \ell = 1$, for simplicity.

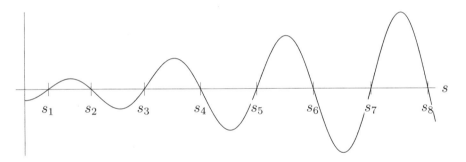

Fig. 2.3 The curve $y = (s - 1/s)\sin s - 2\cos s$

Let's write $s_1, s_2, s_3 \ldots$ for the positive roots, starting with the first one and continuing in ascending order, of the curve in (2.94). (For example, if $h = \ell = 1$, then we find through numerical means that $s_1 \approx 1.31$, $s_2 \approx 3.67$, $s_3 \approx 6.58$, $s_4 \approx 9.63$, $s_5 \approx 12.72$, and so on, cf. Fig. 2.3.) Our situation is then as follows. By (2.52), (2.91), and the definition of our s_n's, we have for each $n \in \mathbb{Z}^+$ a separated solution

$$u_n(x,t) = X(x)T(t) = \left(\cos s_n x + \frac{h}{s_n}\sin s_n x\right)e^{-s_n^2 kt} \tag{2.95}$$

to (2.44) and (2.88). We now seek a solution that will conform to the initial condition (2.47) as well; to this end we apply the usual arguments. That is, we note that equations (2.44) and (2.88) are linear and homogeneous, so an infinite linear combination

$$u(x,t) = \sum_{n=1}^{\infty} D_n \left(\cos s_n x + \frac{h}{s_n}\sin s_n x\right)e^{-s_n^2 kt} \tag{2.96}$$

should, if it converges suitably, itself solve these equations. To this linear combination we then apply the remaining, inhomogeneous condition (2.47), yielding

$$f(x) = \sum_{n=1}^{\infty} D_n \left(\cos s_n x + \frac{h}{s_n} \sin s_n x \right). \tag{2.97}$$

Now, unfortunately, we're stuck!

We have, thus far, neither a method for choosing D_n's that satisfy (2.97) nor even any guarantee that such D_n's exist.

In the next chapter we'll lay some groundwork for a theory that, as we'll see in Chapter 4, provides this guarantee and method and does quite a bit more.

Exercises

2.4.1 Show that the solution, as given by Proposition 2.4.1, to the heat problem of Example 2.4.1 may, alternatively, be written

$$u(x,t) = \sum_{n=-\infty}^{\infty} c_n(f) e^{2\pi i n x/\ell} e^{-4\pi^2 n^2 k t/\ell^2},$$

where $c_n(f)$ is as in (1.213) (with $a = 0$ and $b = \ell$).

2.4.2 Let's suppose, somewhat idealistically, that the temperature at a given location on the surface of the earth is given by $f(t)$, with t in years and f 1-periodic. Then the temperature $u(x,t)$ at a depth of x meters below the surface can be modeled, also rather idealistically and at small enough depths, by the boundary value problem

$$k u_{xx} = u_t \quad (x > 0, \ -\infty < t < \infty),$$
$$u(x,0) = u(x,1) \quad (x > 0),$$
$$u(x,t) \text{ is bounded as } x \to \infty \quad (-\infty < t < \infty),$$
$$u(0,t) = f(t) \quad (-\infty < t < \infty).$$

a. Derive the formal solution

$$u(x,t) = \sum_{n=-\infty}^{\infty} c_n(f) e^{-\sqrt{\pi |n|/k}\,(1+i\,\mathrm{sgn}\,n)x} e^{2\pi i n t},$$

where sgn n denotes the sign of n, cf. (A.12), and $c_n(f)$ is, as usual, the nth Fourier coefficient of the 1-periodic function f. Hint: Write the separated solutions (2.51)/(2.52) to the heat equation $k u_{xx} = u_t$ in complex exponential form. Put the homogeneous, periodic conditions $u(x,0) = u(x,1)$ into these solutions. Doing so should give you eigenvalues $\lambda = -(2\pi n/k)^2$ for $n \in \mathbb{N}$. Now use Exercise A.0.3. You'll be able to restrict the associated eigenfunctions using the boundedness criterion. Plug in the inhomogeneous condition $u(0,t) = f(t)$, in the usual way, to determine the coefficients in your eigenfunction expansion.

b. Assume (simplistically) that surface temperatures vary sinusoidally over the course of the year: $f(t) = T_0 + A \sin 2\pi t$ for some constants T_0, A, say. Show that

$$u(x,t) = T_0 + Ae^{-\sqrt{\pi/k}\,x} \sin(2\pi t - \sqrt{\pi/k}\,x).$$

c. A good depth for a wine cellar is one at which temperatures are *completely out of phase* with the seasons (so that the cellar is cooler in summer and warmer in winter). In the context of part b of this exercise, this means we should choose $\sqrt{\pi/k}\,x = \pi$ (since $\sin(\theta - \pi)$ is completely out of phase with $\sin \theta$). Assuming a thermal diffusivity constant $k = 9\,\text{m}^2/\text{yr}$ (which is fairly realistic) for the earth, what is a good depth for a wine cellar?

2.4.3 Let X_n denote the nth function in the expansion on the right side of (2.97):

$$X_n(x) = \cos s_n x + \frac{h}{s_n} \sin s_n x.$$

Show by direct computation of the relevant integral that the X_n's are orthogonal on $(0, \ell)$: $\int_0^\ell X_k(x)\overline{X_n(x)}\,dx = 0$ for $k \neq n$. (Don't forget that h and the s_n's are real numbers and that, by definition of the s_n's, the right side of (2.94) equals zero when $s = s_n$ for any n.) In Section 4.2, we'll realize this result as an aspect of a more general theory (and we'll evaluate the integral in question when $k = n$).

2.5 CYLINDRICAL AND POLAR COORDINATES

The geometric reality of a given physical situation may mean that it is more readily understood in some coordinate system other than the usual, rectangular one. In the next two sections, we investigate the more common of these other systems. In particular, because the Laplacian ∇_m^2 occurs in so many boundary value problems, we describe its manifestation in such systems. We begin with cylindrical coordinates, according to which a point P in Euclidean xyz space is located by its polar coordinates (ρ, θ) in the xy plane and by its displacement in the z direction (Fig. 2.4).

The cylindrical coordinates (ρ, θ, z) and rectangular coordinates (x, y, z) of P are related by

$$x = \rho \cos \theta, \quad y = \rho \sin \theta, \quad z = z. \tag{2.98}$$

Now suppose u is a sufficiently smooth function of x, y, z. Then by the chain rule and (2.98),

$$\rho u_\rho = \rho(u_x x_\rho + u_y y_\rho + u_z z_\rho) = \rho(u_x \cos \theta + u_y \sin \theta)$$
$$= (\rho \cos \theta) u_x + (\rho \sin \theta) u_y = x u_x + y u_y. \tag{2.99}$$

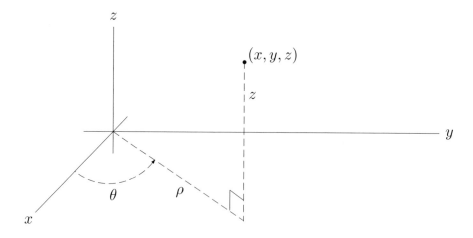

Fig. 2.4 The cylindrical coordinate system

Consequently

$$\rho[\rho u_\rho]_\rho = \rho[xu_x + yu_y]_\rho = \rho[(xu_x + yu_y)_x x_\rho + (xu_x + yu_y)_y y_\rho]$$
$$= \rho[(u_x + xu_{xx} + yu_{yx})\cos\theta + (xu_{xy} + u_y + yu_{yy})\sin\theta]$$
$$= (\rho\cos\theta)(u_x + xu_{xx} + yu_{yx}) + (\rho\sin\theta)(xu_{xy} + u_y + yu_{yy})$$
$$= x(u_x + xu_{xx} + yu_{yx}) + y(xu_{xy} + u_y + yu_{yy}). \quad (2.100)$$

Also,

$$u_\theta = u_x x_\theta + u_y y_\theta = -\rho u_x \sin\theta + \rho u_y \cos\theta = -yu_x + xu_y, \quad (2.101)$$

so that

$$u_{\theta\theta} = (-yu_x + xu_y)_x x_\theta + (-yu_x + xu_y)_y y_\theta$$
$$= (-yu_{xx} + u_y + xu_{yx})(-\rho\sin\theta) + (-u_x - yu_{xy} + xu_{yy})(\rho\cos\theta)$$
$$= -y(-yu_{xx} + u_y + xu_{yx}) + x(-u_x - yu_{xy} + xu_{yy}). \quad (2.102)$$

We add (2.100) to (2.102) and divide the result by ρ^2; we get, after some algebra,

$$\frac{1}{\rho^2} u_{\theta\theta} + \frac{1}{\rho}[\rho u_\rho]_\rho = \frac{x^2 + y^2}{\rho^2}(u_{xx} + u_{yy}) = u_{xx} + u_{yy}. \quad (2.103)$$

We can tack on a u_{zz} to each side if necessary, so we have:

Proposition 2.5.1 **(a)** *The Laplacian in two dimensions has polar coordinate form*

$$\nabla_2^2 u = \frac{1}{\rho^2} u_{\theta\theta} + \frac{1}{\rho}[\rho u_\rho]_\rho. \quad (2.104)$$

(b) *The Laplacian in three dimensions has cylindrical coordinate form*

$$\nabla_3^2 u = \frac{1}{\rho^2} u_{\theta\theta} + \frac{1}{\rho}[\rho u_\rho]_\rho + u_{zz}. \qquad (2.105)$$

Of course

$$[\rho u_\rho]_\rho = u_\rho + \rho u_{\rho\rho}, \qquad (2.106)$$

but often the compact form on the left side is more convenient.

Example 2.5.1 Solve the Dirichlet problem on a disc D of radius ℓ, centered at the origin.

Solution. We seek a function $u = u(\rho, \theta)$ satisfying $\nabla_2^2 u = 0$ on the interior of the disk and equalling a prescribed function f on the boundary. We consider the differential equation

$$u_{\theta\theta} + \rho[\rho u_\rho]_\rho = 0 \quad (0 < \rho < \ell, \ -\pi < \theta < \pi). \qquad (2.107)$$

Note that the left side is, by Proposition 2.5.1(a), just $\rho^2 \nabla_2^2 u$, which will be zero precisely when $\nabla_2^2 u$ is, at least on the domain specified in (2.107), since this domain excludes the origin. The point here is that, in polar coordinates, $\rho^2 \nabla_2^2$ is often easier to deal with than ∇_2^2 itself. We'll worry about what happens at the origin shortly.

The conditions on ρ and θ in (2.107) also exclude those points in D with argument π. We'd like things to "join up" nicely at these points, so we impose the periodic boundary conditions

$$u(\rho, -\pi) = u(\rho, \pi), \quad u_\theta(\rho, -\pi) = u_\theta(\rho, \pi) \quad (0 < \rho < \ell). \qquad (2.108)$$

And finally, we have a Dirichlet boundary condition

$$u(\ell, \theta) = f(\theta) \quad (-\pi < \theta < \pi). \qquad (2.109)$$

We first seek a separated solution $u(\rho, \theta) = R(\rho)\Theta(\theta)$ to our differential equation (2.107) and our periodic conditions (2.108). We note that, applied to such a solution, that differential equation gives

$$R(\rho)\Theta''(\theta) + \rho[\rho R'(\rho)]'\Theta(\theta) = 0 \quad (0 < \rho < \ell, \ -\pi < \theta < \pi). \qquad (2.110)$$

Dividing by $R(\rho)\Theta(\theta)$ and then setting the resulting quotient functions equal to constants in the usual way, we arrive at the ordinary differential equations

$$\Theta''(\theta) + \lambda\Theta(\theta) = 0 \quad (-\pi < \theta < \pi), \quad \rho[\rho R'(\rho)]' - \lambda R(\rho) = 0 \quad (0 < \rho < \ell). \qquad (2.111)$$

We've put the "$+\lambda$" with the equation in Θ, rather than the one in R, because the periodic boundary conditions (2.108) imply linear, homogeneous conditions on Θ. See the discussion following (2.9).

Indeed, applying these periodic conditions to our assumption that $u(\rho, \theta) = R(\rho)\Theta(\theta)$ yields the conditions

$$\Theta(-\pi) = \Theta(\pi), \quad \Theta'(-\pi) = \Theta'(\pi). \tag{2.112}$$

We now observe that these latter conditions, paired with the equation for Θ in (2.111), amount to essentially the same problem as was considered for the function X of Example 2.4.1. So the solution here for Θ is very much like the one there for X. It will, though, be convenient in the present case to write these solutions in complex exponential form. The upshot is this: Our eigenvalue problem in Θ has $\{n^2 : n \in \mathbb{N}\}$ as its set of eigenvalues and has corresponding eigenfunctions

$$\Theta(\theta) = \gamma e^{in\theta} + \delta e^{-in\theta}. \tag{2.113}$$

Putting such an eigenvalue λ into the second of the equations in (2.111) gives

$$\rho[\rho R'(\rho)]' - n^2 R(\rho) = 0 \quad (0 < \rho < \ell). \tag{2.114}$$

We can relate this to a more familiar situation by putting $y = \ln \rho$ and $R(\rho) = Y(y) = Y(\ln \rho)$. (That is, we define Y by $Y(y) = R(e^y)$.) The differential equation in (2.114) then reads

$$\begin{aligned} 0 &= \rho[\rho(Y(\ln \rho))']' - n^2 Y(\ln \rho) = \rho[\rho(\rho^{-1} Y'(\ln \rho))]' - n^2 Y(\ln \rho) \\ &= \rho[Y'(\ln \rho)]' - n^2 Y(\ln \rho) = \rho(\rho^{-1} Y''(\ln \rho)) - n^2 Y(\ln \rho) \\ &= Y''(\ln \rho) - n^2 Y(\ln \rho) = Y''(y) - n^2 Y(y). \end{aligned} \tag{2.115}$$

We already saw, in Example 2.1.1, this equation in Y; from there (see (2.10) and (2.11)) we get the general solution

$$Y(y) = \begin{cases} \sigma + \tau y & \text{if } n = 0, \\ \sigma e^{ny} + \tau e^{-ny} & \text{if not}. \end{cases} \tag{2.116}$$

Consequently, the general solution to (2.114) is

$$R(\rho) = Y(\ln \rho) = \begin{cases} \sigma + \tau \ln \rho & \text{if } n = 0, \\ \sigma \rho^n + \tau \rho^{-n} & \text{if not}. \end{cases} \tag{2.117}$$

But in fact, we can narrow this solution down a bit using a boundary condition *implicit* to our problem. Namely, we want u to be continuous, and therefore bounded, throughout the disk, and in particular at the origin $\rho = 0$, so we'd better not have any negative powers, or logarithms, of ρ appearing. So we'd better have $\tau = 0$ in (2.117). We can summarize and rewrite our result as follows: We have separated solutions

$$u(\rho, \theta) = R(\rho)\Theta(\theta) = \rho^n(\alpha e^{in\theta} + \beta e^{-in\theta}) \tag{2.118}$$

for each $n \in \mathbb{N}$.

We form an infinite linear combination

$$u(\rho,\theta) = D_0 + \sum_{n=1}^{\infty} \rho^n \left(D_n e^{in\theta} + D_{-n} e^{-in\theta} \right) = \sum_{n=-\infty}^{\infty} D_n \rho^{|n|} e^{in\theta}$$

and put into this the inhomogeneous boundary condition (2.109); we get

$$f(\theta) = \sum_{n=-\infty}^{\infty} D_n \ell^{|n|} e^{in\theta} \quad (-\pi < \theta < \pi). \tag{2.119}$$

Theorem 1.11.1 then gives

$$D_n \ell^{|n|} = c_n(f), \tag{2.120}$$

so we have:

Proposition 2.5.2 *If $f \in \mathrm{PSA}(-\pi, \pi)$, then the Dirichlet problem (2.107)–(2.109) has solution*

$$u(\rho,\theta) = \sum_{n=-\infty}^{\infty} \left(\frac{\rho}{\ell}\right)^{|n|} c_n(f) e^{in\theta}. \tag{2.121}$$

To be concrete, let's trot our our old, favorite instance $f(x) = x$, or in this case $f(\theta) = \theta$ on $(-\pi, \pi)$. Then f is just the restriction to $(-\pi, \pi)$ of the 2π-periodic sawtooth function σ of (1.114). So f has the same Fourier coefficients as σ; so by (1.117),

$$c_n(f) = \begin{cases} 0 & \text{if } n = 0, \\ \dfrac{(-1)^{n+1}}{in} & \text{if not.} \end{cases} \tag{2.122}$$

Then by Proposition 2.5.2, we have the solution

$$u(\rho,\theta) = i \sum_{n \in \mathbb{Z}-\{0\}} \frac{1}{n} \left(-\frac{\rho}{\ell}\right)^{|n|} e^{in\theta} = -2 \sum_{n=1}^{\infty} \frac{1}{n} \left(-\frac{\rho}{\ell}\right)^n \sin n\theta \tag{2.123}$$

to our Dirichlet problem. (We've used the fact that $(-1)^{n+1}/i = -(-1)^{|n|}/i = i(-1)^{|n|}$.)

As usual, the conditions stated in the proposition are sufficient to assure that the given solution converges well enough to actually *be* a solution. As usual we omit a proof; we do, though, remark on one of the salient points behind this proof. Namely, the factor $(\rho/\ell)^{|n|}$ in the nth summand of (2.121) decays, as a function of n and for $\rho < \ell$, rapidly enough to ensure satisfactory convergence of the infinite series at hand.

One can always express the solution (2.121) in cosine-sine form, as we did in the particular situation of (2.123). However, the complex exponential form leads more directly to an alternative, integral form, which is sometimes useful:

Proposition 2.5.3 *If f and $u(\rho, \theta)$ are as in Propoosition 2.5.2, then for $\rho < \ell$,*

$$u(\rho, \theta) = \frac{\ell^2 - \rho^2}{2\pi} \int_{-\pi}^{\pi} \frac{f(\phi) \, d\phi}{\ell^2 + \rho^2 - 2\ell\rho \cos(\theta - \phi)}. \tag{2.124}$$

Proof. By the Maclaurin series (1.66) for $z/(1-z)$, we have, for $|r| < 1$ and $\psi \in \mathbb{R}$,

$$\sum_{n=-\infty}^{\infty} r^{|n|} e^{in\psi} = 1 + \sum_{n=1}^{\infty} (re^{i\psi})^n + \sum_{n=1}^{\infty} (re^{-i\psi})^n = 1 + \frac{re^{i\psi}}{1 - re^{i\psi}} + \frac{re^{-i\psi}}{1 - re^{-i\psi}}$$

$$= \frac{(1 - re^{i\psi})(1 - re^{-i\psi}) + re^{i\psi}(1 - re^{-i\psi}) + re^{-i\psi}(1 - re^{i\psi})}{(1 - re^{i\psi})(1 - re^{-i\psi})}$$

$$= \frac{1 - r^2}{1 + r^2 - r(e^{i\psi} + e^{-i\psi})} = \frac{1 - r^2}{1 + r^2 - 2r \cos \psi}. \tag{2.125}$$

We apply this, with $r = \rho/\ell$ and $\psi = \theta - \phi$, to (2.121), as follows:

$$u(\rho, \theta)$$

$$= \sum_{n=-\infty}^{\infty} \left(\frac{\rho}{\ell}\right)^{|n|} c_n(f) e^{in\theta} = \sum_{n=-\infty}^{\infty} \left(\frac{\rho}{\ell}\right)^{|n|} \left(\frac{1}{2\pi} \int_{-\pi}^{\pi} f(\phi) e^{-in\phi} \, d\phi\right) e^{in\theta}$$

$$= \frac{1}{2\pi} \int_{-\pi}^{\pi} f(\phi) \left(\sum_{n=-\infty}^{\infty} \left(\frac{\rho}{\ell}\right)^{|n|} e^{in(\theta - \phi)}\right) d\phi$$

$$= \frac{1}{2\pi} \int_{-\pi}^{\pi} f(\phi) \frac{1 - (\rho/\ell)^2}{1 + (\rho/\ell)^2 - 2(\rho/\ell) \cos(\theta - \phi)} \, d\phi$$

$$= \frac{\ell^2 - \rho^2}{2\pi} \int_{-\pi}^{\pi} \frac{f(\phi) \, d\phi}{\ell^2 + \rho^2 - 2\ell\rho \cos(\theta - \phi)}. \tag{2.126}$$

The manipulations here are seen to be valid under the stated conditions on f and ρ. □

Some interesting consequences of the above proposition are investigated in Exercise 2.5.3.

Exercises

2.5.1 **a.** Sketch the semiannulus

$$T = \{(\rho \cos \theta, \rho \sin \theta) : 1 \le \rho \le b, \, 0 \le \theta \le \pi\} \subset \mathbb{R}^2$$

(where $b > 1$).

b. Formally solve the Dirichlet problem $\nabla_2^2 u = 0$ on T, assuming $u = 0$ on the inner arc $\rho = 1$ of T, $u = f(\theta)$ on the outer arc $\rho = b$, and $u = 0$ on the line segment parts of the boundary of T. ANSWER: $u(\rho, \theta) = \sum_{n=1}^{\infty} D_n(\rho^n - \rho^{-n}) \sin n\theta$, where $D_n = 2/(\pi(b^n - b^{-n})) \int_0^{\pi} f(\theta) \sin n\theta \, d\theta$.

2.5.2   **a.** Sketch the wedge

$$W = \{(\rho \cos \theta, \rho \sin \theta) : 0 \leq \rho \leq a, \ 0 \leq \theta \leq \alpha\} \subset \mathbb{R}^2$$

(where $a > 0$ and $0 < \alpha < \pi$).

   b. Formally solve the problem $\nabla_2^2 u = 0$ on W, assuming $u = f(\theta)$ on the outer arc $\rho = a$ and $u = 0$ on the line segment parts of the boundary of W.

2.5.3   Let ℓ, u, and f be as in the Dirichlet problem of Example 2.5.1. Prove the following.

   a. u has value

$$\frac{1}{2\pi} \int_{-\pi}^{\pi} f(\phi) \, d\phi$$

at the origin. (This is the *mean value theorem for harmonic functions*, which says: If u is harmonic on a disk $\{z \in \mathbb{C} : |z| \leq \ell\}$, meaning u satisfies $\nabla_2^2 u = 0$ there, then the value of u at the center of this disk is the average of its values on the perimeter.) Use either Proposition 2.5.2 or Proposition 2.5.3.

   b. If $f(\theta) \leq M$ for all θ, then $u(\rho, \theta) \leq M$ for all θ and for all $\rho \leq \ell$. (This is the *maximum modulus theorem for harmonic functions*, which says: If u is harmonic on a disk $\{z \in \mathbb{C} : |z| \leq \ell\}$, then the maximum of u on this disk occurs on the perimeter of the disk.) Hint: First show that

$$\frac{\ell^2 - \rho^2}{\ell^2 + \rho^2 - 2\ell\rho\cos(\theta - \phi)}$$

is positive for $\rho < \ell$. Then by (2.124) and the stated assumption on f,

$$u(\rho, \theta) \leq M \frac{\ell^2 - \rho^2}{2\pi} \int_{-\pi}^{\pi} \frac{d\phi}{\ell^2 + \rho^2 - 2\ell\rho\cos(\theta - \phi)}$$

for $\rho < \ell$. Then integrate as follows: First, use Exercise 1.3.1 to show that $\cos(\theta - \phi)$ can be replaced by $\cos \phi$. Into the resulting integral, substitute $u = A \tan(\phi/2)$, where $A = (\ell + \rho)/(\ell - \rho)$. Then $du = (A/2) \sec^2(\phi/2) \, d\phi = (A/2)(1 + \tan^2(\phi/2)) \, d\phi = (A/2)(1 + A^{-2}u^2) \, d\phi$ and (because $\cos 2x = 2\cos^2 x - 1$) $\cos \phi = (A^2 - u^2)/(A^2 + u^2)$ (fill in the details). Simplification should leave you with a constant times $\int_{-\infty}^{\infty} (1 + u^2)^{-1} \, du$, which may be evaluated by standard techniques (put $u = \tan t$).

2.6   SPHERICAL COORDINATES

We now consider the Laplacian ∇_3^2 in spherical coordinates. Let's recall, first, that such coordinates locate a point P in Euclidean xyz space according to its distance r from the origin, its (counterclockwise) inclination θ from the positive x axis, and its (downward) inclination ϕ from the positive z axis (Fig. 2.5).

We have the formulas

$$x = r\cos\theta\sin\phi, \quad y = r\sin\theta\sin\phi, \quad z = r\cos\phi. \tag{2.127}$$

112 FOURIER SERIES AND BOUNDARY VALUE PROBLEMS

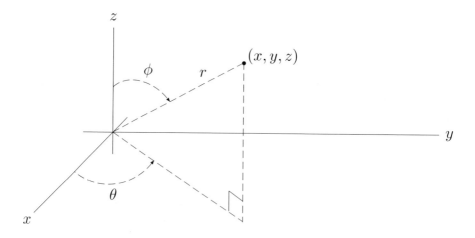

Fig. 2.5 The spherical coordinate system

Here, it's assumed that $0 \leq \phi \leq \pi$. Note also that the θ of Figure 2.5 is the same as that of Figure 2.4.

Our work will be reduced greatly if we exploit some of what we've already seen regarding the cylindrical coordinate situation. Specifically, let's first square each of the first two equations in (2.98), and add the results together to get $x^2 + y^2 = \rho^2$. If we also square and then add together the first two equations in (2.127), we get $x^2 + y^2 = r^2 \sin^2 \phi$. So

$$\rho^2 = r^2 \sin^2 \phi. \tag{2.128}$$

This means $\rho = \pm r \sin \phi$; since ρ, r, and $\sin \phi$ are all nonnegative, we should take the plus sign on the right. In other words, (2.127) implies

$$\rho = r \sin \phi, \quad z = r \cos \phi. \tag{2.129}$$

Now we note the following about equations (2.129): *They are precisely the first two equations in (2.98), but with x, y, ρ, θ replaced by ρ, z, r, ϕ, respectively.* And of course the third equation in (2.98) also remains true under these replacements. Consequently, if we make these *same* replacements in (2.103) then, since this latter result was derived from (2.98), we get a true statement, namely,

$$\frac{1}{r^2} u_{\phi\phi} + \frac{1}{r} [r u_r]_r = u_{\rho\rho} + u_{zz}. \tag{2.130}$$

We make these same replacements also in (2.99) and (2.101); we get

$$r u_r = \rho u_\rho + z u_z, \quad u_\phi = -z u_\rho + \rho u_z. \tag{2.131}$$

Multiplying the second of these last two equations by $-z/\rho$, adding the result to the first of the two, and dividing by $\rho^2 + z^2$, which equals r^2 by (2.129), yields

$$\frac{1}{\rho} u_\rho = \frac{\rho r u_r - z u_\phi}{\rho r^2} = \frac{1}{r} u_r + \frac{1}{r^2 \tan \phi} u_\phi, \tag{2.132}$$

the last step by (2.129) again. Adding together (2.130) and (2.132) gives

$$u_{\rho\rho} + \frac{1}{\rho}u_\rho + u_{zz} = \frac{1}{r}[ru_r]_r + \frac{1}{r}u_r + \frac{1}{r^2}u_{\phi\phi} + \frac{1}{r^2\tan\phi}u_\phi. \tag{2.133}$$

We simplify the left side using (2.106) and the right by observing (DIY) that $[ru_r]_r + u_r = [ru]_{rr}$. We then add $\rho^{-2}u_{\theta\theta}$ to each side of the result, to get

$$\frac{1}{\rho}[\rho u_\rho]_\rho + \frac{1}{\rho^2}u_{\theta\theta} + u_{zz} = \frac{1}{r}[ru]_{rr} + \frac{1}{\rho^2}u_{\theta\theta} + \frac{1}{r^2}u_{\phi\phi} + \frac{1}{r^2\tan\phi}u_\phi$$

$$= \frac{1}{r}[ru]_{rr} + \frac{1}{r^2}\left(\frac{1}{\sin^2\phi}u_{\theta\theta} + u_{\phi\phi} + \frac{1}{\tan\phi}u_\phi\right)$$

$$= \frac{1}{r}[ru]_{rr} + \frac{1}{r^2\sin\phi}\left(\frac{1}{\sin\phi}u_{\theta\theta} + [\sin\phi\, u_\phi]_\phi\right), \tag{2.134}$$

the next to the last step by (2.128), and the last because

$$u_{\phi\phi} + \frac{1}{\tan\phi}u_\phi = \frac{1}{\sin\phi}(\sin\phi\, u_{\phi\phi} + \cos\phi\, u_\phi) = \frac{1}{\sin\phi}[\sin\phi\, u_\phi]_\phi. \tag{2.135}$$

Hope you're having as much fun as we are.

Finally, HERE'S THE RUB: The left side of (2.134) is, by Proposition 2.5.1(b), the Laplacian in cylindrical coordinates. So the right side is the Laplacian in spherical coordinates. In sum:

Proposition 2.6.1 *The three-dimensional Laplacian has spherical coordinate form*

$$\nabla_3^2 u = \frac{1}{r}[ru]_{rr} + \frac{1}{r^2\sin\phi}\left(\frac{1}{\sin\phi}u_{\theta\theta} + [\sin\phi\, u_\phi]_\phi\right). \tag{2.136}$$

Example 2.6.1 A solid ball of radius ℓ and center $(0,0)$ has thermal diffusivity k. Find the temperature u of each point within the ball at time t, assuming that the initial temperature $f(r)$ depends only on distance r from the origin, and that the surface of the ball is kept at temperature zero.

Solution. Since all parameters of the problem depend only on r and t, so clearly will u. So in spherical coordinates, $u_\phi = u_{\theta\theta} = 0$; then by Proposition 2.6.1, our heat equation $k\nabla_3^2 u = u_t$ takes the form

$$\frac{k}{r}[ru]_{rr} = u_t \quad (0 < r < \ell,\ t > 0). \tag{2.137}$$

We also have boundary conditions

$$u(\ell, t) = 0 \quad (t > 0) \tag{2.138}$$

and

$$u(r, 0) = f(r) \quad (0 < r < \ell). \tag{2.139}$$

FOURIER SERIES AND BOUNDARY VALUE PROBLEMS

A trick that works nicely here is to write $u(r,t) = r^{-1}v(r,t)$. Then $u_t = r^{-1}v_t$ and $ru = v$, so that (2.137) yields

$$k\, v_{rr} = v_t \quad (0 < r < \ell,\ t > 0). \tag{2.140}$$

In addition,

$$v(0,t) = 0, \quad v(\ell,t) = 0 \quad (t > 0). \tag{2.141}$$

The first of these equations is because $u = r^{-1}v$ is assumed to be continuous at the origin, which necessitates v being zero there; the second is by (2.138). Finally, from (2.139),

$$v(r,0) = rf(r) \quad (0 < r < \ell). \tag{2.142}$$

It is straightforward enough to solve the boundary value problem (2.140)–(2.142) for v, using the Fourier approach; indeed, this problem is much like that of Example 2.3.1, except that the linear, homogeneous Neumann conditions there are replaced by the corresponding Dirichlet conditions. The upshot is that, in our present situation, we get

$$v(r,t) = \sum_{n=1}^{\infty} D_n e^{-\pi^2 n^2 kt/\ell^2} \sin\frac{\pi n r}{\ell}, \tag{2.143}$$

where the D_n's are the Fourier sine coefficients (see Theorem 1.12.1) of the function g defined by $g(r) = rf(r)$. (Compare this result with Proposition 2.3.1 and with Exercise 2.3.3.)

Finally, putting again $u = r^{-1}v$, we conclude:

Proposition 2.6.2 *If $f \in \text{PSA}(0,\ell)$, then the boundary value problem (2.137)–(2.139) has solution*

$$u(r,t) = \frac{1}{r}\sum_{n=1}^{\infty} D_n e^{-\pi^2 n^2 kt/\ell^2} \sin\frac{\pi n r}{\ell}, \tag{2.144}$$

where

$$D_n = \frac{2}{\ell}\int_0^\ell rf(r)\sin\frac{\pi n r}{\ell}\, dr \quad (n \in \mathbb{Z}^+). \tag{2.145}$$

Let's put $f(r) = r$ for $0 < r < \ell$. Using the integration-by-parts formula

$$\int r^2 \sin\beta r\, dr = \frac{-\beta^2 r^2 \cos\beta r + 2\beta r \sin\beta r + 2\cos\beta r}{\beta^3} + C \quad (\beta \neq 0), \tag{2.146}$$

we get

$$D_n = \frac{2}{\ell}\int_0^\ell r^2 \sin\frac{\pi n r}{\ell}\, dr = \frac{2\ell^2}{\pi^3 n^3}((-1)^n(2 - \pi^2 n^2) - 2). \tag{2.147}$$

(DIY: fill in the details.) We could break this up into cases of odd and even n, but let's not. Then from our above proposition,

$$u(r,t) = \frac{2\ell^2}{\pi^3 r} \sum_{n=1}^{\infty} \frac{(-1)^n(2-\pi^2 n^2)-2}{n^3} e^{-\pi^2 n^2 kt/\ell^2} \sin\frac{\pi n r}{\ell}. \qquad (2.148)$$

As usual, the conditions stated in the proposition are sufficient to assure that the given solution converges well enough to actually *be* a solution; as usual we omit the proof of this.

Exercises

2.6.1 Use Exercise 1.12.3 to solve the heat problem of Example 2.6.1 in the case when $f(r) = \ell - r$. Show that, in this case, there is no heat flow across the sphere concentric with, and of half the radius of, the ball; that is, show $u_r(\ell/2, t) = 0$ for all $t > 0$. (You may take it on faith that it's OK to differentiate term by term.) What is the physical significance of this?

2.6.2 Find the steady state temperatures (that is, the solutions to $\nabla_3^2 u = 0$) between two spheres, the inner one of radius a and constant (that is, θ-, ϕ-, and t-independent) temperature τ_1 and the outer one of radius b and constant temperature τ_2.

2.6.3 Suppose we have temperatures u in the half-ball

$$\{(r,\theta,\phi)\colon 0 \le r \le \ell,\ -\pi \le \theta \le \pi,\ 0 \le \phi \le \pi/2\} \subset \mathbb{R}^3$$

that are independent of the longitude θ: $u = u(r, \phi, t)$. Show that, if the base of the half-ball is insulated, then $u_\phi(r, \pi/2, t) = 0$ for all r, t. Hint: In *cylindrical* coordinates, the base being insulated means $u_z = 0$ when $z = 0$ and $\rho < \ell$. Use the chain rule and the transformations (2.129) between cylindrical and spherical coordinates to express u_ϕ in terms of u_z and u_ρ.

2.6.4 Separate variables to reduce the Laplace equation $\nabla_3^2 u = 0$, in spherical coordinates, to three single-variable ordinary differential equations. (Hint: First, multiply the equation through by r^2.) ANSWER: $\Theta''(\theta) + \lambda\Theta(\theta) = 0$; $r^2 R''(r) + 2rR'(r) + \mu R(r) = 0$; $\sin\phi[\sin\phi\,\Phi'(\phi)]' - \mu\sin^2\phi\,\Phi(\phi) - \lambda\Phi(\phi) = 0$.

2.7 THE WAVE EQUATION I

We now consider a string lying along the x axis, with ends fixed at $x = 0$ and $x = \ell$. Suppose this string is plucked so as to vibrate in the xy plane; this can be assured by stipulating that the initial position and velocity of the string are constrained to that plane. We imagine the string to be perfectly flexible, meaning it offers no resistance to bending. We also ignore the effects of air friction and of the weight of the string itself, and assume there are no other outside forces acting on the string.

All of this amounts to quite a simplistic and idealized model—but for the moment we welcome such a model for various reasons. Namely, it's relatively easy to handle;

it will allow illustration of qualitative and quantitative ideas—relevant even in less idealized, simplistic frameworks—with a minimum of obfuscation and clutter; and it's easier to springboard from the general to the specific than it is to move laterally from one specific to another. (It's easier to complete a reverse 3.5 somersault pike from the high board than it is from the sauna. It's gravity; it's the same idea.)

We have the following fact: The vertical displacement $y(x,t)$ (t denoting time) of a point on such a string satisfies the "wave equation"

$$c^2 y_{xx} = y_{tt} \quad (0 < x < \ell, \ t > 0), \tag{2.149}$$

where the positive number c^2 measures the tension in the string.

We'll not derive the wave equation from physical principles; the interested reader may find such derivations in, for example, Section 7 of [7] or Section 2.5 of [19]. (The uninterested reader may also find them there, but probably won't look for them.)

But we do wish to give a *heuristic* argument for (2.149). What is this equation really telling us? Well, since y_{tt} denotes the *acceleration* (in the y direction) of a point on the string, and y_{xx} denotes the *concavity* of the string at such a point, (2.149) says *acceleration is proportional to concavity* (the constant of proportionality is determined by the tension in the string). In particular, where the string is concave up, the acceleration is positive (so that the string is speeding up if its motion is upward and slowing down if its motion is downward), and where the string is concave down the acceleration is negative (meaning the corresponding thing). Moreover, the greater the concavity, the greater the acceleration, and so on. See Figure 2.6.

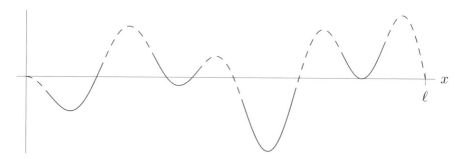

Fig. 2.6 A string vibrating according to the wave equation (2.149). Solid portions have upward acceleration; dashed portions have downward acceleration

In light of this, it's almost possible to visualize a solution to the wave equation. Well okay, maybe that's a stretch, but at any rate the converse, so to speak, is true: if you watch a solution evolve, as we will below (see Section 2.9), you can see each point on the string moving with an acceleration proportional to the concavity there.

Now we note that, since the ends of the string are fixed, we have boundary conditions

$$y(0,t) = 0, \quad y(\ell,t) = 0 \quad (t > 0). \tag{2.150}$$

Also, we'll assume that the starting position and velocity of the string satisfy
$$y(x,0) = f(x), \quad y_t(x,0) = g(x) \quad (0 < x < \ell), \tag{2.151}$$
respectively, where f and g are specified functions.

Example 2.7.1 Solve the boundary value problem (2.149)–(2.151).

Solution. We begin, as usual, by separating variables; that is, we first look for solutions
$$y(x,t) = X(x)T(t) \tag{2.152}$$
to the homogeneous elements of our boundary value problem. For such a solution, equation (2.149) reads
$$c^2 X''(x)T(t) = X(x)T''(t): \tag{2.153}$$
so $c^2 X''/X = T''/T$ or, by the usual process of setting the resulting single-variable functions equal to constants,
$$X''(x) + \lambda X(x) = 0 \quad (0 < x < \ell), \quad T''(t) + \lambda c^2 T(t) = 0 \quad (t > 0). \tag{2.154}$$
These we've seen before, in one form or another. They have solution
$$X(x) = \gamma + \delta x \text{ and } T(t) = \sigma + \tau t \text{ if } \lambda = 0, \tag{2.155}$$
$$X(x) = \gamma \cos \sqrt{\lambda} x + \delta \sin \sqrt{\lambda} x \text{ and } T(t) = \sigma \cos \sqrt{\lambda} ct + \tau \sin \sqrt{\lambda} ct \text{ if } \lambda \neq 0. \tag{2.156}$$

Now imposition of the boundary conditions (2.150) gives $X(0) = X(\ell) = 0$. These are inconsistent with (2.155) except in the trivial case $\gamma = \delta = 0$; so we may assume $\lambda \neq 0$. But $X(0) = 0$ and (2.156) imply $\gamma = 0$, so that
$$X(x) = \delta \sin \sqrt{\lambda} x. \tag{2.157}$$
Putting $X(\ell) = 0$ into this gives $\sin \sqrt{\lambda} \ell = 0$ (since we assume that $\delta \neq 0$), so that
$$\sqrt{\lambda} = \frac{\pi n}{\ell} \tag{2.158}$$
for some integer n, nonzero because λ is. Combining this with (2.157) and with the equation for T in (2.156), we get separated solutions
$$y(x,t) = \sin \frac{\pi n x}{\ell} \left(D_n \cos \frac{\pi n c t}{\ell} + E_n \sin \frac{\pi n c t}{\ell} \right) \tag{2.159}$$
to the linear, homogeneous equations (2.149), (2.150)—here $n \neq 0$, but in fact, since $-n$ yields the same set of solutions as does n, we need only consider $n \in \mathbb{Z}^+$.

To solve the linear, inhomogeneous equation (2.151) as well, we consider infinite linear combinations
$$y(x,t) = \sum_{n=1}^{\infty} \sin \frac{\pi n x}{\ell} \left(D_n \cos \frac{\pi n c t}{\ell} + E_n \sin \frac{\pi n c t}{\ell} \right). \tag{2.160}$$

Substituting the condition $y(x,0) = f(x)$, from (2.151), into this gives

$$f(x) = \sum_{n=1}^{\infty} D_n \sin \frac{\pi n x}{\ell} \quad (0 < x < \ell). \tag{2.161}$$

Theorem 1.12.1 tells us how, at least for reasonable f, to make this last equation true, namely, by putting

$$D_n = \beta_n(f) = \frac{2}{\ell} \int_0^\ell f(x) \sin \frac{\pi n x}{\ell} \, dx. \tag{2.162}$$

And finally, what about the E_n's? We have $y_t(x,0) = g(x)$ by (2.151). So differentiating both sides of our infinite series (2.160) with respect to t and then setting $t = 0$ give

$$g(x) = \sum_{n=1}^{\infty} \left[E_n \cdot \frac{\pi n c}{\ell} \right] \sin \frac{\pi n x}{\ell} \quad (0 < x < \ell). \tag{2.163}$$

And we know how to make this true, for appropriate g, also by Theorem 1.12.1: We put

$$E_n \cdot \frac{\pi n c}{\ell} = \beta_n(g); \tag{2.164}$$

that is,

$$E_n = \frac{\ell}{\pi n c} \beta_n(g) = \frac{2}{\pi n c} \int_0^\ell g(x) \sin \frac{\pi n x}{\ell} \, dx. \tag{2.165}$$

We summarize as follows:

Proposition 2.7.1 *The wave problem* (2.149)–(2.151) *has formal solution given by* (2.160), (2.162), (2.165).

We emphasize, in the above proposition, that the given solution *is* a formal one, meaning convergence questions surrounding that solution have not been addressed. In previous boundary problems we noted, although without proof, that it was sufficient for the relevant "boundary functions" to be piecewise smooth and averaged; things aren't so simple in the present case! The reason they are not is, essentially, this: The summands of the series (2.160) do not, in general, contain factors that decay quickly with n. In previous boundary problems, such factors were in fact present: In Example 2.1.1, there were the e^{-ny} factors; in Examples 2.3.1 and 2.6.1, the $e^{-\pi^2 n^2 k t / \ell^2}$ factors; in Example 2.3.2, the $e^{-\pi^2 (2n-1)^2 k t / (4\ell^2)}$ factors; in Example 2.4.1, the $e^{-4\pi^2 n^2 k t / \ell^2}$ factors; in Example 2.5.1, the $(\rho/\ell)^{|n|}$ factors. The lack of such rapidly decaying terms in the series (2.160) means that series will require a more delicate analysis, and that our above formal solution to the wave problem will be an actual one only under conditions somewhat more stringent than those sufficing in previous situations.

THE WAVE EQUATION I

We'll get to the required analysis, and formulation of appropriate conditions, in the next section. In the meantime, let's look at a specific instance of a formal solution.

Example 2.7.2 Solve (formally) the wave problem (2.149)–(2.151) if $\ell = \pi/2$, $c = 2$,

$$f(x) = \begin{cases} x & \text{if } 0 < x \leq \pi/6, \\ \dfrac{\pi}{4} - \dfrac{x}{2} & \text{if } \pi/6 \leq x < \pi/2 \end{cases} \qquad (2.166)$$

(Fig. 2.7), and

$$g(x) = \begin{cases} 8\sin 8x & \text{if } \pi/8 \leq x \leq 3\pi/8, \\ 0 & \text{elsewhere on } [0, \pi/2]. \end{cases} \qquad (2.167)$$

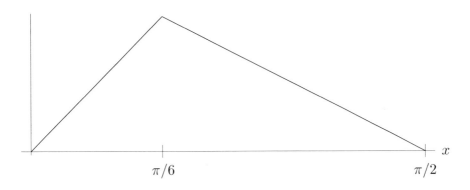

Fig. 2.7 The initial position function f

Solution. By (2.162) we have, for $n \in \mathbb{Z}^+$,

$$\begin{aligned}
D_n &= \frac{4}{\pi} \int_0^{\pi/2} f(x) \sin 2nx \, dx \\
&= \frac{4}{\pi} \left[\int_0^{\pi/6} x \sin 2nx \, dx + \int_{\pi/6}^{\pi/2} \left(\frac{\pi}{4} - \frac{x}{2} \right) \sin 2nx \, dx \right] \\
&= \left(\frac{\sin 2nx}{\pi n^2} - \frac{2x \cos 2nx}{\pi n} \right) \bigg|_0^{\pi/6} - \left(\frac{\cos 2nx}{2n} + \frac{\sin 2nx}{2\pi n^2} - \frac{x \cos 2nx}{\pi n} \right) \bigg|_{\pi/6}^{\pi/2} \\
&= \frac{3 \sin(\pi n/3)}{2\pi n^2}. \qquad (2.168)
\end{aligned}$$

(We integrated by parts for the third equality and canceled all over the place for the last.) Also, by (2.165),

$$E_n = \frac{1}{\pi n}\int_0^{\pi/2} g(x)\sin 2nx\,dx = \frac{8}{\pi n}\int_{\pi/8}^{3\pi/8}\sin 8x\sin 2nx\,dx$$

$$= \begin{cases} \dfrac{1}{4} & \text{if } n = 4, \\ \dfrac{16}{\pi n(n^2-16)}\left(\sin\dfrac{\pi n}{4} - \sin\dfrac{3\pi n}{4}\right) & \text{if } n \neq 4 \end{cases} \qquad (2.169)$$

for $n \in \mathbb{Z}^+$. (Here the integration may be performed using trig identities from Exercise A.0.16.) So by the above proposition,

$$y(x,t) = \sum_{n=1}^\infty \sin 2nx\,(D_n\cos 4nt + E_n\sin 4nt)$$

$$= \frac{3}{2\pi}\sum_{n=1}^\infty \frac{\sin(\pi n/3)}{n^2}\sin 2nx\cos 4nt + \frac{1}{4}\sin 8x\sin 16t$$

$$+ \frac{16}{\pi}\sum_{n\in\mathbb{Z}^+-\{4\}}\frac{1}{n(n^2-16)}\left(\sin\frac{\pi n}{4} - \sin\frac{3\pi n}{4}\right)\sin 2nx\sin 4nt.$$

(2.170)

We could rewrite our answer by noting that $\sin(\pi n/3) = 0$ if n is a multiple of 3, and moreover that $\sin n\pi/4 - \sin 3n\pi/4$ equals zero if n is odd or a multiple of 4, but we won't.

Exercises

2.7.1 Show that the displacement $y(x,t)$ of a vibrating string with ends fixed at $x = 0$ and $x = \pi$, tension constant $c = 2$, initial displacement $y(x,0) = f(x) = x(\pi - x)$, and initial velocity $y_t(x,0) = g(x) = \sin x$ is given by

$$y(x,t) = \frac{1}{2}\sin x\sin 2t + \frac{8}{\pi}\sum_{n=1}^\infty \frac{1}{(2n-1)^3}\sin(2n-1)x\cos(4n-2)t.$$

Hint: Use Exercise 1.12.3.

2.7.2 Show that a vibrating string with ends fixed at $x = 0$ and $x = 3$, tension constant $c = 2$, initial displacement

$$y(x,0) = f(x) = \begin{cases} \sin\pi x & \text{if } 0 \le x \le 1, \\ 0 & \text{if } 1 \le x \le 3, \end{cases}$$

and initial velocity $y_t(x,0) = 0$ has displacement

$$y(x,t) = \frac{1}{3}\sin\pi x\cos 2\pi t + \frac{6}{\pi}\sum_{n\in\mathbb{Z}^+-\{3\}}\frac{1}{9-n^2}\sin\frac{\pi n}{3}\sin\frac{\pi nx}{3}\cos\frac{2\pi nt}{3}.$$

You might want to use trig identities from Exercise A.0.16.

2.7.3 Suppose a string, with ends fixed at $x = 0$ and $x = 1$ and tension constant $c = 1$, is plucked at some point $x = a$ between these endpoints. Suppose this gives the string an initial displacement

$$y(x, 0) = f(x) = \begin{cases} x/a & \text{if } 0 \leq x \leq a, \\ (1-x)/(1-a) & \text{if } a \leq x \leq 1 \end{cases}$$

and initial velocity $y_t(x, 0) = g(x) = 0$.

a. Sketch the graph of $f(x)$ (for some fixed value of a; pick your favorite a between 0 and 1 that's not an endpoint).

b. Show that the displacement $y(x, t)$ of the string is given by

$$y(x, t) = \frac{2}{\pi^2 a(1-a)} \sum_{n=1}^{\infty} \frac{1}{n^2} \sin \pi n a \sin \pi n x \cos \pi n t.$$

c. Show that, for any a with $0 < a < 1$, there is a number M that's independent of n (but depends on a) such that

$$|D_n| \leq \frac{M}{n^2}.$$

d. Show that, as $a \to 1$ from the left, $|D_n|$ approaches $2/(\pi n)$. (Use l'Hôpital's rule, considering D_n to be a function of a.)

e. Compare parts c and d above. Noting that (any constant times) $1/n^2$ decays much faster as $n \to \infty$ than (any constant times) $1/n$, in which type of vibrating string do you think the "overtones," or high-frequency components, will be more prevalent: a string plucked somewhere near the middle or one plucked close to one of the fixed ends? Explain.

2.7.4 (Continuation of the previous exercise) Suppose a string, with ends fixed at $x = 0$ and $x = 1$ and tension constant $c = 1$, is plucked so as to have initial displacement

$$y(x, 0) = f(x) = \begin{cases} x & \text{if } 0 \leq x \leq a, \\ \dfrac{x^2 - x + a^2}{2a - 1} & \text{if } a \leq x \leq 1-a, \\ 1 - x & \text{if } 1-a \leq x \leq 1, \end{cases}$$

where a is some number in the interval $(0, \frac{1}{2})$, and initial velocity $y_t(x, 0) = g(x) = 0$.

a. Sketch the graph of $f(x)$ (for some fixed value of a; pick your favorite a that's strictly between 0 and $\frac{1}{2}$).

b. Show that the displacement $y(x, t)$ of the string is given by

$$y(x, t) = \frac{8}{\pi^3(1-2a)} \sum_{m=1}^{\infty} \frac{1}{(2m-1)^3} \cos \pi(2m-1)a \sin \pi(2m-1)x \cos \pi(2m-1)t.$$

Hint: Use the formula (2.146) as well as Exercise 1.3.5.

   c. Show that, for any a with $0 < a < \frac{1}{2}$, there is a number M that's independent of n (but depends on a) such that

$$|D_n| \leq \frac{M}{n^3}.$$

   d. Show that, as $a \to \frac{1}{2}$ from the left, $|D_n|$ approaches $4/(\pi n)^2$ for n odd. (Use l'Hôpital's rule, considering D_n to be a function of a.) (Of course for even n, all D_n's are zero.)

   e. Compare parts c and d above. Noting that (any constant times) $1/n^3$ decays much faster as $n \to \infty$ than (any constant times) $1/n^2$, in which type of vibrating string do you think the "overtones," or high-frequency components, will be more prevalent: a string plucked *sharply* (with a pick, say), or one plucked more *bluntly* (with a fingertip, say)? Explain.

2.7.5 A somewhat more realistic vibrating string model accounts for damping effects, like those caused by air resistance. In such a model, the differential equation (2.149) is replaced by this one:

$$c^2 y_{xx} = y_{tt} + 2dy_t \quad (0 < x < \ell, \ t > 0), \tag{$*$}$$

where $d > 0$ (the factor of 2 is included simply for convenience).

   a. Show by separation of variables, much as in Example 2.7.1, that the boundary value problem given by $(*)$, (2.150), and (2.151)—assume for simplicity that $g(x) = 0$—has formal solution

$$y(x,t) = e^{-dt} \sum_{n=1}^{\infty} B_n(f) \left(\cos \sqrt{\lambda_n}\, t + \frac{d}{\sqrt{\lambda_n}} \sin \sqrt{\lambda_n}\, t \right) \sin \frac{\pi n x}{\ell},$$

where

$$\lambda_n = \frac{\pi^2 n^2 c^2}{\ell^2} - d^2.$$

Hint: The most general solution to the differential equation $T'' + 2dT' + \mu T = 0$ is $T(t) = e^{-dt}(\gamma \cos \sqrt{\mu - d^2}\, t + \delta \sin \sqrt{\mu - d^2}\, t)$ for $\mu \neq 0$ and $T(t) = \gamma e^{-2dt} + \delta$ for $\mu = 0$ ($\gamma, \delta \in \mathbb{C}$).

   b. Explain why each summand in the above solution is in fact exponentially decaying as time evolves. Warning: For certain values of d, c, ℓ, and n, λ_n may be negative; then the cosine and sine factors will really amount to hyperbolic cosine and sine functions of a real variable, by (1.8). Explain why, when this happens, the exponential growth of these factors is offset by the decay of e^{-dt}.

2.7.6 Suppose we have an elastic string as in Example 2.7.1, except that the ends, instead of being *fixed*, are allowed to *slide up and down* along the lines $x = 0$ and $x = \ell$ in the xy plane. If this sliding is frictionless we get, instead of (2.150), the following boundary conditions:

$$y_x(0,t) = 0, \quad y_x(\ell,t) = 0 \quad (t > 0). \tag{$*$}$$

a. Using, essentially, the method of Example 2.7.1, show that the wave problem consisting of the above boundary conditions (∗), the differential equation (2.149), and the initial conditions (2.151) has formal solution

$$y(x,t) = \frac{1}{2}(\alpha_0(f) + \alpha_0(g)t)$$
$$+ \sum_{n=1}^{\infty} \cos\frac{\pi n x}{\ell}\left(\alpha_n(f)\cos\frac{\pi n c t}{\ell} + \frac{\ell\alpha_n(g)}{\pi n c}\sin\frac{\pi n c t}{\ell}\right),$$

where the $\alpha_n(f)$'s and $\alpha_n(g)$'s are the Fourier cosine coefficients (cf. Theorem 1.12.1) of f and g, respectively.

b. Where does the string "go," as time evolves, if $\alpha_0(g) \neq 0$? Explain qualitatively what's happening here. Hint: $\alpha_0(g)t/2$ is the only term that's not periodic in t. Also, $\alpha_0(g)/2$ is the average initial velocity of the string. (Remark: The above boundary value problem also models, roughly, longitudinal vibrations of air in an organ pipe, for example.)

2.8 THE WAVE EQUATION II: EXISTENCE AND UNIQUENESS OF SOLUTIONS

We promised to supply some details regarding convergence and uniqueness of our solution to the wave problem (2.149)–(2.151). We now make good on that promise. To do so, we'll first need the following, whose proof we omit. (See, for example, Corollary 5.3.4 in [33].)

Lemma 2.8.1 *Suppose f_n is continuously differentiable on an open interval I for each $n \in \mathbb{Z}^+$. If $\sum_{n=1}^{\infty} f_n(x)$ converges pointwise on I and $\sum_{n=1}^{\infty} f'_n(x)$ converges uniformly on I, then the former series is differentiable, and its derivative equals the latter series, throughout I.*

We now have:

Proposition 2.8.1 *The formal solution (2.160), (2.162), (2.165) to the wave problem (2.149)–(2.151) is an actual one, and is in fact the unique one, provided the first three derivatives of f and the first two of g are piecewise smooth and continuous on $(0, \ell)$, and that f and g and all of these derivatives approach zero as $x \to 0^+$ and as $x \to \ell^-$.*

Proof. Let f be as stipulated. It's then readily shown that the first three derivatives of f_{odd}, the odd, 2ℓ-periodic extension of f (see Section 1.12 and especially Figure 1.21), are themselves piecewise smooth and continuous, so by the 2ℓ-periodic analog of Proposition 1.8.1 in the case $m = 4$,

$$c_n(f_{\text{odd}}) = \left(\frac{i\pi n}{\ell}\right)^{-4} c_n(f^{(4)}_{\text{odd}}). \tag{2.171}$$

Now $c_n(f_{\text{odd}}^{(4)})$ is bounded with respect to n (by the 2ℓ-periodic analog of Lemma 1.4.1), and $|b_n(f_{\text{odd}})| \leq |c_n(f_{\text{odd}})| + |c_{-n}(f_{\text{odd}})|$ by the conversion formula (1.7), so we conclude from (2.171) that, for some $L > 0$,

$$|b_n(f_{\text{odd}})| \leq Ln^{-4} \quad (n \in \mathbb{Z}^+). \tag{2.172}$$

On the other hand, $b_n(f_{\text{odd}})$ equals D_n by (2.162) and by (1.230), so

$$|D_n| \leq Ln^{-4} \quad (n \in \mathbb{Z}^+). \tag{2.173}$$

Similarly, by the 2ℓ-periodic analog of Proposition 1.8.1 in the case $m = 3$ and by (2.165), we find that there's a constant M such that

$$|E_n| \leq Mn^{-4} \quad (n \in \mathbb{Z}^+). \tag{2.174}$$

Let's denote by $y(n, x, t)$ the nth summand of the series given by (2.160), (2.162), (2.165). Since $|\cos\theta|, |\sin\theta| \leq 1$ for all real θ, we have $|y(n, x, t)| \leq (L+M)n^{-4}$ for all x, t. Similarly,

$$y_x(n, x, t) = \frac{\pi n}{\ell}\cos\frac{\pi n x}{\ell}\left(D_n \cos\frac{\pi n c t}{\ell} + E_n \sin\frac{\pi n c t}{\ell}\right) \leq \frac{\pi}{\ell}(L+N)n^{-3}, \tag{2.175}$$

and so on. In general, any mth-order partial derivative of $y(n, x, t)$ is seen to be bounded by a constant times n^{m-4}. But $\sum_{n=1}^{\infty} n^{m-4}$ converges for $m = 0, 1$, or 2, whence one shows (much in the way that we showed uniform convergence of the Fourier series for an $f \in \text{PSC}(\mathbb{T})$; see the proof of Theorem 1.7.1) that the series in question converges uniformly, as does the series obtained by differentiating each term once or twice with respect to x or t. By Lemma 2.8.1 above, then,

$$y_{xx}(x, t) + y_{tt}(x, t) = \sum_{n=1}^{\infty}(y_{xx}(n, x, t) + y_{tt}(n, x, t)). \tag{2.176}$$

But $y(n, x, t)$ was chosen to solve the wave equation (2.149), so each summand on the right side of (2.176) is zero, whence so is the left side, as desired.

That the series given by (2.160), (2.162), (2.165) satisfies the conditions (2.150) and (2.151), and the implicit continuity conditions at the boundary (recall the discussions at the end of Section 2.1), follows from the continuity of this series and of the corresponding termwise t-derivative thereof. The stated continuity itself follows, for example, from Lemma 2.8.1 and the fact that differentiability implies continuity.

We now demonstrate uniqueness of our solution. To this end we observe that, if $y = y_1$ and $y = y_2$ are two solutions to our wave problem (2.149)–(2.151), then $y = y_1 - y_2$ satisfies (2.149), (2.150), and, instead of (2.151), the corresponding *homogeneous* conditions

$$y(x, 0) = y_t(x, 0) = 0 \quad (0 < x < \ell). \tag{2.177}$$

THE WAVE EQUATION II: EXISTENCE AND UNIQUENESS OF SOLUTIONS

If we can show that any solution y to (2.149), (2.150), and (2.177) is identically zero, then we'll be done.

We do this as follows. Let y be such a solution; assume y is real-valued (if not, the arguments that follow apply to the real and imaginary parts of y separately). Let

$$V = c^2 y_x^2 + y_t^2. \tag{2.178}$$

Note that

$$V_t = 2c^2 y_x y_{xt} + 2 y_t y_{tt}; \tag{2.179}$$

then by the wave equation (2.149),

$$V_t = 2c^2 y_x y_{xt} + 2c^2 y_t y_{xx} = 2c^2 \frac{\partial}{\partial x}(y_x y_t). \tag{2.180}$$

Therefore, by the fundamental theorem of calculus,

$$\int_0^\ell V_t(x,t)\,dx = 2c^2 \int_0^\ell \frac{\partial}{\partial x}(y_x(x,t) y_t(x,t))\,dx = 2c^2 y_x(x,t) y_t(x,t)\Big|_{x=0}^{x=\ell}$$
$$= 2c^2 (y_x(\ell,t) y_t(\ell,t) - y_x(0,t) y_t(0,t)) = 0 \tag{2.181}$$

since, by the boundary conditions (2.150), $y_t(0,t) = y_t(\ell,t) = 0$ for $t > 0$. On the other hand,

$$\int_0^\ell V_t(x,t)\,dx = \int_0^\ell \frac{\partial}{\partial t} V(x,t)\,dx = \frac{d}{dt} \int_0^\ell V(x,t)\,dx; \tag{2.182}$$

taking the t-derivative outside the integral sign here is readily justified. (See, for example, the proposition on p. 517 of [33].) Comparing (2.181) and (2.182) gives

$$\frac{d}{dt} \int_0^\ell V(x,t)\,dx = 0, \tag{2.183}$$

which implies

$$\int_0^\ell V(x,t)\,dx = C \tag{2.184}$$

for some constant C. We can determine this constant by putting $t = 0$. By definition (2.178) of V, $V(x,0) = c^2 y_x^2(x,0) + y_t^2(x,0)$, which equals zero by (2.177), so by (2.184) with $t = 0$, we find $C = 0$. So by (2.184) again,

$$\int_0^\ell V(x,t)\,dx = 0 \tag{2.185}$$

for any t.

But V is, by its definition (2.178), continuous and nonnegative, so (2.185) implies that V is itself identically zero. This, again by the definition of V, implies that y_x

and y_t are identically zero, so that y must be a constant function. Since $y(x, 0) = 0$, this means y is identically zero, as required. □

It should be noted that the functions f and g of Example 2.7.2 above *do not* satisfy the hypotheses of the proposition just given. Among other things, the function f of that example is not differentiable at $x = \pi/6$. And indeed, neither infinite series on the right side of (2.170) can be twice differentiated, term by term, in either x or t. Such differentiation introduces extra factors of n^2 into corresponding summands, and the resulting series do not converge uniformly.

Still, the setup of that example seems plausible, from a physical point of view. The function f there, for instance, represents a perfectly reasonable initial position for a string plucked sharply at a point one-third of the way along its length. Moreover, the series in (2.170), while not (as just noted) twice differentiable in either variable, *is*, by itself, convergent. And in fact uniformly so. Given all this, shouldn't there be some sense in which this series *can* be considered a bona fide solution to the wave equation, for the designated f, g, ℓ, and c?

Yes there should be, and there is. This is the sense of *weak solutions* to differential equations and boundary value problems. We won't go into detail about this sense, but we do make the following brief remarks about it. First, it may be framed in terms of the theory of *distributions*, of which the *tempered distributions* we discuss in Section 6.9 are a special type. Second, it allows for the kind of thing we'd like to have happen here. That is, it provides a secure mathematical basis for the discussion of boundary value problems that make good physical sense, but have formal solutions that, while not *too* unreasonable, are not amenable to such classical considerations as the one given in Proposition 2.8.1. (Among other things it validates, in many cases, series solutions like (2.170) that converge, but not well enough to be differentiated termwise the requisite number of times.) A more detailed discussion of weak solutions may be found, for example, in Section 9.5 of [19].

Exercises

2.8.1 Here we outline a proof of uniqueness of solutions to the heat equation in a uniform bar, with Dirichlet conditions at the boundary:

$$k u_{xx} = u_t \quad (0 < x < \ell, \; t > 0),$$
$$u(0, t) = 0, \quad u(\ell, t) = 0 \quad (t > 0),$$
$$u(x, 0) = f(x) \quad (0 < x < \ell).$$

(See Exercise 2.3.3.) Let's suppose we have two solutions $u = u_1(x, t)$ and $u = u_2(x, t)$ to this problem. We put $v = u_1 - u_2$, and assume v to be real-valued (if not, we may treat its real and imaginary parts separately). We also define

$$W(t) = \frac{1}{2} \int_0^\ell v^2(x, t) \, dx.$$

a. Show that

$$kv_{xx} = v_t \quad (0 < x < \ell,\ t > 0),$$
$$v(0,t) = 0, \quad v(\ell,t) = 0 \quad (t > 0),$$
$$v(x,0) = 0 \quad (0 < x < \ell).$$

Now suppose that u_1 and u_2 are continuous on $[0,\ell] \times [0,\infty)$ and their derivatives are continuous on $(0,\ell) \times (0,\infty)$, so that the same may be said of v. It may be shown, then, and you may assume, that W is continuous on $[0,\infty)$ and differentiable on $(0,\infty)$, with the derivative obtained by differentiating under the integral sign.

b. Show that

$$W'(t) = k\int_0^\ell v(x,t) v_{xx}(x,t)\, dx.$$

c. Integrate by parts in part b to show that

$$W'(t) = -k\int_0^\ell v_x^2(x,t)\, dx.$$

d. Use the result from part c and the mean value theorem to show that, for any $t \in (0,\infty)$, $W(t) \le 0$.

e. Conclude from the definition of $W(t)$ that $W(t) = 0$ for $t \in [0,\infty)$ and therefore that, for any such t, $v(x,t) = 0$ for all $x \in [0,\ell]$. From this, deduce the uniqueness of solutions to the original heat problem of this exercise.

2.9 THE WAVE EQUATION III: FOURIER VERSUS D'ALEMBERT

We call the series given by (2.160), (2.162), (2.165) "Fourier's form" of the solution to our boundary value problem (2.149)–(2.151). And we observe the following nice feature of this form: It tells us what we *hear*. Specifically, the "temporal frequencies" $nc/2\ell$, coming from the sinusoids $\cos(\pi nct/\ell)$ and $\sin(\pi nct/\ell)$, describe the quality of the sound produced by the string.

The smallest, $c/2\ell$, of these frequencies is the *fundamental frequency*, or *first harmonic*, or *base pitch*. Higher frequencies $nc/2\ell$ ($n > 1$) are *overtones*, or *higher harmonics;* their presence adds richness to the sound. (A single fundamental frequency without any overtones is not much fun to listen to; it sounds like a hearing test.) The sizes of the D_n's and E_n's relative to the sizes of D_1 and E_1 determine the extent to which the overtones are heard relative to the fundamental frequency. Indeed the *amplitude*, or maximum value, of the nth summand in the series (2.160) equals $\sqrt{D_n^2 + E_n^2}$. This follows from our discussions of amplitude in Section 1.10 and from the fact that $|\sin(\pi nx/\ell)| \le 1$.

The *spatial frequencies* $n/2\ell$, corresponding to the $\sin(\pi nx/\ell)$ factors, are harder to relate directly to what's heard: suffice it to say that, because of the dependence

128 FOURIER SERIES AND BOUNDARY VALUE PROBLEMS

of $y(x, t)$ on x, what's heard will vary somewhat with position relative to the string. Quite a bit of acoustics theory is involved here.

But now we consider what solutions to our wave problem *look like*. Consult Figure 2.8: One can see a wave moving to the left, interacting with one moving to the right. (At the fixed ends of the string, each wave reflects and then heads back in the other direction.)

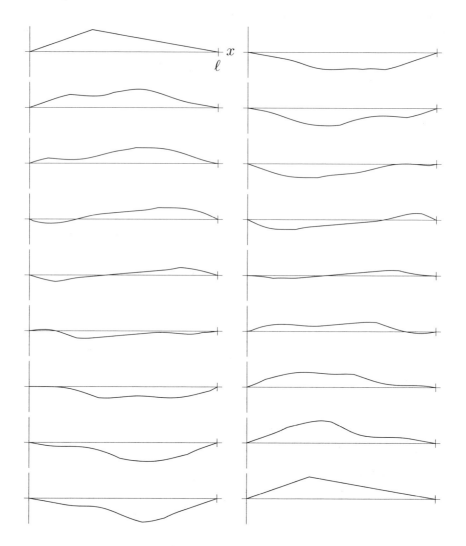

Fig. 2.8 Snapshots of the vibrating string of Example 2.7.2 (here, time evolves from top to bottom, left to right)

An accounting of this behavior is provided by:

THE WAVE EQUATION III: FOURIER VERSUS D'ALEMBERT   129

Proposition 2.9.1 (d'Alembert's form of the solution to the wave problem) *Let f_{odd} and g_{odd} be the odd, 2ℓ-periodic extensions of f and g, respectively, as in Theorem 1.12.1. Let*

$$G(x) = \int_0^x g_{\text{odd}}(u)\, du \qquad (2.186)$$

(an antiderivative of g_{odd}). Then the wave problem (2.149)–(2.151) has formal solution

$$y(x,t) = \frac{1}{2}\left[f_{\text{odd}}(x+ct) + \frac{1}{c}G(x+ct)\right] + \frac{1}{2}\left[f_{\text{odd}}(x-ct) - \frac{1}{c}G(x-ct)\right]. \qquad (2.187)$$

Proof. By the 2ℓ-periodic analog of Proposition 1.8.2, regarding Fourier series of antiderivatives, G has Fourier series

$$-\frac{\ell}{\pi}\sum_{n=1}^{\infty} \frac{b_n(g_{\text{odd}})}{n} \cos\frac{\pi n x}{\ell}. \qquad (2.188)$$

(We've used the fact, clear from the definition of G, that $G(0) = 0$.) Putting this and the Fourier series for f_{odd} into (2.187) gives

$$y(x,t) = \frac{1}{2}\left[\sum_{n=1}^{\infty} b_n(f_{\text{odd}}) \sin\frac{\pi n(x+ct)}{\ell} - \frac{\ell}{\pi c}\sum_{n=1}^{\infty} \frac{b_n(g_{\text{odd}})}{n} \cos\frac{\pi n(x+ct)}{\ell}\right]$$

$$+ \frac{1}{2}\left[\sum_{n=1}^{\infty} b_n(f_{\text{odd}}) \sin\frac{\pi n(x-ct)}{\ell} + \frac{\ell}{\pi c}\sum_{n=1}^{\infty} \frac{b_n(g_{\text{odd}})}{n} \cos\frac{\pi n(x-ct)}{\ell}\right]$$

$$= \sum_{n=1}^{\infty}\left[b_n(f_{\text{odd}}) \cdot \frac{1}{2}\left(\sin\frac{\pi n(x+ct)}{\ell} + \sin\frac{\pi n(x-ct)}{\ell}\right)\right.$$

$$\left. - \frac{\ell b_n(g_{\text{odd}})}{\pi n c} \cdot \frac{1}{2}\left(\cos\frac{\pi n(x-ct)}{\ell} - \cos\frac{\pi n(x+ct)}{\ell}\right)\right]$$

$$= \sum_{n=1}^{\infty} \sin\frac{\pi n x}{\ell}\left[b_n(f_{\text{odd}}) \cos\frac{\pi n c t}{\ell} + \frac{\ell b_n(g_{\text{odd}})}{\pi n c} \sin\frac{\pi n c t}{\ell}\right], \qquad (2.189)$$

the last step by the trig identities of Exercise A.0.16. But by Proposition 2.7.1, the right side of (2.189) solves our wave problem, and we're done. (We've skirted issues of convergence and so on, because we are only concerned with formal solutions. Under appropriate conditions on f and g, though, such as those delineated in Proposition 2.8.1, these solutions are actual ones, and our arguments above are valid.) □

So d'Alembert's formula (2.187) tells us what we *see*. Specifically, the first half

$$\frac{1}{2}\left[f_{\text{odd}}(x+ct) + \frac{1}{c}G(x+ct)\right] \qquad (2.190)$$

of the right side of this formula may be understood as a wave having the shape of the function $\frac{1}{2}[f_{odd} + \frac{1}{c}G]$ and moving to the left with velocity c (units horizontally per unit t). Similarly, the second half,

$$\frac{1}{2}\left[f_{odd}(x - ct) - \frac{1}{c}G(x - ct)\right], \tag{2.191}$$

is a wave with the shape of $\frac{1}{2}[f_{odd} - \frac{1}{c}G]$ moving right with velocity c. This *is* what we see.

It *is* what we saw in Figure 2.8. To understand what's happening a bit more clearly, it's perhaps better to consider an instance in which g is identically zero, so that G is identically zero as well. Then the right side of (2.187) is "half the graph of f_{odd}" (or, more precisely, the graph of $\frac{1}{2}f_{odd}$) moving left with velocity c plus "half the graph of f_{odd}" moving right with velocity c. See Figure 2.9: Here we've dashed in the function (2.187) slightly beyond the "actual" domain $[0, \ell]$. What appears to the left and right of this domain is not part of the physical solution; it's shown simply to indicate how (2.187) explains what *is* happening on the actual domain.

Each of our boundary value problems so far has amounted to a linear, homogeneous second-order differential equation, together with a pair of linear, homogeneous boundary conditions applying along the boundary of one of the variables in question, together with certain linear, inhomogeneous boundary conditions applying along the boundary of the other. Boundary value problems of other natures may require somewhat different approaches. In the next section we consider some variants on the problem (2.149)–(2.151), not necessarily because the latter is the best thing since cup holders (though it would be were it not 250 years precedent to them), but because the resulting boundary value problems provide nice contexts in which to demonstrate some of these other approaches.

Exercises

2.9.1 Consider the displacement $y(x, t)$ of a vibrating string with ends fixed at $x = 0$ and $x = 2$, tension constant $c = 1$, initial displacement

$$y(x, 0) = f(x) = \begin{cases} 0 & \text{if } 0 \le x \le \frac{3}{2}, \\ \sin 2\pi x & \text{if } \frac{3}{2} \le x \le 2, \end{cases}$$

and initial velocity $y_t(x, 0) = g(x) = 0$. You *don't* need to find Fourier's solution for $y(x, t)$. But *do* do the following:

  a. Sketch the graph of the odd periodic extension $f_{odd}(x)$ of $f(x)$.

  b. Using d'Alembert's solution to the wave equation, sketch "snapshots" of $y(x, t)$ at each of the times $t = 1$, $t = \frac{3}{2}$, and $t = 2$.

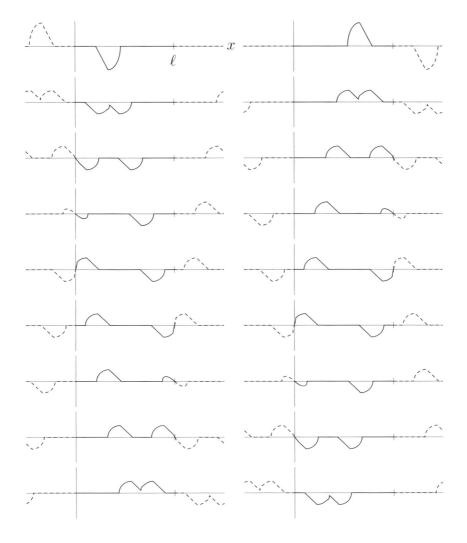

Fig. 2.9 Snapshots of a string vibrating according to (2.149)–(2.151), with initial velocity zero (as before, the chronology is from top to bottom, left to right)

2.9.2 Repeat the above for the string with ends fixed at $x = 0$ and $x = 3$, tension constant $c = 2$, initial displacement

$$y(x,0) = f(x) = \begin{cases} 0 & \text{if } 0 \leq x \leq 1, \\ \sin \pi x & \text{if } 1 \leq x \leq 2, \\ 0 & \text{if } 2 \leq x \leq 3, \end{cases}$$

and initial velocity $y_t(x,0) = g(x) = 0$. This time, sketch snapshots at each of the times $t = \frac{3}{4}, t = 1, t = \frac{5}{4}$, and $t = \frac{3}{2}$.

2.9.3 **a.** Use Exercise 2.7.6a and the method of proof of Proposition 2.9.1 to show that a string with free sliding ends, as in that exercise, and *average* initial velocity equal to zero has displacement

$$y(x,t) = \frac{1}{2}\left[f_{\text{even}}(x+ct) + \frac{1}{c}G(x+ct)\right] + \frac{1}{2}\left[f_{\text{even}}(x-ct) - \frac{1}{c}G(x-ct)\right],$$

where

$$G(x) = \int_0^x g_{\text{even}}(u)\,du.$$

b. What is the analogous result in the case of a nonzero average initial velocity?

2.9.4 Using part a of the previous exercise, repeat Exercise 2.9.2, but this time with free sliding ends instead of fixed ends.

2.10 THE WAVE EQUATION IV: TEMPORALLY CONSTANT INHOMOGENEITY

We now tweak the wave problem considered in the previous three sections in a couple of ways: We account for the effects of gravity and we displace one endpoint vertically. The first of these tweaks renders inhomogeneous the differential equation involved; the second does the same to one of the boundary conditions in x.

We focus on formal rather than actual solutions; the former become the latter under mild conditions (though not as mild, generally, as those sufficing in the context of the heat equation).

Example 2.10.1 Solve the boundary value problem

$$c^2 y_{xx} = y_{tt} - G \quad (0 < x < \ell,\ t > 0), \tag{2.192}$$
$$y(0,t) = 0, \quad y(\ell, t) = 0 \quad (t > 0), \tag{2.193}$$
$$y(x,0) = f(x), \quad y_t(x,0) = g(x) \quad (0 < x < \ell), \tag{2.194}$$

representing a string set up as in the previous section, but now with the acceleration $G\ (\approx -32\ \text{ft/sec}^2)$ due to gravity taken into account.

Solution. Our solution will make use, and will serve as a handy vehicle for illustrating the use, of:

Proposition 2.10.1 (the superposition principle II) *Let* $y = y_h$ *solve the homogeneous equation* $Ly = 0$, *where* L *is a linear operator. If* $y = y_i$ *is a solution to the corresponding inhomogeneous equation* $Ly = f$, *then so is* $y = y_h + y_i$.

Proof. Under the given assumptions,

$$L(y_h + y_i) = Ly_h + Ly_i = 0 + Ly_i = Ly_i = f. \quad \square \tag{2.195}$$

For our purposes, the power of this deceptively simple proposition lies in the following: It may be easy enough to find *one particular* solution $y = y_i$ to the equation $Ly = f$. Further, techniques like those of the previous few sections may yield a whole *slew* of solutions $y = y_h$ to the equation $Ly = 0$. But then putting $y = y_h + y_i$ for each such y_h gives a whole slew of solutions to $Ly = f$. And one can hope that, out of this slew, there's one solution that meets any other requirements of the boundary value problem at hand.

Here's how all this plays out in the context of the present example. We consider first equations (2.192) and (2.193): Note that these are the *same* as equations (2.149) and (2.150) from our earlier wave problem, except that the *homogeneous* differential equation $c^2 y_{xx} - y_{tt} = 0$ there has now been replaced by the inhomogeneous one $c^2 y_{xx} - y_{tt} = -G$. Since the boundary conditions (2.150) of that (and this) problem are linear and homogeneous, and since, as we've seen, the series in (2.160) formally solves (2.149), (2.150), for any choice of the D_n's and E_n's, the superposition principle II says the following: If $y = y_i$ is any solution to the equations (2.192), (2.193), and $D_n, E_n \in \mathbb{C}$, then

$$y(x,t) = y_i(x,t) + \sum_{n=1}^{\infty} \sin\frac{\pi n x}{\ell}\left(D_n \cos\frac{\pi n c t}{\ell} + E_n \sin\frac{\pi n c t}{\ell}\right) \qquad (2.196)$$

formally solves these equations as well.

So, we seek a solution—*any* solution—$y = y_i$ to these equations. And why not make our lives easy by looking for *a solution that does not depend on* t! That is, we look for a solution $y_i(x,t) = Y(x)$.

Since $dY(x)/dt = 0$, (2.192) yields $c^2 Y''(x) = -G$, which means $Y(x) = c^{-2}(K + Hx - Gx^2/2)$ for certain constants K and H. Now what about these constants? Well, the two parts of (2.193) give $0 = Y(0) = c^{-2}K$ and $0 = Y(\ell) = c^{-2}(K + H\ell - G\ell^2/2)$; solving, one finds $K = 0$ and $H = G\ell/2$, so

$$y_i(x,t) = Y(x) = \frac{Gx(\ell - x)}{2c^2}. \qquad (2.197)$$

CONCLUSION: For any choice of the constants D_n and E_n,

$$y(x,t) = \frac{Gx(\ell - x)}{2c^2} + \sum_{n=1}^{\infty} \sin\frac{\pi n x}{\ell}\left(D_n \cos\frac{\pi n c t}{\ell} + E_n \sin\frac{\pi n c t}{\ell}\right) \qquad (2.198)$$

solves (2.192), (2.193).

We'd now like to see if we can get this in line with the inhomogeneous conditions (2.194). Let's try. Putting the first of these conditions into (2.198) gives

$$f(x) = \frac{Gx(\ell - x)}{2c^2} + \sum_{n=1}^{\infty} D_n \sin\frac{\pi n x}{\ell} \qquad (0 < x < \ell). \qquad (2.199)$$

That is, we want $f(x) - Y(x)$, where again $Y(x) = Gx(\ell - x)/(2c^2)$, to equal the above sum on n for $0 < x < \ell$. By Theorem 1.12.1, we can make this so if we choose

D_n to be the nth Fourier sine coefficient of $f - Y$. This means

$$D_n = \beta_n(f - Y) = \beta_n(f) - \beta_n(Y) = \frac{2}{\ell}\int_0^\ell f(x)\sin\frac{\pi n x}{\ell}\,dx - \beta_n(Y)$$

$$= \frac{2}{\ell}\int_0^\ell f(x)\sin\frac{\pi n x}{\ell}\,dx - \frac{2G\ell^2(1-(-1)^n)}{\pi^3 n^3 c^2}; \tag{2.200}$$

the last step is by Exercise 1.12.3.

Similarly, we determine E_n as follows: Putting $y_t(x,0) = g(x)$ into (2.198) gives

$$g(x) = \frac{\pi n c}{\ell}\sum_{n=1}^\infty E_n \sin\frac{\pi n x}{\ell}, \tag{2.201}$$

whence Theorem 1.12.1 gives

$$E_n = \frac{\ell}{\pi n c}\beta_n(g) = \frac{2}{\pi n c}\int_0^\ell g(x)\sin\frac{\pi n x}{\ell}\,dx, \tag{2.202}$$

as in our first wave problem (Example 2.7.1). So we conclude:

Proposition 2.10.2 *The boundary value problem* (2.192)–(2.194) *has formal solution* $y(x,t)$ *given by* (2.198), (2.200), (2.202).

Note that, if we choose an intitial position $f(x) = Y(x) = Gx(\ell - x)/(2c^2)$ and an initial velocity $g(x) = 0$, then all D_n's and E_n's are zero, so (2.198) gives a stationary solution $y(x,t) = Y(x)$. That is, $Y(x)$ describes the natural resting position of the string under the effects of gravity. Note this position is in the shape of a parabola lying below the x axis, as is physically plausible.

The technique employed in the above example will apply even if the inhomogeneity in the differential equation depends on x, as long as it does not depend on t. See Exercise 2.10.1.

Our next example is similar in flavor, except that this time the inhomogeneity is in a boundary condition.

Example 2.10.2 Solve the boundary value problem

$$c^2 y_{xx} = y_{tt} \quad (0 < x < \ell,\ t > 0), \tag{2.203}$$

$$y(0,t) = 0, \quad y(\ell,t) = y_0 \quad (t > 0) \tag{2.204}$$

$$y(x,0) = f(x), \quad y_t(x,0) = g(x) \quad (0 < x < \ell), \tag{2.205}$$

representing a string set up as in Section 2.7, except that its right end is now held at height y_0.

Solution. We note that the above problem is the same as that of Example 2.7.1, except that the homogeneous boundary condition $y(\ell,t) = 0$ there has been replaced by the inhomogeneous one $y(\ell, t) = y_0$. So Proposition 2.10.1 says the following: If $y = y_i$ is any solution to (2.203), (2.204) and $D_n, E_n \in \mathbb{C}$, then

$$y(x,t) = y_i(x,t) + \sum_{n=1}^{\infty} \sin\frac{\pi n x}{\ell}\left(D_n \cos\frac{\pi n c t}{\ell} + E_n \sin\frac{\pi n c t}{\ell}\right) \quad (2.206)$$

also solves (2.203), (2.204), at least formally.

So we seek a particular solution $y = y_i$ to these parts. And again we assume, to make our lives easier, that y_i is independent of t: $y_i(x,t) = Y(x)$.

Now the homogeneous wave equation (2.203) gives $0 = Y''(x)$, which means $Y(x) = \gamma x + \delta$ for certain constants γ and δ. Then by the boundary conditions (2.204), $Y(0) = \gamma \cdot 0 + \delta = 0$ and $Y(\ell) = \gamma \cdot \ell + \delta = y_0$; the first of these latter two results implies $\delta = 0$, so the second yields $\gamma = (y_0 - \delta)/\ell = y_0/\ell$. So

$$y_i(x,t) = Y(x) = \frac{y_0}{\ell}x. \quad (2.207)$$

CONCLUSION: For any choice of the constants D_n and E_n,

$$y(x,t) = Y(x) + \sum_{n=1}^{\infty} \sin\frac{\pi n x}{\ell}\left(D_n \cos\frac{\pi n c t}{\ell} + E_n \sin\frac{\pi n c t}{\ell}\right) \quad (2.208)$$

formally solves (2.203), (2.204).

Into this we put the first equation in (2.205) to get

$$f(x) = Y(x) + \sum_{n=1}^{\infty} D_n \sin\frac{\pi n x}{\ell} \quad (0 < x < \ell). \quad (2.209)$$

By Theorem 1.12.1, we can make this so if we put

$$D_n = \beta_n(f - Y) = \frac{2}{\ell}\int_0^\ell f(x)\sin\frac{\pi n x}{\ell}\,dx - \frac{2y_0}{\ell^2}\int_0^\ell x\sin\frac{\pi n x}{\ell}\,dx$$
$$= \frac{2}{\ell}\int_0^\ell f(x)\sin\frac{\pi n x}{\ell}\,dx - \frac{2y_0(-1)^{n+1}}{\pi n}, \quad (2.210)$$

the last step by (1.237).

It follows just as in our previous example that

$$E_n = \frac{2}{\pi n c}\int_0^\ell g(x)\sin\frac{\pi n x}{\ell}\,dx. \quad (2.211)$$

So:

Proposition 2.10.3 *The boundary value problem (2.203)–(2.205) has formal solution given by (2.208), (2.210), (2.211).*

Exercises

2.10.1 Using the method of Example 2.10.1, solve the boundary value problem

$$c^2 y_{xx} = y_{tt} + a \sin \omega x \quad (0 < x < \ell, \ t > 0),$$
$$y(0, t) = 0, \quad y(\ell, t) = 0 \quad (t > 0),$$
$$y(x, 0) = f(x), \quad y_t(x, 0) = g(x) \quad (0 < x < \ell)$$

$(a, \omega \in \mathbb{R} - \{0\})$, representing a string set up as in Section 2.7, but now with a force proportional to $\sin \omega x$ acting on each point x along the string. (To perform the required integration, you may want to use certain trig identites from Exercise A.0.16.) Warning: The case where w is an integer multiple of π/ℓ will look a bit different from the case where it is not (because $\int \sin \alpha x \sin \beta x \, dx$ takes different forms according to whether $\alpha = \beta$ or $\alpha \neq \beta$).

2.10.2 Suppose a uniform bar, lying along the x axis from $x = 0$ to $x = \ell$, has an internal heat source, generating heat in such a way that the heat flow in the bar satisfies the equation

$$k u_{xx} = u_t - A \quad (0 < x < \ell, \ t > 0),$$

where A is a positive constant. Using the general method of Example 2.10.1, find the temperature $u(x, t)$ in the bar, assuming mixed boundary conditions $u(0, t) = u_x(\ell, t) = 0$ and initial conditions $u(x, 0) = f(x)$. Hints: Refer to Example 2.3.2. Also, Exercise 1.12.4(c) should help.

2.10.3 Suppose a uniform bar, lying along the x axis from $x = 0$ to $x = \ell$, satisfies the usual equation

$$k u_{xx} = u_t \quad (0 < x < \ell, \ t > 0).$$

Using the general method of Example 2.10.2, find the temperature $u(x, t)$ in the bar, assuming the left end is held at constant temperature $u(0, t) = 0$, the right end at constant *nonzero* temperature $u(\ell, t) = \tau_0$, and the initial temperature distribution is $u(x, 0) = f(x)$.

2.10.4 Solve the heat problem as in the previous exercise, but this time with *both* ends held at nonzero temperatures. Hint: By the superposition principle II (Proposition 2.10.1), it suffices to consider the previous exercise and a similar situation with the roles of the two ends reversed.

2.11 THE WAVE EQUATION V: TEMPORALLY VARYING INHOMOGENEITY

In the previous section, we rendered inhomogeneous either the differential equation (2.149) or the boundary conditions (2.150) of our original wave problem, by addition of a t-independent term. This t-independence made it relatively easy to find a particular solution y_i to the relevant inhomogeneous differential equation or boundary condition, and to thereby apply Proposition 2.10.1.

But what if the inhomogeneity is t-dependent? It's sometimes possible, in that case, to make use of the *variation of parameters* technique. The general idea here is as follows: Suppose the inhomogeneity is in the differential equation. (If not one puts it there; see Example 2.11.2.) Apply separation of variables to the corresponding *homogeneous* problem, and solve the resulting eigenvalue problem (or problems; see Sections 2.12 and 2.13). Then put together a series, over the eigenvalues thus obtained, of the corresponding eigenfunctions times as-yet undetermined functions of the remaining variable (or variables). Finally, impose the inhomogeneous differential equation and remaining boundary conditions to determine these functions, as well as the coefficients of the series.

The best way to elucidate this is by example. In the following, we choose our inhomogeneity to be independent of x, though in general one need not — see Exercises 2.11.5–2.11.10.

Example 2.11.1 Solve the boundary value problem

$$c^2 y_{xx} = y_{tt} - K \sin 2\pi s t \quad (0 < x < \ell,\ t > 0), \tag{2.212}$$

$$y(0, t) = 0, \quad y(\ell, t) = 0 \quad (t > 0), \tag{2.213}$$

$$y(x, 0) = 0, \quad y_t(x, 0) = 0 \quad (0 < x < \ell), \tag{2.214}$$

representing a string set up as in Section 2.7, but now with a force $F(t) = K \sin 2\pi s t$ (K and s being constants, with $s > 0$) acting on each point along the string. (Here we've assumed, for simplicity, that the initial position and velocity of the string are zero. See Exercise 2.11.3 below for the more general situation.)

Solution. The variation-of-parameters approach amounts, in the present case, to this: Since the homogeneous parts (2.149), (2.150) of our original wave problem have solution (2.160), we should try to find solutions to the corresponding parts of our new problem that look "the same in x" but different in t. By this we mean solutions of the form

$$y(x, t) = \sum_{n=1}^{\infty} \sin \frac{\pi n x}{\ell} T_n(t), \tag{2.215}$$

where the T_n's are as yet unspecified functions of t. Why should we? Because it works! As we now demonstrate.

We begin by writing

$$1 = \frac{2}{\pi} \sum_{n=1}^{\infty} \frac{1 - (-1)^n}{n} \sin \frac{\pi n x}{\ell} \quad (0 < x < \ell); \tag{2.216}$$

cf. Exercise 1.12.2(a). Such an expansion is desired for the following reason: if we put it, and the series (2.215), into the differential equation (2.212) of our present problem, we get (at least formally)

$$-\left(\frac{\pi nc}{\ell}\right)^2 \sum_{n=1}^{\infty} \sin\frac{\pi nx}{\ell} T_n(t) = \sum_{n=1}^{\infty} \sin\frac{\pi nx}{\ell} T_n''(t)$$

$$-\frac{2K\sin 2\pi st}{\pi} \sum_{n=1}^{\infty} \frac{1-(-1)^n}{n} \sin\frac{\pi nx}{\ell} \quad (2.217)$$

for $0 < x < \ell$ and $t > 0$. We can certainly make this true by equating coefficients of $\sin \pi nx/\ell$ for each n; doing so gives

$$-\left(\frac{\pi nc}{\ell}\right)^2 T_n(t) = T_n''(t) - \frac{2K(1-(-1)^n)\sin 2\pi st}{\pi n} \quad (2.218)$$

for $n \in \mathbb{Z}^+$.

Note that we also have boundary conditions on the T_n's; namely, by putting the initial condition $y(x, 0) = 0$ into the series (2.215), we get

$$\sum_{n=1}^{\infty} T_n(0) \sin\frac{\pi nx}{\ell} = 0 \quad (0 < x < \ell). \quad (2.219)$$

The zero function on the right side has all its Fourier sine coefficients equal to zero, so we arrive at the formula

$$T_n(0) = 0 \quad (n \in \mathbb{Z}^+). \quad (2.220)$$

Similarly, putting $y_t(x, 0) = 0$ into the series (2.215) gives

$$\sum_{n=1}^{\infty} T_n'(0) \sin\frac{\pi nx}{\ell} = 0 \quad (0 < x < \ell); \quad (2.221)$$

so

$$T_n'(0) = 0 \quad (n \in \mathbb{Z}^+) \quad (2.222)$$

as well.

It's shown (see Exercise 2.11.1 below) that (2.218), (2.220), (2.222) have solution

$$T_n(t) = \frac{2K\ell^2(1-(-1)^n)}{\pi^3 n^2 c(4\ell^2 s^2 - n^2 c^2)}\left(2\ell s \sin\frac{\pi nct}{\ell} - nc\sin 2\pi st\right) \quad (2.223)$$

provided $s \neq nc/(2\ell)$. If the two are equal, then we understand (2.223) to denote the limit, as $s \to nc/(2\ell)$, of the quantity given there. That is, by l'Hôpital's rule,

$$T_n(t) = \frac{2K\ell^2(1-(-1)^n)}{\pi^3 n^2 c} \lim_{s \to nc/(2\ell)} \frac{2\ell s \sin(\pi nct/\ell) - nc \sin 2\pi st}{4\ell^2 s^2 - n^2 c^2}$$

$$= \frac{2K\ell^2(1-(-1)^n)}{\pi^3 n^2 c} \lim_{s \to nc/(2\ell)} \frac{2\ell \sin(\pi nct/\ell) - 2\pi nct \cos 2\pi st}{8\ell^2 s}$$

$$= \frac{K\ell(1-(-1)^n)}{\pi^3 n^3 c^2}\left(\ell \sin\frac{\pi nct}{\ell} - \pi nct \cos\frac{\pi nct}{\ell}\right) \quad (2.224)$$

if $s = nc/(2\ell)$.

We summarize:

Proposition 2.11.1 *The boundary value problem (2.212)–(2.214) has formal solution (2.215), with $T_n(t)$ given by (2.223) if $s = nc/(2\ell)$ and by (2.224) if not.*

Let's now consider some physical implications of the problem just solved. Specifically, we note that, if s is a fixed number such that $s \neq nc/(2\ell)$ for any $n \in \mathbb{Z}^+$, then the solution of the above proposition is a *bounded* function of t (and of x, for that matter). Indeed, the proposition gives, for such s,

$$|y(x,t)| \leq \frac{4K\ell^2}{\pi^3 c} \sum_{n=1}^{\infty} \frac{2\ell s + nc}{n^2 |4\ell^2 s^2 - n^2 c^2|} = \frac{4K\ell^2}{\pi^3 c} \sum_{n=1}^{\infty} \frac{1}{n^2 |2\ell s - nc|} \quad (2.225)$$

(we've used the facts that $\sin\theta \leq 1$ for all real θ and $4\ell^2 s^2 - n^2 c^2 = (2\ell s - nc)(2\ell s + nc)$); the sum on the right is just a constant with respect to x and t. INTERPRETATION: In the "generic" case, meaning the one where s does not equal nc/ℓ for any n, the vibrations of the string remain bounded as time evolves.

But NOW, suppose s *does* equal one of the $nc/(2\ell)$'s: say $s = n_0 c/(2\ell)$ for some *odd* positive integer n_0. (The quantity in (2.224) is zero for n even.) In this case $T_{n_0}(t)$ looks like Figure 2.10.

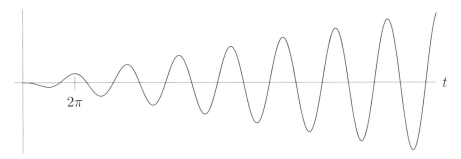

Fig. 2.10 The graph of $T_{n_0}(t)$ for $s = n_0 c/(2\ell)$. (Here $n_0 = c = 1$ and $K = \ell = \pi$)

Note that $T_{n_0}(t)$ *does* blow up as t increases. (This is because of the factor of t preceding the cosine term in (2.224).) Consequently so does $y(x,t)$.

The moral of the story is this. If a vibrating string, or a physical system admitting a similar mathematical model, is subject to an external, sinusoidal force of an appropriate frequency s, then that system will experience oscillations that grow unboundedly in magnitude as time elapses. (In "real life," nothing (?) is infinite: The oscillations will either be tempered by damping forces—see Exercise 2.7.5—or will cease when the system in question breaks! Still, as the above model suggests, the vibrations can become rather large.)

The phenomenon just described is called *resonance*; an input frequency s that causes resonance is called a *resonant frequency*. So in fact, our above vibrating string has infinitely many resonant frequencies, one for each odd $n \in \mathbb{Z}^+$.

More generally, by the resonant frequency of a physical system we mean one at which the system "likes" to vibrate or oscillate; it's a frequency that, when imposed on such a system from without, induces very large oscillations or vibrations.

Many systems besides vibrating strings behave, at least in part, in a manner dictated by a differential equation of the form (2.212). Consequently, many have resonant frequencies. Suspension bridges do, for example. This is why they can sway rather unpleasantly when traversed by a large group of people walking in synch. See [38]. (Even more dramatic is the famous Tacoma Narrows Bridge collapse, which has been attributed to resonance, although there is lively debate as to whether this was actually the cause.) So do certain electronic circuits; without this we would not have feedback, without which we would not have had Jimi Hendrix, without whom civilization as we know it would be severely compromised, and unquestionably a bit too civilized.

So do electrons in an atom and bonds in a molecule, whence all sorts of things to which Fourier analysis is relevant. One of these things is spectroscopy; see Section 7.10. And so on.

We now consider another boundary value problem—representing a vibrating string again, but this time under somewhat different circumstances—whose solution has a resonant frequency. The key difference between this problem and the previous one is that, now, the inhomogeneity resides not in the differential equation, but in the conditions at the boundary in the x direction.

Example 2.11.2 Solve the boundary value problem

$$c^2 y_{xx} = y_{tt} \quad (0 < x < \ell,\ t > 0), \tag{2.226}$$
$$y(0, t) = 0, \quad y(\ell, t) = A \sin 2\pi s t \quad (t > 0), \tag{2.227}$$
$$y(x, 0) = 0, \quad y_t(x, 0) = 0 \quad (0 < x < \ell), \tag{2.228}$$

representing a string set up as in Section 2.7, but now with the right endpoint being driven in such a way as to have displacement $A \sin 2\pi s t$ at time t. (Here we've again stipulated, for simplicity, a null initial position and velocity. See Exercise 2.11.4 below for the more general case.)

Solution. We use:

Proposition 2.11.2 *Let $v(x, t)$ be any function satisfying a boundary condition of the form $B_1 v = G$, where G is a function of t. If $u(x, t)$ satisfies*

$$Lu = -Lv, \quad B_1 u = 0, \quad B_2 u = H - B_2 v, \tag{2.229}$$

where H is a function of x, then $y = u + v$ satisfies

$$Ly = 0, \quad B_1 y = G, \quad B_2 y = H. \tag{2.230}$$

Proof. For such u, v, y, we have

$$Ly = Lu + Lv = 0, \quad B_1 y = B_1 u + B_1 v = G, \quad B_2 y = B_2 u + B_2 v = H. \quad \square \tag{2.231}$$

What this proposition does is allow us to trade off inhomogeneity in one of the boundary conditions for inhomogeneity in the differential equation. Such trade-off requires determination of an auxiliary function v that satisfies certain boundary conditions, but such a v will often be relatively easy to find.

So for our present boundary value problem, the situation is this: Suppose we can find a v so that

$$v(0,t) = 0, \quad v(\ell, t) = A \sin 2\pi s t \quad (t > 0). \tag{2.232}$$

Since our differential operator here is $Lv = c^2 v_{xx} - v_{tt}$, the above proposition then says we should solve

$$c^2 u_{xx} - u_{tt} = -c^2 v_{xx} + v_{tt} \quad (0 < x < \ell,\ t > 0), \tag{2.233}$$
$$u(0,t) = 0, \quad u(\ell, t) = 0 \quad (t > 0), \tag{2.234}$$
$$u(x,0) = -v(x,0), \quad u_t(x,0) = -v_t(x,0) \quad (0 < x < \ell) \tag{2.235}$$

for u. Doing so and putting $y = u + v$, we get a solution to (2.226)–(2.228).

The first step is to find a suitable v, but this is easy! Indeed, to obtain a v satisfying (2.232), we need only multiply $A \sin 2\pi s t$ by a function of x that's equal to 0 at $x = 0$ and 1 at $x = \ell$. There are lots of such functions, but x/ℓ is the simplest one we can imagine. So we should put

$$v(x,t) = \frac{Ax}{\ell} \sin 2\pi s t. \tag{2.236}$$

Let's.

Then (2.233) reads

$$c^2 u_{xx} - u_{tt} = -\frac{(2\pi s)^2 A x}{\ell} \sin 2\pi s t \quad (0 < x < \ell,\ t > 0). \tag{2.237}$$

But this differential equation in u together with the boundary conditions (2.234) amount to a situation much like that of equations (2.212) and (2.213) from our previous example: We have t-dependent (and also, in this case, x-dependent) inhomogeneity in the differential equation and homogeneous boundary conditions in x. So we proceed similarly: We look for solutions u of the form

$$u(x,t) = \sum_{n=1}^{\infty} \sin \frac{\pi n x}{\ell} T_n(t). \tag{2.238}$$

By Example 1.12.1,

$$x = \frac{2\ell}{\pi} \sum_{n=1}^{\infty} \frac{(-1)^{n+1}}{n} \sin \frac{\pi n x}{\ell} \quad (0 < x < \ell). \tag{2.239}$$

Putting this and the series (2.238) for u into the differential equation (2.237) gives

$$-\left(\frac{\pi n c}{\ell}\right)^2 T_n(t) - \sum_{n=1}^{\infty} \sin \frac{\pi n x}{\ell} T_n''(t) = \frac{2(2\pi s)^2 A \sin 2\pi s t}{\pi} \sum_{n=1}^{\infty} \frac{(-1)^n}{n} \sin \frac{\pi n x}{\ell}$$

$$\tag{2.240}$$

for $0 < x < \ell$ and $t > 0$.

We solve much as we did in the previous example; that is, we equate coefficients of the trig functions of x. We get

$$-\left(\frac{\pi nc}{\ell}\right)^2 T_n(t) - T_n''(t) = \frac{2(2\pi s)^2 A(-1)^n \sin 2\pi st}{\pi n}. \tag{2.241}$$

We determine boundary conditions for T_n as follows: By the first equation in (2.235), by (2.236), and by (2.238), we have

$$\sum_{n=1}^{\infty} \sin \frac{\pi nx}{\ell} T_n(0) = 0 \quad (0 < x < \ell); \tag{2.242}$$

similarly, by the second equation in (2.235), by (2.236), by (2.238), and by (2.239), we have

$$\sum_{n=1}^{\infty} \sin \frac{\pi nx}{\ell} T_n'(0) = -\frac{2\pi s Ax}{\ell} = 4sA \sum_{n=1}^{\infty} \frac{(-1)^n}{n} \sin \frac{\pi nx}{\ell} \quad (0 < x < \ell). \tag{2.243}$$

We can guarantee (2.242) and (2.243) by requiring that

$$T_n(0) = 0, \quad T_n'(0) = \frac{4sA(-1)^n}{n}. \tag{2.244}$$

The boundary value problem (2.241), (2.244) has solution

$$T_n(t) = \begin{cases} \dfrac{A(-1)^n}{\pi n \ell} \left(\ell \sin \dfrac{\pi nct}{\ell} + \pi nct \cos \dfrac{\pi nct}{\ell} \right) & \text{if } s = \dfrac{nc}{2\ell}, \\ \dfrac{4A\ell s(-1)^n}{\pi n \left(4\ell^2 s^2 - n^2 c^2\right)} \left(2\ell s \sin 2\pi st - nc \sin \dfrac{\pi nct}{\ell} \right) & \text{if not.} \end{cases} \tag{2.245}$$

(See Exercise 2.11.1f.) Finally, we recall that our desired function y is the function u given by the series (2.238), with $T_n(t)$ as just described, *plus* the function v of (2.236). So we have:

Proposition 2.11.3 *The boundary value problem (2.226)–(2.228) has formal solution*

$$y(x,t) = \frac{Ax}{\ell} \sin 2\pi st + \sum_{n=1}^{\infty} \sin \frac{\pi nx}{\ell} T_n(t), \tag{2.246}$$

with $T_n(t)$ given by (2.245).

Note the resonance when $s = nc/(2\ell)$ for some positive integer n.

Exercises

EXERCISES

2.11.1 By the superposition principle II (Proposition 2.10.1), the differential equation
$$T''(t) + a^2 T(t) = b \sin \omega t \qquad (*)$$
has solution $T(t) = T_h(t) + T_i(t)$, where T_h is the general solution to the corresponding homogeneous equation and T_i any particular solution to $(*)$.

  a. What is T_h? Hint: See the Introduction.

  b. Find a particular solution T_i to $(*)$ in the case where $\omega \neq \pm a$, by plugging in $T_i = A \cos \omega t + B \sin \omega t$ and solving for A and B.

  c. Use your above results to derive the solution (2.223) to (2.218), (2.220), (2.222) in the case $s \neq nc/(2\ell)$. (Remember that $s, n, c, \ell > 0$.)

  d. Find a particular solution T_i to $(*)$ in the case where ω^2 and a^2 are equal, by plugging in $T_i = At \cos \omega t + Bt \sin \omega t$ and solving for A and B.

  e. Use your above results to derive the solution (2.224) to (2.218), (2.220), (2.222) in the case $s = nc/(2\ell)$.

  f. Similarly, derive the solution (2.245) to (2.240), (2.244).

2.11.2 Formally solve the wave problem of Example 2.11.1, but with the "fixed ends" boundary conditions (2.213) replaced by the "free sliding ends" conditions
$$y_x(0,t) = 0, \quad y_x(\ell, t) = 0 \quad (t > 0).$$
See Exercise 2.7.6a. The idea is that, because of the $(\cos \pi nx/\ell)$'s encountered in that exercise, you should proceed here as we did in Example 2.11.1, but using $(\cos \pi nx/\ell)$'s instead of $(\sin \pi nx/\ell)$'s throughout. In particular, you'll want a cosine series for the constant function 1 on $(0, \ell)$; see Exercise 1.12.2(b).

2.11.3 What does the formal solution to the wave problem of Example 2.11.1 look like if the homogeneous initial conditions (2.214) are replaced by the more general conditions
$$y(x,0) = f(x), \quad y_t(x,0) = g(x) \quad (0 < x < \ell)?$$
Hint: Use the superposition principle II (Proposition 2.10.1) to reduce the problem to one of solving two simpler boundary value problems (both of which we've already solved).

2.11.4 Formally solve the wave problem of Example 2.11.2, but with the homogeneous initial conditions (2.228) replaced by the general conditions
$$y(x,0) = f(x), \quad y_t(x,0) = g(x) \quad (0 < x < \ell).$$

2.11.5 **a.** Show, by direct differentiation of the relevant quantities, that the single-variable boundary value problem
$$\tau''(t) + a^2 \tau(t) = \varphi(t) \quad (t > 0), \quad \tau(0) = 0, \quad \tau'(0) = 0$$
has formal solution
$$\tau(t) = \begin{cases} a^{-1} \int_0^t \sin a(t-s) \varphi(s)\, ds & \text{if } a \neq 0, \\ \int_0^t (t-s) \varphi(s)\, ds & \text{if } a = 0. \end{cases}$$

144 FOURIER SERIES AND BOUNDARY VALUE PROBLEMS

Hint: Expand the sine term in the integrand using a trig identity; use the fact that

$$\frac{d}{dt}\int_0^t f(s)\,ds = f(t)$$

for nice enough f.

  b. Show that the wave problem of Example 2.11.1, but with the differential equation (2.212) replaced by the more general equation

$$c^2 y_{xx} = y_{tt} - F(x,t) \quad (0 < x < \ell,\ t > 0),$$

has formal solution

$$y(x,t) = \sum_{n=1}^{\infty}\left[\frac{\ell}{\pi n c}\int_0^t \sin\frac{\pi n c(t-s)}{\ell}\varphi_n(s)\,ds\right]\sin\frac{\pi n x}{\ell},$$

where the $\varphi_n(t)$'s are the Fourier sine coefficients of $F(x,t)$, considered as a function of x:

$$F(x,t) = \sum_{n=1}^{\infty}\varphi_n(t)\sin\frac{\pi n x}{\ell} \quad (0 < x < \ell).$$

Hint: Proceed much as in that example, and use part a of this exercise.

  c. Suppose $F(x,t)$ is a product of a function of x with a function of t: $F(x,t) = U(x)V(t)$, say. Show that the above formula for $y(x,t)$ becomes

$$y(x,t) = \sum_{n=1}^{\infty}\beta_n(U)\left[\frac{\ell}{\pi n c}\int_0^t \sin\frac{\pi n c(t-s)}{\ell}V(s)\,ds\right]\sin\frac{\pi n x}{\ell},$$

the $\beta_n(U)$'s being the Fourier sine coefficients of U.

2.11.6 Write down the general, formal solution to the wave problem

$$c^2 y_{xx} = y_{tt} - F(x,t) \quad (0 < x < \ell,\ t > 0),$$
$$y(0,t) = 0, \quad y(\ell,t) = 0 \quad (t > 0),$$
$$y(x,0) = f(x), \quad y_t(x,0) = g(x) \quad (0 < x < \ell).$$

Hint: Use the superposition principle II and Exercises 2.11.3 and 2.11.5 above.

2.11.7 a. Show that Example 2.10.1 agrees with Exercise 2.11.6 above.
  b. Show that Exercise 2.10.1 also agrees with this exercise.

2.11.8 a. Show, by direct differentiation of the relevant quantities, that the single-variable boundary value problem

$$\sigma'(t) + a\sigma(t) = \psi(t) \quad (t > 0), \quad \sigma(0) = 0$$

has formal solution

$$\sigma(t) = \begin{cases} e^{-at}\int_0^t e^{as}\psi(s)\,ds & \text{if } a \neq 0, \\ \int_0^t \psi(s)\,ds & \text{if } a = 0. \end{cases}$$

b. Show that the mixed boundary condition heat problem of Example 2.3.2, but with the heat equation (2.44) replaced by the more general equation

$$k\,u_{xx} = u_t - F(x,t) \quad (0 < x < \ell,\ t > 0),$$

and assuming initial temperature equal to zero, has formal solution

$$u(x,t) = \sum_{n=1}^{\infty} \left[\int_0^t e^{\pi^2(2n-1)^2 ks/(4\ell^2)} \psi_n(s)\,ds \right] e^{-\pi^2(2n-1)^2 kt/(4\ell^2)} \sin \frac{\pi(2n-1)x}{2\ell},$$

where the $\psi_n(t)$'s are the Fourier halfsine coefficients (cf. Theorem 1.12.2) of $F(x,t)$, considered as a function of x:

$$F(x,t) = \sum_{n=1}^{\infty} \psi_n(t) \sin \frac{\pi(2n-1)x}{2\ell} \quad (0 < x < \ell).$$

Hint: Proceed much as in that example and use part a of this exercise.

2.11.9 Write down the general, formal solution to the heat problem

$$k\,u_{xx} = u_t - F(x,t) \quad (0 < x < \ell,\ t > 0),$$
$$u(0,t) = 0, \quad u_x(\ell,t) = 0 \quad (t > 0),$$
$$u(x,0) = f(x) \quad (0 < x < \ell).$$

2.11.10 Show that Exercise 2.10.2 agrees with Exercise 2.11.9.

2.12 THE WAVE EQUATION VI: DRUMMING UP SOME INTEREST

We now consider some boundary problems in three variables, which give rise to Fourier series in two variables. We discuss the particulars of such series as they come up.

Example 2.12.1 The boundary value problem

$$c^2(z_{xx} + z_{yy}) = z_{tt} \quad (0 < x < \ell_1,\ 0 < y < \ell_2,\ t > 0), \tag{2.247}$$
$$z(0,y,t) = 0, \quad z(\ell_1,y,t) = 0 \quad (0 < y < \ell_2,\ t > 0), \tag{2.248}$$
$$z(x,0,t) = 0, \quad z(x,\ell_2,t) = 0 \quad (0 < x < \ell_1,\ t > 0), \tag{2.249}$$
$$z(x,y,0) = f(x,y), \quad z_t(x,y,0) = g(x,y) \quad (0 < x < \ell_1,\ 0 < y < \ell_2) \tag{2.250}$$

represents (an idealized model for) the vibrations in a drumhead stretched across a rectangular frame (of length ℓ_1 and width ℓ_2), with initial position $f(x,y)$ and initial velocity $g(x,y)$. Solve this problem for $z = z(x,y,t)$.

Solution. We begin, as usual, by separating variables. If

$$z(x, y, t) = X(x)Y(y)T(t), \qquad (2.251)$$

then our above differential equation in z gives

$$c^2(X''(x)Y(y)T(t) + X(x)Y''(y)T(t)) = X(x)Y(y)T''(t) \qquad (2.252)$$

for the relevant values of x, y, t. There are three variables here instead of two; to handle this situation, we introduce two unspecified constants instead of one. That is, we note that, certainly, this last equation will be satisfied if $X'' = -\lambda X, Y'' = -\mu Y$, and $T'' = -(\lambda + \mu)c^2 T$. So it suffices to consider the equations

$$X''(x) + \lambda X(x) = 0 \quad (0 < x < \ell_1), \quad Y''(y) + \mu Y(y) = 0 \quad (0 < y < \ell_2), \qquad (2.253)$$

$$T''(t) + (\lambda + \mu)c^2 T(t) = 0 \quad (t > 0). \qquad (2.254)$$

The boundary conditions (2.248) yield the conditions $X(0) = X(\ell_1) = 0$, which together with the above equation for X imply, as we've seen a number of times, the following eigenvalues and eigenfunctions:

$$\lambda = \left(\frac{\pi n_1}{\ell_1}\right)^2, \quad X(x) = \sin \sqrt{\lambda}\, x = \sin \frac{\pi n_1 x}{\ell_1} \quad (n_1 \in \mathbb{Z}^+). \qquad (2.255)$$

In the same way, the given equation for Y and the conditions (2.249) give

$$\mu = \left(\frac{\pi n_2}{\ell_2}\right)^2, \quad Y(y) = \sin \sqrt{\mu}\, y = \sin \frac{\pi n_2 y}{\ell_2} \quad (n_2 \in \mathbb{Z}^+). \qquad (2.256)$$

The above equation for T then has solution

$$T(t) = \gamma \cos \sqrt{\lambda + \mu}\, ct + \delta \sin \sqrt{\lambda + \mu}\, ct$$

$$= \gamma \cos \pi \sqrt{\frac{n_1^2}{\ell_1^2} + \frac{n_2^2}{\ell_2^2}}\, ct + \delta \sin \pi \sqrt{\frac{n_1^2}{\ell_1^2} + \frac{n_2^2}{\ell_2^2}}\, ct \qquad (2.257)$$

for any $\gamma, \delta \in \mathbb{C}$, $n_1, n_2 \in \mathbb{Z}^+$.

Forming the product solutions $z = XYT$ and then summing, we get the formal solutions

$$z(x, y, t) = \sum_{n_1, n_2 = 1}^{\infty} \sin \frac{\pi n_1 x}{\ell_1} \sin \frac{\pi n_2 y}{\ell_2}$$

$$\cdot \left(D_{n_1, n_2} \cos \pi \sqrt{\frac{n_1^2}{\ell_1^2} + \frac{n_2^2}{\ell_2^2}}\, ct + E_{n_1, n_2} \sin \pi \sqrt{\frac{n_1^2}{\ell_1^2} + \frac{n_2^2}{\ell_2^2}}\, ct \right)$$

$$\qquad (2.258)$$

to the homogeneous parts (2.247)–(2.249) of our boundary value problem.

We now impose the inhomogeneous conditions (2.250): The first of these conditions gives

$$f(x,y) = \sum_{n_2=1}^{\infty} \left(\sum_{n_1=1}^{\infty} D_{n_1,n_2} \sin \frac{\pi n_1 x}{\ell_1} \right) \sin \frac{\pi n_2 y}{\ell_2}. \tag{2.259}$$

We've grouped terms as we have for the following reason: Imagine for the moment that x is fixed. Then we may consider (2.259) a Fourier sine series, in the index n_2 of summation, for $f(x,y)$, considered as a function of y. Then Theorem 1.12.1 gives us, under appropriate conditions, the following formula for the coefficients of this series:

$$\sum_{n_1=1}^{\infty} D_{n_1,n_2} \sin \frac{\pi n_1 x}{\ell_1} = \frac{2}{\ell_2} \int_0^{\ell_2} f(x,y) \sin \frac{\pi n_2 y}{\ell_2} dy. \tag{2.260}$$

But now we note that the left side of this last equation itself represents a Fourier sine series for the function of x on the right. So again by Theorem 1.12.1,

$$D_{n_1,n_2} = \frac{2}{\ell_1} \int_0^{\ell_1} \left[\frac{2}{\ell_2} \int_0^{\ell_2} f(x,y) \sin \frac{\pi n_2 y}{\ell_2} dy \right] \sin \frac{\pi n_1 x}{\ell_1} dx$$

$$= \frac{4}{\ell_1 \ell_2} \int_0^{\ell_2} \int_0^{\ell_1} f(x,y) \sin \frac{\pi n_1 x}{\ell_1} \sin \frac{\pi n_2 y}{\ell_2} dx\, dy. \tag{2.261}$$

A similar argument yields

$$E_{n_1,n_2} = \frac{4}{\pi \sqrt{n_1^2 \ell_2^2 + n_2^2 \ell_1^2}\, c} \int_0^{\ell_2} \int_0^{\ell_1} g(x,y) \sin \frac{\pi n_1 x}{\ell_1} \sin \frac{\pi n_2 y}{\ell_2} dx\, dy. \tag{2.262}$$

So:

Proposition 2.12.1 *The boundary value problem (2.247)–(2.250) has formal solution (2.258), (2.261), (2.262).*

What's really behind the above example, and especially the determination of the D_{n_1,n_2}'s and E_{n_1,n_2}'s, is the following two-dimensional generalization of Theorem 1.12.1: *If f is a reasonable function on $(0,\ell_1) \times (0,\ell_2)$, then*

$$f(x,y) = \sum_{n_1,n_2=1}^{\infty} \beta_{n_1,n_2}(f) \sin \frac{\pi n_1 x}{\ell_1} \sin \frac{\pi n_2 y}{\ell_2} \quad (0 < x < \ell_1,\ 0 < x < \ell_2), \tag{2.263}$$

where

$$\beta_{n_1,n_2}(f) = \frac{4}{\ell_1 \ell_2} \int_0^{\ell_2} \int_0^{\ell_1} f(x,y) \sin \frac{\pi n_1 x}{\ell_1} \sin \frac{\pi n_2 y}{\ell_2} dx\, dy \quad (n_1, n_2 \in \mathbb{Z}^+). \tag{2.264}$$

148 FOURIER SERIES AND BOUNDARY VALUE PROBLEMS

For suitable "reasonability" conditions on f, see, for example, Chapter 7, Section 3, of [47]. It's worth noting that these conditions are somewhat elaborate, and do not generalize in particularly obvious ways the corresponding one-variable criteria of Theorem 1.12.1, or of any of our other convergence results from Chapter 1. Discussions of pointwise or uniform convergence of Fourier series in several variables can be messy. Fortunately, things are considerably cleaner if one focuses instead on *norm convergence*, as we will in the next chapter. See, in particular, Section 3.8 for some elegantly simple results regarding norm convergence of multivariable Fourier series.

Leaving aside, again, convergence questions, let's remark on an interesting qualitative feature of our drum, which is nicely reflected by our above quantitative analysis. Namely, we observe from the series (2.258) that this drum has temporal frequencies of vibration

$$\frac{1}{2}\sqrt{\frac{n_1^2}{\ell_1^2} + \frac{n_2^2}{\ell_2^2}}\, c \quad (n_1, n_2 \in \mathbb{Z}^+). \tag{2.265}$$

These are not all integer multiples of some fundamental frequency. For example, taking $\ell_1 = \ell_2 = c/2$ for simplicity, we get, for $(n_1, n_2) = (1,1), (2,1), (2,2), (3,1),$ and $(3,2)$, the frequencies $\sqrt{2}, \sqrt{5}, \sqrt{8}, \sqrt{10},$ and $\sqrt{13}$, respectively. This is to be contrasted with the one-dimensional wave problem of Section 2.7; there, if $\ell = c/2$, say, we have a base pitch of $n = 1$ and overtones $n = 2, 3, 4, \ldots$.

Thus our drum will not sound as "tuned" as does a vibrating string or similar instrument. (It might for certain special choices of initial position and velocity—see Exercise 2.12.1—but not in general.) Some would say the drum is less "musical" than these other instruments, though the drummers among us would dispute this characterization.

Of course the drum of the above example is quite nonstandard, since it's rectangular. What happens with a more familiar, circular drum? Let's see.

Example 2.12.2 Solve the boundary value problem analogous to that of the previous example, but where this time the drum is circular of radius ℓ.
Solution. We solve for $z = z(\rho, \theta, t)$, where

$$c^2\left(\rho^{-2} z_{\theta\theta} + \rho^{-1}[\rho z_\rho]_\rho\right) = z_{tt} \quad (0 < \rho < \ell,\ -\pi < \theta < \pi,\ t > 0), \tag{2.266}$$

$$z(\rho, -\pi, t) = z(\rho, \pi, t), \quad z_\theta(\rho, -\pi, t) = z_\theta(\rho, \pi, t) \quad (0 < \rho < \ell,\ t > 0), \tag{2.267}$$

$$z(\ell, \theta, t) = 0 \quad (-\pi < \theta < \pi,\ t > 0), \tag{2.268}$$

$$z(\rho, \theta, 0) = f(\rho, \theta), \quad z_t(\rho, \theta, 0) = g(\rho, \theta) \quad (0 < \rho < \ell,\ -\pi < \theta < \pi). \tag{2.269}$$

The left side of (2.266) is just $c^2 \nabla_2^2 z$ in polar coordinates; see Proposition 2.5.1(a).
As usual, we separate variables. We put

$$z(\rho, \theta, t) = R(\rho)\Theta(\theta)T(t); \tag{2.270}$$

then the differential equation (2.266) gives

$$c^2\{\rho^{-2}R(\rho)\Theta''(\theta)T(t) + \rho^{-1}[\rho R'(\rho)]'\Theta(\theta)T(t)\} = R(\rho)\Theta(\theta)T''(t) \quad (2.271)$$

for the ρ, θ, t in question. To reduce the problem of solving this equation to one involving three ordinary differential equations, we proceed essentially as usual, though just a bit more care is required. We begin by putttting $\Theta'' = -\lambda\Theta$; then the left side of (2.271) becomes

$$c^2\,\Theta(\theta)T(t)\{-\lambda\rho^{-2}R(\rho) + \rho^{-1}[\rho R'(\rho)]'\}. \quad (2.272)$$

We can now, certainly, make this and the right side of (2.271) equal to each other if we let

$$-\lambda\rho^{-2}R(\rho) + \rho^{-1}[\rho R'(\rho)]' = -\mu R(\rho), \quad T''(t) = -\mu c^2\,T(t). \quad (2.273)$$

So first of all, we get the eigenvalue problem

$$\Theta''(\theta) + \lambda\Theta(\theta) = 0 \quad (-\pi < \theta < \pi), \quad \Theta(-\pi) = \Theta(\pi), \quad \Theta'(-\pi) = \Theta'(\pi) \quad (2.274)$$

(the boundary conditions coming from (2.267)) in Θ. Note this is the same problem as we had for λ and Θ as in Example 2.5.1. So we have the same eigenvalues and corresponding eigenfunctions

$$\lambda = n^2, \quad \Theta(\theta) = \gamma e^{i\sqrt{\lambda}\theta} + \delta e^{-i\sqrt{\lambda}\theta} = \gamma e^{in\theta} + \delta e^{-in\theta} \quad (\gamma, \delta \in \mathbb{C}, n \in \mathbb{N}). \quad (2.275)$$

Then in light of (2.273), we have this eigenvalue problem in μ and R:

$$\rho^{-1}[\rho R'(\rho)]' - n^2\rho^{-2}R(\rho) + \mu R(\rho) = 0 \quad (0 < \rho < \ell), \quad R(\ell) = 0. \quad (2.276)$$

(Here the boundary condition is from (2.268).) But now we're stuck: The differential equation for R is unfamiliar.

Its nonzero solutions turn out to be not sinusoids, but certain less elementary entities known as *Bessel functions*. We'll discuss these functions in Sections 4.3–4.5, where we'll also solve the problem (2.276).

We'll then have, for each eigenvalue μ, solutions

$$T(t) = \sigma \cos\sqrt{\mu}\,ct + \tau\sin\sqrt{\mu}\,ct \quad (2.277)$$

($\sigma, \tau \in \mathbb{C}$) to our above equation in T. (We could of course write these solutions in terms of complex exponentials, but for our eventual imposition of initial conditions, cosines and sines will be more convenient.) We'll put everything together to get our separated solutions $z = R\Theta T$ to the homogeneous parts of our boundary value problem; we'll form an infinite linear combination of these; finally, we'll impose the inhomogeneous conditions (2.269) to try to determine the necessary coefficients of this combination. Note that doing so will leave us with a doubly infinite linear combination of sinusoids times Bessel functions. So we'll certainly need to know something about infinite series of Bessel functions. In other words, we'll be in a situation something like that of Example 2.4.2, where separation of variables led us to a certain unfamiliar infinite series of functions, and lacking any general theory of such series, we were unable to complete that problem.

But again, we will, and we'll finish this one as well, in Chapter 4.

Exercises

2.12.1 Solve the rectangular drum problem of Example 2.12.1, assuming for simplicity the dimensions $\ell_1 = \ell_2 = \pi$ and initial velocity $g(x, y) = 0$, for each of the following initial position functions:

a. $f(x, y) = 2 \sin x (\sin y + 3 \sin 2y) - \sin 2x (\sin 3y + \sin 7y)$;
b. $f(x, y) = \sin x (\sin y - 2 \sin 7y) + 5 \sin 2x (\sin 2y - \sin 14y)$.

Which of the above initial positions yields a more "tuned" sound? Explain.

2.12.2 a. Solve the heat problem

$$\begin{aligned} k(u_{xx} + u_{yy}) &= u_t \quad (0 < x < \ell_1,\ 0 < y < \ell_2,\ t > 0), \\ u(0, y, t) &= 0, \quad u(\ell_1, y, t) = 0 \quad (0 < y < \ell_2,\ t > 0), \\ u(x, 0, t) &= 0, \quad u(x, \ell_2, t) = 0 \quad (0 < x < \ell_1,\ t > 0), \\ u(x, y, 0) &= f(x, y) \quad (0 < x < \ell_1,\ 0 < y < \ell_2) \end{aligned}$$

for an infinite rectangular prism with all four faces held at temperature zero.

b. In the case where $f(x, y)$ is a product $f_1(x) f_2(y)$ of single-variable functions, express your answer as a product of solutions to single-variable heat problems.

2.12.3 Solve the heat problem of part a of the previous exercise, but with the condition $u(\ell_1, y, t) = 0$ replaced with $u(\ell_1, y, t) = T_0$. You may want to recall the techniques of Section 2.10; see especially Example 2.10.2 and Exercise 2.10.3.

2.13 TRIPLE FOURIER SERIES

The big step is from singly to doubly infinite Fourier series; any additional index of summation after that is considerably less daunting. Still, let's not underestimate the value of triple, or greater, Fourier series: A variety of boundary value problems of intrinsic natural interest entail such series.

Example 2.13.1 Solve the heat equation $k \nabla_3^2 u = u_t$ in a solid rectangular box

$$B = \{(x, y, z) \in \mathbb{R}^3 : 0 < x < \ell_1,\ 0 < y < \ell_2,\ 0 < z < \ell_3\} \qquad (2.278)$$

given an initial temperature distribution $f(x, y, z)$ and the following boundary assumptions: The faces at $x = 0, x = \ell_1$, and $z = 0$ are held at temperature zero; those at $y = 0, y = \ell_2$, and $z = \ell_3$ are insulated.

Solution. Our boundary value problem is

$$k(u_{xx} + u_{yy} + u_{zz}) = u_t \quad (0 < x < \ell_1,\ 0 < y < \ell_2,\ 0 < z < \ell_3,\ t > 0), \tag{2.279}$$

$$u(0, y, z, t) = 0, \quad u(\ell_1, y, z, t) = 0 \quad (0 < y < \ell_2,\ 0 < z < \ell_3,\ t > 0), \tag{2.280}$$

$$u_y(x, 0, z, t) = 0, \quad u_y(x, \ell_2, z, t) = 0 \quad (0 < x < \ell_1,\ 0 < z < \ell_3,\ t > 0), \tag{2.281}$$

$$u(x, y, 0, t) = 0, \quad u_z(x, y, \ell_3, t) = 0 \quad (0 < x < \ell_1,\ 0 < y < \ell_2,\ t > 0), \tag{2.282}$$

$$u(x, y, z, 0) = f(x, y, z) \quad (0 < x < \ell_1,\ 0 < y < \ell_2,\ 0 < z < \ell_3). \tag{2.283}$$

Putting $u = XYZT$ into the differential equation here gives

$$k(X''(x)Y(y)Z(z)T(t) + X(x)Y''(y)Z(z)T(t) + X(x)Y(y)Z''(z)T(t))$$
$$= X(x)Y(y)Z(z)T'(t), \tag{2.284}$$

which will hold if $X'' = -\lambda X$, $Y'' = -\mu Y$, $Z'' = -\nu Z$, and $T' = -(\lambda+\mu+\nu)kT$. So we consider the equations

$$X''(x) + \lambda X(x) = 0 \quad (0 < x < \ell_1), \quad Y''(y) + \mu Y(y) = 0 \quad (0 < y < \ell_2),$$
$$Z''(z) + \nu Z(z) = 0 \quad (0 < z < \ell_3), \quad T'(t) + (\lambda + \mu + \nu)kT(t) = 0 \quad (t > 0). \tag{2.285}$$

The boundary conditions (2.280)–(2.282) imply

$$X(0) = X(\ell_1) = Y'(0) = Y'(\ell_2) = Z(0) = Z'(\ell_3) = 0. \tag{2.286}$$

We therefore have an eigenvalue problem in X and μ, one in Y and ν, and one in Z and ν. Each of these we've encountered before, perhaps in other contexts and with quantities having different names. And we've seen that they have the following eigenvalues and corresponding eigenfunctions:

$$\lambda = \left(\frac{\pi n_1}{\ell_1}\right)^2, \quad X(x) = \sin\sqrt{\lambda}\,x = \sin\frac{\pi n_1 x}{\ell_1} \quad (n_1 \in \mathbb{Z}^+), \tag{2.287}$$

$$\mu = \left(\frac{\pi n_2}{\ell_2}\right)^2, \quad Y(y) = \cos\sqrt{\mu}\,y = \cos\frac{\pi n_2 y}{\ell_2} \quad (n_2 \in \mathbb{N}), \tag{2.288}$$

$$\nu = \left(\frac{\pi(2n_3 - 1)}{2\ell_3}\right)^2, \quad Z(z) = \sin\sqrt{\nu}\,z = \sin\frac{\pi(2n_3 - 1)z}{2\ell_3} \quad (n_3 \in \mathbb{Z}^+). \tag{2.289}$$

(See, for instance, Examples 2.7.1, 2.3.1, and 2.3.2, respectively.) Also, our above equation for T has solution

$$T(t) = e^{-(\lambda+\mu+\nu)kt}. \tag{2.290}$$

152   FOURIER SERIES AND BOUNDARY VALUE PROBLEMS

So we get separated solutions $u = u_{n_1,n_2,n_3}$ given by

$$u_{n_1,n_2,n_3}(x,y,z,t) = \sin\frac{\pi n_1 x}{\ell_1} \cos\frac{\pi n_2 y}{\ell_2} \sin\frac{\pi(2n_3-1)z}{2\ell_3}$$
$$\cdot \exp\left(-\pi^2\left(\frac{n_1^2}{\ell_1^2} + \frac{n_2^2}{\ell_2^2} + \frac{(2n_3-1)^2}{4\ell_3^2}\right)kt\right), \qquad (2.291)$$

where $n_1 \in \mathbb{Z}^+$, $n_2 \in \mathbb{N}$, $n_3 \in \mathbb{Z}^+$.

We should be a bit careful about forming, from these, formal infinite series solutions u. Specifically, the cosine terms in y suggest that, when we impose our initial conditions, we'll bring Fourier cosine series into play. And we know that, in such series, a factor of $\frac{1}{2}$ appears with the constant term. Let's account for this. That is, let's write our formal solution as follows:

$$u(x,y,z,t) = \sum_{n_1=1}^{\infty}\sum_{n_2=0}^{\infty}\sum_{n_3=1}^{\infty}\left(1 - \frac{\delta_{0,n_2}}{2}\right)D_{n_1,n_2,n_3}\sin\frac{\pi n_1 x}{\ell_1}\cos\frac{\pi n_2 y}{\ell_2}$$
$$\cdot \sin\frac{\pi(2n_3-1)z}{2\ell_3}\exp\left(-\pi^2\left(\frac{n_1^2}{\ell_1^2} + \frac{n_2^2}{\ell_2^2} + \frac{(2n_3-1)^2}{4\ell_3^2}\right)kt\right).$$
$$(2.292)$$

The factor $1 - \delta_{0,n_2}/2$ is equal to 1 if $n_2 \neq 0$ and $\frac{1}{2}$ if $n_2 = 0$; this factor will allow for nice formulas for the D_{n_1,n_2,n_3}'s.

These formulas are now obtained by applying the initial conditions (2.283) and invoking Theorems 1.12.1 and 1.12.2. (Both series described by the former theorem are relevant here.) We leave the details to the reader; the result is:

Proposition 2.13.1 *The boundary value problem (2.279)–(2.283) has formal solution (2.292), where*

$$D_{n_1,n_2,n_3}$$
$$= \frac{8}{\ell_1\ell_2\ell_3}\int_0^{\ell_3}\int_0^{\ell_2}\int_0^{\ell_1} f(x,y,z)\sin\frac{\pi n_1 x}{\ell_1}\cos\frac{\pi n_2 y}{\ell_2}\sin\frac{\pi(2n_3-1)z}{2\ell_3}\,dx\,dy\,dz$$

for $n_1 \in \mathbb{Z}^+$, $n_2 \in \mathbb{N}$, $n_3 \in \mathbb{Z}^+$.

In each boundary value problem considered so far, the number of indices in our series solution has been 1 less than the number of independent variables present. Our final example shows that this needn't always be the case.

Example 2.13.2 Under certain circumstances and assumptions, electrostatic potential in a crystal can be modeled by a function $u(x,y,z)$ of the following sort: u has period P in each variable, P being the length of a "fundamental cube" of the crystal lattice, and $\nabla_3^2 u = -4\pi\rho(x,y,z)$, where ρ denotes *charge density* within the crystal.

Solve for u in such a situation, assuming the net charge within a fundamental cube to be zero.

Solution. We have the following boundary value problem:

$$u_{xx} + u_{yy} + u_{zz} = -4\pi\rho(x,y,z) \quad (0 < x,y,z < P), \quad (2.293)$$
$$u(0,y,z) = u(P,y,z), \quad u_x(0,y,z) = u_x(P,y,z) \quad (0 < y,z < P), \quad (2.294)$$
$$u(x,0,z) = u(x,P,z), \quad u_y(x,0,z) = u_y(x,P,z) \quad (0 < x,z < P), \quad (2.295)$$
$$u(x,y,0) = u(x,y,P), \quad u_z(x,y,0) = u_z(x,y,P) \quad (0 < x,y < P). \quad (2.296)$$

How might we solve this problem?

We might use variation of parameters, in the manner discussed at the beginning of Section 2.11. That is, we might first consider the corresponding homogeneous problem; separate variables there; solve any resulting eigenvalue problems; put together a series involving these obtained eigenfunctions; and then get this series to conform to the inhomogeneous differential equation.

Not only might we, but we should, because variation of parameters works particularly well when *all* boundary conditions are homogenous, as they are here. In such a case one may often, by separating variables and applying the corresponding homogeneous differential equation, arrive at an eigenvalue problem in *any* of the variables under consideration. One may then multiply together the resulting eigenfunctions and sum over corresponding eigenvalues to obtain a series whose nature in each variable is known; all that remains is to determine the coefficients of these series.

Let's see how this works in our present context. If we replace the differential equation (2.293) with the homogeneous equation $\nabla_3^2 u = 0$, then we may arrive, through separation of variables, at the following periodic eigenvalue problem in X:

$$X''(x) + \lambda X(x) = 0 \quad (0 < x < P), \quad X(0) = X(P), \quad X'(0) = X'(P). \quad (2.297)$$

This problem has solutions

$$X(x) = \gamma e^{in_1 x} + \delta e^{-in_1 x} \quad (n_1 \in \mathbb{N}). \quad (2.298)$$

One treats the variables y and z in the same way. So the variation-of-parameters method leads us to consideration of the series solutions

$$u(x,y,z) = \sum_{n_1=-\infty}^{\infty} \sum_{n_2=-\infty}^{\infty} \sum_{n_3=-\infty}^{\infty} D_{n_1,n_2,n_3} e^{2\pi i(n_1 x + n_2 y + n_3 z)/P} \quad (2.299)$$

to our boundary value problem (2.293)–(2.296). To determine the coefficients of this series, we do as we did in Section 2.11: We expand the inhomogeneity in our differential equation into a Fourier series of the appropriate form; plug this expansion, as well as the one for u, into the inhomogeneous differential equation; and match up summands to solve. That is, we write, here,

$$\rho(x,y,z) = \sum_{n_1=-\infty}^{\infty} \sum_{n_2=-\infty}^{\infty} \sum_{n_3=-\infty}^{\infty} c_{n_1,n_2,n_3}(\rho) e^{2\pi i(n_1 x + n_2 y + n_3 z)/P}, \quad (2.300)$$

where

$$c_{n_1,n_2,n_3}(\rho) = \frac{1}{P^3} \int_0^P \int_0^P \int_0^P \rho(x,y,z) e^{-2\pi i(n_1 x + n_2 y + n_3 z)/P} \, dx \, dy \, dz.$$

(2.301)

We plug the above series for u and for ρ into our differential equation (2.293); we get

$$-\frac{4\pi^2}{P^2}(n_1^2 + n_2^2 + n_3^2) D_{n_1,n_2,n_3} = -4\pi \, c_{n_1,n_2,n_3}(\rho).$$

(2.302)

We've assumed that the net charge in a fundamental cube is zero; this means

$$0 = \frac{1}{P^3} \int_0^P \int_0^P \int_0^P \rho(x,y,z) \, dx \, dy \, dz = c_{0,0,0}(\rho).$$

(2.303)

So (2.302) will hold for $n_1 = n_2 = n_3 = 0$ whatever our choice of $D_{0,0,0}$. (It's evident from equations (2.293)–(2.296) themselves that solutions to them are determined at most up to addition of a constant.) For any other triple (n_1, n_2, n_3), we divide (2.302) by $-4\pi^2(n_1^2 + n_2^2 + n_3^2)/P^2$ to determine D_{n_1,n_2,n_3}.

So we have:

Proposition 2.13.2 *The boundary value problem (2.293)–(2.296) has formal solution (2.299), where $D_{0,0,0}$ is arbitrary, while for n_1, n_2, n_3 not all zero,*

$$D_{n_1,n_2,n_3} = \frac{P^2}{\pi(n_1^2 + n_2^2 + n_3^2)} c_{n_1,n_2,n_3}(\rho) = \frac{1}{P\pi(n_1^2 + n_2^2 + n_3^2)}$$

$$\cdot \int_0^P \int_0^P \int_0^P \rho(x,y,z) e^{-2\pi i(n_1 x + n_2 y + n_3 z)/P} \, dx \, dy \, dz.$$

Exercises

2.13.1 Solve the problem of Example 2.13.1 with the condition $u_y(x, \ell_2, z, t) = 0$ there replaced by $u_y(x, \ell_2, z, t) = z$. (Recall Example 2.10.2 and Exercise 2.10.3.)

2.13.2 Solve the problem of Example 2.13.1 with Dirichlet conditions (that is, temperature equal to zero) on all faces, and the homogeneous differential equation replaced by $k(u_{xx} + u_{yy} + u_{zz}) = u_t - A$. (Recall Exercise 2.10.2.)

2.13.3 Consider the potential problem of Example 2.13.2 in the case where $P = 2\pi$, $\rho(x,y,z) = \cos x \cos 2y + \sin y \cos 3z$.
    a. Show that the net charge in the fundamental cube *is* zero.
    b. Solve the problem; write your answer out in terms of cosines and sines.
    c. Verify by direct substitution into equations (2.293)–(2.296) that your function u really does solve the problem.

3

L^2 Spaces: Optimal Contexts for Fourier Series

In this chapter, we study *norm convergence of functions* and *vector spaces of square integrable functions*. We demonstrate that these notions provide an especially elegant and fruitful framework in which to study not only frequency, but also other, more general decompositions.

3.1 THE MEAN SQUARE NORM AND THE INNER PRODUCT ON $C(\mathbb{T})$

Let f be a function on $\mathbb{T}$. In Section 1.7, the issue of convergence of the Fourier series for f was phrased in terms of the quantity

$$\lim_{N\to\infty} \sup_{x\in\mathbb{T}} |S_N^f(x) - f(x)|. \tag{3.1}$$

Here's a justification for this choice of phrasing: The quantity $\sup_{x\in\mathbb{T}} |f(x)|$ is a useful, intuitively plausible indicator of the "magnitude" of a function f on $\mathbb{T}$. So the quantity $\sup_{x\in\mathbb{T}} |f(x) - g(x)|$ is a useful, intuitively plausible indicator of the *distance* between functions f and g on $\mathbb{T}$. So the quantity (3.1) is a useful, intuitively plausible indicator of whether *the distance between S_N^f and f is approaching zero as $N \to \infty$*.

All of which suggests that, should we produce a *more* useful, intuitively plausible indicator of the magnitude of a function, we'd also have a more useful, intuitively plausible way of discussing distance between functions and in turn a more useful, intuitively plausible way of discussing convergence of Fourier series.

The following definition affords us all three of these things.

Definition 3.1.1 Denote by $C(\mathbb{T})$ the space of continuous functions on $\mathbb{T}$.

(a) The *mean square magnitude*, or *mean square norm*, or simply *norm*, denoted $||f||$ (or sometimes $||f||_2$), of $f \in C(\mathbb{T})$ is given by

$$||f|| = \left(\int_{\mathbb{T}} |f(x)|^2 \, dx \right)^{1/2}. \tag{3.2}$$

(Here and forever after, "$\int_{\mathbb{T}}$" denotes integration over any interval of length 2π.)

(b) By *the mean square distance*, or simply the *distance*, between functions $f, g \in C(\mathbb{T})$, we mean the quantity

$$||f - g|| = \left(\int_{\mathbb{T}} |f(x) - g(x)|^2 \, dx \right)^{1/2}. \tag{3.3}$$

(c) If $f, f_1, f_2, \ldots \in C(\mathbb{T})$ satisfy

$$\lim_{N \to \infty} ||f_N - f|| = 0, \tag{3.4}$$

then we say the sequence $f_1, f_2, \ldots$ *converges to* f, or *the f_N's converge to* f, *in mean square norm*, or simply *in norm*.

Why is the above so intuitively plausible? Because the given prescription for $||f||$ is entirely analogous to the definition

$$||a|| = \left(\sum_{m=1}^{M} |a_m|^2 \right)^{1/2} \tag{3.5}$$

of the length, or norm, $||a||$ of a vector $a = (a_1, a_2, \ldots, a_M)$ in M-dimensional real Euclidean space $\mathbb{R}^M$ or complex Euclidean space $\mathbb{C}^M$. (In the former case, the absolute values on the right side of (3.5) are redundant.) Or to put it another way, using $\sup_{x \in \mathbb{T}} |f(x)|$ to measure a function f on $\mathbb{T}$ is like using $\sup_{1 \leq m \leq M} |a_m|$ to measure a vector $(a_1, a_2, \ldots, a_M)$ in $\mathbb{R}^M$ or $\mathbb{C}^M$. One doesn't, typically, do the latter; we won't, generally (from now on), do the former.

Consideration of the *utility* of $||f||$, as an indicator of the size of f, will occupy much of our attention in this chapter (and thereafter). It is, in fact, our intention to show that use of this indicator leads to an exceptionally clean, complete, symmetric, aesthetically pleasing, convenient, and useful theory of frequency and other decompositions.

We begin with some basic properties of the norm.

Proposition 3.1.1 *Let $f \in C(\mathbb{T})$. Then:*

(a) $||f|| \geq 0$; *moreover*, $||f|| = 0$ *if and only if* $f(x) = 0$ *for all* $x \in \mathbb{T}$.

(b) $||\alpha f|| = |\alpha|\, ||f||$ *for any* $\alpha \in \mathbb{C}$.

Proof. It's clear from Definition 3.1.1(a) that $||f|| \geq 0$ always. Further, the Riemann integral of a continuous, nonnegative function over any interval of positive length is zero if and only if that function is identically zero on that interval. So part (a) of our proposition is proved.

Part (b) is demonstrated as follows: We have

$$||\alpha f||^2 = \int_\mathbb{T} |\alpha f(x)|^2\, dx = |\alpha|^2 \int_\mathbb{T} |f(x)|^2\, dx = |\alpha|^2 ||f||^2 \qquad (3.6)$$

for f and α as specified. Taking square roots gives the desired result. □

We make several observations regarding norm versus uniform and norm versus pointwise convergence. First:

Proposition 3.1.2 *Let $f_1, f_2, \ldots$ be a sequence of functions in $C(\mathbb{T})$; also let $f \in C(\mathbb{T})$. If this sequence converges uniformly to f, then it converges in norm to f.*

Proof. By basic properties of the Riemann integral,

$$||f_N - f||^2 = \int_\mathbb{T} |f_N(x) - f(x)|^2\, dx \leq \sup_{x \in \mathbb{T}} |f_N(x) - f(x)|^2 \int_\mathbb{T} dx \qquad (3.7)$$

$$= 2\pi \left(\sup_{x \in \mathbb{T}} |f_N(x) - f(x)| \right)^2,$$

the last step because $\mathbb{T}$ has length 2π, and because "the sup of the square equals the square of the sup." If the f_N's converge uniformly to f, then the right side of the above goes to zero as $N \to \infty$. By the squeeze law, then, so does $||f_N - f||^2$, and therefore so does $||f_N - f||$. So the f_N's converge in norm to f, as required. □

On the other hand, *pointwise* convergence does not, in general, imply norm convergence:

Example 3.1.1 For $N \in \mathbb{Z}^+$, let f_N be the function in $C(\mathbb{T})$ defined by

$$f_N(x) = N\left(\frac{x}{2\pi}\right)^N \sqrt{2\pi - x} \quad (0 \leq x \leq 2\pi) \qquad (3.8)$$

(Fig. 3.1). Also let $f(x) = 0$ for all $x \in [0, 2\pi]$. Show that the f_N's converge pointwise to f but do not converge to f in norm.

Solution. Let's first take care of the cases $x = 0$ and $x = 2\pi$, which are easy: $f_N(0) = f_N(2\pi) = 0 \to 0 = f(0) = f(2\pi)$ as $N \to \infty$, as required.

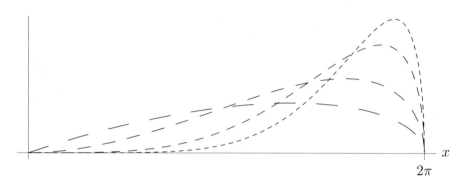

Fig. 3.1 A few f_N's (shorter dashes connote larger N)

Next, for any fixed element x of $(0, 2\pi)$, we use l'Hôpital's rule as follows:

$$\lim_{N \to \infty} f_N(x) = \lim_{N \to \infty} N \left(\frac{x}{2\pi}\right)^N \sqrt{2\pi - x} = \sqrt{2\pi - x} \lim_{N \to \infty} \frac{N}{(2\pi/x)^N}$$

$$= \sqrt{2\pi - x} \lim_{N \to \infty} \frac{1}{(2\pi/x)^N \ln(2\pi/x)}$$

$$= \frac{\sqrt{2\pi - x}}{\ln(2\pi/x)} \lim_{N \to \infty} \left(\frac{x}{2\pi}\right)^N = 0. \qquad (3.9)$$

(The limit on the right is zero because $0 < x < 2\pi$.) So the f_N's converge pointwise to f on $[0, 2\pi]$, as required.

On the other hand,

$$\lim_{N \to \infty} \|f_N - f\|^2 = \lim_{N \to \infty} \int_0^{2\pi} |f_N(x) - f(x)|^2 \, dx$$

$$= \lim_{N \to \infty} N^2 \int_0^{2\pi} \left(\frac{x}{2\pi}\right)^{2N} (2\pi - x) \, dx$$

$$= 4\pi^2 \lim_{N \to \infty} N^2 \int_0^1 u^{2N}(1-u) \, dx$$

$$= 4\pi^2 \lim_{N \to \infty} N^2 \left[\frac{u^{2N+1}}{2N+1} - \frac{u^{2N+2}}{2N+2}\right]_0^1$$

$$= 4\pi^2 \lim_{N \to \infty} N^2 \frac{1}{(2N+1)(2N+2)} = \pi^2 \neq 0 \qquad (3.10)$$

(we substituted $x = 2\pi u$), so $\|f_N - f\| \not\to 0$, so the f_N's *do not* converge to f in norm. (Even though $\|f_N - f\|$ *does* converge to something!)

Nor does norm convergence imply pointwise convergence. For a rather extreme example of this, see Figure 3.2: Here one has a sequence of "plateau" functions g_N that become more and more narrow, so that their norms approach zero as $N \to \infty$. So the g_N's converge in norm to the zero function. Yet the plateaus keep sliding along $\mathbb{T}$ in such a way that, for any given $x \in \mathbb{T}$, $g_N(x) = 1$ for infinitely many N. So

$\lim_{N\to\infty} g_N(x)$ does not equal zero for *any* x. In fact, since $g_N(x)$ is, for any $x \in \mathbb{T}$, also equal to 0 for infinitely many N, $\lim_{N\to\infty} g_N(x)$ *does not exist* for any x!

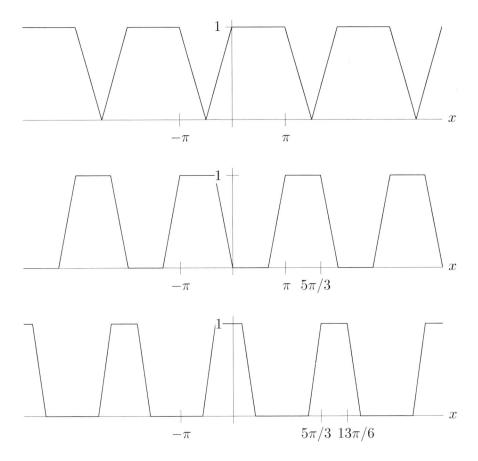

Fig. 3.2 The g_N's converge in norm, but pointwise nowhere (top: g_1; middle: g_2; bottom: g_3)

(The reader is encouraged to flesh out the details of this example. The key idea is that the sum of the reciprocals of the positive integers diverges. Consequently, any given $x \in [-\pi, \pi]$ will find a plateau directly above it infinitely often, so that $g_N(x)$ will equal 1 for infinitely many $N \in \mathbb{Z}^+$. For similar reasons, $g_N(x)$ will, for any given x, vanish infinitely often.)

In the next section, we'll reexamine relationships among various modes of convergence. For the present, we wish to leave aside questions of convergence and discuss an entity closely associated with the norm $||f||$. We have:

Definition 3.1.2 The *inner product* of two functions $f, g \in C(\mathbb{T})$ is the complex number $\langle f, g \rangle$ given by

$$\langle f, g \rangle = \int_{\mathbb{T}} f(x)\overline{g(x)}\, dx. \tag{3.11}$$

We now document some basic properties of this inner product.

Proposition 3.1.3 (*Throughout, $f, g, h \in C(\mathbb{T})$.*)

(a) *The inner product on $C(\mathbb{T})$ is:*

  (i) conjugate symmetric, *meaning*

  $$\overline{\langle f, g \rangle} = \langle g, f \rangle; \tag{3.12}$$

  (ii) linear in the first variable, *meaning*

  $$\langle \alpha f + \beta g, h \rangle = \alpha \langle f, h \rangle + \beta \langle g, h \rangle \quad \text{for } \alpha, \beta \in \mathbb{C}; \tag{3.13}$$

  (iii) conjugate linear in the second variable, *meaning*

  $$\langle f, \alpha g + \beta h \rangle = \overline{\alpha} \langle f, g \rangle + \overline{\beta} \langle f, h \rangle \quad \text{for } \alpha, \beta \in \mathbb{C}. \tag{3.14}$$

(b) *The norm on $C(\mathbb{T})$ satisfies:*

  (i) $||f|| = \langle f, f \rangle^{1/2}$;

  (ii) *the Cauchy-Schwarz inequality*

  $$|\langle f, g \rangle| \leq ||f|| \cdot ||g||; \tag{3.15}$$

  (iii) *the triangle inequality*

  $$||f + g|| \leq ||f|| + ||g||; \tag{3.16}$$

  (iv) *the Pythagorean theorem: if $\langle f, g \rangle = 0$, then*

  $$||f + g||^2 = ||f||^2 + ||g||^2. \tag{3.17}$$

Proof. Part (a)(i) goes like this:

$$\overline{\langle f, g \rangle} = \overline{\int_{\mathbb{T}} f(x)\overline{g(x)}\, dx} = \int_{\mathbb{T}} \overline{f(x)}\left(\overline{\overline{g(x)}}\right) dx = \int_{\mathbb{T}} g(x)\overline{f(x)}\, dx = \langle g, f \rangle. \tag{3.18}$$

For part (a)(ii), we have

$$\langle \alpha f + \beta g, h \rangle = \int_{\mathbb{T}} (\alpha f(x) + \beta g(x))\overline{h(x)}\, dx$$
$$= \alpha \int_{\mathbb{T}} f(x)\overline{h(x)}\, dx + \beta \int_{\mathbb{T}} g(x)\overline{h(x)}\, dx = \alpha \langle f, h \rangle + \beta \langle g, h \rangle. \quad (3.19)$$

Part (a)(iii) may be proved similarly or, as is at least as much fun, may be deduced from parts (i) and (ii), as follows:

$$\langle f, \alpha g + \beta h \rangle = \overline{\langle \alpha g + \beta h, f \rangle} = \overline{\alpha \langle g, f \rangle + \beta \langle h, f \rangle}$$
$$= \overline{\alpha}\, \overline{\langle g, f \rangle} + \overline{\beta}\, \overline{\langle h, f \rangle} = \overline{\alpha} \langle f, g \rangle + \overline{\beta} \langle f, h \rangle. \quad (3.20)$$

Part (b)(i) is because

$$\langle f, f \rangle = \int_{\mathbb{T}} f(x)\overline{f(x)}\, dx = \int_{\mathbb{T}} |f(x)|^2\, dx = ||f||^2. \quad (3.21)$$

We prove part (b)(ii) as follows. In the case $||g|| = 0$, the function g is identically zero by Proposition 3.1.1(a), so $f\,\overline{g}$ is too, and therefore so is $\langle f, g \rangle$. So the Cauchy-Schwarz inequality certainly holds in this case. Now let's assume $||g|| \neq 0$. For any complex number γ, we have

$$||f + \gamma g||^2 = \langle f + \gamma g, f + \gamma g \rangle = \langle f, f + \gamma g \rangle + \gamma \langle g, f + \gamma g \rangle$$
$$= \langle f, f \rangle + \overline{\gamma} \langle f, g \rangle + \gamma \langle g, f \rangle + \gamma \overline{\gamma} \langle g, g \rangle$$
$$= ||f||^2 + \left(\overline{\gamma} \langle f, g \rangle + \gamma \overline{\langle f, g \rangle} \right) + |\gamma|^2 ||g||^2$$
$$= ||f||^2 + 2\operatorname{Re}(\overline{\gamma} \langle f, g \rangle) + |\gamma|^2 ||g||^2 \quad (3.22)$$

(we've used parts of the present proposition and parts of Proposition A.0.3). Since the left side of (3.22) is nonnegative, we conclude that

$$-2\operatorname{Re}(\overline{\gamma} \langle f, g \rangle) - |\gamma|^2 ||g||^2 \leq ||f||^2. \quad (3.23)$$

Moreover, since $||g|| \neq 0$, we may put $\gamma = -\langle f, g \rangle / ||g||^2$ into (3.23). We get

$$2\frac{|\langle f, g \rangle|^2}{||g||^2} - \frac{|\langle f, g \rangle|^2}{||g||^4}||g||^2 \leq ||f||^2 \quad (3.24)$$

or, after some algebra,

$$|\langle f, g \rangle| \leq ||f||\, ||g||, \quad (3.25)$$

as required.

To get the triangle inequality, we put $\gamma = 1$ into (3.22). The result is

$$||f + g||^2 = ||f||^2 + 2\operatorname{Re}\langle f, g \rangle + ||g||^2 \leq ||f||^2 + 2|\langle f, g \rangle| + ||g||^2, \quad (3.26)$$

whence, by the Cauchy-Schwarz inequality,

$$||f+g||^2 \leq ||f||^2 + 2||f|| \cdot ||g|| + ||g||^2 = (||f|| + ||g||)^2. \qquad (3.27)$$

Taking square roots gives the desired result.

Finally, the Pythagorean theorem is just the equality given in (3.26), in the case $\langle f, g \rangle = 0$. □

The above proposition can perhaps be better understood, at an intuitive level, by considering its analog in the real vector space $\mathbb{R}^2$. There, we have an inner product

$$\langle a, b \rangle = a_1 b_1 + a_2 b_2 \qquad (3.28)$$

for vectors $a = (a_1, a_2)$ and $b = (b_1, b_2)$, and a norm $||a|| = \sqrt{\langle a, a \rangle}$. Moreover, it may be shown that

$$\langle a, b \rangle = ||a|| \cdot ||b|| \cos \theta_{a,b}, \qquad (3.29)$$

where $\theta_{a,b}$ is the angle between a and b. Note in particular that, by (3.29), $a, b \in \mathbb{R}^2$ are perpendicular if and only if $\langle a, b \rangle = 0$.

If we replace, in Proposition 3.1.3, every $C(\mathbb{T})$ function with a point in $\mathbb{R}^2$ and assume α, β to be real, then all parts of that proposition continue to hold. In particular, the Cauchy-Schwarz inequality in $\mathbb{R}^2$ follows immediately from (3.29); the triangle inequality there is just Proposition A.0.3(c) (if we identify $\mathbb{R}^2$ with $\mathbb{C}$, whereby the norm of $(a_1, a_2) \in \mathbb{R}^2$ is just $|a_1 + ia_2|$), and the Pythagorean theorem there is the one familiar from elementary geometry.

The inner product on $C(\mathbb{T})$, and on analogous vector spaces to be defined shortly, will prove instrumental throughout the remainder of our studies. The relationship between this inner product and the mean square norm is one of the main advantages of the latter as a measure of magnitude of functions.

Exercises

3.1.1 For $N = 1, 2, 3, \ldots$, let $f_N \in C(\mathbb{T})$ be defined by

$$f_N(x) = \begin{cases} \sqrt{N}(1 - Nx) & \text{if } 0 < x \leq \dfrac{1}{N}, \\ 0 & \text{if } -\pi < x \leq 0 \text{ or } \dfrac{1}{N} < x \leq \pi. \end{cases}$$

Also let $f(x) = 0$ for all x.

    a. Sketch, on the same set of axes, the graphs of a couple of f_N's.

    b. Show that f_N converges pointwise to f on $(-\pi, \pi]$ (and therefore on $\mathbb{R}$). Hint: The case $x \in (-\pi, 0]$ is easy. For $x \in (0, \pi]$, note that $1/N$ is eventually (meaning for N large enough) smaller than x.

    c. What is

$$\lim_{N \to \infty} ||f_N - f||?$$

Does f_N converge in norm to f?

d. Show that f_N does not converge uniformly to f on $\mathbb{R}$.

3.1.2 Show that the Cauchy-Schwarz inequality (3.15) is an equality if and only if f is a constant multiple of g. Hint: The "if" part is straightforward. For the "only if" part, first suppose $||g|| = 0$; this case is also straightforward. For the case $||g|| \neq 0$, put $\gamma = -\langle f, g\rangle / ||g||^2$ into the right side of (3.22); then put the assumption $|\langle f, g\rangle| = ||f|| \, ||g||$ into the result and consider the left side of (3.22).

3.1.3 Show that, if $f, g \in C(\mathbb{T})$, then

$$\big| \, ||f|| - ||g|| \, \big| \leq ||f - g||.$$

Hint: First consider the case where $||f|| \geq ||g||$; write $f = (f - g) + g$ and apply the triangle inequality. Then consider the case $||f|| \leq ||g||$ and make a similar argument.

3.1.4 Using the previous exercise show that, if f_N converges to f in the norm on $C(\mathbb{T})$, then $||f_N|| \to ||f||$. Is the converse true?

3.2 THE VECTOR SPACE $L^2(\mathbb{T})$

(Much of the material in this section derives from the edifying and highly readable treatment of integration and related matters in [31], Appendix A.)

Whenever one is studying convergence, one hopes to be working in a space that is *complete*. By this we mean, informally speaking, a space in which any sequence that looks like it *should* converge to an element of that space actually *does* converge to an element of that space. Unfortunately, $C(\mathbb{T})$ is *not* complete with respect to norm convergence.

Let's formalize this: To do so, we recall some notions from general topology and real analysis. First, a *pseudometric* on a set X is a function d that associates, to each pair x, y of elements of X, a nonnegative real number $d(x, y)$ such that (i) $x = y \Rightarrow d(x, y) = 0$, (ii) $d(x, y) = d(y, x)$, and (iii) $d(x, z) \leq d(x, y) + d(y, z)$ for all $z \in X$. If a pseudometric d exists on X, then X is called a *pseudometric space* under d, or simply (when d is understood) a *pseudometric space*.

If, in the above situation, d also satisfies the condition

$$d(x, y) = 0 \Rightarrow x = y, \tag{3.30}$$

then d is called a *metric* on X and X a *metric space* (under d).

Next, a *Cauchy sequence* in a pseudometric space X is a sequence $x_1, x_2, \ldots$ of elements of X such that

$$\lim_{N \to \infty} \sup_{M > N} d(x_N, x_M) = 0. \tag{3.31}$$

(Throughout, we'll use $\sup_{M>N}$ to denote $\sup_{M \in \{N+1, N+2, N+3, \ldots\}}$.) And we say X is *complete* (with respect to d) if every Cauchy sequence in X converges to an

164 L^2 SPACES: OPTIMAL CONTEXTS FOR FOURIER SERIES

element of X. In other words, X is by definition complete if, for any sequence $x_1, x_2, \ldots$ in X, (3.31) implies the existence of an $x \in X$ such that

$$\lim_{N \to \infty} d(x_N, x) = 0. \tag{3.32}$$

Otherwise, X is called *incomplete*.

Intuitively, the terms in a Cauchy sequence all bunch up together as you go farther out in that sequence. (Indeed, (3.31) says that you can make the Nth term x_N as close as you want to all subsequent terms if you choose N large enough.) So a complete pseudometric space is, intuitively, one in which terms that *bunch up against each other* will necessarily *bunch up against some fixed element of the space*. Or to put it another way, a complete pseudometric space has no holes!

One shows (see Exercise 3.2.2) that $C(\mathbb{T})$ is a metric space under the metric $d(f, g) = ||f - g||$. But it's not complete! To see this, consider the sequence $f_1, f_2, \ldots$ in $C(\mathbb{T})$ defined as follows:

$$f_N(x) = \begin{cases} N^{1/3} & \text{if } |x| \leq N^{-1}, \\ |x|^{-1/3} & \text{if } N^{-1} < |x| \leq \pi. \end{cases} \tag{3.33}$$

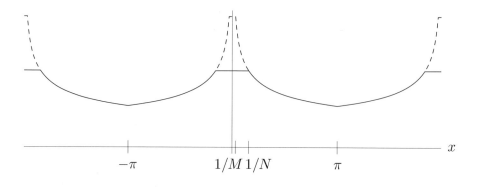

Fig. 3.3 f_N (solid) and f_M (dashed) for $M > N$

Note (see Fig. 3.3) that, if $M > N$, then $|f_N(x) - f_M(x)|$ is less than $M^{1/3}$ for $|x| \leq M^{-1}$, is less than $|x^{-1/3}|$ for $M^{-1} \leq |x| \leq N^{-1}$, and equals zero elsewhere on $[-\pi, \pi]$. So, since all f_N's are even functions on $[-\pi, \pi]$,

$$||f_N - f_M||^2 = 2 \int_0^\pi |f_N(x) - f_M(x)|^2 \, dx$$
$$< 2 \left(\int_0^{M^{-1}} M^{2/3} \, dx + \int_{M^{-1}}^{N^{-1}} x^{-2/3} \, dx \right)$$
$$= 2(M^{-1/3} + 3(N^{-1/3} - M^{-1/3})) = 6N^{-1/3} - 4M^{-1/3}. \tag{3.34}$$

The least upper bound over all $M > N$ of the right side equals $6N^{-1/3}$, which indeed goes to zero as $N \to \infty$. So the corresponding least upper bound of the left side does too, and thus the f_N's form a Cauchy sequence in $C(\mathbb{T})$.

But this sequence does not converge to anything in $C(\mathbb{T})$, for the following reason. Let $f \colon \mathbb{T} \to \mathbb{C}$ be defined by

$$f(x) = |x|^{-1/3} \quad (0 < |x| \leq \pi) \tag{3.35}$$

(we may define $f(0)$ any way we like, or not at all). Then $f \notin C(\mathbb{T})$; in fact, $f \notin PC(\mathbb{T})$, since $\lim_{x \to 0} f(x) = \infty$. On the other hand, one shows (see Exercise 3.2.3) that the improper Riemann integral of $|f_N(x) - f(x)|^2$ over $\mathbb{T}$ exists for all $N \in \mathbb{Z}^+$ and approaches zero as $N \to \infty$. So the f_N's converge in norm to a function $f \notin C(\mathbb{T})$ and consequently cannot converge in norm to anything *in* $C(\mathbb{T})$. (The argument embodied by this last sentence is a bit incomplete (pun intended); we'll complete it at the end of this section. But the basic idea is correct.)

Let's not be discouraged, though. While the above example does highlight a problem with $C(\mathbb{T})$, it also suggests a *solution* to this problem! Here's what we mean. The function f, which "fills in the hole at the end of the sequence of f_N's," may be obtained directly from those f_N's by taking their *pointwise* limit. That is (as is readily checked),

$$f(x) = \lim_{N \to \infty} f_N(x) \tag{3.36}$$

for all real numbers x except the integer multiples of 2π. So why not fill in *all* holes in $C(\mathbb{T})$ by augmenting this space with all pointwise limits, defined wherever they converge, of Cauchy sequences therein?

Why not? Because, as we saw in the previous section, there exist Cauchy sequences $g_1, g_2, \ldots$ in $C(\mathbb{T})$ such that the $g_N(x)$'s fails to converge for *any* $x \in \mathbb{T}$!

This seems at first like a major difficulty, but there is a way to resolve it while retaining the merit of the original idea. This is accomplished essentially by *restricting our attention* to certain Cauchy sequences in $C(\mathbb{T})$ that converge pointwise "almost everywhere." Let's make this precise.

Definition 3.2.1 (a) Let $\varepsilon > 0$. A subset S of $\mathbb{R}$ is said to be *ε-small* if there is a collection $\{I_k\}_{k=1}^{\infty}$ of open intervals in $\mathbb{R}$ whose union contains S and the sum of whose lengths is at most ε.

(b) A subset S of $\mathbb{R}$ is called *negligible* if S is ε-small for every positive number ε.

(c) Let $A(x)$ be a statement regarding a real variable x; let I be either an interval or a torus. If the set $\{x \in I : A(x) \text{ is false}\}$ is negligible, then we say $A(x)$ is true *almost everywhere on I* or *for almost all x in I*. (Technically, by "almost everywhere on $\mathbb{T}_P$" we mean "almost everywhere on any interval of length P.")

We sometimes drop the "on I" or the "in I" when I is understood.

(d) A function f defined almost everywhere on I is called a *realization* of a sequence $f_1, f_2, \ldots$ of functions defined almost everywhere on I if $f(x) = \lim_{N \to \infty} f_N(x)$ almost everywhere.

If S is ε-small then, clearly, it is ε'-small for any $\varepsilon' \geq \varepsilon$.

We think of a negligible set (also called a *set of measure zero*) as being "very small." Indeed, according to the definition, a negligible set is one that may be *completely covered* by a collection of open intervals $I_1, I_2, \ldots$ the sum of whose lengths is as small as is desired.

But just how small is very small? Well, any finite subset of $\mathbb{R}$ is negligible, as is any countably infinite one (meaning, as the reader may recall, one whose elements may be put into one-to-one correspondence with the positive integers). See Exercise 3.2.1. So is any subset of a negligible set. One also checks that any countable — that is, finite or countably infinite — union of negligible sets is negligible. (Note, then, that "almost everywhere on $\mathbb{T}_P$" is equivalent to "almost everywhere on $\mathbb{R}$," since for any $P > 0$, $\mathbb{R}$ is a countable union of intervals of length P.)

On the other hand, one shows that no interval of positive length is negligible. As intervals of positive length are perhaps the most basic and familiar examples of uncountable sets, we might conjecture that negligible subsets of $\mathbb{R}$ must be countable. We might, but we'd be wrong: A counterexample is the famous "Cantor set," due to Cantor, hence the name. See Example 4, p. 53, in [29]. The behavior of this set should serve as a warning that, with respect to negligible sets, intuition will take us only so far. On the other hand, we shouldn't let this warning deter us from imagining, intuitively, that such sets *are* very small.

We are now just about ready to fill in the holes in $C(\mathbb{T})$. After the following lemma, we *will be* ready.

Lemma 3.2.1 *Suppose the sequence $h_1, h_2, \ldots$ in $C(\mathbb{T})$ is ultraCauchy, meaning, for some constant $c > 0$,*

$$\|h_k - h_{k+1}\| \leq \frac{c}{k^4} \qquad (3.37)$$

for each $k \in \mathbb{Z}^+$. Then this sequence is Cauchy; moreover, for any $\varepsilon > 0$, it converges uniformly except on an ε-small subset of $\mathbb{R}$. In particular, it has a realization.

Proof. Let the h_k's be as stated. Then for $M > N$, we have

$$h_N - h_M = (h_N - h_{N+1}) + (h_{N+1} - h_{N+2}) + \cdots + (h_{M-1} - h_M)$$
$$= \sum_{k=N}^{M-1} (h_k - h_{k+1}). \qquad (3.38)$$

So by the triangle inequality (3.16),

$$\sup_{M>N} \|h_N - h_M\| \leq \sup_{M>N} \sum_{k=N}^{M-1} \|h_k - h_{k+1}\| = \sum_{k=N}^{\infty} \|h_k - h_{k+1}\| \leq \sum_{k=N}^{\infty} \frac{c}{k^4}. \qquad (3.39)$$

THE VECTOR SPACE $L^2(\mathbb{T})$ 167

The right side, being the tail end of a convergent series, goes to zero as $N \to \infty$; so the left side does too, and the h_k's therefore form a Cauchy sequence.

We now claim that the set

$$G_k = \{x \in \mathbb{T} : |h_k(x) - h_{k+1}(x)| \geq k^{-2}\} \qquad (3.40)$$

is $(\sqrt{2\pi}\, c/k^2)$-small for each $k \in \mathbb{Z}^+$. We do so by invoking *Tchebyschev's inequality for continuous functions* (see (9) and the surrounding discussions in Section A.2 of [31]), which says: If g is continuous and nonnegative on an interval I, then the set $\{x \in I : g(x) \geq a\}$ is δ-small, where δ equals a^{-1} times the integral of g over I. But this and the definition of G_k mean the latter is δ-small, where

$$\delta = (k^{-2})^{-1} \int_{\mathbb{T}} |h_k(x) - h_{k+1}(x)|\, dx \leq k^2 \|h_k - h_{k+1}\| \cdot \|1\|$$

$$\leq k^2 \cdot \frac{c}{k^4} \cdot \sqrt{2\pi} = \frac{\sqrt{2\pi}\, c}{k^2} \qquad (3.41)$$

(the first inequality by the Cauchy-Schwarz inequality (3.15) with $f = |h_k - h_{k+1}|$ and $g = 1$), as required.

Finallly, we use the claim just verified to establish the desired uniform and pointwise convergence properties of the h_M's, as follows. Let $\varepsilon > 0$, and pick $N \in \mathbb{Z}^+$ such that

$$\sqrt{2\pi}\, c \sum_{k=N}^{\infty} \frac{1}{k^2} < \varepsilon; \qquad (3.42)$$

such N exists because $\sum_{k=1}^{\infty} k^{-2}$ is convergent, and therefore its tail ends get small. Then $H_N = \cup_{k=N}^{\infty} G_k$ is ε-small. Moreover, suppose $x \in \mathbb{T} - H_N$; then $x \in \mathbb{T} - G_k$ for all $k \geq N$, so by definition of the G_k's,

$$\sum_{k=N}^{\infty} |h_k(x) - h_{k+1}(x)| < \sum_{k=N}^{\infty} \frac{1}{k^2} < \infty. \qquad (3.43)$$

That is, the series on the left converges; therefore (since absolute convergence implies convergence), the limit

$$\lim_{M \to \infty} h_M(x) = -h_N(x) + \lim_{M \to \infty} \sum_{k=N}^{M-1} (h_k(x) - h_{k+1}(x))$$

$$= -h_N(x) + \sum_{k=N}^{\infty} (h_k(x) - h_{k+1}(x)) \qquad (3.44)$$

(we've used (3.38) again) exists on $\mathbb{T} - H_N$. So the $h_M(x)$'s have a finite limit on $\mathbb{T} - H_N$. Since H_N is ε-small and ε is arbitrary, the h_M's have a realization.

Finally, if we denote this realization by h, then again for $x \in \mathbb{T} - H_N$, we have

$$|h_M(x) - h(x)| = \left|\sum_{k=M}^{\infty}(h_k(x) - h_{k+1}(x))\right| \le \sum_{k=M}^{\infty}|h_k(x) - h_{k+1}(x)|$$
$$< \sum_{k=M}^{\infty}\frac{1}{k^2} \qquad (3.45)$$

as long as $M \ge N$. The right side, being the tail end of a convergent series, goes to zero as $M \to \infty$. But then the sup over $x \in \mathbb{T} - H_N$ of the left side does too, which gives us the stated uniform convergence of the h_M's. $\square$

Now we're ready for:

Proposition 3.2.1 (a) *Given a realization f of an ultraCauchy sequence $f_1, f_2, \ldots$ in $C(\mathbb{T})$, define the norm $||f||$ of f by*

$$||f|| = \lim_{N \to \infty}||f_N||. \qquad (3.46)$$

This limit exists and is well defined: If f is a realization of another ultra-Cauchy sequence $g_1, g_2, \ldots$ in $C(\mathbb{T})$, then the right side of (3.46) equals $\lim_{N \to \infty}||g_N||$.

(b) *Denote by $L^2(\mathbb{T})$ the collection of all realizations of ultraCauchy sequences in $C(\mathbb{T})$. Then $C(\mathbb{T}) \subset L^2(\mathbb{T})$; moreover, $L^2(\mathbb{T})$ is a complex vector space, closed under complex conjugation and absolute values.*

(c) *For $f \in L^2(\mathbb{T})$, $||f|| = 0$ if and only if $f(x) = 0$ almost everywhere.*

(d) *The formula $\rho(f, g) = ||f - g||$ defines a pseudometric on $L^2(\mathbb{T})$. With respect to ρ, $C(\mathbb{T})$ is dense in $L^2(\mathbb{T})$, meaning: Given $f \in L^2(\mathbb{T})$, there is a sequence $f_1, f_2, \ldots$ in $C(\mathbb{T})$ converging to f in norm (in the sense of (3.4)). In fact we may take, for this sequence, any Cauchy sequence in $C(\mathbb{T})$ of which f is a realization.*

(e) $PSC(\mathbb{T})$ *is also dense in $L^2(\mathbb{T})$.*

(f) *Suppose a sequence $f_1, f_2, \ldots$ in $L^2(\mathbb{T})$ converges in norm to $f \in L^2(\mathbb{T})$. Then this sequence converges in norm to a function $g \in L^2(\mathbb{T})$ if and only if $f(x) = g(x)$ almost everywhere.*

(g) $L^2(\mathbb{T})$ *is complete with respect to ρ. More specifically, suppose $f_1, f_2, \ldots$ is a Cauchy sequence in $L^2(\mathbb{T})$. Then some subsequence of this sequence has a realization $f \in L^2(\mathbb{T})$, and the original sequence of f_N's converges in norm to f.*

(h) *If a sequence in $L^2(\mathbb{T})$ has both a limit in norm and a realization, then the former equals the latter almost everywhere.*

Proof. We begin with part (a): Let all quantities be as stipulated there. To show that the limit in (3.46) exists, we note that, for $M, N \in \mathbb{Z}^+$,

$$\big| ||f_M|| - ||f_N|| \big| \le ||f_M - f_N|| \tag{3.47}$$

(see Exercise 3.1.3). By assumption on the f_N's, the limit as $N \to \infty$ of the sup over $N > M$ of the entities on the right is zero. But then the same is true of the entities on the left. This means the $||f_N||$'s form a Cauchy sequence of real numbers with respect to the metric $d(x, y) = |x - y|$ on $\mathbb{R}$. Since $\mathbb{R}$ is complete under this metric, the $||f_N||$'s have a limit in $\mathbb{R}$, as required.

To show that this limit is well defined, we observe that, for $f_N, g_N \in C(\mathbb{T})$,

$$\big| ||f_N|| - ||g_N|| \big| \le ||f_N - g_N|| \tag{3.48}$$

(by Exercise 3.1.3, again). We wish to show that the left side of (3.48) goes to zero as $N \to \infty$; it will suffice to show that the right side does. But note that, since the f_N's and the g_N's constitute Cauchy sequences in $C(\mathbb{T})$ convergent pointwise almost everywhere to f, the formula $\ell_N = f_N - g_N$ defines a Cauchy sequence in $C(\mathbb{T})$ convergent pointwise almost everywhere to zero. So it will suffice to show that, for any such sequence $\ell_1, \ell_2, \dots$,

$$\lim_{N \to \infty} ||\ell_N|| = 0. \tag{3.49}$$

To show this, we note that, for $J, N \in \mathbb{Z}^+$ and $J > N$,

$$\int_\mathbb{T} |\ell_J(x) - \ell_N(x)|^2 \, dx = ||\ell_J - \ell_N||^2 \le \sup_{M > N} ||\ell_M - \ell_N||^2, \tag{3.50}$$

the inequality by definition of the supremum. Since $\lim_{J \to \infty} \ell_J(x) = 0$ almost everywhere, we may, by established properties of Riemann integrals (see Section 8 in [5]), replace ℓ_J by zero in (3.50); we get

$$\int_\mathbb{T} |-\ell_N(x)|^2 \, dx = ||\ell_N||^2 \le \sup_{M > N} ||\ell_M - \ell_N||^2. \tag{3.51}$$

Since the ℓ_N's form, by assumption, a Cauchy sequence, the right and therefore the left side of (3.51) approaches zero as $N \to \infty$, as required.

For part (b), we note first that, if $f \in C(\mathbb{T})$ and we define $f_N = f$ for $N \in \mathbb{Z}^+$, then f is a realization of the ultraCauchy sequence $f_1, f_2, \dots$. So indeed $C(\mathbb{T}) \subset L^2(\mathbb{T})$. Next, to show closure of $L^2(\mathbb{T})$ under linear combinations, complex conjugation, and absolute values, let's suppose that elements f and g in $L^2(\mathbb{T})$ are realizations of ultraCauchy sequences $f_1, f_2, \dots$ and $g_1, g_2, \dots$, respectively, in $C(\mathbb{T})$ and that $\alpha, \beta \in \mathbb{C}$. Then one checks that $\alpha f + \beta g$ is a realization of the ultraCauchy sequence of $\alpha f_N + \beta g_N$'s, $\overline{f}$ is a realization of the ultraCauchy sequence of $\overline{f_N}$'s, and $|f|$ is a realization of the ultraCauchy sequence of $|f_N|$'s. So $\alpha f + \beta g, \overline{f}, |f| \in L^2(\mathbb{T})$ as well.

For part (c), let $f \in L^2(\mathbb{T})$. We first assume $f(x) = 0$ almost everywhere. Then, defining $f_N(x) = 0$ for all $N \in \mathbb{Z}^+$ and $x \in \mathbb{R}$, we have

$$f(x) = 0 = \lim_{N \to \infty} 0 = \lim_{N \to \infty} f_N(x) \tag{3.52}$$

almost everywhere. So by the definition (3.46),

$$||f|| = \lim_{N \to \infty} ||f_N|| = \lim_{N \to \infty} 0 = 0, \tag{3.53}$$

as required.

Conversely, suppose $||f|| = 0$. Let's pick an ultraCauchy sequence $f_1, f_2, \ldots$ in $C(\mathbb{T})$ of which f is a realization. Since $||f_N|| \to ||f|| = 0$ as $N \to \infty$, we may extract a subsequence $g_1, g_2, \ldots$ (here $g_k = f_{N_k}$ for certain N_k's with $N_1 < N_2 < \cdots$) with $||g_k|| \leq k^{-4}$ for all $k \in \mathbb{Z}^+$. Now consider the sequence $h_1, h_2, \ldots$ defined by

$$h_k(x) = \begin{cases} 0 & \text{if } k \text{ is odd,} \\ g_{k/2}(x) & \text{if } k \text{ is even.} \end{cases} \tag{3.54}$$

The h_k's are ultraCauchy: If k is odd, then $||h_k - h_{k+1}|| = ||-g_{(k+1)/2}|| \leq ((k+1)/2)^{-4} < (2/k)^4$; if k is even, then $||h_k - k_{k+1}|| = ||g_{k/2}|| \leq (2/k)^4$. So by Lemma 3.2.1, the $h_k(x)$'s converge on $\mathbb{T} - S$, where S is negligible.

Let $x \in \mathbb{T} - (R \cup S)$, where R is the negligible set on which $f(x)$ is unequal to $f_N(x)$. Write $h(x) = \lim_{k \to \infty} h_k(x)$. Since any subsequence of a convergent sequence converges to the same limit, we find that the subsequence $g_1(x), g_2(x), \ldots$ of the $h_k(x)$'s converges to $h(x)$, as does the subsequence $0, 0, \ldots$. But the former subsequence, being a subsequence of the $f_N(x)$'s, converges to $f(x)$; the latter converges to zero for all x.

So

$$0 = \lim_{k \to \infty} 0 = \lim_{k \to \infty} h_k(x) = \lim_{k \to \infty} g_k(x) = f(x) \tag{3.55}$$

on $\mathbb{T} - (R \cup S)$, which is to say almost everywhere. Therefore, part (c) is proved.

We now consider part (d). The fact that $L^2(\mathbb{T})$ is a pseudometric space under ρ follows from limiting arguments applied to the fact that $C(\mathbb{T})$ is a metric space under d. To prove the remaining claims, we suppose that f is a realization of the Cauchy sequence $f_1, f_2, \ldots$ in $C(\mathbb{T})$. Note then that, for any fixed N, $f_N - f \in L^2(\mathbb{T})$ is a realization of the Cauchy sequence $f_N - f_1, f_N - f_2, \ldots$. Then, by definition of the norm on $L^2(\mathbb{T})$ and by properties of the supremum,

$$||f_N - f|| = \lim_{M \to \infty} ||f_N - f_M|| \leq \sup_{M > N} ||f_N - f_M||. \tag{3.56}$$

The right side goes to zero as $N \to \infty$, since the f_N's form a Cauchy sequence in $C(\mathbb{T})$. So the left side does too, whence the f_N's converge to f in norm. So $C(\mathbb{T})$ is dense in $L^2(\mathbb{T})$.

We move on to part (e). By one version of the Stone-Weierstrass theorem (see Theorem 5.8.1 in [33]), the space of polynomials p on $[-\pi, \pi]$ such that $p(-\pi) = p(\pi)$

is dense in the space of continuous functions on $[-\pi, \pi]$. Note that the 2π-periodic extension of any such polynomial is in PSC($\mathbb{T}$): so certainly PSC($\mathbb{T}$) is dense in $C(\mathbb{T})$. The latter is dense in $L^2(\mathbb{T})$ so, by transitivity of norm density (see Exercise 3.2.6), the former is too, and part (e) is proved.

For the remainder, it will be important to note that the triangle inequality (3.16) holds in $L^2(\mathbb{T})$ as in $C(\mathbb{T})$. Indeed, because ρ is a pseudometric,

$$||F + G|| = \rho(F, -G) \le \rho(F, 0) + \rho(0, -G) = ||F|| + ||G|| \tag{3.57}$$

for $F, G \in L^2(\mathbb{T})$. So let's prove part (f): Suppose $\lim_{N \to \infty} ||f_N - f|| = 0$ for $f_1, f_2, \ldots$ and f in $L^2(\mathbb{T})$. If also $\lim_{N \to \infty} ||f_N - g|| = 0$ for $g \in L^2(\mathbb{T})$ then, because

$$||f - g|| \le ||f - f_N|| + ||f_N - g||, \tag{3.58}$$

we find that $||f - g|| = 0$, so $f(x) - g(x) = 0$ almost everywhere, by part (c) of this proposition. Coversely, if $f(x) - g(x) = 0$ almost everywhere, then $||f - g|| = 0$ again by part (c) above, so the inequality

$$||f_N - g|| \le ||f_N - f|| + ||f - g|| = ||f_N - f|| \tag{3.59}$$

implies $\lim_{N \to \infty} ||f_N - g|| = 0$, as required.

To prove part (g), let $f_1, f_2, \ldots$ be a Cauchy sequence in $L^2(\mathbb{T})$. Since the terms in this sequence are bunching up as we go further out, there is a subsequence $g_1, g_2, \ldots$ (as before $g_k = f_{N_k}$ for certain N_k's with $N_1 < N_2 < \cdots$) such that

$$||g_k - g_{k+1}|| \le \frac{1}{k^4} \tag{3.60}$$

for $k \in \mathbb{Z}^+$. (See Exercise 3.2.5.) We wish to show that the g_k's have a realization f to which they converge in norm; this will suffice because, if a subsequence of a Cauchy sequence converges, then the original sequence does too, and to the same limit.

Now for a given $k \in \mathbb{Z}^+$ there exists, by definition of $L^2(\mathbb{T})$, an ultraCauchy sequence $g_{1,k}, g_{2,k}, \ldots$ of functions in $C(\mathbb{T})$ such that g_k is a realization of the $g_{j,k}$'s. By part (d) of the present proposition, the $g_{j,k}$'s converge to g_k in norm; by Lemma 3.2.1, they converge uniformly to g_k outside of an ε-small set for any $\varepsilon > 0$. We may therefore pick a $j \in \mathbb{Z}^+$ (depending on k) such that the function $h_k = g_{j,k}$ satisfies

$$||h_k - g_k|| \le \frac{1}{k^4}, \tag{3.61}$$

and

$$|h_k(x) - g_k(x)| \le \frac{1}{k} \text{ for all } x \in \mathbb{T} - S_k, \text{ where } S_k \text{ is } k^{-2}\text{-small.} \tag{3.62}$$

Let's do so: We claim that the h_k's have a realization.

To see this we observe that, by the triangle inequality and the identity

$$h_k - h_{k+1} = (h_k - g_k) + (g_k - g_{k+1}) + (g_{k+1} - h_{k+1}), \qquad (3.63)$$

we have

$$\begin{aligned}\|h_k - h_{k+1}\| &\leq \|h_k - g_k\| + \|g_k - g_{k+1}\| + \|g_{k+1} - h_{k+1}\| \\ &\leq \frac{1}{k^4} + \frac{1}{k^4} + \frac{1}{(k+1)^4} < \frac{3}{k^4},\end{aligned} \qquad (3.64)$$

the next to the last step by (3.60) and (3.61). By Lemma 3.2.1, then, the h_k's have a realization f.

We show that the g_k's converge in norm, and pointwise almost everywhere, to f, as follows. First, by the triangle inequality and (3.61),

$$\|f - g_k\| \leq \|f - h_k\| + \|h_k - g_k\| \leq \|f - h_k\| + \frac{1}{k^4}; \qquad (3.65)$$

the right side goes to zero as $k \to \infty$ by part (d) of this proposition. Hence the requisite norm convergence. Next, suppose R is the negligible subset of $\mathbb{T}$ on which $f(x) \neq \lim_{k \to \infty} h_k(x)$. If $x \in \mathbb{T}$ is neither in R nor in the set S_k of (3.62), then

$$|f(x) - g_k(x)| \leq |f(x) - h_k(x)| + |h_k(x) - g_k(x)| \leq |f(x) - h_k(x)| + \frac{1}{k}. \qquad (3.66)$$

Now let $\varepsilon > 0$. Pick $N \in \mathbb{Z}^+$ such that $\sum_{k=N}^{\infty} k^{-2} < \varepsilon$; then

$$T_N = R \cup \left(\bigcup_{k=N}^{\infty} S_k\right) \qquad (3.67)$$

is ε-small. Moreover, for $x \in \mathbb{T} - T_N$, (3.66) holds for all $k \geq N$. For such x, certainly $x \notin R$, so $f(x) - h_k(x)$ and consequently the entire right side of (3.66) goes to zero as $k \to \infty$. The same then holds for the left side. So the g_k's converge pointwise to f, outside of the ε-small set T_N. Since ε was arbitrary, they converge to f pointwise almost everywhere, as required.

Finally, for part (h), suppose the sequence $f_1, f_2, \ldots$ in $L^2(\mathbb{T})$ converges in norm to f and pointwise almost everywhere to h. By the previous part of this proposition, we have a subsequence $g_1, g_2, \ldots$ of the f_N's converging both in norm and pointwise almost everywhere to some $g \in L^2(\mathbb{T})$.

But again, any subsequence of a convergent sequence (of complex numbers or of functions in norm) converges to the same limit. So, on the one hand, the g_N's converge in norm to f; consequently $f(x) = g(x)$ almost everywhere, by part (e) of this proposition. On the other hand, they converge pointwise almost everywhere to h; consequently $g(x) = h(x)$ amost everywhere. So $f(x) = h(x)$ almost everywhere, as claimed. □

Elements of $L^2(\mathbb{T})$ are said to be *Lebesque square integrable*, or sometimes simply *square integrable*, on $\mathbb{T}$. The logic behind this terminology will be made clear shortly.

We note that $L^2(\mathbb{T})$ contains not only $C(\mathbb{T})$ but also PC($\mathbb{T}$), and more. Indeed, if f is 2π-periodic and has only finitely many discontinuities on $\mathbb{T}$, then $f \in L^2(\mathbb{T})$ if and only if the (perhaps improper) Riemann integral of $|f|^2$ over $\mathbb{T}$ is defined. (For example, the function f of (3.35) is in $L^2(\mathbb{T})$.) And when this Riemann integral *is* defined, its square root equals the quantity $||f||$ of (3.46). (In other words, the latter definition of $||f||$ agrees, in this situation, with the one given by (3.2).) This may be seen by expressing such an f as a realization of an appropriate ultraCauchy sequence in $C(\mathbb{T})$.

But this is not all that $L^2(\mathbb{T})$ contains. Indeed, consider a "2π-periodic Dirichlet function" $Q_{2\pi}$; namely, define $Q_{2\pi}(x) = 1$ for x a *rational* number in $(-\pi, \pi]$ and $Q_{2\pi}(x) = 0$ for x an *irrational* number in $(-\pi, \pi]$; then extend $Q_{2\pi}$ periodically. (Compare with the function Q of Exercise 1.1.1.) Because $\mathbb{Q}$ is countable, $Q_{2\pi}(x) = 0$ almost everywhere. So $||Q_{2\pi}|| = 0$ by Proposition 3.2.1(c). On the other hand, one shows that $|Q_{2\pi}|^2$ (which equals $Q_{2\pi}$, since $|0|^2 = 0$ and $|1|^2 = 1$) is discontinuous *everywhere* (and therefore is not Riemann integrable on $\mathbb{T}$).

There are even stranger creatures lurking in the depths of $L^2(\mathbb{T})$; for example, there exists a function g there that's unbounded on *every positive-length subinterval* of $\mathbb{R}$. See Chapter IV in [52]. Still, this space is not *so* bizarre: Everything therein is a limit, in the sense described by Proposition 3.2.1, of a sequence of perfectly "ordinary" continuous functions on $\mathbb{T}$. We will do well, both mathematically and philosophically, to keep in mind that elements of $L^2(\mathbb{T})$ are, in this sense, "not so different" from such perfectly ordinary functions.

Let's now flesh out our rather imprecise argument, above, regarding the nonconvergence in norm of the f_N's of (3.33) to any $g \in C(\mathbb{T})$. Here's how: We've already seen that these f_N's are Cauchy, so by part (g) of the above proposition, they have a norm limit in $L^2(\mathbb{T})$. But we've also seen that they have the function f of (3.35) as a realization; so by part (h) of the proposition, they in fact converge in norm to f. Now were they also to converge to a $g \in C(\mathbb{T})$, then, by part (f) of the proposition, we'd have $g(x) = f(x)$ almost everywhere. But functions in $C(\mathbb{T})$ are bounded, and f is not equal almost everywhere to any bounded function (see Exercise 3.2.4). So we have a contradiction to our assumption on g, and we're done.

Exercises

3.2.1 Show that any countably infinite set $S = \{s_k : k \in \mathbb{Z}^+\} \subset \mathbb{R}$ is negligible. Hint: Given $\varepsilon > 0$, consider, for each k, the interval $I_k = (s_k - \varepsilon/2^{k+1}, s_k + \varepsilon/2^{k+1})$.

3.2.2 Show that $\rho(f, g) = ||f - g||$ defines a metric on $C(\mathbb{T})$. Parts of Propositions 3.1.1 and 3.1.3 should be of use here.

3.2.3 Let f_N be as in (3.33) and f as in (3.35).
  a. Compute $||f_N - f||$.
  b. Use part a to show directly that the f_N's converge in norm to f.

3.2.4 Show that the function f of (3.35) is not equal almost everywhere to any bounded function. That is, let g be bounded, say $g(x) \leq M$ always; find a nonnegligible set S such that $f(x) > M$ for all x is S. (Recall that nonempty open intervals are nonnegligible.)

3.2.5 Show that every Cauchy sequence $f_1, f_2, \ldots$ in $L^2(\mathbb{T})$ has a subsequence $g_1, g_2, \ldots$, where $g_k = f_{N_k}$ and $N_1 < N_2 < \cdots$, satisfying (3.60) (so that, in particular, the g_k's are ultraCauchy). Hint: Using the definition of Cauchy sequence and mathematical induction, show there are integers $N_1, N_2, \ldots$ with $N_1 < N_2 < \cdots$ and such that, for any k, $||f_{N_k} - f_M|| \leq 1/k^4$ whenever $M > N_k$.

3.2.6 Prove *transitivity of norm density*, meaning: If $A \subset B \subset C \subset L^2(\mathbb{T})$, A is dense in B, and B is dense in C, then A is dense in C. Hint: Let $f \in C$; choose a sequence $g_1, g_2, \ldots$ in B converging in norm to f. Then for each $N \in \mathbb{Z}^+$, choose an $f_N \in A$ such that $||f_N - g_N|| < 1/N$. (Why is this possible?) Use the triangle inequality in $L^2(\mathbb{T})$ and the squeeze law to examine $\lim_{N\to\infty} ||f_N - f||$.

3.2.7 For $N \in \mathbb{Z}^+$, let $f_N \in L^2(\mathbb{T})$ be defined by

$$f_N(x) = \begin{cases} N^{1/3} & \text{if } |x| < \dfrac{1}{N}, \\ 0 & \text{if } \dfrac{1}{N} \leq |x| \leq \pi. \end{cases}$$

Also let f be the zero function in $L^2(\mathbb{T})$.
   a. Sketch, on the same set of axes, the graphs of a few f_N's.
   b. Show that f_N converges to f in the norm on $L^2(\mathbb{T})$.
   c. Show that f_N does not converge pointwise to f on $\mathbb{T}$.
   d. Does f_N converge uniformly to f on $\mathbb{T}$? Explain.

3.2.8 For $N \in \mathbb{Z}^+$, let $g_N \in L^2(\mathbb{T})$ be defined by

$$g_N(x) = \begin{cases} 1 & \text{if } -\pi < x \leq 0, \\ 1 + x(N^{5/4} - N) & \text{if } 0 < x \leq 1/N, \\ x^{-1/4} & \text{if } 1/N < x \leq \pi. \end{cases}$$

Also, let $g \in L^2(\mathbb{T})$ be defined by

$$g(x) = \begin{cases} 1 & \text{if } -\pi < x \leq 0, \\ x^{-1/4} & \text{if } 0 < x \leq \pi. \end{cases}$$

   a. Sketch the graph of a couple of g_N's, and of g, all on the same set of axes.
   b. Show g_N converges to g in the norm on $L^2(\mathbb{T})$.
   c. Show g_N converges pointwise to g on $\mathbb{T}$.
   d. Show g_N does not converge uniformly to g. You can either do this directly, by the definition of uniform convergence, or you can use the following result from advanced calculus: If g_N converges uniformly to g and each g_N is continuous, then so is g.

3.2.9 Repeat the above problem, but this time with $g_N \in L^2(\mathbb{T})$ given by

$$g_N(x) = \begin{cases} x + \pi & \text{if } -\pi < x \leq -1/N, \\ (1 - N\pi)x & \text{if } -1/N < x \leq 1/N, \\ x - \pi & \text{if } 1/N < x \leq \pi \end{cases}$$

and $g \in L^2(\mathbb{T})$ by

$$g(x) = \begin{cases} x + \pi & \text{if } -\pi < x \leq 0, \\ 0 & \text{if } x = 0, \\ x - \pi & \text{if } 0 < x \leq \pi. \end{cases}$$

3.2.10 Let $f = 1 - 2Q_{2\pi}$, where $Q_{2\pi}$ is the 2π-periodic Dirichlet function described above. Show that $|f|$ is Riemann integrable on $\mathbb{T}$ but f is not. Hint for the second part: if f were Riemann integrable on $\mathbb{T}$, what would that say about $Q_{2\pi}$?

3.3 MORE ON $L^2(\mathbb{T})$; THE VECTOR SPACE $L^1(\mathbb{T})$

The containment of $C(\mathbb{T})$ in $L^2(\mathbb{T})$, together with our notion of inner product on the former space, provide us with a natural notion of inner product on the latter. Namely, we have:

Proposition 3.3.1 *For $f, g \in L^2(\mathbb{T})$, define the* inner product $\langle f, g \rangle$ *by*

$$\langle f, g \rangle = \lim_{N \to \infty} \langle f_N, g_N \rangle = \lim_{N \to \infty} \int_{\mathbb{T}} f_N(x) \overline{g_N(x)} \, dx, \qquad (3.68)$$

where $f_1, f_2, \ldots$ and $g_1, g_2, \ldots$ are any Cauchy sequences in $C(\mathbb{T})$ of which f and g, respectively, are realizations. This limit exists and is well defined, meaning independent of the choices of f_N's converging to f and g_N's converging to g.

Proof. For $M, N \in \mathbb{Z}^+$,

$$|\langle f_M, g_M \rangle - \langle f_N, g_N \rangle| = |\langle f_M, g_M - g_N \rangle + \langle f_M - f_N, g_N \rangle|$$
$$\leq |\langle f_M, g_M - g_N \rangle| + |\langle f_M - f_N, g_N \rangle|$$
$$\leq ||f_M|| \cdot ||g_M - g_N|| + ||f_M - f_N|| \cdot ||g_N||; \qquad (3.69)$$

the first inequality is by the triangle inequality on $\mathbb{C}$ and the second by the Cauchy-Schwarz inequality (3.15). The $||f_M||$'s and the $||g_N||$'s are convergent and therefore bounded sequences of real numbers, by Proposition 3.2.1. So we may conclude that

$$|\langle f_M, g_M \rangle - \langle f_N, g_N \rangle| \leq A||g_M - g_N|| + B||f_M - f_N|| \qquad (3.70)$$

for some $A, B > 0$. Since the f_N's and g_N's form Cauchy sequences in $C(\mathbb{T})$, the limit as $N \to \infty$ of the sup over $M > N$ of the right side, and thus the left side, of

(3.70) is zero. But this means the inner products $\langle f_N, g_N \rangle$ form a Cauchy and hence convergent sequence in $\mathbb{R}$, as required.

That $\langle f, g \rangle$ is well defined on $L^2(\mathbb{T})$ follows from the fact that $||f||$ is; see Exercise 3.3.1. □

As we would expect, the above inner product on $L^2(\mathbb{T})$ coincides with the one given by the usual Riemann integral of $f\bar{g}$ over $\mathbb{T}$, in cases where this Riemann integral is defined. Further, and as we would also expect, *Proposition 3.1.3 holds on $L^2(\mathbb{T})$ as it does on $C(\mathbb{T})$*.

It will be useful to consider, in addition to $L^2(\mathbb{T})$, an analogous pseudometric space $L^1(\mathbb{T})$, and a notion of *integration* on this space. This space and notion are suggested by the above proposition.

Definition 3.3.1 By $L^1(\mathbb{T})$ we mean the collection $\{fg\colon f, g \in L^2(\mathbb{T})\}$. And by the *Lebesgue integral*

$$\int_{\mathbb{T}} h(x)\, dx \tag{3.71}$$

of h over $\mathbb{T}$, for $h \in L^1(\mathbb{T})$, we mean the inner product $\langle f, \bar{g} \rangle$, where f and g are any elements of $L^2(\mathbb{R})$ such that $h = fg$.

Elements of $L^1(\mathbb{T})$ are said to be *Lebesgue integrable*, or sometimes simply *integrable*, on $\mathbb{T}$.

Henri Lebesgue was motivated to develop the theory of the Lebesgue integral (hence the name), in the early 1900's, in large part because the Riemann theory was imperfectly suited to the analysis of Fourier series. That is, the Lebesgue integral is an eminently "Fourier" thing; its relevance, to be demonstrated shortly, to Fourier analysis is no accident. (Lebesgue's original approach was substantially diffferent from the one we've taken above.)

It may be shown that the Lebesgue integral over $\mathbb{T}$ of an $h \in L^1(\mathbb{T})$ equals the corresponding (proper or improper) Riemann integral when the latter is defined. Also, the Lebesgue integral inherits from the Riemann integral all of the usual properties. That is, the Lebesgue integral of a linear combination equals the corresponding linear combination of Lebesgue integrals, the Lebesgue integral of a nonnegative function is nonnegative, the absolute value of a Lebesgue integral is less than or equal to the Lebesgue integral of the absolute value, and so on. In light of all this, it's safe to adopt—so let's adopt—the following convention: *From now on all integrals over $\mathbb{T}$ are, unless otherwise indicated, Lebsegue.*

We note that $L^1(\mathbb{T})$ may, alternatively, be constructed using the general idea of Proposition 3.2.1 but with a different pseudometric. Specifically:

Proposition 3.3.2 *Define the mean norm* $||f||_1$ *on* $C(\mathbb{T})$ *by*

$$||f||_1 = \int_{\mathbb{T}} |f(x)|\, dx. \tag{3.72}$$

Then $C(\mathbb{T})$ is a metric space under the metric $d(f,g) = ||f-g||_1$.

Proposition 3.2.1(a) holds with the mean norm in place of the mean square norm (and with all sequences in question being Cauchy with respect to this mean norm). The collection of all realizations of Cauchy sequences with respect to this mean norm is the pseudometric space $L^1(\mathbb{T})$ defined above; moreover, the remainder of Proposition 3.2.1 holds if we again replace $||\cdot||$ with $||\cdot||_1$ and also replace $L^2(\mathbb{T})$ with $L^1(\mathbb{T})$.

The proof of Proposition 3.3.2 is quite similar to that of Proposition 3.2.1.

Note that $L^2(\mathbb{T}) \subset L^1(\mathbb{T})$: If f is in the former space then, since $f = f \cdot 1$ (by 1 we mean the constant function equal to 1 everywhere), f is a product of two functions in $L^2(\mathbb{T})$ and is therefore in $L^1(\mathbb{T})$. (That the constant function 1 is integrable over $\mathbb{T}$ is crucial here. The same does not hold with $\mathbb{T}$ replaced by an unbounded interval, since a rectangle of height 1 and infinite length has infinite area.) The reverse inclusion does not hold: If $k \colon \mathbb{T} \to \mathbb{C}$ is defined by $k(x) = |x|^{-2/3}$ ($0 < |x| < \pi$), then $k \in L^1(\mathbb{T})$ ($k = f^2$, with f as in (3.35)), but

$$\int_{\mathbb{T}} |k(x)|^2 \, dx = 2 \int_0^\pi x^{-4/3} = -6x^{-1/3}\Big|_0^\pi = \infty, \tag{3.73}$$

so $k \notin L^2(\mathbb{T})$.

The triangle inequality (3.16) holds on $L^1(\mathbb{T})$ as it does on $L^2(\mathbb{T})$; one sees this by noting simply that $|f(x) + g(x)| \leq |f(x)| + |g(x)|$. We must, though, remark on a glaring difference between these two spaces: Namely, $L^1(\mathbb{T})$ does not possess any reasonable kind of *inner product*. That $L^2(\mathbb{T})$ *does* possess one, behaving as in Proposition 3.1.3, is one of its more attractive features, Fourier-wise (and otherwise), as we'll see.

Our last major observation for this section is this. It will be convenient, in what follows, to consider certain "collapsings" of the spaces $L^2(\mathbb{T})$ and $L^1(\mathbb{T})$ onto themselves. The idea here is that two functions differing only on a "small" set are, from many perspectives, best treated as *equal*. Indeed, it might be argued that, intuitively, if a subset S of $\mathbb{R}$ is "too small to see," then the difference between two functions that differ only on S is also too small to see. It might be argued further that, intuitively, the subsets of $\mathbb{R}$ that are too small to see are precisely the *negligible* ones. All of this suggests it might make good sense to identify, in our minds, functions equal almost everywhere.

Here is a mathematically rigorous way of doing so.

Definition 3.3.2 Let $p = 1$ or 2. Define an *equivalence relation* "$\sim$" on $L^p(\mathbb{T})$ as follows: $f \sim g$, for $f, g \in L^p(\mathbb{T})$, if and only if $f(x) = g(x)$ almost everywhere. Then by $\mathcal{L}^p(\mathbb{T})$ we mean the set of *equivalence classes* $[f]$ of elements of $L^p(\mathbb{T})$.

The names $\mathcal{L}^1(\mathbb{T})$ and $\mathcal{L}^2(\mathbb{T})$ won't stick; in a mere few paragraphs, we'll replace them with new (but old!) names.

Recall that an equivalence relation on a set X is a relation R thereon such that, for all $x, y, z \in X$, (i) xRx, (ii) $xRy \Rightarrow yRx$, and (iii) xRy and $yRz \Rightarrow xRz$. The above relation $\sim$ is an equivalence relation on either of our spaces $L^p(\mathbb{T})$; see Exercise 3.3.2.

Also recall that, in a set X on which there is defined an equivalence relation R, the *equivalence class* $[x]$ of an $x \in X$ just the subset

$$[x] = \{y \in X : yRx\}. \tag{3.74}$$

That is, $[x]$ is the collection of all elements related to x under R. So, in particular, in constructing $\mathcal{L}^p(\mathbb{T})$ from $L^p(\mathbb{T})$, we are indeed grouping together—in a sense, identifying—functions equal almost everywhere.

Note that $\mathcal{L}^p(\mathbb{T})$ may be given a natural complex vector structure. Namely, we define $\alpha[f] + \beta[g]$ to be the class $[\alpha f + \beta g]$, for $f, g \in \mathcal{L}^p(\mathbb{T})$ and $\alpha, \beta \in \mathbb{C}$. This operation is well defined: If $[f] = [f^*]$ and $[g] = [g^*]$, then f and f^* agree almost everywhere, as do g and g^*, whence it follows that $\alpha f + \beta g$ and $\alpha f^* + \beta g^*$ agree almost everywhere, meaning $[\alpha f + \beta g] = [\alpha f^* + \beta g^*]$.

Similarly, we may define an inner product and mean square norm on $\mathcal{L}^2(\mathbb{T})$, as well as an integral and mean norm on $\mathcal{L}^1(\mathbb{T})$, in terms of the corresponding entities on $L^2(\mathbb{T})$ and $L^1(\mathbb{T})$:

$$\langle [f], [g] \rangle = \langle f, g \rangle, \quad \|[f]\| = \|f\| \quad ([f], [g], \in \mathcal{L}^2(\mathbb{T})),$$

$$\int_{\mathbb{T}} [f] = \int_{\mathbb{T}} f(x)\, dx, \quad \|[f]\|_1 = \|f\|_1 \quad ([f], [g], \in \mathcal{L}^1(\mathbb{T})).$$

Because, for $h \in L^1(\mathbb{T})$, $|\int_{\mathbb{T}} h(x)\, dx| \leq \int_{\mathbb{T}} |h(x)|\, dx = \|h\|_1$, and because the mean norm of h is zero precisely when $h(x) = 0$ almost everywhere, all the above entities are well defined (independent of the choice of representative of each given class).

Observe further that, by part (c) of Proposition 3.2.1 and the analogous part of Proposition 3.3.2, $\mathcal{L}^2(\mathbb{T})$ and $\mathcal{L}^1(\mathbb{T})$ are *metric spaces* under their corresponding norms. That is, (3.30) holds for $[f], [g] \in \mathcal{L}^2(\mathbb{T})$ and $d([f], [g]) = \|[f] - [g]\|$, as it does for $[f], [g] \in \mathcal{L}^1(\mathbb{T})$ and $d([f], [g]) = \|[f] - [g]\|_1$.

It follows from Proposition 3.2.1(f) and the $L^1(\mathbb{T})$ analog thereof that a sequence in $\mathcal{L}^1(\mathbb{T})$ or in $\mathcal{L}^2(\mathbb{T})$ can have at most one limit in norm.

Most of the time, it's not such a bad thing to *confuse* $\mathcal{L}^p(\mathbb{T})$ with its precursor $L^p(\mathbb{T})$. Most of the time, especially when that time is spent doing Fourier analysis, it's a *good* thing: Fourier coefficients are, after all, integrals, and therefore neither they nor the resulting Fourier series recognize distinctions between equivalent functions. So we will, most of the time, be thus confused. In particular, from now on, *we will simply write $L^p(\mathbb{T})$ for $\mathcal{L}^p(\mathbb{T})$ (and for $L^p(\mathbb{T})$)*!

Of course, when discussing *values* $f(x)$ of a function f, we are tacitly agreeing that f is a *true* function, and not an equivalence class of such. It makes no sense to evaluate an equivalence class at a real number; equivalent functions certainly need not agree at a given point x. On the other hand, note that, if f and g are *continuous* at x, or even if both are piecewise continuous and averaged on $\mathbb{R}$ (see Definition 1.4.2), then $f \sim g$ *does* imply $f(x) = g(x)$. See Exercise 3.3.3.

Exercises

3.3.1 Show that the inner product on $L^2(\mathbb{T})$ is well defined. In other words show that, if we have sequences f_N and f_N^* in $C(\mathbb{T})$ converging in norm to f and sequences g_N and g_N^* in $C(\mathbb{T})$ converging in norm to g, then

$$\lim_{N\to\infty} \langle f_N, g_N \rangle = \lim_{N\to\infty} \langle f_N^*, g_N^* \rangle.$$

Hint: Consider (3.69), (3.70), and the surrounding argument, with f_N^* in place of f_M and g_N^* in place of g_M.

3.3.2 Show that the relation "$\sim$" of Definition 3.3.2 *is* an equivalence relation, meaning it satisfies the requirements (i),(ii),(iii) described just below that definition.

3.3.3 Suppose f and g are piecewise continuous and averaged on $\mathbb{R}$ (again see Definition 1.4.2). Show that $f \sim g \Rightarrow f(x) = g(x)$ for all x. Hint: Under the given assumptions, $|f - g|$ is piecewise continuous and averaged. Now use the fact, already cited in the proof of Proposition 3.1.1, that the integral of a *continuous*, nonnegative function over an interval is zero only if that function is identically zero on that interval.

3.4 NORM CONVERGENCE OF FOURIER SERIES: A THEOREM

As a prelude to our main result for this section, we supply the following "two-stage" squeeze law.

Proposition 3.4.1 (the squeeze law II)

(a) *Suppose we have real numbers $a_N, b_{M,N}$, and c_M ($M, N \in \mathbb{N}$) such that*

$$0 \leq a_N \leq b_{M,N} + c_M \tag{3.75}$$

for all M, N sufficiently large (larger than some fixed integer K). If

$$\lim_{M\to\infty} c_M = 0 \tag{3.76}$$

and, for any fixed M,

$$\lim_{N\to\infty} b_{M,N} = 0, \tag{3.77}$$

then

$$\lim_{N\to\infty} a_N = 0. \tag{3.78}$$

(b) *Suppose we have real-valued functions $a(s)$ and $b_M(s)$ and real numbers c_M ($M \in \mathbb{N}$, $s \in \mathbb{R}$) such that*

$$0 \leq a(s) \leq b_M(s) + c_M \tag{3.79}$$

180 L^2 SPACES: OPTIMAL CONTEXTS FOR FOURIER SERIES

for all M, s sufficiently large (larger than some fixed number R). If

$$\lim_{M \to \infty} c_M = 0 \tag{3.80}$$

and, for any fixed M,

$$\lim_{s \to \infty} b_M(s) = 0, \tag{3.81}$$

then

$$\lim_{s \to \infty} a(s) = 0. \tag{3.82}$$

The squeeze law II is perhaps less familiar and less intuitively manifest than the standard squeeze law. The reader is encouraged to draw some pictures, to illustrate and illuminate the two-stage law. (See Exercise 3.4.4 regarding a proof.)

We're now ready for the following, which reigns supreme among convergence-of-Fourier-series results.

Theorem 3.4.1 $f \in L^2(\mathbb{T})$ *if and only if S_N^f converges in norm to f.*

Proof. Assume $f \in L^2(\mathbb{T})$. To show that S_N^f converges in norm to f, we first show that S_N^f makes sense for all N. This will be the case if the Fourier coefficients $c_n(f)$ make sense—in other words, if the integrals (1.18) are defined for all n. But they are: The function ρ_n defined by $\rho_n(x) = e^{inx}$ is certainly in $C(\mathbb{T})$, and therefore in $L^2(\mathbb{T})$, for all n, whence the product $f \overline{\rho_n}$ is in $L^1(\mathbb{T})$ (by Definition 3.3.1) for all n. (The $a_n(f)$'s and $b_n(f)$'s are defined for similar reasons.)

Not only is S_N^f defined, but in fact it's in $L^2(\mathbb{T})$, since it's clearly in $C(\mathbb{T})$.

We now need to show that $\|S_N^f - f\| \to 0$ as $N \to \infty$. To do so, we choose a sequence of f_M's in PSC$(\mathbb{T})$ converging in norm to f. Such a sequence exists by Proposition 3.2.1(e). We write

$$S_N^f - f = (S_N^f - S_N^{f_M}) + (S_N^{f_M} - f_M) + (f_M - f) \tag{3.83}$$

and apply the triangle inequality (3.16) to get

$$\|S_N^f - f\| \le \|S_N^f - S_N^{f_M}\| + \|S_N^{f_M} - f_M\| + \|f_M - f\|. \tag{3.84}$$

Now note that

$$\|S_N^f - S_N^{f_M}\| = \|S_N^{f - f_M}\| \le \|f - f_M\|; \tag{3.85}$$

the equality here follows readily from Definition 1.4.1 for partial sums of Fourier series, and the inequality is by (1.74) and (1.77) in the case $g = f_M - f$. (These latter two results were proved under the assumption $g \in$ PC$(\mathbb{T})$, but note that all the relevant arguments are valid whenever $g \in L^2(\mathbb{T})$.) So (3.84) gives

$$\|S_N^f - f\| \le \|S_N^{f_M} - f_M\| + 2\|f_M - f\|. \tag{3.86}$$

Now $f_M \in \text{PSC}(\mathbb{T})$, so by Theorem 1.7.1, the $S_N^{f_M}$'s converge uniformly to f_M as $N \to \infty$. But then by Proposition 3.1.2, $||S_N^{f_M} - f_M|| \to 0$ as $N \to \infty$ for any fixed M. Further, $||f_M - f|| \to 0$ as $M \to \infty$, by choice of the f_M's. So by the above squeeze law II, part (a), the left side of (3.86) approaches zero as $N \to \infty$, as desired.

Conversely, suppose S_N^f converges in norm to f. We write $f = S_N^f + (f - S_N^f)$. Again, S_N^f is in $L^2(\mathbb{T})$. So is $f - S_N^f$ for N large enough, since we're assuming that $||f - S_N^f|| \to 0$ and therefore tacitly that $||f - S_N^f||$ is eventually defined. So f is a sum of two elements of the complex vector space $L^2(\mathbb{T})$, and is therefore in $L^2(\mathbb{T})$ itself. □

Exercises

3.4.1 Let $f(x) = -\ln\left|2\sin\frac{x}{2}\right|$.
   a. Show that f is 2π-periodic.
   b. Sketch the graph of f.
   c. Show that $f \notin \text{PC}(\mathbb{T})$, but that $f \in L^2(\mathbb{T})$. Hint for the second part: Compare f to $g(x) = -\ln x$. Integrate $|g|^2$ over $[-\pi, \pi]$ directly.
   d. Show that
   $$f(x) \sim \sum_{n=1}^{\infty} \frac{\cos nx}{n}.$$

Hints: First, f is even. Second, to compute $a_0(f)$, first note that $\sin(x/2) = 2\cos(x/4)\sin(x/4)$ and then use properties of the logarithm to break up the integral defining $a_0(f)$. When you do this, you should be able to relate the original integral you started with to the ones you end up with, via some changes of variable. (The fact that $\cos(\pi/2 - x) = \sin x$ may help.) Third, to compute $a_n(f)$ for $n > 0$, integrate by parts with $u = -\ln\left|2\sin\frac{x}{2}\right|$ and $dv = \cos nx\, dx$. The integral that results from this can be done using the trig identity

$$\sin nx \cos\frac{x}{2} = \frac{1}{2}(\sin(n+\frac{1}{2})x + \sin(n-\frac{1}{2})x)$$

combined with the integral formula

$$\frac{1}{\pi}\int_0^\pi \frac{\sin(m+1/2)x}{\sin(x/2)}\, dx = 1 \quad (m \in \mathbb{N}).$$

To see where this last formula comes from:
   e. *Prove* this last formula. Hint: See (1.83) and (1.122).

3.4.2 a. Does $S_N^f(x)$, for f as in the previous exercise, converge for all x?
   b. Assuming that $S_N^f(\pi)$ converges to $f(\pi)$ (it's true), evaluate
   $$\sum_{n=1}^{\infty} \frac{(-1)^n}{n}.$$

3.4.3 The human ear is essentially "phase deaf" at low frequencies: Typically, it can distinguish only amplitudes and not phases of sinuosoids of frequencies below a couple of hundred hertz. (This is why you only need one subwoofer, and it doesn't much matter where you put it.) Given this, what *piecewise polynomial*, 2π-periodic function does the function f of the previous example "sound like," at least in the low-frequency range? (It's a function we've encountered often already.)

3.4.4 (For those familiar with "$\varepsilon - N$" and "$\varepsilon - \delta$" arguments) Prove Proposition 3.4.1.

3.5 MORE ON INTEGRATION

Let D be an interval of some finite length $\ell > 0$ or a torus $\mathbb{T}_P$ for some $P > 0$. The discussions and results of Sections 3.1 and 3.2 carry over with little change if "$\mathbb{T}$" is replaced by "D" throughout. Indeed, all the definitions there continue to apply, and all the propositions there remain true, under this replacement. (The examples given in those sections must be, and may be, modified suitably to illustrate analogous behavior of functions on D.) We may therefore work with $C(D)$; $L^2(D)$ and $L^1(D)$; norms, inner products, and integrals on D; and so on, and have a pretty decent idea what we're doing.

We have to be a bit careful, though, with *unbounded* intervals; the problem here is that *a continuous function on such an interval need not be integrable there*. As a simple example of this, consider the constant functions: Any such is certainly continuous, whatever its domain, but no such, except for the zero function, is integrable if that domain is unbounded.

Fortunately, the situation may be remedied without too much difficulty. Here's how: If I is an unbounded interval, then instead of $C(I)$, we begin with the space $C_c(I)$ of *compactly supported* continuous functions on I. Here, by "compactly supported" we mean "supported on some bounded interval"; by "supported on" we in turn mean "vanishing almost everywhere outside of." So by definition, $C_c(I)$ is the collection of $f \in C(I)$ such that, for some $R > 0$, $f(x) = 0$ for almost all $x \notin [-R, R]$.

Continuous functions are integrable over bounded intervals, so if $f \in C_c(I)$, then the Riemann integral of f, or of $\overline{f}$ or $|f|$ or f^2 or $|f|^2$, is defined.

One therefore has a mean square norm and an inner product on $C_c(I)$; Definitions 3.1.1 and 3.1.2 as well as Propositions 3.1.1 and 3.1.3 apply to this space as they do to $C(\mathbb{T})$. But BEWARE: The same may *not* be said of Proposition 3.1.2, as the following example illustrates.

Example 3.5.1 For $N \in \mathbb{Z}^+$, let $f_N \colon \mathbb{R} \to \mathbb{C}$ be defined by

$$f_N(x) = \begin{cases} N^{-1/3}(1 - N^{-1}|x|) & \text{if } |x| \leq N, \\ 0 & \text{if not} \end{cases} \quad (3.87)$$

(see Fig. 3.4).

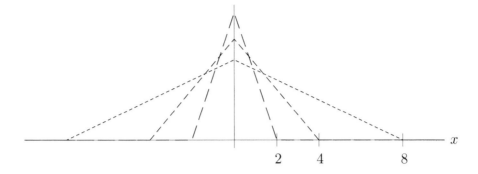

Fig. 3.4 f_2, f_4, and f_8

(Note that each f_N is compactly supported and continuous on $\mathbb{R}$.) Also let f be the zero function on $\mathbb{R}$. Show that the f_N's converge uniformly to f on $\mathbb{R}$ but do *not* converge to f in the mean square norm on $C_c(\mathbb{R})$.

Solution. We have

$$\lim_{N\to\infty}\sup_{x\in\mathbb{R}}|f_N(x)-f(x)| = \lim_{N\to\infty}\sup_{x\in\mathbb{R}}|f_N(x)| = \lim_{N\to\infty} N^{-1/3} = 0, \quad (3.88)$$

so the f_N's converge uniformly to f on $\mathbb{R}$.

On the other hand,

$$||f_N - f||^2 = ||f_N||^2 = \int_{\mathbb{R}} |f_N(x)|^2\, dx = 2N^{-2/3}\int_0^N (1 - N^{-1}x)^2\, dx$$

$$= 2N^{1/3}\int_0^1 u^2\, du = \frac{2N^{1/3}}{3} \quad (3.89)$$

(we used the fact that f_N is even as well as the substitution $u = 1 - N^{-1}x$), so the f_N's do not converge to f in the mean square norm on $C_c(\mathbb{R})$.

This example notwithstanding, we may define $L^2(I)$, for I an unbounded interval, as the collection of (equivalence classes of) all functions obtainable as realizations of ultraCauchy sequences (with respect to the mean square norm) in $C_c(I)$. Almost all of the ideas of Section 3.2 then carry over nicely. The definitions there continue to apply, and the propositions continue to hold, under the replacement of "PSC($\mathbb{T}$)," "$C(\mathbb{T})$," and "$\mathbb{T}$" by "PSC$_c(I)$," "$C_c(I)$," and "I," respectively. About the only part of that section that does *not* translate so directly is the obervation that $L^2(\mathbb{T}) \subset L^1(\mathbb{T})$; it's definitely *not* the case that $L^2(I) \subset L^1(I)$ when I is unbounded. For example, let $f(x) = (1 + |x|)^{-1}$ for all real x. Then

$$\int_{\mathbb{R}} |f(x)|^2\, dx = 2\int_0^\infty \frac{dx}{(1+x)^2} = -\frac{1}{1+x}\bigg|_0^\infty = 1 \quad (3.90)$$

so $f \in L^2(\mathbb{R})$, but

$$\int_{\mathbb{R}} |f(x)|\,dx = 2\int_0^\infty \frac{dx}{1+x} = \ln(1+x)\Big|_0^\infty = \infty \tag{3.91}$$

so $f \notin L^1(\mathbb{R})$. (On the other hand, see Corollary 3.5.1(c).)

At any rate, we've established the notions of Lebesgue integration, norms, inner products, and so on, on quite general domains D: D may, so far, be any torus or (bounded or unbounded) interval. We'd now like to push the theory a bit further: We'd like to make sense of these notions on *product domains*.

We achieve this by relatively familiar methods. Specifically, suppose $D = D_1 \times D_2 \times \cdots \times D_m$, where $m \in \mathbb{Z}^+$ and each D_j is either an interval or a torus. By an ε-small subset of D, we now mean one that may be covered by a countable collection of *open boxes* the sum of whose *volumes* is at most ε. (An open box in D is a product $I_1 \times I_2 \times \cdots \times I_m$ of open intervals; the volume of such a box is the product of the lengths of the I_j's.) The notions of negligible set and of almost everywhere then make sense as before, so in particular we may consider the space $C_c(D)$ of compactly supported, continuous functions on D, where by "compactly supported" we now mean "vanishing almost everywhere outside of some product of bounded intervals." We define the mean square norm, distance, and convergence, as well as the inner product, on $C_c(D)$ as on $C(\mathbb{T})$, and then define $L^2(D)$ as the collection of all (equivalence classes of) realizations of ultraCauchy sequences in $C_c(D)$. The definition of $L^1(D)$ and of the Lebesgue integral thereon then follow, and the usual, consequent results apply, under a few modifications and with a few caveats. To wit; first, any generalization to more than one variable of Proposition 3.2.1(e) would require a notion of piecewise smoothness in several dimensions. Rather than specify such a notion, let's simply not worry about such generalizations: We won't need them. Second, neither Proposition 3.1.2 nor the fact that $L^2(\mathbb{T}) \subset L^1(\mathbb{T})$ carries over to D if any factor D_j of D is an unbounded interval.

We now record, mostly without proof, some particularly elegant and powerful results in the theory of the Lebesgue integral. (Here and hereafter, D denotes a product domain with one or more factors, as described above.)

Proposition 3.5.1 (the Lebesgue dominated convergence theorem) *Suppose f_1, $f_2, \ldots$ is a sequence of functions in $L^1(D)$ such that $\lim_{N\to\infty} f_N(x)$ exists for almost all $x \in D$; suppose also that there's a function $g \in L^1(D)$ such that, for almost all $x \in D$,*

$$|f_N(x)| \leq g(x) \tag{3.92}$$

for each $N \in \mathbb{Z}^+$. Then

$$\int_D \left(\lim_{N\to\infty} f_N(x)\right) dx = \lim_{N\to\infty} \int_D f_N(x)\,dx \tag{3.93}$$

(in particular, the pointwise limit on the left defines a function in $L^1(D)$, and the limit on the right exists).

Moreover, the result continues to hold if, instead of the sequence $f_1, f_2, \ldots$, we have a collection $\{f_\delta \colon \delta \in (a, b)\}$ of functions in $L^1(D)$ (here $-\infty \leq a < b \leq \infty$) and we replace "N" with "δ"; "$\mathbb{Z}^+$" with "(a,b)"; and "$N \to \infty$" with "$\delta \to c$ for some $c \in (a,b)$," or with "$\delta \to a^+$," or with "$\delta \to b^-$"; throughout.

A proof may be found in Section 30.1 of [29].

To properly state our next few results, we need a new notion, that of *measurable function*. The technical definition is this: The function $f \colon D \to \mathbb{C}$ is measurable (on D) if, given any $\varepsilon > 0$, there is a continuous function $f_\varepsilon \colon D \to \mathbb{C}$ such that the set

$$\{x \in D \colon f(x) \neq f_\varepsilon(x)\} \tag{3.94}$$

is ε-small. (This definition is perhaps not the standard one, but is equivalent to the standard one. See Chapter 2, especially Theorem 2.19 and Exercise 11 of that chapter, in [24].)

All "familiar" functions—functions with only finitely many discontinuities per bounded interval, integrable and square integrable functions, functions zero almost everywhere—are seen to be measurable. Moreover, linear combinations and products of measurable functions, reciprocals of measurable functions nonzero almost everywhere, continuous functions of measurable functions, and realizations of sequences of measurable functions are measurable.

Nonmeasurable functions do exist; see, for example, Section 44, Chapter VI of [35]. But they're a bit hard to produce and are generally considered rare, pathological, and of little use (except as pathological, rare examples). At any rate, the reader should rest assured that, in any application of any of the following results requiring measurability, the relevant functions are indeed measurable.

Proposition 3.5.2 *Suppose $h \in L^1(D)$; let $p = 1$ or 2. If f is measurable on D and $|f(x)|^p \leq h(x)$ for almost all $x \in D$, then $f \in L^p(D)$, and*

$$\left| \int_D f(x)^p \, dx \right| \leq \int_D h(x) \, dx. \tag{3.95}$$

For a proof, see Section 29.1 in [29].

We next have:

Corollary 3.5.1 (a) *Let f be measurable. Then $|f| \in L^p(D) \Rightarrow f \in L^p(D)$.*

(b) *If $f \in L^1(D)$ and f is equal almost everywhere to a bounded function, then $f \in L^2(D)$.*

(c) *If $f \in L^2(D)$ and f is compactly supported, then $f \in L^1(D)$.*

Proof. For part (a), suppose f is measurable and in $L^p(D)$. Then $|f| \in L^p(D)$ by the generalizations to D of Proposition 3.2.1(b) and its L^1 analog. So $h = |f|^p \in L^1(D)$; since certainly $|f|^p \leq h$, we have $f \in L^p(D)$ by the above proposition.

To prove (b), suppose $f \in L^1(D)$ and, for some $M \geq 0$, $|f(x)| \leq M$ for almost all $x \in D$. Then f is measurable (being integrable), and $|f(x)|^2 \leq M|f(x)|$ for almost all $x \in D$. Since $M|f|$ is integrable by assumption on f, we have $f \in L^2(D)$ by the above proposition.

For part (c) we simply note that, if $f \in L^2(D)$ is supported on the bounded box R, then we can write $f(x) = f(x)\chi_R(x)$ for almost all x, where by definition χ_R is the "characteristic function of R":

$$\chi_R(x) = \begin{cases} 1 & \text{if } x \in R, \\ 0 & \text{if not.} \end{cases} \tag{3.96}$$

So f is equal almost everywhere to a product of two functions in $L^2(D)$ (χ_R is square integrable because R is bounded). That product belongs to $L^1(D)$ by definition of this space, so f does too. □

Proposition 3.5.2 has the following interpretation. A function can fail to be integrable for one (or both) of only two reasons: because it's "too big" (not bounded in absolute value than any integrable function) or because it's "too strange" (not measurable). In light of this, one often writes $\int_D h(x)\,dx = \infty$, if h is measurable and nonnegative, but not integrable, on D. And in light of this and Corollary 3.5.1(a), one also often writes

$$L^p(D) = \left\{ \text{measurable functions } f \text{ on } D : \int_D |f(x)|^p\,dx < \infty \right\}. \tag{3.97}$$

(Sometimes, one even leaves out the word "measurable"; however, see Exercise 3.5.1.)

Finally, we cite a tremendously important result regarding "iterated" integrals.

Proposition 3.5.3 (Fubini's theorem) *Let D_1 be either an interval or a torus, and similarly for D_2. Suppose h is a measurable function on $D_1 \times D_2$. If any one of the integrals*

$$\int_{D_1 \times D_2} |h(x)|\,dx, \quad \int_{D_2}\left(\int_{D_1} |h(u,v)|\,du\right)dv, \quad \int_{D_1}\left(\int_{D_2} |h(u,v)|\,dv\right)du \tag{3.98}$$

exists, then they all do, and they are equal to each other, and moreover the integrals

$$\int_{D_1 \times D_2} h(x)\,dx, \quad \int_{D_2}\left(\int_{D_1} h(u,v)\,du\right)dv, \quad \int_{D_1}\left(\int_{D_2} h(u,v)\,dv\right)du \tag{3.99}$$

all exist and are equal to each other.

See Section 35.3 in [29] for a proof.

Fubini's theorem generalizes in the expected way to m-dimensional domains D.

Exercises

3.5.1 Show that $|f|$ can be measurable, and even integrable, even when f is non-measurable (and therefore nonintegrable). Hint: There exists a set $S \subset [0,1]$ such that χ_S is not measurable. (See [29]. A set with a nonmeasurable characteristic function is, itself, said to be nonmeasurable.) Now consider $1 - 2\chi_S$ as a function on $[0, 1]$.

3.5.2 Let
$$h(u, v) = \frac{uv}{(u^2 + v^2)^2}.$$

Show that
$$\int_{-\infty}^{\infty} \left(\int_{-\infty}^{\infty} h(u,v) \, du \right) dv = \int_{-\infty}^{\infty} \left(\int_{-\infty}^{\infty} h(u,v) \, dv \right) du = 0$$

but that
$$\int_{-\infty}^{\infty} \left(\int_{-\infty}^{\infty} |h(u,v)| \, du \right) dv$$

does not exist. Hint for the last part: $|h(u, v)|$ is an even function of u or v.

3.5.3 Let f be the function on $(0, 1]$ defined by
$$f(x) = (-1)^k k \quad \text{if} \quad \frac{1}{k+1} < x \le \frac{1}{k} \quad (k = 1, 2, 3, \dots).$$

a. Sketch the graph of f.

b. Show that the improper Riemann integral
$$\int_0^1 f(x) \, dx$$

converges. Hint: Since $(0, 1]$ is the union of the nonoverlapping intervals $(1/2, 1]$, $(1/3, 1/2], (1/4, 1/3], (1/5, 1/4], \dots$, we can write
$$\int_0^1 f(x) \, dx = \lim_{N \to \infty} \int_{1/N}^1 f(x) \, dx = \lim_{N \to \infty} \sum_{k=1}^N \int_{1/(k+1)}^{1/k} f(x) \, dx. \qquad (*)$$

c. Show that f is not Lebesgue integrable on $(0, 1]$. Hint: Use Corollary 3.5.1(a), and $(*)$ above with $|f|$ in place of f.

3.6 ORTHOGONALITY, ORTHONORMALITY, AND FOURIER SERIES

Here and in much of what follows, we'll be interested primarily with $L^2(D)$, and with regard to this space, it will be convenient to adopt some conventions. Namely, suppose $C \subset \mathbb{Z}$ and, for each $n \in C$, we have numbers $d_n \in \mathbb{C}$ and functions $\sigma_n \in L^2(D)$. Then by
$$\sum_{n \in C} d_n \sigma_n \qquad (3.100)$$

we will mean, unless otherwise specified, the mean square norm limit of the sequence of partial sums

$$\sum_{\substack{n \in C \\ |n| \leq N}} d_n \sigma_n. \tag{3.101}$$

As observed at the end of Section 3.3 (in the case $D = \mathbb{T}$), this sequence can have at most one limit; since $L^2(D)$ is complete, the limit exists whenever the sequence is Cauchy.

Let's now go way back to Section 1.1, where we introduced the idea of *orthogonality*. (See Lemma 1.1.2 and the discussions at the end of that section.) This idea has had vast consequences: It suggested to us the nature of frequency decompositions of periodic functions, and from this suggestion a rich, varied, and tremendously powerful theory has sprung.

It's still springing. And at this point, to keep it bouncing along, it will be beneficial to *revisit* the issue of orthogonality. We reformulate this issue and expand on it using the language of inner products on $L^2(D)$, as follows.

Definition 3.6.1 Let C be a (perhaps infinite) subset of $\mathbb{Z}$; let $S = \{\sigma_n : n \in C\}$ be a subset of $L^2(D)$.

(a) We call S an *orthogonal set* if none of its elements is zero and if $\langle \sigma_n, \sigma_k \rangle = 0$ whenever $n, k \in C$ and $n \neq k$.

(b) We call S an *orthonormal set* if it's an orthogonal set and each of its elements has norm 1. That is, S is an orthonormal set if

$$\langle \sigma_n, \sigma_k \rangle = \delta_{n,k} \quad (n, k \in C). \tag{3.102}$$

Elements of an orthogonal set are themselves said to be *orthogonal* (to each other); similarly for orthonormal sets.

(c) Let $V \subset L^2(D)$ be a vector space. We call S an *orthogonal* (respectively, *orthonormal*) *basis* for V if it's an orthogonal (respectively, orthonormal) subset of V and, given $f \in V$, there is a unique sequence $(d_n)_{n \in C}$ of complex numbers such that

$$f = \sum_{n \in C} d_n \sigma_n. \tag{3.103}$$

Example 3.6.1 Let

$$\sigma_n(x) = \frac{1}{\sqrt{2\pi}} e^{inx} \tag{3.104}$$

for $n \in \mathbb{Z}$ and $x \in \mathbb{R}$. Show that $\mathcal{S} = \{\sigma_n \colon n \in \mathbb{Z}\}$ is an orthonormal set in $L^2(\mathbb{T})$. Do the same for $\mathcal{T} = \{\tau_n \colon n \in \mathbb{Z}\}$, where

$$\tau_n(x) = \begin{cases} \dfrac{1}{\sqrt{\pi}} \cos nx & \text{if } n < 0, \\ \dfrac{1}{\sqrt{2\pi}} & \text{if } n = 0, \\ \dfrac{1}{\sqrt{\pi}} \sin nx & \text{if } n > 0. \end{cases} \qquad (3.105)$$

Solution. We have

$$\langle \sigma_n, \sigma_k \rangle = \frac{1}{\sqrt{2\pi}} \overline{\left(\frac{1}{\sqrt{2\pi}}\right)} \int_{-\pi}^{\pi} e^{inx} \overline{e^{ikx}} \, dx = \frac{1}{2\pi} \int_{-\pi}^{\pi} e^{inx} e^{-ikx} \, dx = \delta_{n,k}, \qquad (3.106)$$

by Lemma 1.1.2. So $\mathcal{S}$ is an orthonormal set. The orthonormality of $\mathcal{T}$ follows in a similar way from Exercise 1.1.5.

We'll have more to say about $\mathcal{S}$ and $\mathcal{T}$ shortly. First, we need some additional, general results concerning $L^2(D)$.

Lemma 3.6.1 *"Inner products may be brought inside norm limits." That is, if* (3.103) *holds, with all quantities as above, then for any* $g \in L^2(D)$,

$$\langle f, g \rangle = \sum_{n \in C} d_n \langle \sigma_n, g \rangle, \qquad (3.107)$$

the right side denoting

$$\lim_{N \to \infty} \sum_{\substack{n \in C \\ |n| \leq N}} d_n \langle \sigma_n, g \rangle. \qquad (3.108)$$

Proof. Under the given assumptions,

$$\left| \langle f, g \rangle - \sum_{\substack{n \in C \\ |n| \leq N}} d_n \langle \sigma_n, g \rangle \right| = \left| \left\langle f - \sum_{\substack{n \in C \\ |n| \leq N}} d_n \sigma_n, g \right\rangle \right| \leq \left\| f - \sum_{\substack{n \in C \\ |n| \leq N}} d_n \sigma_n \right\| \|g\|, \qquad (3.109)$$

the last step by the Cauchy-Schwarz inequality (3.15) on $L^2(D)$. The right side of (3.109) goes to zero as $N \to \infty$, by assumption; so by the squeeze law, the left side does too, as claimed. □

To bring an inner product inside a norm limit in the *second* variable—that is, to compute $\langle g, f \rangle$, with all quantities as above—we may simply apply to the above result the identity $\langle g, f \rangle = \overline{\langle f, g \rangle}$.

The next proposition tells us (among other things) that, if S is an orthonormal basis for $V \subset L^2(D)$, then the d_n's in Definition 3.6.1(c) are readily determined by way of the inner product.

Proposition 3.6.1 *Let C and V be as in Definition 3.6.1(c); suppose $S = \{\sigma_n : n \in C\}$ is an orthonormal set in V. Then S is an orthonormal basis for V if and only if*

$$f = \sum_{n \in C} \langle f, \sigma_n \rangle \sigma_n \tag{3.110}$$

for each $f \in V$.

Proof. Suppose S is an orthonormal basis for V, so that, given $f \in V$, equation (3.103) holds for certain d_n's. Then by the above lemma and the orthonormality of S,

$$\langle f, \sigma_k \rangle = \sum_{n \in C} d_n \langle \sigma_n, \sigma_k \rangle = \sum_{n \in C} d_n \delta_{n,k} = d_k \tag{3.111}$$

for any $k \in C$. That is, $d_n = \langle f, \sigma_n \rangle$ for $n \in C$, so (3.103) yields (3.110).

Conversely, suppose (3.110) holds for each $f \in V$. Then certainly any such f has an expansion of type (3.103). But we just showed that, if f has such an expansion and the σ_n's constitute an orthonormal set, then necessarily $d_n = \langle f, \sigma_n \rangle$ for all n. That is, the d_n's are unique, so S *is* a basis, and we're done. □

In contemplating the meaning and import of the above proposition, it is helpful, we think, to reflect on the corresponding picture in the Euclidean plane $\mathbb{R}^2$ (Fig. 3.5).

We now consider some $L^2(D)$ analogs of results from Chapter 1. We begin with the following consequence of Theorem 3.4.1.

Theorem 3.6.1 *Each of the sets S and T of Example 3.6.1 is an orthonormal basis for $L^2(\mathbb{T})$.*

Proof. In Example 3.6.1, we saw that S is an orthonormal set. Further, for $f \in L^2(\mathbb{T})$ and σ_n as in (3.104), we have

$$\langle f, \sigma_n \rangle = \frac{1}{\sqrt{2\pi}} \int_{-\pi}^{\pi} f(x) \overline{e^{inx}} \, dx = \sqrt{2\pi} \, c_n(f), \tag{3.112}$$

so

$$S_N^f(x) = \sum_{n=-N}^{N} c_n(f) e^{inx} = \sum_{n=-N}^{N} \langle f, \sigma_n \rangle \frac{e^{inx}}{\sqrt{2\pi}} = \sum_{n=-N}^{N} \langle f, \sigma_n \rangle \sigma_n(x), \tag{3.113}$$

so by Theorem 3.4.1,

$$f = \sum_{n=-\infty}^{\infty} \langle f, \sigma_n \rangle \sigma_n. \tag{3.114}$$

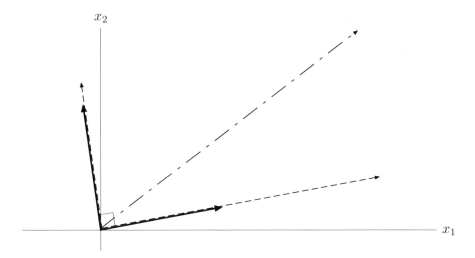

Fig. 3.5 Geometric interpretation of the formula $x = \sum_{n=1}^{2} \langle x, v_k \rangle v_k$ for $x \in \mathbb{R}^2$ and $\{v_1, v_2\}$ an orthonormal basis for $\mathbb{R}^2$. The vector x is dot-dashed; v_1 and v_2 are solid; the "projections" $\langle x, v_1 \rangle v_1$ and $\langle x, v_2 \rangle v_2$ of x in the directions of v_1 and v_2, respectively, are dashed

By Proposition 3.6.1, then, the σ_n's form an orthonormal basis for $L^2(\mathbb{T})$, as claimed.

We also saw in Example 3.6.1 that $\mathcal{T}$ is an orthonormal set. Now for $f \in L^2(\mathbb{T})$ and τ_n as in (3.105), we check easily that

$$\langle f, \tau_n \rangle = \begin{cases} \sqrt{\pi}\, a_{-n}(f) & \text{if } n < 0, \\ \sqrt{\dfrac{\pi}{2}}\, a_0(f) & \text{if } n = 0, \\ \sqrt{\pi}\, b_n(f) & \text{if } n > 0. \end{cases} \tag{3.115}$$

So

$$\sum_{n=-N}^{N} \langle f, \tau_n \rangle \tau_n(x) = \langle f, \tau_0 \rangle \tau_0(x) + \sum_{n=1}^{N} \big(\langle f, \tau_{-n} \rangle \tau_{-n}(x) + \langle f, \tau_n \rangle \tau_n(x) \big)$$

$$= \frac{1}{\sqrt{2\pi}} \cdot \sqrt{\frac{\pi}{2}}\, a_0(f)$$

$$+ \frac{1}{\sqrt{\pi}} \cdot \sqrt{\pi} \sum_{n=1}^{N} (a_n(f) \cos nx + b_n(f) \sin nx) = S_N^f(x). \tag{3.116}$$

Then by Theorem 3.4.1 again,

$$f = \sum_{n=-\infty}^{\infty} \langle f, \tau_n \rangle \tau_n, \tag{3.117}$$

so by Proposition 3.6.1 again, $\mathcal{T}$ is an orthonormal basis for $L^2(\mathbb{T})$. □

From the above results and simple change of variable arguments, much as were employed in Section 1.9, we now deduce the following. (Here S_N^f is as defined in Theorem 1.9.1.)

Theorem 3.6.2 (a) $f \in L^2(\mathbb{T}_P)$ if and only if S_N^f converges in norm to f.

(b) If

$$\sigma_n(x) = \frac{1}{\sqrt{P}} e^{2\pi i n x/P} \qquad (3.118)$$

for $n \in \mathbb{Z}$, then $\{\sigma_n : n \in \mathbb{Z}\}$ is an orthonormal basis for $L^2(\mathbb{T}_P)$.

(c) If

$$\tau_n(x) = \begin{cases} \sqrt{\dfrac{2}{P}} \cos \dfrac{2\pi n x}{P} & \text{if } n < 0, \\ \dfrac{1}{\sqrt{P}} & \text{if } n = 0, \\ \sqrt{\dfrac{2}{P}} \sin \dfrac{2\pi n x}{P} & \text{if } n > 0, \end{cases} \qquad (3.119)$$

then $\{\tau_n : n \in \mathbb{Z}\}$ is an orthonormal basis for $L^2(\mathbb{T}_P)$.

We reiterate that any expansion in terms of a basis is unique. So as a consequence of Theorem 3.6.2, we have:

Corollary 3.6.1 Let $f \in L^2(\mathbb{T}_P)$ and $\rho_n(x) = e^{2\pi i n x/P}$. If there are complex numbers z_n such that

$$f = \sum_{n=-\infty}^{\infty} z_n \rho_n, \qquad (3.120)$$

then $z_n = c_n(f)$ for all n. (An analogous result holds for expansions into sines and cosines.)

Our next step is to move from the periodic situation to that of functions on bounded intervals. Let $-\infty < a < b < \infty$. Since all of the spaces $L^2([a,b])$, $L^2([a,b))$, $L^2((a,b])$, and $L^2((a,b))$ are really the same (as the sets $\{a\}$, $\{b\}$, and $\{a,b\}$ are all negligible), we may and will safely denote any of these spaces by $L^2(a,b)$.

If $f \in L^2(a,b)$, then the $(b-a)$-periodic extension f_{per} of f is in $L^2(\mathbb{T}_{b-a})$. So, arguing much as in Section 1.11, we arrive at the following analog of Theorem 1.11.1.

Theorem 3.6.3 Let $-\infty < a < b < \infty$.

(a) $f \in L^2(a,b)$ if and only if the partial sums of the series in (1.210) converge in norm to f.

(b) If
$$\sigma_n(x) = \frac{1}{\sqrt{b-a}} e^{2\pi i n x/(b-a)} \tag{3.121}$$
for $n \in \mathbb{Z}$, then $\{\sigma_n : n \in \mathbb{Z}\}$ is an orthonormal basis for $L^2(a,b)$.

(c) If
$$\tau_n(x) = \begin{cases} \sqrt{\dfrac{2}{b-a}} \cos \dfrac{2\pi n x}{b-a} & \text{if } n < 0, \\ \dfrac{1}{\sqrt{b-a}} & \text{if } n = 0, \\ \sqrt{\dfrac{2}{b-a}} \sin \dfrac{2\pi n x}{b-a} & \text{if } n > 0, \end{cases} \tag{3.122}$$
then $\{\tau_n : n \in \mathbb{Z}\}$ is an orthonormal basis for $L^2(a,b)$.

Finally, we consider the space $L^2(0,\ell)$, for $\ell > 0$. An element f of this space has both an even 2ℓ-periodic extension and an odd one, as well as an odd 2ℓ-antiperiodic extension, much as in Section 1.12. Of course there are infinitely many other ways of periodizing f, but as in Section 1.12 we content ourselves, here, with the three ways just described. We have the following analog of Theorems 1.12.1 and 1.12.2.

Theorem 3.6.4 *Let $\ell > 0$.*

(a) $f \in L^2(0,\ell)$ if and only if the partial sums of each of the three series in Theorems 1.12.1 and 1.12.2 converge in norm to f.

(b) If
$$\tau_n(x) = \begin{cases} \dfrac{1}{\sqrt{\ell}} & \text{if } n = 0, \\ \sqrt{\dfrac{2}{\ell}} \cos \dfrac{\pi n x}{\ell} & \text{if } n > 0, \end{cases} \tag{3.123}$$
then $\{\tau_n : n \in \mathbb{N}\}$ is an orthonormal basis for $L^2(0,\ell)$.

(c) If
$$\omega_n(x) = \sqrt{\frac{2}{\ell}} \sin \frac{\pi n x}{\ell} \tag{3.124}$$
for $n \in \mathbb{Z}^+$, then $\{\omega_n : n \in \mathbb{Z}^+\}$ is an orthonormal basis for $L^2(0,\ell)$.

(d) *If*

$$\xi_n(x) = \sqrt{\frac{2}{\ell}} \sin \frac{\pi(2n-1)x}{2\ell} \qquad (3.125)$$

for $n \in \mathbb{Z}^+$, then $\{\xi_n : n \in \mathbb{Z}^+\}$ is an orthonormal basis for $L^2(0, \ell)$.

Let's return to the general setting of $L^2(D)$. Sometimes, in this setting, we'll want to work directly with orthogonal, rather than necessarily orthonormal, sets. The following results will aid us in that work.

Proposition 3.6.2 *Let $\mathcal{R} = \{\rho_n : n \in C\}$ be an an orthogonal set in a vector space $V \subset L^2(D)$. Then $\mathcal{R}$ is an orthogonal basis for V if and only if $\mathcal{S} = \{\rho_n/\|\rho_n\| : n \in C\}$ is an orthonormal basis for V.*

Proof. Certainly $\mathcal{R}$ is a basis if and only if $\mathcal{S}$ is. Moreover, the orthogonality of $\mathcal{R}$ yields $\langle \rho_n, \rho_k \rangle = \|\rho_n\|\|\rho_k\|\delta_{n,k}$ (either side equals zero if $n \neq k$ and $\|\rho_n\|^2$ if $n = k$), which yields

$$\left\langle \frac{\rho_n}{\|\rho_n\|}, \frac{\rho_k}{\|\rho_k\|} \right\rangle = \frac{\langle \rho_n, \rho_k \rangle}{\|\rho_n\|\|\rho_k\|} = \delta_{n,k}, \qquad (3.126)$$

which yields the orthonormality of $\mathcal{S}$. □

Consequently:

Corollary 3.6.2 *Let $\mathcal{R} = \{\rho_n : n \in C\}$ be an orthogonal set in a vector space $V \subset L^2(D)$. Then $\mathcal{R}$ is an orthogonal basis for V if and only if*

$$f = \sum_{n \in C} \frac{\langle f, \rho_n \rangle}{\|\rho_n\|^2} \rho_n \qquad (3.127)$$

for each $f \in V$.

Proof. By Propositions 3.6.1 and 3.6.2, $\mathcal{R}$ is an orthogonal basis if and only if, for any $f \in V$,

$$f = \sum_{n \in C} \left\langle f, \frac{\rho_n}{\|\rho_n\|} \right\rangle \frac{\rho_n}{\|\rho_n\|} = \sum_{n \in C} \frac{\langle f, \rho_n \rangle}{\|\rho_n\|^2} \rho_n. \qquad \square \qquad (3.128)$$

Each of the explicit bases produced in this section has *infinitely many* elements. From standard "uniqueness-of-dimension" results in linear algebra, it follows that *any* basis for *any* of the L^2 spaces considered herein is infinite. We say that these spaces are *infinite dimensional*.

Not surprisingly, $L^2(D)$ is also infinite dimensional when D is an *unbounded* interval or any product of intervals and tori. See Sections 3.8 and 7.1.

Exercises

3.6.1 Find real constants $a_0, b_0, b_1, c_0, c_1, c_2$ such that the polynomials $\rho_0 = a_0$, $\rho_1(x) = b_0 + b_1 x$, and $\rho_2(x) = c_0 + c_1 x + c_2 x^2$ form an orthogonal set in $L^2(0,1)$.

3.6.2

a. Use Theorem 3.6.4(c) and a change of variable to show that, if

$$\sigma_n(x) = \sqrt{\frac{8}{\pi(4+x^2)}} \sin\left(2n \arctan \frac{x}{2}\right),$$

then $\{\sigma_n : n \in \mathbb{Z}^+\}$ is an orthonormal *set* in $L^2(0, \infty)$.

b. Show these σ_n's form an orthonormal *basis* for $L^2(0, \infty)$. Hint: Use Theorem 3.6.4(b) to show that, for any $f \in L^2(0, \pi/2)$, the function g defined by $g(x) = (4+x^2)^{-1/2} f(\arctan(x/2))$ can be written as an infinite linear combination of the σ_n's. Now show *any* $g \in L^2(0, \infty)$ can be expressed in terms of an $f \in L^2(0, \pi/2)$ in the manner just described.

3.6.3 Use Corollary 3.6.1 and the Maclaurin series (A.20) to evaluate

$$\int_{-\pi}^{\pi} e^{e^{i\theta}} e^{-in\theta} \, d\theta$$

for $n \in \mathbb{Z}$.

3.7 MORE ON THE INNER PRODUCT

We catalog some additional useful results concerning orthogonal and orthonormal sets and bases in $L^2(D)$. Throughout, for the sake of concreteness, we also detail exactly what these results look like when $D = \mathbb{T}$, and the sets and bases in question comprise complex exponentials. We leave to the reader explicit formulation of these results in some other familiar cases.

Proposition 3.7.1 *Let $S = \{\sigma_n : n \in C\}$ be an orthonormal basis for a vector space $V \subset L^2(D)$; let $f \in V$. Then:*

(a) *For any $g \in L^2(D)$,*

$$\langle f, g \rangle = \sum_{n \in C} \langle f, \sigma_n \rangle \overline{\langle g, \sigma_n \rangle}. \tag{3.129}$$

(b) (Parseval's equation)

$$||f||^2 = \sum_{n \in C} |\langle f, \sigma_n \rangle|^2. \tag{3.130}$$

Proof. By Lemma 3.6.1 we have, for $g \in L^2(D)$,

$$\langle f, g \rangle = \sum_{n \in C} \langle f, \sigma_n \rangle \langle \sigma_n, g \rangle = \sum_{n \in C} \langle f, \sigma_n \rangle \overline{\langle g, \sigma_n \rangle}. \tag{3.131}$$

So we've proved part (a); to get part (b), we put $g = f$ into part (a) and use the fact that $||f||^2 = \langle f, f \rangle$. $\square$

Corollary 3.7.1 (a) *If $f, g \in L^2(\mathbb{T})$, then*

$$\langle f, g \rangle = 2\pi \sum_{n=-\infty}^{\infty} c_n(f)\overline{c_n(g)}. \tag{3.132}$$

(b) (Parseval's equation). *If $f \in L^2(\mathbb{T})$, then*

$$||f||^2 = 2\pi \sum_{n=-\infty}^{\infty} |c_n(f)|^2. \tag{3.133}$$

Proof. Put (3.112) into Proposition 3.7.1. $\square$

Note Corollary 3.7.1(b) says that Bessel's inequality (Lemma 1.7.2) is actually an equality, for $g \in L^2(\mathbb{T})$.

One application of the above proposition or its corollary is to the evaluation of infinite series. Such as:

Example 3.7.1 Evaluate

$$\sum_{m=-\infty}^{\infty} \frac{1}{(2m-1)^4}. \tag{3.134}$$

Solution. By Example 1.3.1 with $r = \pi/2$ we find that, if $g \in C(\mathbb{T})$ is defined by

$$g(x) = \begin{cases} x & \text{if } -\dfrac{\pi}{2} < x \le \dfrac{\pi}{2}, \\ \pi - x & \text{if } \dfrac{\pi}{2} < x \le \dfrac{3\pi}{2}, \end{cases} \tag{3.135}$$

then

$$g(x) \sim \frac{2}{\pi i} \sum_{n \ne 0} \frac{\sin(\pi n/2)}{n^2} e^{inx} = \frac{2}{\pi i} \sum_{m=-\infty}^{\infty} \frac{(-1)^{m+1}}{(2m-1)^2} e^{i(2m-1)x} \tag{3.136}$$

(the last step by (1.63) and the fact that the sine of an even multiple of $\pi/2$ is zero). So by Corollary 3.7.1,

$$2\pi \left|\frac{2}{\pi i}\right|^2 \sum_{m=-\infty}^{\infty} \left|\frac{(-1)^{m+1}}{(2m-1)^2}\right|^2 = \int_{-\pi/2}^{3\pi/2} |g(x)|^2 \, dx$$

$$= \int_{-\pi/2}^{\pi/2} x^2 \, dx + \int_{\pi/2}^{3\pi/2} (\pi - x)^2 \, dx = \frac{\pi^3}{6}. \tag{3.137}$$

Solving gives

$$\sum_{m=-\infty}^{\infty} \frac{1}{(2m-1)^4} = \frac{1}{2\pi} \cdot \frac{\pi^2}{4} \cdot \frac{\pi^3}{6} = \frac{\pi^4}{48}. \tag{3.138}$$

We next consider a problem of supreme practical import, namely, that of *approximating a function in $L^2(D)$ by linear combinations of elements of an orthogonal set*. Part (a) of the following proposition tells us how to do this; part (b) generalizes Proposition 3.7.1(b).

Proposition 3.7.2 *Let $\mathcal{R} = \{\rho_n \colon n \in C\}$ be an orthogonal set in $L^2(D)$; also let $f \in L^2(D)$. Then the following are true.*

(a) (the best approximation theorem)

$$\left\| f - \sum_{n \in C} \frac{\langle f, \rho_n \rangle}{||\rho_n||^2} \rho_n \right\| \leq \left\| f - \sum_{n \in C} d_n \rho_n \right\| \tag{3.139}$$

for any choice of coefficients $d_n \in \mathbb{C}$ (we agree that the norm on the right equals $+\infty$ if the series there does not converge in norm), with equality holding if and only if

$$d_n = \frac{\langle f, \rho_n \rangle}{||\rho_n||^2} \quad \text{for all } n \in C. \tag{3.140}$$

In other words,

$$\sum_{n \in C} \frac{\langle f, \rho_n \rangle}{||\rho_n||^2} \rho_n \tag{3.141}$$

provides the best approximation in norm to f, among all linear combinations of the ρ_n's.

(b) *Moreover,*

$$\left\| \sum_{n \in C} \frac{\langle f, \rho_n \rangle}{||\rho_n||^2} \rho_n \right\| = \sqrt{\sum_{n \in C} \frac{|\langle f, \rho_n \rangle|^2}{||\rho_n||^2}} \leq ||f||. \tag{3.142}$$

(c) If $C' \subset C$, then

$$\left\|f - \sum_{n \in C} \frac{\langle f, \rho_n\rangle}{\|\rho_n\|^2} \rho_n \right\| \leq \left\|f - \sum_{n \in C'} \frac{\langle f, \rho_n\rangle}{\|\rho_n\|^2} \rho_n \right\|. \tag{3.143}$$

Proof. For general $f, g \in L^2(D)$,

$$\|f - g\|^2 = \|f\|^2 - 2\operatorname{Re}\langle f, g\rangle + \|g\|^2 \tag{3.144}$$

(cf. (3.26) with $-g$ in place of g). So, by various parts of Proposition 3.1.3,

$$\left\|f - \sum_{n \in C} d_n \rho_n \right\|^2 = \|f\|^2 - 2\operatorname{Re}\left\langle f, \sum_{n \in C} d_n \rho_n \right\rangle + \left\|\sum_{n \in C} d_n \rho_n\right\|^2$$

$$= \|f\|^2 - 2\operatorname{Re}\left(\sum_{n \in C} \overline{d_n} \langle f, \rho_n\rangle\right) + \sum_{n \in C} \|d_n \rho_n\|^2$$

$$= \|f\|^2 + \sum_{n \in C} \left(-2\operatorname{Re}\left(\overline{d_n} \langle f, \rho_n\rangle\right) + |d_n|^2 \|\rho_n\|^2\right). \tag{3.145}$$

Now we compute easily that

$$-2\operatorname{Re}\overline{d}z + |d|^2 = |d - z|^2 - |z|^2 \tag{3.146}$$

for complex numbers d and z. Putting this, with $d = d_n\|\rho_n\|$ and $z = \langle f, \rho_n\rangle / \|\rho_n\|$, into (3.145) yields

$$\left\|f - \sum_{n \in C} d_n \rho_n \right\|^2 = \|f\|^2 + \sum_{n \in C} \left(\left|d_n\|\rho_n\| - \frac{\langle f, \rho_n\rangle}{\|\rho_n\|}\right|^2 - \frac{|\langle f, \rho_n\rangle|^2}{\|\rho_n\|^2}\right). \tag{3.147}$$

But $\left|d_n\|\rho_n\| - \langle f, \rho_n\rangle / \|\rho_n\|\right| \geq 0$, with equality holding if and only if $d_n = \langle f, \rho_n\rangle / \|\rho_n\|^2$. So the sum on the right side of (3.147), and hence the square root of the quantity on the left, attains its smallest value if (3.140) holds for each $n \in C$, and is strictly larger than this value if not. This completes the proof of part (a).

To prove part (b), we actually *put* $d_n = \langle f, \rho_n\rangle / \|\rho_n\|^2$ into (3.147); we get

$$\|f\|^2 - \sum_{n \in C} \frac{|\langle f, \rho_n\rangle|^2}{\|\rho_n\|^2} = \left\|f - \sum_{n \in C} \frac{\langle f, \rho_n\rangle}{\|\rho_n\|^2} \rho_n\right\|^2 \geq 0, \tag{3.148}$$

which gives us the inequality in (3.142). The equality there is by the Pythagorean theorem (3.17) on $L^2(D)$, as follows:

$$\left\|\sum_{n \in C} \frac{\langle f, \rho_n\rangle}{\|\rho_n\|^2} \rho_n\right\|^2 = \sum_{n \in C} \left\|\frac{\langle f, \rho_n\rangle}{\|\rho_n\|^2} \rho_n\right\|^2 = \sum_{n \in C} \frac{|\langle f, \rho_n\rangle|^2}{\|\rho_n\|^4} \|\rho_n\|^2$$

$$= \sum_{n \in C} \frac{|\langle f, \rho_n\rangle|^2}{\|\rho_n\|^2}. \tag{3.149}$$

For part (c), we simply put

$$d_n = \begin{cases} \dfrac{\langle f, \rho_n \rangle}{||\rho_n||^2} & \text{if } n \in C', \\ 0 & \text{if not} \end{cases} \qquad (3.150)$$

into (3.139). We're done. □

We catalog a number of happy consequences of the above. The first of these will be of considerable aid in our studies of *wavelets* in Chapter 8.

Corollary 3.7.2 *Let $\mathcal{R} = \{\rho_n : n \in C\}$ be an orthogonal set in a vector space $V \subset L^2(D)$. Then the following are equivalent.*

(i) *$\mathcal{R}$ is an orthogonal basis for V.*

(ii) *For all $f \in V$,*

$$||f||^2 = \sum_{n \in C} \frac{|\langle f, \rho_n \rangle|^2}{||\rho_n||^2}. \qquad (3.151)$$

(iii) *The set $\mathrm{span}(\mathcal{R})$ of all finite linear combinations of elements of $\mathcal{R}$ is dense in V.*

Proof. We have (i)⇒(ii) by Parseval's equation (Proposition 3.7.1(b)), so it suffices to show (ii)⇒(i) and (i)⇔(iii).

That (ii)⇒(i) is immediate: If (ii) holds then, by (3.148),

$$\left\| f - \sum_{n \in C} \frac{\langle f, \rho_n \rangle}{||\rho_n||^2} \rho_n \right\|^2 = 0 \qquad (3.152)$$

for all $f \in V$, so by Corollary 3.6.2, (i) holds.

Now suppose (i) is true. Given $f \in V$, define, for $N \in \mathbb{Z}^+$,

$$f_N = \sum_{\substack{n \in C \\ |n| \leq N}} \frac{\langle f, \rho_n \rangle}{||\rho_n||^2} \rho_n. \qquad (3.153)$$

Then by Corollary 3.6.2, the f_N's converge to f in norm. So f is a norm limit of elements of $\mathrm{span}(\mathcal{R})$. Thus (iii) follows.

Conversely, suppose (iii) holds. Let $f \in V$: We wish to show that $||f_N - f||$ converges to zero for f_N as above. But by Proposition 3.7.2(c), $||f_N - f||$ is a monotone decreasing sequence of nonnegative numbers; so by the monotone sequence property, this sequence converges. Therefore, to show that it converges to zero, we need only show that some subsequence $f_{M_1}, f_{M_2}, \ldots$ ($M_1 < M_2 < \cdots$) does.

We do this as follows. By assumption on span($\mathcal{R}$), we have a sequence

$$g_N = \sum_{k \in C_N} a_{k,N} \rho_k \quad (N \in \mathbb{Z}^+), \tag{3.154}$$

where C_N is a finite subset of C and $a_{k,N} \in \mathbb{C}$, such that $\lim_{N \to \infty} \|g_N - f\| = 0$. By defining $a_{k,N} = 0$ where necessary, we may as well assume that $C_N = \{-M_N, -M_N + 1, \ldots, M_N - 1, M_N\}$, where $0 < M_1 < M_2 < \cdots$. Then by Proposition 3.7.2(a), we have $\|f_{M_N} - f\| \le \|g_N - f\|$ for all N; so by the squeeze law and the nature of the g_N's, the f_{M_N}'s converge in norm to f, as required. □

The best approximation theorem is of great value in "real life" where, often, one is working not with an entire orthogonal basis (infinite sets typically being rather unwieldy in practice), but with a finite subset C thereof. The theorem tells us just how to get the most out of such a subset.

In particular, in the context of Fourier series in $L^2(\mathbb{T})$, we have the following.

Corollary 3.7.3 *Let $f \in L^2(\mathbb{T})$ and $N \in \mathbb{Z}^+$. The partial sum S_N^f of the Fourier series for f gives the best approximation to f in mean square distance, among all linear combinations of the complex exponentials*

$$\rho_{-N}, \rho_{-N+1}, \ldots, \rho_{-1}, \rho_0, \rho_1, \ldots, \rho_{N-1}, \rho_N, \tag{3.155}$$

where $\rho_n(x) = e^{inx}$.

Proof. We have

$$S_N^f = \sum_{n=-N}^{N} \frac{\langle f, \rho_n \rangle}{\|\rho_n\|^2} \rho_n \tag{3.156}$$

by (3.113) and the relation between the σ_n's there and the ρ_n's here. Applying Proposition 3.7.2(a) gives the desired result. □

The *gestalt* of Corollary 3.7.3 is as follows. Because of Theorem 3.4.1, we know that, for $f \in L^2(\mathbb{T})$, S_N^f serves as a *good* approximation to f, in the sense that the mean square distance between the former and the latter can be made arbitrarily small by choosing N large enough. But does it provide the *best* approximation? Is it, for any given N, *closer* to f than any other superposition of the complex exponentials (3.155)? Yes, says Corollary 3.7.3, it is, and it does.

For a somewhat philosophical application of the best approximation theorem, let's now consider the case where $C = \{-n_0, n_0\}$ for some given $n_0 \in \mathbb{Z}^+$. We still assume, for simplicity, that $D = \mathbb{T}$, and take ρ_n to be as in the above corollary. In this situation, the theorem tells us the following: The linear combination of ρ_{-n_0} and ρ_{n_0}—in other words, *the sinusoid of frequency n_0*—that best approximates a given $f \in L^2(\mathbb{T})$, in mean square norm, is *the frequency component*

$$c_{-n_0}(f)\rho_{-n_0}(x) + c_{n_0}(f)\rho_{n_0}(x) = a_{n_0}(f)\cos n_0 x + b_{n_0}(f)\sin n_0 x. \tag{3.157}$$

We've argued, since the Introduction, for thinking of the sinusoid (3.157) as "that part of f with frequency n_0." The fact, just observed, that this sinusoid "looks more like f" (in mean square distance) than anything *else* with this frequency certainly supports our argument. (Likewise, $c_0(f) = a_0(f)/2$ is the "constant part of f.")

Analogously, we understand $c_n(f)\rho_{n_0}$ as "that part of f that looks like the complex exponential ρ_{n_0}," for $f \in L^2(\mathbb{T})$. And in general, for *any* space $L^2(D)$ and functions f and $\rho \neq 0$ in that space, we think of

$$\frac{\langle f, \rho \rangle}{||\rho||^2} \rho \tag{3.158}$$

as "that part of f that looks like ρ." (Note that, for $\rho \in L^2(D) - \{0\}$, $\mathcal{R} = \{\rho\}$ is an orthogonal set in $L^2(D)$, so that Proposition 3.7.2(a) applies.) And we think of the coefficient $\langle f, \rho \rangle / ||\rho||^2$ itself as gauging "the extent to which ρ goes into the makeup of f."

Exercises

3.7.1 Use Corollary 3.7.1(b) and the result of Exercise 1.2.14 to show that, for $a \in \mathbb{R}$,

$$\sum_{n=-\infty}^{\infty} \frac{\sin^2 \pi(a+n)}{\pi^2 (a+n)^2} = 1.$$

3.7.2 Use Corollary 3.7.1(b) and the result of Exercise 1.3.16 to show that, for $a \in \mathbb{R}$,

$$\sum_{n=-\infty}^{\infty} \frac{\sin^4 \pi(a+n)}{\pi^4 (a+n)^4} = \frac{1}{3}(2 + \cos 2\pi a).$$

3.7.3 Use Corollary 3.7.1(b) and results from Exercise 1.8.6 to evaluate

$$\zeta(s) = \sum_{n=1}^{\infty} \frac{1}{n^s}$$

for $s = 8$ and $s = 10$.

3.7.4 (For this exercise and the next one, it may help to recall Example 1.2.1.) Let $f \in L^2(\mathbb{T})$ be defined by $f(x) = e^x$ ($-\pi < x \leq \pi$).

    a. Find the best approximation g in norm to f among all linear combinations of the complex exponentials $\rho_{-2}, \rho_0,$ and ρ_2, where $\rho_n(x) = e^{inx}$.

    b. Sketch the graphs of f and g on the same set of axes. (The g you found should be real-valued, like f.)

3.7.5 Repeat the above exercise, but this time using $\rho_{-2}, \rho_0,$ and ρ_3. Your best approximation g will not, in this case, be real-valued, so you'll have to graph f versus $\operatorname{Re} g$ on one set of axes and $\operatorname{Im} f$, which is zero, versus $\operatorname{Im} g$ on another.

3.7.6 Consider the functions $\varphi_1, \varphi_2, \varphi_3$ in $L^2(0,1)$ defined by
$$\varphi_1(x) = \chi_{[0,\frac{1}{3}]}(x); \quad \varphi_2(x) = \chi_{[\frac{1}{3},\frac{1}{2}]}(x); \quad \varphi_3(x) = \chi_{[\frac{1}{2},1]}(x).$$
(Recall the definition (3.96) of χ_R.)

a. Show that $\{\varphi_1, \varphi_2, \varphi_3\}$ is an orthogonal set in $L^2(0,1)$.
b. Find $\|\varphi_1\|, \|\varphi_2\|, \|\varphi_3\|$.
c. Among all linear combinations of the functions $\varphi_1, \varphi_2, \varphi_3$, find the best approximation in norm to the function $f(x) = x - \pi \sin 4\pi x$ in $L^2(0,1)$. Sketch this best approximation and f on the same set of axes. (Your best approximation should be a *step function*—a function that's piecewise constant.)

3.7.7 Consider the functions ψ_k ($k = 1, 2, 3, \ldots$) in $L^2(0,1)$ defined by
$$\psi_k(x) = \begin{cases} \sqrt{k(k+1)} & \text{if } \frac{1}{k+1} < x \leq \frac{1}{k}; \\ 0 & \text{otherwise.} \end{cases}$$

a. Show that $\{\psi_k : k \in \mathbb{Z}^+\}$ is an orthonormal set in $L^2(0,1)$.
b. Show that $\{\psi_k\}$ is an orthonormal *set* in $L^2(0,1)$.
c. Among all linear combinations of the form
$$\sum_{k=1}^{\infty} c_k \psi_k,$$
find the best approximation in norm to the function $f(x) = x$ in $L^2(0,1)$. Sketch this best approximation and f on the same set of axes.

3.7.8 **a.** Show that, if $\rho_1(x) = x$, $\rho_2(x) = 3x^2 - 1$, and $\rho_3(x) = 1$ for all $x \in [-1, 1]$, then $\{\rho_1, \rho_2, \rho_3\}$ is an orthogonal set in $L^2(-1, 1)$.
b. Compute $\|\rho_1\|, \|\rho_2\|$, and $\|\rho_3\|$.
c. Find the best approximation in norm to the function $f(x) = 5x^3 - 3x^2$ in $L^2(-1, 1)$ among all linear combinations of the functions ρ_1, ρ_2, ρ_3. Sketch this best approximation and f on the same set of axes.

3.7.9 Show that, in $L^2(\mathbb{T})$, the partial sums S_N^f do *not* necessarily provide the best *uniform* approximation to f among all linear combinations of complex exponentials $\rho_n(x) = e^{inx}$. That is, find an f in $L^2(\mathbb{T})$, a nonnegative integer N, and complex numbers d_n ($-N \leq n \leq N$) such that
$$\sup_{x \in \mathbb{R}} |S_N^f(x) - f(x)| > \sup_{x \in \mathbb{R}} \left| \sum_{n=-N}^{N} d_n \rho_n(x) - f(x) \right|.$$
Hint: Take $f(x) = x^2$ on $(-\pi, \pi]$, and $N = 0$.

3.8 ORTHONORMAL BASES FOR PRODUCT DOMAINS

The following result, of stunning elegance, simplicity, and power, tells us that orthonormal bases may be constructed "one real variable at a time."

Proposition 3.8.1 *Let A be a torus or interval, and likewise for B; let $D = A \times B$. If $\Phi = \{\phi_m \colon m \in \mathbb{Z}\}$ and $\Psi = \{\psi_n \colon n \in \mathbb{Z}\}$ are orthonormal bases for $L^2(A)$ and $L^2(B)$, respectively, then*

$$\Lambda = \{\phi_m \psi_n \colon m, n \in \mathbb{Z}\} \tag{3.159}$$

is an orthonormal basis for $L^2(D)$.

Proof. Let all quantities be as stipulated; we first show that Λ is an orthonormal *set* in $L^2(D)$. This is straightforward: If $(m, n), (k, \ell) \in \mathbb{Z}^2$, then

$$\begin{aligned}
\langle \phi_m \psi_n, \phi_k \psi_\ell \rangle &= \int_D (\phi_m \psi_n)(x) \overline{(\phi_k \psi_\ell)(x)} \, dx \\
&= \int_A \int_B \phi_m(u) \overline{\phi_k(u)} \psi_n(v) \overline{\psi_\ell(v)} \, dv \, du \\
&= \int_A \phi_m(u) \overline{\phi_k(u)} \, du \int_B \psi_n(v) \overline{\psi_\ell(v)} \, dv \\
&= \delta_{m,k} \delta_{n,\ell}, \tag{3.160}
\end{aligned}$$

which equals 1 if $m = k$ and $n = \ell$ and 0 otherwise. We've used Fubini's theorem (Proposition 3.5.3); its application is valid here because $\phi_m, \psi_n, \phi_k, \psi_\ell \in L^2(D)$, and consequently

$$\int_A \int_B \left| \phi_m(u) \overline{\phi_k(u)} \psi_n(v) \overline{\psi_\ell(v)} \right| \, dv \, du \tag{3.161}$$

$$= \int_A \left| \phi_m(u) \overline{\phi_k(u)} \right| du \int_B \left| \psi_n(v) \overline{\psi_\ell(v)} \right| dv < \infty. \tag{3.162}$$

Also, we've used and will continue to use, throughout this section, the symbol $\langle \cdot, \cdot \rangle$ to denote the inner product on $L^2(D)$ (rather than $L^2(A)$ or $L^2(B)$).

To show Λ is a basis, it will suffice, by Proposition 3.6.1, to show that

$$\lim_{M \to \infty} \lim_{N \to \infty} \left\| \sum_{m=-M}^{M} \sum_{n=-N}^{N} \langle f, \phi_m \psi_n \rangle \phi_m \psi_n - f \right\| = 0 \tag{3.163}$$

for $f \in L^2(D)$, the symbol $\|\cdot\|$ denoting, throughout this section, the norm on that space. (Actually, some care needs to be taken here. To apply Proposition 3.6.1 directly, we should, strictly speaking, "line up" the elements of Λ in an infinite list $\lambda_1, \lambda_2, \ldots$, and then show that $\sum_{|j|<J} \langle f, \lambda_j \rangle \lambda_j$ approaches f, in norm, as $J \to \infty$. It is not *a priori* clear that this is equivalent to showing that the above double limit is zero for such f. Fortunately, though, the two *are* equivalent. This follows from Parseval's equation (Proposition 3.7.1(b)) above: All summands on the right side of

that equation are nonnegative, so they may be summed in any order; therefore so may the building blocks $\langle f, \sigma_n \rangle \sigma_n$ of the function f on the left.)

So let's pick $f \in L^2(D)$ and demonstrate (3.163). To do so, we produce some auxiliary functions f_v in $L^2(A)$, and some additional auxiliary functions $g_m \in L^2(B)$, as follows. First, let $v \in B$ be fixed, and denote by f_v the function on A defined by

$$f_v(u) = f(u, v) \tag{3.164}$$

whenever the right side makes sense. We show that $f_v \in L^2(A)$ for almost all $v \in B$, as follows. Since $f \in L^2(D)$, we have

$$\int_B \left(\int_A |f(u,v)|^2 \, du \right) dv < \infty; \tag{3.165}$$

this is possible only if the integral in parentheses defines a function on B, meaning this integral makes sense—in other words $f_v \in L^2(A)$—for almost all $v \in B$, as claimed.

Let's also define, for each $m \in \mathbb{Z}$, a function g_m on B by

$$g_m(v) = \langle f_v, \phi_m \rangle_A, \tag{3.166}$$

where now we subscript an inner product, or a norm, with an A or a B to indicate the domain in question. We check that $g_m \in L^2(B)$ for all $m \in \mathbb{Z}$ as follows. We note that

$$|g_m(v)|^2 = |\langle f_v, \phi_m \rangle_A|^2 \leq \|f_v\|_A^2 \|\phi_m\|_A^2 = \|f_v\|_A^2 \tag{3.167}$$

by the Cauchy-Schwarz inequality on $L^2(A)$ and the fact that the ϕ_m's form an orthonormal set in this space. But then, since

$$\int_B \|f_v\|_A^2 \, dv = \int_B \int_A |f_v(u)|^2 \, du \, dv = \int_A \int_B |f(u,v)|^2 \, du \, dv < \infty, \tag{3.168}$$

Proposition 3.5.2 implies that $g_m \in L^2(B)$, as required.

We return now to the verification of (3.163). We write

$$\sum_{m=-M}^{M} \sum_{n=-N}^{N} \langle f, \phi_m \psi_n \rangle \phi_m \psi_n - f$$

$$= \sum_{m=-M}^{M} \phi_m \left(\sum_{n=-N}^{N} \langle f, \phi_m \psi_n \rangle \psi_n - g_m \right) + \left(\sum_{m=-M}^{M} g_m \phi_m - f \right), \tag{3.169}$$

so that, by the triangle inequality,

$$\left\| \sum_{m=-M}^{M} \sum_{n=-N}^{N} \langle f, \phi_m \psi_n \rangle \phi_m \psi_n - f \right\|$$

$$\leq \sum_{m=-M}^{M} \left\| \phi_m \left(\sum_{n=-N}^{N} \langle f, \phi_m \psi_n \rangle \psi_n - g_m \right) \right\| + \left\| \sum_{m=-M}^{M} g_m \phi_m - f \right\|. \tag{3.170}$$

Thus, for any fixed M,

$$\lim_{N\to\infty} \left\| \sum_{m=-M}^{M} \sum_{n=-N}^{N} \langle f, \phi_m \psi_n \rangle \phi_m \psi_n - f \right\|$$
$$\leq \sum_{m=-M}^{M} \lim_{N\to\infty} \left\| \phi_m \left(\sum_{n=-N}^{N} \langle f, \phi_m \psi_n \rangle \psi_n - g_m \right) \right\| + \left\| \sum_{m=-M}^{M} g_m \phi_m - f \right\|,$$
(3.171)

provided the limits in question exist.

The one on the left side does for the following reason. Since Λ is an orthonormal set in $L^2(D)$, Proposition 3.7.2(c) tells us that the norm on the left side of (3.171) is a decreasing function of N. It's also bounded below, by zero, so by the bounded monotone sequence property of $\mathbb{R}$, it does indeed have a limit.

We now show that each limit on the right not only exists but is zero. To accomplish this, we first observe that

$$\langle f, \phi_m \psi_n \rangle = \int_B \int_A f(u,v) \overline{\phi_m(u) \psi_n(v)} \, du \, dv$$
$$= \int_B \left(\int_A f_v(u) \overline{\phi_m(u)} \, du \right) \overline{\psi_n(v)} \, dv$$
$$= \int_B \langle f_v, \phi_m \rangle_A \overline{\psi_n(v)} \, dv = \int_B g_m(v) \overline{\psi_n(v)} \, dv = \langle g_m, \psi_n \rangle_B.$$
(3.172)

So

$$\left\| \phi_m \left(\sum_{n=-N}^{N} \langle f, \phi_m \psi_n \rangle \psi_n - g_m \right) \right\| = \left\| \phi_m \left(\sum_{n=-N}^{N} \langle g_m, \psi_n \rangle_B \psi_n - g_m \right) \right\|$$
$$= \int_B \int_A \left| \phi_m(u) \left(\sum_{n=-N}^{N} \langle g_m, \psi_n \rangle_B \psi_n(v) - g_m(v) \right) \right|^2 du \, dv$$
$$= \left(\int_B \left| \sum_{n=-N}^{N} \langle g_m, \psi_n \rangle_B \psi_n(v) - g_m(v) \right|^2 dv \right) \left(\int_A |\phi_m(u)|^2 du \right)$$
$$= \left\| \sum_{n=-N}^{N} \langle g_m, \psi_n \rangle_B \psi_n - g_m \right\|_B^2 \|\phi_m\|_A^2 = \left\| \sum_{n=-N}^{N} \langle g_m, \psi_n \rangle_B \psi_n - g_m \right\|_B^2.$$
(3.173)

The second equality is by by definition of the norm on $L^2(D)$, the third because the double integral breaks up into a product of two single ones, the fourth by definition of the norms on $L^2(A)$ and $L^2(B)$, and the fifth because, by assumption, the ϕ_m's have norm 1. The right side of (3.173) does go to zero as $N \to \infty$, because $g_m \in L^2(B)$ and the ψ_n's form an orthonormal basis for that space.

So by (3.171),

$$\lim_{N\to\infty}\left\|\sum_{m=-M}^{M}\sum_{n=-N}^{N}\langle f,\phi_m\psi_n\rangle\,\phi_m\psi_n - f\right\| \le \left\|\sum_{m=-M}^{M} g_m\phi_m - f\right\|. \tag{3.174}$$

To prove our proposition, it suffices to show that the right side of (3.174) has limit zero as $M \to \infty$. Let's. We have

$$\left\|\sum_{m=-M}^{M} g_m\phi_m - f\right\|^2 = \int_B\int_A\left|\sum_{m=-M}^{M} g_m(v)\phi_m(u) - f(u,v)\right|^2 du\,dv$$

$$= \int_B\int_A\left|\sum_{m=-M}^{M}\langle f_v,\phi_m\rangle_A\,\phi_m(u) - f_v(u)\right|^2 du\,dv$$

$$= \int_B\left\|\sum_{m=-M}^{M}\langle f_v,\phi_m\rangle_A\,\phi_m - f_v\right\|_A^2 dv. \tag{3.175}$$

The integrand approaches zero as $M \to \infty$ for almost all $v \in B$, since the ϕ_m's form an orthonormal basis for $L^2(A)$ and $f_v \in L^2(A)$ almost everywhere on B. So, by Proposition 3.5.1, the integral on the right side of (3.175) goes to zero as $M \to \infty$ as well, *provided* there's some function $H(v)$ that's integrable on B and bounds the integrand independently of M. But there is such an H: If we let $H(v) = \|f_v\|_A^2$ then, by Proposition 3.7.2(c) with C' equal to the empty set,

$$\left\|\sum_{m=-M}^{M}\langle f_v,\phi_m\rangle_A\,\phi_m - f_v\right\|_A^2 \le \|-f_v\|_A^2 = H(v). \tag{3.176}$$

And H *is* integrable on B, by (3.168). So the right side of (3.175) does go to zero as $M \to \infty$, and we're done. □

As a basic example we have, by the above proposition and Theorem 3.6.2, the following for $f \in L^2(\mathbb{T}_P \times \mathbb{T}_Q)$:

$$f = \sum_{m=-\infty}^{\infty}\sum_{n=-\infty}^{\infty} \langle f,\phi_m\psi_n\rangle\,\phi_m\psi_n, \tag{3.177}$$

where

$$\phi_m(u) = \frac{e^{2\pi imu/P}}{\sqrt{P}}, \quad \psi_n(v) = \frac{e^{2\pi inv/Q}}{\sqrt{Q}}, \tag{3.178}$$

and consequently

$$\langle f,\phi_m\psi_n\rangle = \frac{1}{\sqrt{PQ}}\int_{-P/2}^{P/2}\int_{-Q/2}^{Q/2} f(u,v)e^{-2\pi i((mu/P)+(nv/Q))}\,du\,dv. \tag{3.179}$$

We note that Proposition 3.8.1 generalizes in the expected way to the context of domains D with any finite number of factors.

Exercises

3.8.1 Define $\phi_m \in L^2(0,1)$, for $m \in \mathbb{Z}^+$, by $\phi_m(u) = u^m$; define $\psi_n \in L^2(-\pi, \pi)$, for $n \in \mathbb{Z}^+$, by $\psi_n(v) = e^{inv}$.
  a. Is $\{\phi_m : m \in \mathbb{Z}^+\}$ an orthogonal set in $L^2(0,1)$?
  b. Is $\{\psi_n : n \in \mathbb{Z}^+\}$ an orthogonal set in $L^2(-\pi, \pi)$?
  c. Is $\{\phi_m \psi_n : m, n \in \mathbb{Z}^+\}$ an orthogonal set in $L^2((0,1) \times (-\pi, \pi))$? Explain.
  d. Is $\{\phi_m \psi_m : m \in \mathbb{Z}^+\}$ an orthogonal set in $L^2((0,1) \times (-\pi, \pi))$? Explain.

3.8.2 Show that $f_n(x, y) = (x + iy)^n$ ($n \in \mathbb{Z}^+$) defines an orthogonal set in $L^2(D)$, where D is the disk $\{(x, y) : x^2 + y^2 \leq 1\} \subset \mathbb{R}^2$, and compute $\|f_n\|$ for each n. Hint: polar coordinates.

3.8.3 Let A and B be as in Proposition 3.8.1. Suppose $\Phi = \{\phi_m : m \in \mathbb{Z}\}$ is an orthonormal *set* in $L^2(A)$, $\Psi = \{\psi_n : n \in \mathbb{Z}\}$ is an orthonormal *set* in $L^2(B)$, and $\Lambda = \{\phi_m \psi_n : m, n \in \mathbb{Z}\}$ is an orthonormal *basis* for $L^2(A \times B)$. Are Φ and Ψ necessarily orthonormal bases for $L^2(A)$ and $L^2(B)$, respectively? Hint: given $f \in L^2(A)$, consider the expansion of $f\psi_k$, where ψ_k is any element of Ψ, in terms of the basis Λ. Similarly for $g \in L^2(B)$.

3.9 AN APPLICATION: THE ISOPERIMETRIC PROBLEM

If you have a fixed length of bordering material and you want to bound as much area as possible with it, into what shape should you make it? Fourier series provide the answer.

Proposition 3.9.1 *Let Γ be a curve of length 2π in the pq plane. Suppose Γ is:*

(a) parametrized counterclockwise by arc length—*this means the curve is swept out by a point moving counterclockwise along it, with position $(p(t), q(t))$ at time t, and at constant speed 1 (that is, time elapsed equals arc length traversed);*

(b) piecewise smooth and continuous—*this means $p, q \in \mathrm{PSC}(\mathbb{T})$;*

(c) closed, *meaning $(p(t_1), q(t_1)) = (p(t_2), q(t_2))$ if t_1 and t_2 differ by an integer multiple of 2π, and* simple, *meaning the converse.*

(See Fig. 3.6.)
Let A be the area of the region enclosed by Γ. Then $A < \pi$ unless Γ is a circle (in which case $A = \pi$).

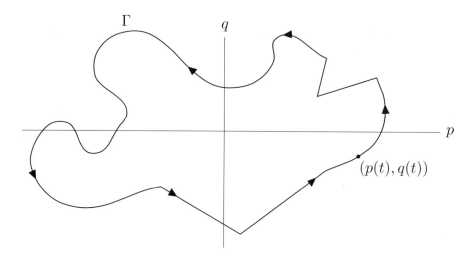

Fig. 3.6 A curve Γ satisfying (a)(b)(c) above

Proof. By translating and rotating our curve Γ if necessary, we may assume that it intersects the horizontal axis at $t = 0$ and at $t = \pi$—in other words, that $q(0) = q(\pi) = 0$.

Now by Green's theorem from vector calculus (see Section 14.4 in [16]), the area A of the region bounded by Γ satisfies

$$A = \frac{1}{2} \int_0^{2\pi} [p(t)q'(t) - p'(t)q(t)]\, dt = -\int_0^{2\pi} p'(t)q(t)\, dt. \tag{3.180}$$

(The second equality is obtained from the first through integration by parts.) Applying the Cauchy-Schwarz inequality (3.15) and equation (1.142), we get

$$A \leq ||p'|| \, ||q|| \leq \frac{1}{2}(||p'||^2 + ||q||^2). \tag{3.181}$$

We now invoke *Wirtinger's inequality*, which states: *If $q \in \mathrm{PSC}(\mathbb{T})$ and $q(0) = q(\pi) = 0$, then $||q|| \leq ||q'||$, with equality holding if and only if q is a constant times $\sin t$.* (See Exercise 3.9.1.) Then (3.181) reads

$$A \leq \frac{1}{2}(||p'||^2 + ||q'||^2) = \frac{1}{2} \int_0^{2\pi} [(p'(t))^2 + (q'(t))^2]\, dt. \tag{3.182}$$

(Observe that p' and q' are real-valued.) But $(p'(t))^2 + (q'(t))^2$ is just the length of the velocity vector $(p'(t), q'(t))$; since the speed of the point $(p(t), q(t))$ is assumed constant and equal to 1, this length equals 1. So the right side of (3.182) equals π, and we have our desired upper bound on A.

To finish, we must determine exactly when this upper bound *equals* A. Tracing through our arguments beginning with (3.180), we see that this will be the case if and only if all of the following are true: The integral on the right side of (3.180) equals $||p'||\,||q||$; the latter quantity equals $(||p'||^2+||q||^2)/2$; and $||q||=||q'||=\sqrt{\pi}$. Let's work backward: by Wirtinger's inequality, the last string of equalities will obtain if and only if $q(t)=\pm\sin t$. Next, by simple algebra, $||p'||\,||q||=(||p'||^2+||q||^2)/2$ if and only if $||q||=||p'||$; so $A=\pi$ if and only if $q(t)=\pm\sin t$ and $||p'||=||q||=\sqrt{\pi}$. Finally, the absolute value of the integral over $[0,2\pi]$ of $-p'q$ equals $||p'||\,||q||$ if and only if p' is a constant multiple of q, by Exercise 3.1.2. In sum, $A=\pi$ if and only if $q(t)=\pm\sin t$ and $p(t)=\pm\cos t$. Then certainly $(p(t),q(t))$ traces out the circle $\{(x,y)\colon x^2+y^2=1\}$, and we're done. □

The requirement of piecewise smoothness, in the above proposition, may be weakened, by applying limiting arguments to the result just demonstrated. (That is, one approximates nonsmooth curves by smooth ones.) Also, the assumption of simplicity is not really necessary.

Exercises

3.9.1 Prove Wirtinger's inequality, stated just below equation (3.181), as follows.
   a. Let w be the 2π-periodic, odd extension of the restriction to $[0,\pi]$ of q. (It might help to draw a picture.) Explain why $w \in \mathrm{PSC}(\mathbb{T})$.
   b. Explain why $c_0(w) = c_0(w') = 0$.
   c. Use Parseval's equation (Corollary 3.7.1(b)) and the fact that $|nc_n(w)| \geq |c_n(w)|$ when $|n| \geq 1$ to show that $||w'|| \geq ||w||$.
   d. Now repeat the above arguments with the 2π-periodic, odd extension v of the restriction to $[-\pi, 0]$ of q. Add your results together to deduce that $||q'|| \geq ||q||$.
   e. By looking back over your above proof and determining when the inequalities there become equalities, prove that $||q'|| = ||q||$ if and only if q is a multiple of $\sin t$.

3.10 WHAT IS $L^2(\mathbb{T})$?

$L^2(\mathbb{T})$ is a somewhat subtle, but incredibly beautiful, complex vector space. Here, to strengthen our understanding of this space, we discuss it in seven different but related ways, by presenting seven different but related "things that $L^2(\mathbb{T})$ is."

Some of what we have to say is familiar; some is new. Most of it applies, or extends, to the context of more general spaces $L^2(D)$. We warn, though, that Thing 3.10.7 is a distinctly one-dimensional thing; while it does carry over with little change to Fourier series on tori $\mathbb{T}_P$ or on bounded intervals, it definitely does *not* generalize nicely to product domains D.

Thing 3.10.1 $L^2(\mathbb{T})$ is the *completion* of $C(\mathbb{T})$ with respect to the mean square norm. That is, $L^2(\mathbb{T})$ is the unique, complete metric space that contains $C(\mathbb{T})$ as a dense subspace.

Actually, we're being a bit imprecise here. The metric space $L^2(\mathbb{T})$ is, strictly speaking, a space of *equivalence classes* of functions (see the end of Section 3.3), and $C(\mathbb{T})$ a space of functions *per se*. Nevertheless, we may *think* of $C(\mathbb{T})$ as sitting inside $L^2(\mathbb{T})$ in a natural way. Namely, we identify $f \in C(\mathbb{T})$ with its equivalence class $[f]$.

We claim that this identification is an *isometry*; this means it preserves norms ($||f|| = ||[f]||$) and is one-to-one (if $[f] = [g]$, then $f = g$). The first of these claims is true by the definition (cf. the end of Section 3.3) of the norm on $L^2(\mathbb{T})$. The second is true by Propositions 3.1.1(a) and 3.2.1(c): The second of these propositions tells us that $[f] = [g] \Rightarrow ||f - g|| = 0$, while the first says $||f - g|| = 0 \Rightarrow f(x) = g(x)$ everywhere, provided f and g are continuous.

Also imprecise is our claim of uniqueness of $L^2(\mathbb{T})$: It's only unique up to isometry. That is, if W is any complete metric space containing $C(\mathbb{T})$ as a dense subspace, then there is an isometry from W onto $L^2(\mathbb{T})$.

Isometric spaces "look the same" from a metric space point of view.

It should be noted that every metric space X has a completion, meaning a complete metric space W containing (an isometric image of) X as a dense subset, and that this completion is unique (up to isometry). See Section 7.4 in [29] as well as Exercise 3.10.3 below.

Thing 3.10.2 $L^2(\mathbb{T})$ *is* the the space of (equivalence classes of) 2π-periodic, measurable functions f such that

$$\int_{\mathbb{T}} |f(x)|^2 \, dx < \infty, \tag{3.183}$$

the integral being Lebesgue. See (3.97) and the surrounding discussions in Section 3.5.

Next we have the following, which is particularly relevant to the issue of convergence of Fourier series:

Thing 3.10.3 $L^2(\mathbb{T})$ *is* the space of (equivalence classes of) 2π-periodic functions f such that S_N^f converges in norm to f.

The above thing *is* the upshot of Theorem 3.4.1.
Somewhat more abstractly:

Thing 3.10.4 $L^2(\mathbb{T})$ *is* a complex Hilbert space.

By this we mean:

Definition 3.10.1 A *complex Hilbert space* is a complex vector space V with the following four properties.

(a) V is equipped with an inner product $\langle \cdot, \cdot \rangle$ and a norm $|| \cdot ||$ satisfying all conclusions of Propositions 3.1.1 and 3.1.3.

(b) V is *separable*. That is, there is some countable subset W of V that's dense in V with respect to the norm.

(c) V is complete with respect to the norm.

(d) V is infinite dimensional.

Some commentary is merited here. First, $V = L^2(\mathbb{T})$ does indeed satisfy property (b) above if we let W be the vector space of equivalence classes of 2π-periodic extensions of all polynomials $p: (-\pi, \pi] \to \mathbb{C}$ with rational coefficients. That this space W really is dense in $L^2(\mathbb{T})$ follows from a variant on the Stone-Weierstrass theorem (Theorem 5.8.1 in [33]). Second, we noted in Section 3.6 that $L^2(\mathbb{T})$ is infinite dimensional.

According to a major but standard theorem in the theory of complex Hilbert spaces (see, for example, Theorems 11 and 11', Chapter 4, in [29]), any two such spaces V_1 and V_2 are *isomorphic* in the following sense.

Definition 3.10.2 (a) A mapping $\mathcal{I}: V_1 \to V_2$, where V_1 and V_2 are vector spaces equipped with inner products $\langle \cdot, \cdot \rangle_1$ and $\langle \cdot, \cdot \rangle_2$, respectively, is called an *isomorphism* from V_1 to V_2 if $\mathcal{I}$ is:

(i) *bijective*, meaning one-to-one and onto V_2;

(ii) *linear*, meaning

$$\mathcal{I}(\alpha_1 f_1 + \alpha_2 f_2) = \alpha_1 \mathcal{I}(f_1) + \alpha_2 \mathcal{I}(f_2) \tag{3.184}$$

for any $\alpha_1, \alpha_2 \in \mathbb{C}$ and $f_1, f_2 \in V_1$;

(iii) *inner product preserving*, meaning

$$\langle \mathcal{I}(f_1), \mathcal{I}(f_2) \rangle_2 = \langle f_1, f_2 \rangle_1 \tag{3.185}$$

for $f_1, f_2 \in V_1$.

(b) A vector space V_1 is *isomorphic* to a vector space V_2 if there's an isomorphism $\mathcal{I}: V_1 \to V_2$.

If V_1 is isomorphic to V_2, then V_2 is isomorphic to V_1, since if $\mathcal{I}$ is an isomorphism from V_1 to V_2 then the inverse map $\mathcal{I}^{-1}$ is an isomorphism from V_2 to V_1. We call either of these maps an isomorphism *between* the spaces.

That all complex Hilbert spaces are isomorphic means they all "look the same" in many ways (much as isometric spaces do, only more so).

The above definition brings us to:

Thing 3.10.5 $L^2(\mathbb{T})$ *is* isomorphic to the vector space l^2 of *square summable* complex sequences, meaning infinite sequences

$$(r_n)_{n \in \mathbb{Z}} = (\ldots, r_{-2}, r_{-1}, r_0, r_1, r_2, \ldots) \tag{3.186}$$

of complex numbers r_n such that

$$\sum_{n=-\infty}^{\infty} |r_n|^2 < \infty. \tag{3.187}$$

(We previously encountered l^2 in Exercise 2.2.5.)

It will be convenient to denote the sequence (3.186) simply by (r_n).

It's readily shown that l^2 *is a vector space.* (By the sum of two elements (r_n) and (s_n) of l^2, we mean the sequence $(r_n + s_n)$; by $\alpha(r_n)$ we mean (αr_n), for $\alpha \in \mathbb{C}$.) What's less immediate is that it's a Hilbert space, with inner product

$$\langle (r_n), (s_n) \rangle = \sum_{n=-\infty}^{\infty} r_n \overline{s_n}. \tag{3.188}$$

But it's true. In particular, if ϕ_n is, for any $n \in \mathbb{Z}$, the infinite sequence with nth coordinate equal to 1 and all others 0, then $\mathcal{S} = \{\phi_n : n \in \mathbb{Z}\}$ *is* an orthonormal basis for l^2. We leave the details to the reader.

By the authority of the "major but standard" theorem cited just before Definition 3.10.2, $L^2(\mathbb{T})$ *is* isomorphic to l^2. But in order to understand more concretely what's going on here, let's invoke our inalienable right to question authority, and demonstrate the isomorphism explicitly. We have:

Proposition 3.10.1 *Given $f \in L^2(\mathbb{T})$, define $\mathcal{I}f \in l^2$ by*

$$\mathcal{I}f = \left(\sqrt{2\pi}\, c_n(f)\right). \tag{3.189}$$

Then $\mathcal{I}$ defines an isomorphism between $L^2(\mathbb{T})$ and l^2.

Proof. First we must check that $\mathcal{I}f$ really is in l^2 when $f \in L^2(\mathbb{T})$. This is so because

$$\sum_{n=-\infty}^{\infty} |\sqrt{2\pi}\, c_n(f)|^2 \tag{3.190}$$

converges by Parseval's equation (Corollary 3.7.1(b)).

We leave it to the reader to check that $\mathcal{I}$ is linear, injective, and inner product preserving. (See Exercise 3.10.1.) It then remains only to check surjectivity, so we need to show that, given $r = (r_n) \in l^2$, there is an $f \in L^2(\mathbb{T})$ such that $\mathcal{I}f = r$.

We do this as follows: Given such an r, we define $f_N \in L^2(\mathbb{T})$ ($N \in \mathbb{N}$) by

$$f_N = \sum_{n=-N}^{N} r_n \sigma_n, \tag{3.191}$$

with σ_n as in (3.104). Note that, for $M > N$,

$$\|f_M - f_N\|^2 = \left\| \sum_{N < |n| \leq M} r_n \sigma_n \right\|^2 = \sum_{N < |n| \leq M} |r_n|^2 \|\sigma_n\|^2 = \sum_{N < |n| \leq M} |r_n|^2; \tag{3.192}$$

the second equality is by the Pythagorean theorem (3.17) and the orthogonality of the σ_n's and the last is by orthonormality of the σ_n's. Now because the sum of the squares of the $|r_n|$'s is, by assumption, convergent, and because convergent series have tail ends going to zero, we have

$$\lim_{N \to \infty} \sup_{M > N} \sum_{N < |n| \leq M} |r_n|^2 = \lim_{N \to \infty} \sum_{|n| > N} |r_n|^2 = 0. \qquad (3.193)$$

This and (3.192) tell us that the f_N's form a Cauchy sequence in $L^2(\mathbb{T})$. So by completeness of this space, it contains an element f to which these f_N's converge in norm.

That is,

$$f = \sum_{n=-\infty}^{\infty} r_n \sigma_n. \qquad (3.194)$$

So by Theorem 3.4.1 and the fact that expansions in terms of bases are unique, we find that $r_n = \langle f, \sigma_n \rangle = \sqrt{2\pi}\, c_n(f)$ (the last step by (3.112)) for all n. In other words, $(r_n) = \mathcal{I}f$, as required. □

Thing 3.10.6 $L^2(\mathbb{T})$ *is the vector space of (equivalence classes of) all 2π-periodic functions f with square summable sequences of Fourier coefficients.*

Note that, here, we are saying something even more substantial than we did in Thing 3.10.5. There we certainly *did* say that, if $f \in L^2(\mathbb{T})$, then the sequence $(c_n(f))$ of Fourier coefficients of f is in l^2. We said some other things too, but if you think about it (or even if you don't), we have *not* yet said whether $(c_n(f)) \in l^2$ assures that $f \in L^2(\mathbb{T})$. Does it?

The following, whose proof we omit (see Section 4.3 in [4]), implies that yes, in fact, it does.

Proposition 3.10.2 *If $f, g \in L^1(\mathbb{T})$ and $c_n(f) = c_n(g)$ for all $n \in \mathbb{Z}$, then $f = g$ (in $L^1(\mathbb{T})$; that is, $f(x) = g(x)$ for almost all x).*

How does this proposition yield Thing 3.10.6? As follows. Let $f \in L^1(\mathbb{T})$, and suppose $r = (c_n(f))$ is in l^2. Then, by Proposition 3.10.1, there's a $g \in L^2(\mathbb{T})$ with $(c_n(g)) = r$. But this means $c_n(f) = c_n(g)$ for all n. And since $c_0(g) = \int_\mathbb{T} g(x)\,dx$ is by assumption defined, necessarily $g \in L^1(\mathbb{T})$. So by Proposition 3.10.2, $f = g$. And again g is in $L^2(\mathbb{T})$, so f is too, as promised.

Thing 3.10.7 $L^2(\mathbb{T})$ *is a space of (equivalence classes of) 2π-periodic functions f such that f is a realization of S_N^f.*

That $L^2(\mathbb{T})$ really is Thing 3.10.7 is rightly considered a HUGE RESULT—large enough that we proclaim it again.

Theorem 3.10.1 (Carleson, 1964) *If $f \in L^2(\mathbb{T})$, then*

$$\lim_{N \to \infty} S_N^f(x) = f(x) \tag{3.195}$$

for almost all x.

(See [9].)

We should explain why we used the article "a," rather than "the," in presenting Thing 3.10.7. It's because $L^2(\mathbb{T})$ is *not* THE space of (all) equivalence classes of 2π-periodic functions f such that S_N^f converges pointwise almost everywhere to f. For example, consider the function f constructed as follows. For any $N \in \mathbb{Z}^+$, let

$$f_N(x) = \sum_{n=1}^{N} \frac{\sin nx}{\sqrt{n}}. \tag{3.196}$$

It may be shown (see Exercise 9, Section 2.5, in [4]) that $\lim_{N \to \infty} f_N(x)$ exists for any $x \in \mathbb{R}$.

Let's denote this limit by $f(x)$. Then certainly f is a realization of the f_N's. It may further be shown that $f \in L^1(\mathbb{T})$, and that $a_n(f) = 0$, $b_n(f) = n^{-1/2}$, and consequently $S_N^f = f_N$, for all n and N. So f is a realization of the S_N^f's.

But $f \notin L^2(\mathbb{T})$ because, again, square integrable functions have square summable sequences of Fourier coefficients. And f does not since, by (1.5),

$$\sum_{n=-\infty}^{\infty} |c_n(f)|^2 = \frac{1}{2} \sum_{n=1}^{\infty} |b_n(f)|^2 = \frac{1}{2} \sum_{n=1}^{\infty} \frac{1}{n}, \tag{3.197}$$

and the rightmost series diverges.

No nice characterization of the space

$$\{2\pi\text{-periodic functions } f \colon S_N^f \text{ converges pointwise almost everywhere to } f\} \tag{3.198}$$

is known, unless (3.198) counts as a nice characterization.

Exercises

3.10.1 Show that the map $\mathcal{I} : L^2(\mathbb{T}) \to l^2$ of Proposition 3.10.1 is:
   a. linear;
   b. injective (meaning $f_1 = f_2 \Rightarrow \mathcal{I} f_1 = \mathcal{I} f_2$; use Corollary 3.7.1(b) and Proposition 3.2.1(c), and recall that equality in $L^2(\mathbb{T})$ means equality almost everywhere);
   c. inner product preserving.

3.10.2 Use Proposition 3.10.1 and the Cauchy-Schwarz inequality on $L^2(\mathbb{T})$ to show that, for any sequences (r_n) and (s_n) in l^2,

$$\left| \sum_{n=-\infty}^{\infty} r_n \overline{s_n} \right| \le \left(\sum_{n=-\infty}^{\infty} |r_n|^2 \right)^{1/2} \left(\sum_{n=-\infty}^{\infty} |s_n|^2 \right)^{1/2}.$$

3.10.3 Let $\mathcal{R} = \{\rho_n : n \in \mathbb{Z}\}$ be an orthogonal set in $L^2(D)$. Show that the *completion*, call it $\overline{\mathrm{span}(\mathcal{R})}$, of the metric space $\mathrm{span}(\mathcal{R})$ is given by

$$\overline{\mathrm{span}(\mathcal{R})} = \left\{ \sum_{n=-\infty}^{\infty} d_n \rho_n : \sum_{n=-\infty}^{\infty} |d_n|^2 < \infty \right\}.$$

(Recall, cf. Corollary 3.7.2, that $\mathrm{span}(\mathcal{R})$ denotes the collection of all finite linear combinations of elements of $\mathcal{R}$. Also recall, cf. our discussions of Thing 3.10.1, that the completion of a metric space V is by definition the unique complete metric space in which V is dense.) Tips: You first need to check that elements of $\overline{\mathrm{span}(\mathcal{R})}$ are defined—that is, that the infinite series used to define this space really do converge in norm to elements of $L^2(D)$. See the proof of Proposition 3.10.1. You also need to show that $\overline{\mathrm{span}(\mathcal{R})}$ is complete and that $\mathrm{span}(\mathcal{R})$ is dense therein; see Corollary 3.7.2. (Don't worry about uniqueness of $\overline{\mathrm{span}(\mathcal{R})}$.)

4
Sturm-Liouville Problems

Here we see how the theory developed in the previous chapter yields a framework that accommodates a large, important class of eigenvalue problems. We observe that this class includes all eigenvalue problems encountered in Chapter 2. We then use this observation to complete the study of some boundary value problems considered but only partially solved there.

We also examine a number of new eigenvalue and boundary value problems, arising in various mathematical and physical contexts, and solve them using our new methods.

4.1 DEFINITIONS AND BASIC PROPERTIES

Recall Definition 2.2.2 of an eigenvalue problem. In Chapter 2 we considered a variety of such problems, each one resulting from application of the Fourier method to a boundary value problem in several variables.

We'll revisit those eigenvalue and boundary value problems shortly; first, we introduce a notion that will unify, and facilitate further discussion and application of, many of the key ideas.

Definition 4.1.1 A *Sturm-Liouville problem* on an interval $[a, b]$ ($-\infty < a < b < \infty$) is an eigenvalue problem on $[a, b]$ consisting of the following.

(a) A differential equation of the form

$$[r(x)X'(x)]' + q(x)X(x) + \lambda w(x)X(x) = 0 \quad (a < x < b), \qquad (4.1)$$

where r, q, and w are specified, *real-valued* functions, with r and r' continuous on $[a, b]$, q and w continuous on (a, b), and r and w positive on (a, b).

(b) Boundary conditions that are either *separated* or *periodic* with respect to r, by which we mean this.

(i) Conditions are *separated* with respect to r if they're of the form

$$\sin\alpha\, X(a) - \cos\alpha\, r(a)X'(a) = 0 = \sin\beta\, X(b) + \cos\beta\, r(b)X'(b) \tag{4.2}$$

for some specified numbers $\alpha, \beta \in [0, \pi)$, or of an equivalent form.

(ii) Conditions are *periodic* with respect to r if they're of the form

$$X(b) - X(a) = 0 = r(b)X'(b) - r(a)X'(a), \tag{4.3}$$

or of an equivalent form.

We assume that either type of condition carries along with it the implicit requirement that X and rX' be *continuous* on $[a, b]$.

We make a few notes. First, in the above context, w is called a *weight function*. Second, the Sturm-Liouville differential equation (4.1) is as in Definition 2.2.2(a), with

$$LX(x) = \frac{[r(x)X'(x)]' + q(x)X(x)}{w(x)}. \tag{4.4}$$

Third, the use of sines and cosines in the formulation (4.2) of separated boundary conditions will simplify some equations and proofs. In particular, because of the assumption $\alpha \in [0, \pi)$, specification of the coefficients $\sin\alpha$ and $\cos\alpha$ will require only that of $\tan\alpha$, where we define $\tan\pi/2 = \infty$. Similarly for β. Fourth, as with any eigenvalue problem on $[a, b]$, eigenfunctions of a Sturm-Liouville problem are, by definition, nonzero.

We remark that *every eigenvalue problem encountered in Chapter 2 is in fact a Sturm-Liouville problem*. Indeed, as is well worth checking, all but one of the eigenvalue problems arising there comprises (in one form or another) the differential equation

$$X''(x) + \lambda X(x) = 0 \quad (a < x < b), \tag{4.5}$$

together with one of the following five sets of boundary conditions:

$$X(a) = X(b) = 0, \quad X'(a) = X'(b) = 0, \quad X(a) = X'(b) = 0,$$
$$X(b) - X(a) = X'(b) - X'(a) = 0, \quad hX(a) - X'(a) = hX(b) + X'(b) = 0. \tag{4.6}$$

Note that (4.5) satisfies the criteria of Definition 4.1.1(a), with r and w identically 1, and q identically 0. Observe also that the first, second, third, and fifth sets of

DEFINITIONS AND BASIC PROPERTIES 219

boundary conditions in (4.6) are separated, with $(\tan\alpha, \tan\beta)$ equal to (∞, ∞), $(0,0)$, $(\infty, 0)$, and (h, h), respectively, and that the fourth set of boundary conditions there is periodic.

The remaining eigenvalue problem from Chapter 2 is this one:

$$[\rho R'(\rho)]' - n^2 \rho^{-1} R(\rho) + \mu \rho R(\rho) = 0 \quad (0 < \rho < \ell), \quad R(\ell) = 0. \tag{4.7}$$

(See Section 2.12, in particular (2.276).) This is a Sturm-Liouville problem on $[0, \ell]$, with $r(\rho) = w(\rho) = \rho$ and $q(\rho) = -n^2/\rho$; its boundary conditions are separated, with $(\tan\alpha, \tan\beta) = (0, \infty)$. (Note that, with $a = \alpha = 0$ and $r(\rho) = \rho$, the boundary condition at a, in (4.2), reads $0 = 0$. We interpret this as a *lack*—as there is in (4.7)—of explicit restrictions on $X(a)$ or $X'(a)$.) In sum, all six of our eigenvalue problems *are* of Sturm-Liouville type.

It's highly recommended that the reader reflect on the examples of Chapter 2 and on how they lead to these Sturm-Liouville problems. It should be noted especially that, because of the physical contexts from which these problems arose, it's more than reasonable to assume that each carries along with it the *implicit* continuity requirements of Definition 4.1.1(b).

Of the above six problems, there are two that we have not yet solved: the one consisting of the differential equation (4.5) and the fifth set of conditions in (4.6), and the one given by (4.7). In this chapter, we will develop the Sturm-Liouville theory sufficiently to obtain these solutions, to complete the study of the corresponding boundary value problems, and to solve a variety of other Sturm-Liouville and associated boundary value problems. To these ends, we first need two lemmas and two definitions.

Lemma 4.1.1 *Boundary conditions of any Sturm-Liouville problem are* self-adjoint. *This means the following: If we define the* Wronskian $W_r[X, Y]$ *by*

$$W_r[X, Y](x) = r(x)\left(X(x)\overline{Y'(x)} - X'(x)\overline{Y(x)}\right) \tag{4.8}$$

(with r as in Definition 4.1.1), then for any functions X and Y satisfying such boundary conditions,

$$W_r[X, Y](a) = W_r[X, Y](b). \tag{4.9}$$

Proof. We begin with the case of separated conditions, as described in Definition 4.1.1(b)(i). We assume that X and Y satisfy such conditions; we'll show that, in fact, this implies

$$W_r[X, Y](a) = 0 = W_r[X, Y](b). \tag{4.10}$$

By the inherent symmetry of the situation, we need only demonstrate the first equality. We do so as follows:

$$W_r[X, Y](a) = X(a)\left[r(a)\overline{Y'(a)}\right] - \overline{Y(a)}\left[r(a)X'(a)\right]$$
$$= X(a)\left[\tan\alpha\,\overline{Y(a)}\right] - \overline{Y(a)}\left[\tan\alpha\,X(a)\right] = 0. \tag{4.11}$$

The first step here is just a rewrite of the definition (4.8) at $x = a$; the second is by the assumptions that X and Y satisfy the first equation in (4.2) and that r is real-valued; the third is algebra. Actually, the second step is justified only if $\alpha \neq \pi/2$, but as the reader should check, a similar argument applies when $\alpha = \pi/2$.

The case of periodic conditions is straightforward: If X and Y satisfy such conditions, as described in Definition 4.1.1(b)(ii), then each of the four quantities

$$X(x),\ Y(x),\ r(x)X'(x),\ r(x)\overline{Y'(x)} \tag{4.12}$$

has the same value at $x = b$ as it does at $x = a$. So by the first equality in (4.11), $W_r[X,Y](a) = W_r[X,Y](b)$, and we're done. $\square$

We now describe a useful variant on the notion of L^2 space.

Definition 4.1.2 Let w be as in Definition 4.1.1(a).

(a) We denote by $L^2_w(a,b)$ the vector space of equivalence classes of measurable functions $f: (a,b) \to \mathbb{C}$ such that

$$\int_a^b |f(x)|^2 w(x)\, dx < \infty. \tag{4.13}$$

(In keeping with the discussions of Sections 3.3 and 3.5, it's generally okay to think of elements of $L^2_w(a,b)$ as functions *per se* and not to worry much about equivalence classes.)

(b) The *inner product* $\langle \cdot, \cdot \rangle_w$ and *norm* $||\cdot||_w$ on $L^2_w(a,b)$ are defined by

$$\langle f, g \rangle_w = \int_a^b f(x)\overline{g(x)}\, w(x)\, dx, \quad ||f||_w = \left(\int_a^b |f(x)|^2 w(x)\, dx \right)^{1/2}. \tag{4.14}$$

(Note that $||f||_w = \langle f, f \rangle_w^{1/2}$.)

(c) *Orthogonal and orthonormal sets and bases* in $L^2_w(a,b)$ are defined as in Definition 3.6.1, but with the inner product and norm there replaced by the ones just described.

Note that $L^2_1(a,b)$ (the subscript "1" denotes the constant function 1) is just the usual space $L^2(a,b)$ of Chapter 3; that $f \in L^2_w(a,b)$ if and only if $\sqrt{w}f \in L^2(a,b)$; and that for f, g in the former space, $\langle f, g \rangle_w = \langle \sqrt{w}f, \sqrt{w}g \rangle$ (so $||f||_w = ||\sqrt{w}f||$). Because of this, most of our earlier constructs and results concerning $L^2(a,b)$ generalize in straightforward ways to $L^2_w(a,b)$. In particular, we have and will make frequent use of the following generalization of Corollary 3.6.2.

Lemma 4.1.2 *Let $\mathcal{R} = \{\rho_n : n \in C\}$ be an orthogonal set in $L_w^2(a,b)$. Then $\mathcal{R}$ is an orthogonal basis for $L_w^2(a,b)$ if and only if*

$$f = \sum_{n \in C} \frac{\langle f, \rho_n \rangle_w}{||\rho_n||_w^2} \rho_n \tag{4.15}$$

for each $f \in L_w^2(a,b)$.

The proof of the above is left to the exercises.

Next, we have a convenient further classification of Sturm-Liouville problems.

Definition 4.1.3 A Sturm-Liouville problem on $[a,b]$ is:

(a) *regular* if the functions r, r', q, and w in Definition 4.1.1 are continuous on all of $[a,b]$ and the functions r are w there are positive on all of $[a,b]$;

(b) *singular* otherwise.

The problems delineated in (4.5) and (4.6) are regular, since (again), for each of these problems, $r = w = 1$ and $q = 0$. But (4.7) is singular: $r(\rho) = \rho$ and $w(\rho) = \rho$ vanish at $\rho = 0$; also $q(\rho) = -n^2/\rho$ is discontinuous there if $n \neq 0$.

We present finally our main result regarding Sturm-Liouville problems.

Theorem 4.1.1 *Let a Sturm-Liouville problem on $[a,b]$ be given.*

(a) *Let λ be an eigenvalue. Then the eigenspace for λ has dimension at most 2; that is, there are eigenfunctions Y and Z corresponding to λ such that any eigenfunction X corresponding to λ is a linear combination of Y and Z.*

Now suppose that the function w of Definition 4.1.1(a) is integrable on (a,b). Then the following are also true.

(b) *All eigenvalues λ are real.*

(c) *Eigenfunctions X and Y corresponding to distinct eigenvalues λ and μ are orthogonal in $L_w^2(a,b)$.*

(d) *Suppose the boundary conditions are separated, as in Definition 4.1.1(b)(i). Then:*

  (i) *If at least one of the quantities $\alpha, \beta, r(a), r(b)$ is not equal to zero, then the eigenspace for any eigenvalue λ has dimension 1: that is, all eigenfunctions corresponding to λ are in fact constant multiples of a fixed one Y. Moreover, we may choose Y to be real-valued.*

  (ii) *Suppose $\alpha, \beta \in [0, \pi/2]$. If $q(x) \leq 0$ on (a,b), then any eigenvalue λ is nonnegative; if $q(x) < 0$ on (a,b), then any eigenvalue λ is positive.*

(e) *Suppose the problem is regular. Then the set of all eigenvalues λ is countably infinite, and we also have the following. Suppose we choose a single eigenfunction for each eigenvalue λ whose eigenspace has dimension 1, and two* eigenfunctions orthogonal in $L_w^2(a,b)$ for each other eigenvalue λ. The set of eigenfunctions thus chosen is an orthogonal basis for $L_w^2(a,b)$.

Proof. Let our Sturm-Liouville problem be as in Definition 4.1.1.

To prove part (a) of our proposition, we divide equation (4.1) through by $r(x)$ to get the equivalent equation

$$X''(x) + \frac{r'(x)}{r(x)} X'(x) + \frac{q(x) + \lambda w(x)}{r(x)} X(x) = 0 \quad (a < x < b). \tag{4.16}$$

Now r'/r and $(q + \lambda w)/r$ are continuous on (a,b). So by a standard result in the theory of second-order differential equations, the space of solutions to (4.16) has, for any given λ, dimension 2. Consequently, any eigenspace of our Sturm-Liouville problem has dimension *at most* 2.

For the remainder of our proof, we suppose w to be integrable on $[a,b]$. To prove parts (b) and (c), we take (for the moment, not necessarily distinct) eigenvalues λ and μ of our problem, with corresponding eigenfunctions X and Y, respectively. Then on (a,b), we have

$$\lambda w X = -[rX']' - qX, \quad \mu w Y = -[rY']' - qY. \tag{4.17}$$

We multiply the first equation through by $\overline{Y}$; we take complex conjugates throughout the second equation and then multiply the result through by X. Since r, q, and w are real-valued, we get

$$\lambda w X \overline{Y} = -[rX']'\overline{Y} - qX\overline{Y}, \quad \overline{\mu} w X \overline{Y} = -[r\overline{Y'}]'X - qX\overline{Y}; \tag{4.18}$$

subtracting gives

$$(\lambda - \overline{\mu}) w X \overline{Y} = [r\overline{Y'}]'X - [rX']'\overline{Y}. \tag{4.19}$$

Now, by adding and subtracting $(r\overline{Y'})X'$ on the right side, we find

$$(\lambda - \overline{\mu}) w X \overline{Y} = \left([r\overline{Y'}]'X + (r\overline{Y'})X'\right) - \left([rX']'\overline{Y} + (r\overline{Y'})X'\right)$$
$$= [(r\overline{Y'})X]' - [(rX')\overline{Y}]' = [r(X\overline{Y'} - X'\overline{Y})]' = W_r[X,Y]'. \tag{4.20}$$

The left, and therefore the right, side is integrable on $[a,b]$, by assumption on w, X, and Y. (On $[a,b]$, $|wX\overline{Y}|$ is bounded by $|w|$ times the maximum of $|X\overline{Y}|$, which exists by the continuity of X and Y there. The integrability of $wX\overline{Y}$ then follows from Proposition 3.5.2.) So we integrate (4.20); we get

$$(\lambda - \overline{\mu}) \langle X, Y \rangle_w = \int_a^b W_r[X,Y]'(x)\,dx = W_r[X,Y](x)\Big|_a^b$$
$$= W_r[X,Y](b) - W_r[X,Y](a) = 0 \tag{4.21}$$

by the fundamental theorem of calculus and Lemma 4.1.1.

Now suppose $\lambda = \mu$ and $X = Y$; then (4.21) gives

$$(\lambda - \overline{\lambda})||X||_w^2 = 0. \tag{4.22}$$

Dividing by $||X||_w^2$ (which we may do by the fact that eigenfunctions are nonzero) gives $\lambda = \overline{\lambda}$, so that λ must be real. Thus part (b) is proved.

This result, applied to (4.21), then gives

$$(\lambda - \mu)\langle X, Y\rangle_w = 0. \tag{4.23}$$

If we assume $\lambda \neq \mu$, then we conclude $\langle X, Y\rangle_w = 0$. So part (c) is also proved.

We now assume our boundary conditions to be separated as in Definition 4.1.1(b)(i). If either α or $r(a)$ is nonzero, then the first equation in (4.2) is a "linear constraint" on $X(a)$ and $X'(a)$. (Again, if $\alpha = r(a) = 0$ then this equation reads $0 = 0$, meaning there's no constraint on the values $X(a)$ and $X'(a)$.) By essentially the same standard result cited in the above proof of part (a), nonzero solutions to (4.1) satisfying such a constraint are, in fact, multiples of each other. A similar argument applies if $(\beta, r(b)) \neq (0, 0)$. Hence, the first statement in part (d)(i).

For the second statement there, we observe the following. Since r, q, w, λ, α, and β are real (λ by part (a) of this proposition; all other quantities by assumption), we may take complex conjugates of the differential equation (4.1) and the boundary conditions (4.2) to conclude that $\overline{X}$ is an eigenfunction corresponding to λ whenever X is. Then by linearity and homogeneity of that equation and those conditions, so is $(X + \overline{X})/2 = \operatorname{Re} X$. So we're done with part (d)(i) by letting the function Y there equal $\operatorname{Re} X$.

Part (d)(ii) goes like this. Putting $Y = X$ into the first equation in (4.18) gives

$$\lambda w|X|^2 = -[rX']'\overline{X} - q|X|^2. \tag{4.24}$$

The left side is integrable on $[a, b]$ for reasons recently cited. Also, integration by parts with $u = \overline{X}$ and $dv = (rX')'\, dx$ gives

$$-\int [r(x)X'(x)]'\,\overline{X}(x)\, dx = -r(x)X'(x)\overline{X(x)} + \int r(x)X'(x)\overline{X'}(x)\, dx$$

$$= -r(x)X'(x)\overline{X(x)} + \int r(x)|X'(x)|^2\, dx, \tag{4.25}$$

so (4.24) yields

$$\lambda||X||_w^2 = -r(b)X'(b)\overline{X(b)} + r(a)X'(a)\overline{X(a)}$$
$$+ \int_a^b \left(r(x)|X'(x)|^2 - q(x)|X(x)|^2\right)\, dx. \tag{4.26}$$

Now $||X||_w > 0$; also, if $q(x) \leq 0$ on (a, b), then the integral on the right side of (4.26) is nonnegative. If in fact $q(x) < 0$ on (a, b), then $r|X'|^2 - q|X|^2$ is nonnegative, continuous, and not identically zero on that interval, which implies that the integral is in fact positive. So we'll be done with part (d)(ii) if we can show that, when $\alpha, \beta \in [0, \pi/2]$, we have

$$-r(b)X'(b)\overline{X(b)} + r(a)X'(a)\overline{X(a)} \geq 0. \tag{4.27}$$

Indeed, we'll prove that $\alpha \in [0, \pi/2]$ implies $r(a)X'(a)\overline{X(a)} \geq 0$ and that $\beta \in [0, \pi/2]$ implies $r(b)X'(b)\overline{X(b)} \leq 0$; this will certainly suffice.

If $\alpha = \pi/2$, then the first equation in (4.2) gives $X(a) = 0$, so of course $r(a)X'(a)\overline{X(a)} = 0$. If $\alpha \in [0, \pi/2)$, then $\tan \alpha$ is defined and nonnegative, so that same equation gives

$$r(a)X'(a)\overline{X(a)} = [\tan \alpha X(a)]\overline{X(a)} = \tan \alpha |X(a)|^2 \geq 0. \tag{4.28}$$

We've demonstrated the first of the two required implications; the proof of the second is nearly identical (because the second equation in (4.2) has a "+" where the first has a "−"). So part (d)(ii) is proved.

Regarding part (e), we observe first that the choice described there is always possible. That is, suppose Y and Z are eigenfunctions that correspond to the same eigenvalue λ but are *not* constant multiples of each other (so that the eigenspace for λ *does not* have dimension 1). Then, letting

$$c = -\frac{\langle Z, Y \rangle_w}{||Y||_w^2}, \tag{4.29}$$

we find that $cY + Z$ is also an eigenfunction corresponding to λ ($cY + Z$ is nonzero by assumption on Y and Z), and that

$$\langle cY + Z, Y \rangle_w = c \langle Y, Y \rangle_w + \langle Z, Y \rangle_w = -\langle Z, Y \rangle_w + \langle Z, Y \rangle_w = 0. \tag{4.30}$$

The *proof* of part (e) is considerably more difficult; it involves the theory of functions of a complex variable. We omit this proof (see Chapter 1 in [46]). □

We consider an example that will be useful in the next section.

Example 4.1.1 Solve the Sturm-Liouville problem

$$[\rho R'(\rho)]' + \lambda \rho^{-1} R(\rho) = 0 \quad (a < \rho < 1), \quad R(a) = R(1) = 0, \tag{4.31}$$

where a is a constant in the interval $(0, 1)$. Also write down the expansion of an arbitrary element of the appropriate "L_w^2" space in terms of the eigenfunctions of the problem.

Solution. This problem fits the mold of Definition 4.1.1, with $r(\rho) = \rho$, $q(\rho) = 0$, and $w(\rho) = 1/\rho$, all of this on $[a, 1]$. Note in particular that, according to part (b)(i) of that definition, our problem is separated, with $(\alpha, \beta) = (\pi/2, \pi/2)$; also, according to Definition 4.1.3(a), it's regular. So from part (d) of the above proposition, we find that any eigenvalue λ is nonnegative and has eigenspace of dimension 1. From part (e) of that proposition we then conclude that, if we choose one eigenfunction R for each such λ, we get an orthogonal basis for $L^2_w(a,1)$. Elements of that space then have expansions prescribed by Lemma 4.1.2.

So let's actually, explicitly, solve (4.31). We approach it much as we did the differential equation (2.114) of Example 2.5.1: We put $y = \ln \rho$ and $R(\rho) = Y(y) = Y(\ln \rho)$. As we saw in the course of that example, this yields $[\rho R'(\rho)]' = \rho^{-1} Y''(y)$. So the differential equation in (4.31) yields

$$Y''(y) + \lambda Y(y) = 0, \tag{4.32}$$

whose general solution is

$$Y(y) = \begin{cases} \gamma + \delta y & \text{if } \lambda = 0, \\ \gamma \cos\sqrt{\lambda}\, y + \delta \sin\sqrt{\lambda}\, y & \text{if not.} \end{cases} \tag{4.33}$$

Consequently, our original differential equation has general solution

$$R(\rho) = Y(\ln \rho) = \begin{cases} \gamma + \delta \ln \rho & \text{if } \lambda = 0, \\ \gamma \cos(\sqrt{\lambda} \ln \rho) + \delta \sin(\sqrt{\lambda} \ln \rho) & \text{if not.} \end{cases} \tag{4.34}$$

Now for the boundary conditions. We note first that zero is not an eigenvalue, for the following reason: Putting the boundary condition $R(1) = 0$ into the equation $R(\rho) = \gamma + \delta \ln \rho$ gives $\gamma = 0$; putting $R(a) = 0$ into the resulting equation $R(\rho) = \delta \ln \rho$ then gives $\delta = 0$, so that $R(\rho)$ is identically zero.

So we assume $\lambda \neq 0$. Then our boundary conditions on R, applied to (4.34), yield

$$\gamma \cos(\sqrt{\lambda} \ln a) + \delta \sin(\sqrt{\lambda} \ln a) = 0,$$
$$\gamma \cos(\sqrt{\lambda} \ln 1) + \delta \sin(\sqrt{\lambda} \ln 1) = 0. \tag{4.35}$$

The second equation gives $\gamma = 0$; the first then gives $\sin(\sqrt{\lambda} \ln a) = 0$, meaning $\sqrt{\lambda}$ must be a nonzero integer multiple of $\pi/\ln a$.

We may as well take this integer to be positive; negative n's give nothing extra. In sum, our problem (4.31) has the following eigenvalues and corresponding eigenfunctions:

$$\lambda_n = \left(\frac{\pi n}{\ln a}\right)^2, \quad R_n(\rho) = \sin(\sqrt{\lambda} \ln \rho) = \sin\left(\frac{\pi n \ln \rho}{\ln a}\right) \quad (n \in \mathbb{Z}^+); \tag{4.36}$$

for each of these λ_n's, *any* corresponding eigenfunction is a multiple of R_n.

So again, $\{R_n : n \in \mathbb{Z}^+\}$ is an orthogonal basis for $L^2_w(a,1)$. We compute

$$\|R_n\|^2_w = \int_a^1 |R_n(\rho)|^2 w(\rho)\, d\rho = \int_a^1 \sin^2\left(\frac{\pi n \ln \rho}{\ln a}\right) \frac{d\rho}{\rho}$$
$$= \frac{\ln a}{\pi n} \int_{\pi n}^0 \sin^2 u\, du = \frac{\ln a}{2\pi n} \int_{\pi n}^0 (1 - \cos 2u)\, du$$
$$= \frac{\ln a}{2\pi n} \left[u - \frac{\sin 2u}{2}\right]_{\pi n}^0 = -\frac{\ln a}{2} \tag{4.37}$$

(we put $u = \pi n \ln \rho / \ln a$). So finally we have, by Lemma 4.1.2, the following expansion of an $f \in L_w^2(a, 1)$:

$$f(\rho) = \sum_{n=1}^{\infty} d_n(f) \sin\left(\frac{\pi n \ln \rho}{\ln a}\right), \qquad (4.38)$$

where

$$d_n(f) = \frac{\langle f, R_n \rangle_w}{\|R_n\|_w^2} = -\frac{2}{\ln a} \int_a^1 f(\rho) \sin\left(\frac{\pi n \ln \rho}{\ln a}\right) \frac{d\rho}{\rho}. \qquad (4.39)$$

Exercises

4.1.1 Let $b > 1$. Using results from Example 4.1.1 where necessary:
  a. Show that the Sturm-Liouville problem

$$[\rho R'(\rho)]' + \lambda \rho^{-1} R(\rho) = 0 \quad (1 < \rho < b), \quad R'(1) = 0, \quad R(b) = 0$$

has the following eigenvalues and associated eigenfunctions:

$$\lambda_n = \left(\frac{\pi(2n-1)}{2 \ln b}\right)^2, \quad R_n(\rho) = \cos(\sqrt{\lambda_n} \ln \rho) \quad (n \in \mathbb{Z}^+)$$

(and no others, except for constant multiples of the R_n's).
  b. Show that the problem is regular.
  c. For an appropriate weight function w, find $\|R_n\|_w$ for each n and write down the expansion of an arbitrary element of $L_w^2(1, b)$ in terms of the eigenfunctions of this problem.

4.1.2 Let $b > 1$.
  a. Solve the Sturm-Liouville problem

$$[\rho^2 Q'(\rho)]' + \mu Q(\rho) = 0 \quad (1 < \rho < b), \quad Q(1) = 0, \quad Q(b) = 0.$$

Hint: Substitute $Q(\rho) = R(\rho)/\sqrt{\rho}$ and $\mu = \frac{1}{4} + \lambda$; use Example 4.1.1.
  b. Show that the problem is regular.
  c. For an appropriate weight function w, find $\|Q_n\|_w$ for each n and write down the expansion of an arbitrary element of $L_w^2(1, b)$ in terms of the eigenfunctions of this problem.

4.1.3 Let $\ell > 0$.
  a. Show that the Sturm-Liouville problem

$$[xP'(x)]' - \frac{1}{4x}P(x) + \lambda x P(x) = 0 \quad (0 < x < \ell), \quad P(\ell) = 0$$

has the following eigenvalues and associated eigenfunctions:

$$\lambda_n = \left(\frac{\pi n}{\ell}\right)^2, \quad P_n(x) = \frac{\sin \sqrt{\lambda_n}\, x}{\sqrt{x}} \quad (n \in \mathbb{Z}^+).$$

(Do so by plugging the given eigenvalues and eigenfunctions directly into the given problem.)

  b. Show that the problem is singular.

  c. For an appropriate weight function w, find $||P_n||_w$ for each n.

Show that, in spite of the singularity of this problem, every element of the appropriate "L_w^2" space *can* be expanded into eigenfunctions of this problem. Hint: Theorem 3.6.4(c). Also see the paragraph following Definition 4.1.2.

(The Sturm-Liouville problem of this exercise should be compared with (4.7).)

4.2 SOME BOUNDARY VALUE PROBLEMS

Here are an old one and a new one.

Example 4.2.1 Complete the (formal) solution of the Newton's law of cooling boundary value problem of Example 2.4.2.

Solution. Let's recall the story so far. What we found in that example (see especially (2.96) and (2.97)) is, in essence, this: *If* there are constants D_n so that

$$f = \sum_{n=1}^{\infty} D_n X_n, \tag{4.40}$$

where

$$X_n(x) = \cos s_n x + \frac{h}{s_n} \sin s_n x \tag{4.41}$$

and the s_n's are the positive roots of (2.94), *then* the boundary value problem in question has formal solution

$$u(x,t) = \sum_{n=1}^{\infty} D_n X_n(x) e^{-s_n^2 kt}. \tag{4.42}$$

So *are* there such D_n's, and if so, what are they?

To answer, we observe that the X_n of (4.41) is, for each $n \in \mathbb{Z}^+$, an eigenfunction of the regular, separated Sturm-Liouville problem consisting of (4.5) and the fifth set of conditions in (4.6), with eigenvalue $\lambda_n = s_n^2$. The reader may want to review Example 2.4.2 to see that this is so. Indeed, what we saw in that example is that these λ_n's are the *only* positive eigenvalues of that problem. But this problem only *has* positive eigenvalues: This follows from Theorem 4.1.1(d)(ii); from the facts that, for the problem in question, $(\tan \alpha, \tan \beta) = (h, h)$ and q is identically zero; and also from the fact, shown in the course of Example 2.4.2, that zero is not an eigenvalue for this problem.

The upshot is that our λ_n's give *all* eigenvalues for this problem. Now by Theorem 4.1.1(d)(i), every eigenspace has dimension 1. So from part (e) of that proposition and the facts that, here, $[a, b] = [0, \ell]$ and w is identically 1, we conclude that $\{X_n: n \in \mathbb{Z}^+\}$ is an orthogonal basis for $L^2(0, \ell)$.

But then Lemma 4.1.2 tells us that, for f in the latter space, (4.40) *does* hold, with

$$D_n = \frac{\langle f, X_n \rangle}{||X_n||^2}. \tag{4.43}$$

We compute that

$$||X_n||^2 = \frac{\ell(s_n^2 + h^2) + 2h}{2s_n^2} \tag{4.44}$$

(see Exercise 4.2.1), so we finally have the following.

Proposition 4.2.1 *The Newton's law of cooling problem given by (2.44), (2.47), and (2.88) has formal solution*

$$u(x, t) = \sum_{n=1}^{\infty} D_n \left(\cos s_n x + \frac{h}{s_n} \sin s_n x \right) e^{-s_n^2 kt}, \tag{4.45}$$

where

$$D_n = \frac{2s_n^2}{\ell(s_n^2 + h^2) + 2h} \int_0^\ell f(x) \left(\cos s_n x + \frac{h}{s_n} \sin s_n x \right) dx, \tag{4.46}$$

and again the s_n's are the positive solutions to $(sh^{-1} - s^{-1}h) \sin s\ell - 2 \cos s\ell = 0$.

Example 4.2.2 Solve the Dirichlet problem $\nabla_2^2 u = 0$ on the semiannulus

$$S = \{(\rho \cos \theta, \rho \sin \theta) : a \leq \rho \leq 1, -\pi/2 \leq \theta \leq \pi/2\} \subset \mathbb{R}^2 \tag{4.47}$$

(where $0 < a < 1$), assuming that $u = 0$ on the inner and outer arcs of S and (for simplicity) that the values of u on the upper line segment of the boundary are the mirror image of those on the lower one (Fig. 4.1).

Solution. By Proposition 2.5.1(a), giving the Laplacian in polar coordinates, we have the boundary value problem

$$\rho^{-1} u_{\theta\theta} + [\rho u_\rho]_\rho = 0 \quad (a < \rho < 1, -\pi/2 < \theta < \pi/2), \tag{4.48}$$
$$u(a, \theta) = u(1, \theta) = 0 \quad (-\pi/2 < \theta < \pi/2), \tag{4.49}$$
$$u(\rho, -\pi/2) = u(\rho, \pi/2) = f(\rho) \quad (a < \rho < 1). \tag{4.50}$$

We put $u(\rho, \theta) = R(\rho)\Theta(\theta)$ into the differential equation (4.48) to get

$$\rho^{-1} R(\rho) \Theta''(\theta) + [\rho R'(\rho)]' \Theta(\theta) = 0 \quad (a < \rho < 1, -\pi/2 < \theta < \pi/2), \tag{4.51}$$

which will be true if

$$[\rho R'(\rho)]' + \lambda \rho^{-1} R(\rho) = 0 \quad (a < \rho < 1), \tag{4.52}$$
$$\Theta''(\theta) - \lambda \Theta(\theta) = 0 \quad (-\pi/2 < \theta < \pi/2). \tag{4.53}$$

SOME BOUNDARY VALUE PROBLEMS

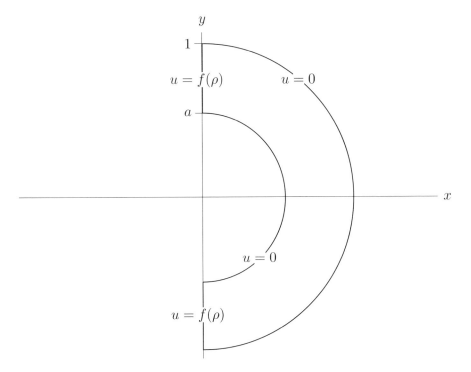

Fig. 4.1 The region S

Note that we've proceeded as we did in Example 2.5.1, with one notable exception: Here, we put the "$+\lambda$" with the ordinary differential equation in R, and the "$-\lambda$" with the one in Θ; there, it was the other way around. Why make this switch? Because the *homogeneous* boundary conditions in our present example lie at the boundary in the "ρ direction," (see (4.49)), whereas in that earlier example they were in the "θ direction" (see (2.108)). Consequently, while previously we ended up with a Sturm-Liouville problem in Θ, here it will be in R. (And again, we write the Sturm-Liouville differential equation (4.1) with a "$+\lambda$," not a "$-\lambda$.")

Indeed we arrive, in light of the conditions (4.49) and the differential equation (4.52), at *precisely* the problem of Example 4.1.1. We wrote down the eigenvalues λ_n and eigenfunctions R_n for that problem in (4.36). All of the λ_n's there are positive, and since the equation for Θ in (4.53) has, for $\lambda \neq 0$,

$$\Theta_\lambda(\theta) = \alpha e^{\sqrt{\lambda}\theta} + \beta e^{-\sqrt{\lambda}\theta} \tag{4.54}$$

as its general solution, we get the following formal series solution to (4.48), (4.49):

$$u(\rho,\theta) = \sum_{n=1}^{\infty} R_n(\rho)\left(D_n e^{\sqrt{\lambda_n}\theta} + E_n e^{-\sqrt{\lambda_n}\theta}\right)$$

$$= \sum_{n=1}^{\infty} \sin\left(\frac{\pi n \ln \rho}{\ln a}\right)\left(D_n e^{\pi n \theta/\ln a} + E_n e^{-\pi n \theta/\ln a}\right). \tag{4.55}$$

Into this, we now put the inhomogeneous conditions (4.50). We get

$$f(\rho) = \sum_{n=1}^{\infty} \sin\left(\frac{\pi n \ln \rho}{\ln a}\right)\left(D_n e^{-\pi^2 n/(2\ln a)} + E_n e^{\pi^2 n/(2\ln a)}\right)$$

$$= \sum_{n=1}^{\infty} \sin\left(\frac{\pi n \ln \rho}{\ln a}\right)\left(D_n e^{\pi^2 n/(2\ln a)} + E_n e^{-\pi^2 n/(2\ln a)}\right). \quad (4.56)$$

We match these up with (4.38) and (4.39); we get

$$D_n e^{-\pi^2 n/(2\ln a)} + E_n e^{\pi^2 n/(2\ln a)} = D_n e^{\pi^2 n/(2\ln a)} + E_n e^{-\pi^2 n/(2\ln a)}$$
$$= d_n(f)$$
$$= -\frac{2}{\ln a}\int_a^1 f(\rho)\sin\left(\frac{\pi n \ln \rho}{\ln a}\right)\frac{d\rho}{\rho}. \quad (4.57)$$

Solving this for D_n and E_n gives

$$D_n = E_n = -\frac{2}{\ln a (e^{\pi^2 n/(2\ln a)} + e^{-\pi^2 n/(2\ln a)})}\int_a^1 f(\rho)\sin\left(\frac{\pi n \ln \rho}{\ln a}\right)\frac{d\rho}{\rho}. \quad (4.58)$$

We summarize: To do so concisely, we invoke the hyperbolic cosine function $\cosh x = (e^x + e^{-x})/2$. We have:

Proposition 4.2.2 *The boundary value problem (4.48)–(4.50) has formal solution*

$$u(\rho,\theta) = -\frac{2}{\ln a}\sum_{n=1}^{\infty}\left[\int_a^1 f(v)\sin\left(\frac{\pi n \ln v}{\ln a}\right)\frac{dv}{v}\right]\frac{\sin\left(\frac{\pi n \ln \rho}{\ln a}\right)\cosh\left(\frac{\pi n \theta}{\ln a}\right)}{\cosh\left(\frac{\pi^2 n}{2\ln a}\right)}. \quad (4.59)$$

(Here we've written v for our variable of integration so as not to confuse this variable with the independent variable ρ of $u(\rho,\theta)$.)

The above result should be compared with that of Exercise 2.5.1.

Now imagine that $a \to 0^+$, so our semiannulus above becomes a semidisk, and the problem (4.48)–(4.50) becomes

$$\rho^{-1}u_{\theta\theta} + [\rho u_\rho]_\rho = 0 \quad (0<\rho<1,\ -\pi/2<\theta<\pi/2), \quad (4.60)$$
$$u(1,\theta) = 0 \quad (-\pi/2<\theta<\pi/2), \quad (4.61)$$
$$u(\rho,-\pi/2) = u(\rho,\pi/2) = f(\rho) \quad (0<\rho<1). \quad (4.62)$$

The condition on the inner arc is now gone because that arc itself is. The associated regular Sturm-Liouville problem (4.31) thus becomes the problem

$$[\rho R'(\rho)]' + \lambda \rho^{-1}R(\rho) = 0 \quad (0<\rho<1),\quad R(1)=0, \quad (4.63)$$

which is *singular* because $r(\rho) = \rho$ is zero at $\rho = 0$ and $w(\rho) = 1/\rho$ is discontinuous there.

It turns out that (4.63) has *no* eigenfunctions in our Sturm-Liouville sense of the word. More specifically, the function

$$R_\lambda(\rho) = \begin{cases} \ln \rho & \text{if } \lambda = 0, \\ \sin(\sqrt{\lambda} \ln \rho) & \text{if not} \end{cases} \qquad (4.64)$$

does, for *any* $\lambda \in \mathbb{C}$, satisfy both the differential equation and the boundary condition in (4.63) and is, up to constant multiples, the only function to do so. All of this is readily shown. But for no λ is $R_\lambda(\rho)$ continuous on $[0, 1]$ ($\lim_{\rho \to 0+} R_\lambda(\rho)$ does not exist).

Still, for $\lambda \in \mathbb{C}$, R_λ as in (4.64), and Θ_λ as in (4.54), $u_\lambda = R_\lambda \Theta_\lambda$ *does* solve equations (4.60), (4.61) in the literal sense, meaning in the absence of implicit continuity conditions at the origin. Moreover, solutions to the complete boundary value problem (4.60)–(4.62) *may* be obtained as superpositions of these u_λ's. But these superpositions are *definite integrals*, over $\lambda \in (0, \infty)$, rather than series. We'll see how this goes in Section 7.2. For the present, the moral is that singular Sturm-Liouville problems can behave quite unlike regular ones.

Or not. In the next section, we'll investigate some singular problems with rather pronounced delusions of regularity.

Exercises

4.2.1 Show that, for X_n as in (4.41) and the s_n's as in Example 4.2.1 (and Example 2.4.2), the formula (4.44) for $||X_n||^2$ holds. Hints: Compute $||X_n||^2$ using the facts that $\cos^2 \theta$ has antiderivative $(\theta + \sin \theta \cos \theta)/2$, $\sin \theta \cos \theta$ has antiderivative $\sin^2 \theta/2$, and $\sin^2 \theta$ has antiderivative $(\theta - \sin \theta \cos \theta)/2$. Simplify using the fact that the s_n's make the right side of (2.94) zero.

4.2.2 Into Example 4.2.1 and Proposition 4.2.1, put $h = \ell = 1$. (Just to make life less complicated.) This means that the numbers s_n there are precisely the positive solutions, listed in ascending order, to the equation

$$(s_n - s_n^{-1}) \sin s_n - 2 \cos s_n = 0. \qquad (*)$$

a. Show that, for each $n \in \mathbb{Z}^+$, $s_n \sin s_n - \cos s_n = \pm 1$. Hint: First show, using basic trig identities (for example, $\cos^2 \theta - 1 = -\sin^2 \theta$), that

$$(s_n \sin s_n - \cos s_n)^2 - 1 = s_n \sin s_n((s_n - s_n^{-1}) \sin s_n - 2 \cos s_n).$$

Then use the fact that the s_n's satisfy $(*)$.

b. Actually one may show, though we won't (draw yourself some graphs if you really need convincing), that $s_n \sin s_n - \cos s_n$ equals -1 precisely when n is even and equals 1 precisely when n is odd. Using this information, show that, in each of the following cases, the Newton's law of cooling heat problem (2.44), (2.47), (2.88) has the stated (formal) solution. (We still assume $h = \ell = 1$; the given function f is defined on $(0, 1)$.)

(i) $f(x) = 1$. ANSWER:

$$u(x,t) = 4 \sum_{m=1}^{\infty} \frac{\cos s_{2m-1}x + s_{2m-1}^{-1} \sin s_{2m-1}x}{s_{2m-1}^2 + 3} e^{-s_{2m-1}^2 kt}.$$

(ii) $f(x) = \cos \pi x$. ANSWER:

$$u(x,t) = 4 \sum_{m=1}^{\infty} \frac{s_{2m}^2 \cos s_{2m}x + s_{2m} \sin s_{2m}x}{(s_{2m}^2 + 3)(s_{2m}^2 - \pi^2)} e^{-s_{2m}^2 kt}.$$

4.2.3 Formally solve the Dirichlet problem of Example 4.2.2 explicitly in the case where:

a. $f(\rho) = 1$;
b. $f(\rho) = \rho$;
c. $f(\rho) = \ln \rho$.

Hint: Substitute $u = \ln v / \ln a$.

4.2.4 Find the general formal solution to the Dirichlet problem of Example 4.2.2 if the condition $u(\rho, -\pi/2) = f(\rho)$ there is replaced by $u(\rho, -\pi/2) = 0$. (We still assume $u(\rho, \pi/2) = f(\rho)$.)

4.3 BESSEL FUNCTIONS I: BESSEL'S EQUATION OF ORDER n

The Sturm-Liouville problem (4.7) is, again, singular. We wish to show that, nonetheless, it behaves in many ways as if it were regular, and to reveal that behavior explicitly.
To do so, we first examine the differential equation

$$\rho[\rho R'(\rho)]' + (\mu \rho^2 - n^2) R(\rho) = 0, \tag{4.65}$$

which is just the one in (4.7) multiplied through by ρ. We assume $\mu > 0$; this will suffice for our purposes. Also, as before, $n \in \mathbb{N}$.
How might we solve (4.65)? Well, first suppose J_n solves the equation

$$x[xJ_n'(x)]' + (x^2 - n^2) J_n(x) = 0, \tag{4.66}$$

called *Bessel's equation of order* n. If we write $x = \sqrt{\mu}\rho$ and $R(\rho) = J_n(x) = J_n(\sqrt{\mu}\rho)$, then by the chain rule,

$$\rho[\rho R'(\rho)]' = \rho[\sqrt{\mu}\rho J_n'(\sqrt{\mu}\rho)]' = \rho[\sqrt{\mu} J_n'(\sqrt{\mu}\rho) + \mu\rho J_n''(\sqrt{\mu}\rho)]$$
$$= (\sqrt{\mu}\rho)^2 J_n''(\sqrt{\mu}\rho) + \sqrt{\mu}\rho J_n'(\sqrt{\mu}\rho) = x^2 J_n''(x) + x J_n'(x)$$
$$= x[xJ_n'(x)]' = -(x^2 - n^2) J_n(x) = -(\mu\rho^2 - n^2) R(\rho), \tag{4.67}$$

so R solves (4.65). That is, it suffices to consider (4.66).

How might we solve *that* equation? We might look for a power series solution. Let's: We note that such a series will begin with, say, the Kth power of x, where $K \in \mathbb{N}$ is to be determined. So we can write that series in the form

$$J_n(x) = \sum_{k=0}^{\infty} B_k x^{k+K}, \tag{4.68}$$

where B_0 is assumed *nonzero*—we'll choose a convenient value for B_0 shortly—and the remaining B_k's are also to be determined.

We have

$$x(x(x^{k+K})')' = x((k+K)x^{k+K})' = (k+K)^2 x^{k+K}, \tag{4.69}$$

so putting (4.68) into (4.66) and rearranging a bit, we get

$$\sum_{k=0}^{\infty} \left((k+K)^2 - n^2\right) B_k x^{k+K} = -\sum_{k=0}^{\infty} B_k x^{2+k+K}. \tag{4.70}$$

Substituting $j = 2 + k$ into the expression on the right turns that expression into

$$-\sum_{j=2}^{\infty} B_{j-2} x^{j+K}. \tag{4.71}$$

But the name of the index of summation does not affect what the given sum equals, so (4.70) can be rewritten as

$$\sum_{k=0}^{\infty} \left((k+K)^2 - n^2\right) B_k x^{k+K} = -\sum_{k=2}^{\infty} B_{k-2} x^{k+K}. \tag{4.72}$$

We make the latter true by equating like powers of x. We first note that, since the sum on the right starts at $k = 2$, the coefficients of x^K and x^{1+K} on that side are zero. For the same to be true on the left, we need

$$[K^2 - n^2] B_0 = [(K+1)^2 - n^2] B_1 = 0. \tag{4.73}$$

To make this so, we choose

$$K = n, \quad B_1 = 0. \tag{4.74}$$

We put this K into (4.72) and then match up coefficients of x^{k+n} for $k \geq 2$ there. We get

$$k(2n+k) B_k = -B_{k-2} \quad (k \geq 2). \tag{4.75}$$

Note that $k(2n+k)$ is, under our assumptions on n and k, nonzero, so from (4.75) and the fact that $B_1 = 0$, we conclude that $B_k = 0$ for k *odd*. So by (4.68),

$$J_n(x) = \sum_{\text{even } k \in \mathbb{N}} B_k x^{k+n} = \sum_{m=0}^{\infty} B_{2m} x^{2m+n}. \tag{4.76}$$

We next consider the B_{2m}'s. We observe that, by (4.75),

$$B_{2\cdot 1} = -\frac{B_0}{2(2n+2)} = \frac{B_0(-1)^1}{1!(n+1)\,4^1},$$

$$B_{2\cdot 2} = -\frac{B_{2\cdot 1}}{4(2n+4)} = -\frac{B_{2\cdot 1}}{2(n+2)\cdot 4} = \frac{B_0(-1)^2}{2!(n+1)(n+2)\,4^2},$$

$$B_{2\cdot 3} = -\frac{B_{2\cdot 2}}{6(2n+6)} = -\frac{B_{2\cdot 2}}{3(n+3)\cdot 4} = \frac{B_0(-1)^3}{3!(n+1)(n+2)(n+3)\,4^3},$$

$$\tag{4.77}$$

and so on. A pattern emerges:

$$B_{2m} = \frac{B_0(-1)^m}{m!(n+1)(n+2)\cdots(n+m)\,4^m}. \tag{4.78}$$

(See Exercise 4.3.1.) Now since any constant multiple of a solution to the differential equation (4.66) is still a solution, we can choose B_0 however we want. Let's choose $B_0 = 1/(2^n n!)$; then

$$B_{2m} = \frac{(-1)^m}{m!(1\cdot 2\cdots n)(n+1)(n+2)\cdots(n+m)\,4^m 2^n} = \frac{(-1)^m}{m!(n+m)!\,2^{2m+n}}. \tag{4.79}$$

Putting this into the series (4.76) gives

$$J_n(x) = \sum_{m=0}^{\infty} \frac{(-1)^m}{m!(n+m)!}\left(\frac{x}{2}\right)^{2m+n}. \tag{4.80}$$

This *is* our series solution $J_n(x)$; we've not yet discussed its convergence, but it and its termwise derivatives do in fact converge quite nicely. Indeed, by the ratio test and Lemma 2.8.1, say, it's differentiable any number of times term by term, and satisfies the requisite differential equation (4.66), on all of $\mathbb{R}$.

So $R(\rho) = J_n(\sqrt{\mu}\rho)$ converges just as nicely and, as we recall, solves (4.65). We therefore have part (a) of the following result, part (b) of which we don't prove. For that proof, the reader may consult the classic text of Watson [49], which in fact is an excellent source for any Bessel function details that we omit (and any others).

Proposition 4.3.1 *Let* $n \in \mathbb{N}$, *and define the* Bessel function J_n *of the first kind, of order* n, *by* (4.80). *Then for* $\mu > 0$:

(a) *The function* $R(\rho) = J_n(\sqrt{\mu}\rho)$ *is differentiable term by term any number of times, and solves* (4.65), *on all of* $\mathbb{R}$.

(b) *Any* continuous *solution* $R(\rho)$ *to* (4.65) *on any interval* $[0,\ell]$ *is a constant multiple of* $J_n(\sqrt{\mu}\rho)$.

Part (b) has this useful consequence: If a Sturm-Liouville problem on $[0, \ell]$ entails the differential equation (4.65), then, for any eigenvalue μ, any eigenfunction of that problem *must* be a constant times $J_n(\sqrt{\mu}\rho)$.

We'll consider such problems in the next section. There and for some later computations the following result will be of aid.

Proposition 4.3.2 *For $x \in \mathbb{R}$ and $n \in \mathbb{N}$,*

$$x J_n'(x) = n J_n(x) - x J_{n+1}(x). \quad (4.81)$$

Proof. We have

$$x \frac{d}{dx} \left(\frac{x}{2}\right)^{2m+n} = x \cdot \frac{2m+n}{2} \left(\frac{x}{2}\right)^{2m+n-1} = n \left(\frac{x}{2}\right)^{2m+n} + mx \left(\frac{x}{2}\right)^{2m+n-1}. \quad (4.82)$$

So termwise differentiation of (4.80), justified by earlier observations, gives

$$x J_n'(x) = n \sum_{m=0}^{\infty} \frac{(-1)^m}{m!(n+m)!} \left(\frac{x}{2}\right)^{2m+n} + x \sum_{m=0}^{\infty} \frac{(-1)^m \, m}{m!(n+m)!} \left(\frac{x}{2}\right)^{2m+n-1}$$

$$= n J_n(x) + x \sum_{m=1}^{\infty} \frac{(-1)^m}{(m-1)!(n+m)!} \left(\frac{x}{2}\right)^{2m+n-1}; \quad (4.83)$$

to get the sum on the far right, we used the fact that $m/m!$ equals 0 if $m = 0$ and $1/(m-1)!$ if $m \in \mathbb{Z}^+$. We'll be done if we can show that this sum equals $-J_{n+1}(x)$. We do so by letting $k = m - 1$; we get

$$\sum_{m=1}^{\infty} \frac{(-1)^m}{(m-1)!(n+m)!} \left(\frac{x}{2}\right)^{2m+n-1} = \sum_{k=0}^{\infty} \frac{(-1)^{k+1}}{k!((n+1)+k)!} \left(\frac{x}{2}\right)^{2k+(n+1)}$$

$$= -J_{n+1}(x), \quad (4.84)$$

as promised. □

Our discussions so far of J_n have been rather technical, as some below also promise to be. To navigate the technicalities, we'd do well to bear in mind the following general principle, which provides a more *intuitive* sense of Bessel functions: $J_n(x)$ behaves, for large x, quite a bit like

$$f_n(x) = \sqrt{\frac{2}{\pi x}} \sin\left(x - \frac{(2n-1)\pi}{4}\right). \quad (4.85)$$

What do we mean by this? There are many mathematically rigorous ways to answer, but again we're concerned now with heuristics; for these, we refer to Figure 4.2.

Pictures for other values of n are similar, although the larger n is, the further out along the positive x axis one must go before J_n "looks like" f_n. (For any $n \in \mathbb{N}$,

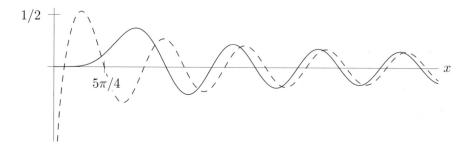

Fig. 4.2 The Bessel function J_5 (solid), and f_5 for f_n as in (4.85) (dashed)

as in Figure 4.2 for $n = 5$, $J_n(x)$ and $f_n(x)$ look quite *different* for x small enough. Again, $J_n(x)$ is bounded as $x \to 0^+$, but as is clear from its definition, $f_n(x)$ is not.)

We make a few closing remarks. First, we came across J_1 already in Exercise 1.8.9, where it took a form quite different from the one given by (4.80). See Exercise 5.6.9 for the equivalence of these forms.

Second, for *any* complex number n, (4.65) has a solution J_n that looks in many ways like the series (4.80). See the exercises for this section for the case where n is half of an odd positive integer; see the exercises for Section 5.6 for more general values of n.

Finally, for each $n \in \mathbb{C}$ there is a *Bessel function of the second kind* Y_n that solves (4.65) but is *not* a constant multiple of J_n. But our close encounters with Bessel functions will be of the first kind only.

Exercises

4.3.1 Show, by plugging the B_{2m}'s of (4.78) directly into the recurrence relation (4.75), that these coefficients do indeed satisfy this relation.

4.3.2   **a.** Multiply equation (4.80) by x^n, differentiate, and rearrange to show that

$$\frac{d}{dx}(x^n J_n(x)) = x^n J_{n-1}(x)$$

for $n \in \mathbb{Z}^+$.

     b. Conclude that, for such n, $xJ'_n(x) = -nJ_n(x) + xJ_{n-1}(x)$.

     c. Use this result and (4.81) to express J_{n+1} in terms of J_n and J_{n-1} for $n \in \mathbb{Z}^+$.

4.3.3 Use part a of the previous exercise to show that

$$\int_0^t x^{n+1} J_n(x)\, dx = t^{n+1} J_{n+1}(t)$$

for $n \in \mathbb{N}$.

4.3.4 Let $k \in \mathbb{N}$. The *Bessel function* $J_{k+1/2}$ *of the first kind, of order* $k + \frac{1}{2}$, is defined by

$$J_{k+1/2}(x) = \frac{2^{k+1/2}}{\sqrt{\pi}} \sum_{m=0}^{\infty} \frac{(-1)^m (m+k)!}{m!(2m+2k+1)!} x^{2m+k+1/2}.$$

(The factor in front will be explained in Exercise 5.6.8.)
Show that $J_{k+1/2}$ satisfies:

  a. *Bessel's equation of order* $k + \frac{1}{2}$, meaning the differential equation (4.66) with $k + \frac{1}{2}$ in place of n;
  b. the recurrence relation (4.81) with $k + \frac{1}{2}$ in place of n.
Hint for both parts: Differentiate term by term.

4.3.5 Let $J_{k+1/2}$ be as in the previous exercise.

  a. Show that $J_{1/2}(x)$ is a constant times $x^{-1/2} \sin x$ (Use the Maclaurin series for $\sin x$, cf. (A.21).)
  b. Use the relation (4.81) to recursively find explicit furmulas, in terms of cosines, sines, and half-integer powers of x (and constants), for $J_{3/2}(x)$ and $J_{5/2}(x)$.
  c. We remarked, just before Figure 4.2, that $J_n(x)$ behaves quite a bit like the function f_n of (4.85), for nonnegative integers n. Is this also true for $n = \frac{1}{2}$? If so, how much is "quite a bit"? Answer these same questions for $n = \frac{3}{2}, \frac{5}{2}$.

4.4 BESSEL FUNCTIONS II: FOURIER-BESSEL SERIES

Here is our second major result concerning solutions to Sturm-Liouville problems.

Theorem 4.4.1 *Let* $n \in \mathbb{N}$; *let* $\beta \in [0, \pi/2]$.

(a) *The singular Sturm-Liouville problem*

$$[\rho R'(\rho)]' - n^2 \rho^{-1} R(\rho) + \mu \rho R(\rho) = 0 \quad (0 < \rho < \ell), \tag{4.86}$$
$$\sin \beta\, R(\ell) + \cos \beta\, \ell R'(\ell) = 0 \tag{4.87}$$

has an eigenvalue $\mu = s^2$ *for each positive solution* s *to the equation*

$$\sin \beta\, J_n(s\ell) + \cos \beta\, s\ell J_n'(s\ell) = 0. \tag{4.88}$$

These are the only eigenvalues unless $\beta = n = 0$, *in which case there is exactly one more, namely* $\mu = 0$.

(b) *For each of the eigenvalues* μ *described in part (a), any corresponding eigenfunction is a constant multiple of*

$$R(\rho) = J_n(\sqrt{\mu}\rho) = \begin{cases} J_0(0) = 1 & \text{if } \beta = n = \mu = 0, \\ J_n(s\rho) & \text{if not.} \end{cases} \tag{4.89}$$

(*That $J_0(0) = 1$ follows from* (4.80).)

(c) *The set of positive solutions s to* (4.88) *forms an infinite, unbounded sequence* $s_{n,1}, s_{n,2}, s_{n,3}, \ldots$. *Moreover, let's write* $v(\rho) = \rho$;

$$R_{n,j}(\rho) = J_n(s_{n,j}\rho) \tag{4.90}$$

for $j \in \mathbb{Z}^+$; $R_{0,0}(\rho) = 1$; *and* $\mathcal{R} = \{R_{n,j} : j \geq \delta_{\beta+n,0}\}$. *(That is, $\mathcal{R}$ includes the constant function $R_{0,0} = 1$ if $\beta = n = 0$; otherwise it does not.) Then $\mathcal{R}$ is an orthogonal basis for* $L_v^2(0, \ell)$.

(d) *Let all notation be as above. Then* $\|R_{0,0}\|_v^2 = \ell^2/2$, *while for* $j \in \mathbb{Z}^+$,

$$\|R_{n,j}\|_v^2 = \begin{cases} \dfrac{\ell^2}{2} J_{n+1}^2(s_{n,j}\ell) & \text{if } \beta = \dfrac{\pi}{2}, \\ \dfrac{s_{n,j}^2 \ell^2 - n^2 + \tan^2\beta}{2s_{n,j}^2} J_n^2(s_{n,j}\ell) & \text{if not.} \end{cases} \tag{4.91}$$

Proof. The problem (4.86), (4.87) conforms to Definition 4.1.1(a) with $a = 0, b = \ell$, $r(\rho) = \rho$, $q(\rho) = -n^2/\rho$, and $w(\rho) = v(\rho) = \rho$; and to Definition 4.1.1(b)(i) with $\alpha = 0$ and $\beta \in [0, \pi/2]$. So by Theorem 4.1.1(d)(ii), any eigenvalue μ is nonnegative and is in fact positive except perhaps in the case $n = 0$, in which case $\mu = 0$ may be an eigenvalue. We claim that it is, in this case, if and only if $\beta = 0$. Proof: To say that R solves the differential equation (4.86), with $\mu = n = 0$, on $(0, \ell)$ is to say that $[\rho R'(\rho)]'$ is zero there. This means $\rho R'(\rho)$ is a constant C, and therefore $R'(\rho) = C/\rho$. So R is either constant (if $C = 0$) or a constant plus $\ln \rho$ times another constant (if $C \neq 0$). But $\ln(0^+)$ is undefined, so any eigenfunction R must in fact be a nonzero constant. For such R, of course, $R' = 0$, but then the boundary condition (4.87) will be met if and only if $\beta = 0$. So our claim is proved.

To summarize, we've verified parts (a) and (b) of our proposition in the case of the eigenvalue $\mu = 0$, and have shown that any nonzero eigenvalues μ are positive. So now, let's say a given $\mu > 0$ actually *is* an eigenvalue, with corresponding eigenfunction R. This means precisely these four things: First, R is not identically zero on $(0, \ell)$; second, R solves (4.86)—or equivalently, R solves the differential equation (4.65) on $(0, \ell)$; third, $R(\rho)$ and $\rho R'(\rho)$ are continuous on $[0, \ell]$; fourth, $\sin \beta\, R(\ell) + \cos \beta\, \ell R'(\ell) = 0$. The first three items are equivalent, by Proposition 4.3.1, to $R(\rho)$ being equal to $\delta J_n(\sqrt{\mu}\rho)$ for some $\delta \neq 0$. The fourth item then becomes the condition

$$\sin \beta\, J_n(\sqrt{\mu}\ell) + \cos \beta\, \ell \left[\frac{d}{d\rho} J_n(\sqrt{\mu}\rho)\right]_{\rho=\ell}$$
$$= \sin \beta\, J_n(\sqrt{\mu}\ell) + \cos \beta\, \sqrt{\mu}\ell J_n'(\sqrt{\mu}\ell) = 0; \tag{4.92}$$

that is, $\mu = s^2$, where s is a positive solution to (4.88). So we've proved completely parts (a) and (b).

BESSEL FUNCTIONS II: FOURIER-BESSEL SERIES 239

We do not prove part (c). (A proof may be found in [49].) Note that it's similar in spirit to Theorem 4.1.1(e), but does not follow from that result, because (4.86), (4.87) is a singular problem.

We now consider part (d). First, we compute

$$||R_{0,0}||_v^2 = \int_0^\ell |R_{0,0}(\rho)|^2 v(\rho)\, d\rho = \int_0^\ell |1|^2 \rho\, d\rho = \left.\frac{\rho^2}{2}\right|_0^\ell = \frac{\ell^2}{2}. \tag{4.93}$$

The case $j > 0$ requires a bit more work; for this case we note that, since $R_{n,j}$ by its definition solves (4.86) with $\mu = s_{n,j}^2$, we have

$$[\rho R'_{n,j}(\rho)]' - n^2 \rho^{-1} R_{n,j}(\rho) + s_{n,j}^2 \rho R_{n,j}(\rho) = 0 \tag{4.94}$$

on $(0, \ell)$. We multiply through by $\rho R'_{n,j}(\rho)/s_{n,j}^2$ and rearrange a bit to get

$$-\rho^2 R_{n,j}(\rho) R'_{n,j}(\rho) = \frac{1}{s_{n,j}^2}\left[\rho R'_{n,j}(\rho)[\rho R'_{n,j}(\rho)]' - n^2 R_{n,j}(\rho) R'_{n,j}(\rho)\right]. \tag{4.95}$$

We integrate both sides of this over $[0, \ell]$. Doing so on the right side is easy, because in general $g^2/2$ is an antiderivative of gg'; so

$$\int_0^\ell \left[\rho R'_{n,j}(\rho)[\rho R'_{n,j}(\rho)]' - n^2 R_{n,j}(\rho) R'_{n,j}(\rho)\right] d\rho$$

$$= \left[\frac{\rho^2(R'_{n,j}(\rho))^2 - n^2(R_{n,j}(\rho))^2}{2}\right]_0^\ell = \frac{\ell^2(R'_{n,j}(\ell))^2 - n^2(R_{n,j}(\ell))^2}{2}. \tag{4.96}$$

The last step is because, by (4.80), $R_{n,j}(0) = J_n(0) = 0$ for $n \geq 1$, and because for $n = 0$, certainly $n^2(R_{n,j}(0))^2 = 0$. The left side of (4.95) is now integrated by parts, with $u = \rho^2/2$ and $dv = 2R_{n,j}(\rho) R'_{n,j}(\rho)\, d\rho = d((R_{n,j}(\rho))^2)$. The result is

$$-\int_0^\ell \rho^2 R_{n,j}(\rho) R'_{n,j}(\rho)\, d\rho = -\left.\frac{\rho^2 (R_{n,j}(\rho))^2}{2}\right|_0^\ell + \int_0^\ell \rho (R_{n,j}(\rho))^2\, d\rho$$

$$= -\frac{\ell^2 (R_{n,j}(\ell))^2}{2} + ||R_{n,j}||_v^2. \tag{4.97}$$

(We've used the fact that $R_{n,j}$ is real valued.) Combining equations (4.95), (4.95), and (4.97), we get

$$||R_{n,j}||_v^2 = \frac{\ell^2 (R_{n,j}(\ell))^2}{2} + \frac{1}{s_{n,j}^2}\left[\frac{\ell^2 (R'_{n,j}(\ell))^2 - n^2 (R_{n,j}(\ell))^2}{2}\right]$$

$$= \frac{1}{2}\left[\ell^2 (J'_n(s_{n,j}\ell))^2 + \frac{s_{n,j}^2 \ell^2 - n^2}{s_{n,j}^2}(J_n(s_{n,j}\ell))^2\right], \tag{4.98}$$

the last step by some algebra and the fact that

$$R_{n,j}(\ell) = J_n(s_{n,j}\ell), \quad R'_{n,j}(\ell) = \left[\frac{d}{d\rho} J_n(s_{n,j}\rho)\right]_{\rho=\ell} = s_{n,j} J'_n(s_{n,j}\ell). \tag{4.99}$$

To finish, we consider (4.98), in the separate cases $\beta \neq \pi/2$ and $\beta = \pi/2$. The first is straightforward: Since $s_{n,j}$ solves (4.88), we have

$$J_n'(s_{n,j}\ell) = \frac{\tan \beta}{s_{n,j}\ell} J_n(s_{n,j}\ell), \tag{4.100}$$

so that (4.98) reads

$$\|R_{n,j}\|_v^2 = \frac{s_{n,j}^2 \ell^2 - n^2 + \tan^2 \beta}{2s_{n,j}^2} J_n^2(s_{n,j}\ell) \tag{4.101}$$

as promised. Finally, if $\beta = \pi/2$, then (4.88) gives $J_n(s_{n,j}\ell) = 0$, so by (4.98) and Proposition 4.3.2,

$$\|R_{n,j}\|_v^2 = \frac{\ell^2}{2} (J_n'(s_{n,j}\ell))^2 = \frac{\ell^2}{2} \left(\frac{nJ_n(s_{n,j}\ell) - s_{n,j}\ell J_{n+1}(s_{n,j}\ell)}{s_{n,j}\ell} \right)^2$$

$$= \frac{\ell^2}{2} J_{n+1}^2(s_{n,j}\ell), \tag{4.102}$$

as claimed. We're done. □

For any particular choice of β, an expansion of an $f \in L_v^2(0, \ell)$ into the basis $\mathcal{R}$ described above is called a *Fourier-Bessel series for f*.

Example 4.4.1 Solve the Sturm-Liouville problem (4.7) in the general case of $n \in \mathbb{N}$. Write down the corresponding Fourier-Bessel series for an $f \in L_v^2(0, \ell)$. Compute this series explicitly in the case $n = 0$, $f(\rho) = \ell^2 - \rho^2$.

Solution. By parts (a) and (b) of the above proposition, with $\beta = \pi/2$, we have the following complete list of eigenvalues of (4.7), with associated eigenfunctions:

$$\mu_{n,j} = s_{n,j}^2, \quad R_{n,j}(\rho) = J_n(\sqrt{\mu_{n,j}}\rho) = J_n(s_{n,j}\rho) \quad (j \in \mathbb{Z}^+), \tag{4.103}$$

where the $s_{n,j}$'s are the positive zeroes of $J_n(s\ell)$. By parts (c) and (d) of the proposition and by Lemma 4.1.2,

$$f(\rho) = \sum_{j=1}^{\infty} d_{n,j}(f) J_n(s_{n,j}\rho), \tag{4.104}$$

where

$$d_{n,j}(f) = \frac{\langle f, R_{n,j} \rangle_v}{\|R_{n,j}\|_v^2} = \frac{2}{\ell^2 J_{n+1}^2(s_{n,j}\ell)} \int_0^\ell f(\rho) J_n(s_{n,j}\rho) \rho \, d\rho. \tag{4.105}$$

That is, the partial sums on the right side of (4.104) converge, in the norm $\|\cdot\|_v$ on $L^2(0, \ell)$, to the function on the right, although under appropriate additional assumptions on f, one also has pointwise, or even uniform, convergence. (Fourier-Bessel series behave, convergence-wise, quite a bit like standard Fourier series. We omit the details; see [49].)

If $n = 0$ and $f(\rho) = \ell^2 - \rho^2$, the coefficients

$$d_{0,j}(f) = \frac{2}{\ell^2 J_1^2(s_{0,j}\ell)} \int_0^\ell (\ell^2 - \rho^2) J_0(s_{0,j}\rho) \rho \, d\rho \qquad (4.106)$$

may be evaluated explicitly in terms of Bessel functions using the following formulas:

$$\int x J_0(x) \, dx = x J_1(x) + C, \qquad (4.107)$$

$$\int x^3 J_0(x) \, dx = (x^3 - 4x) J_1(x) + 2x^2 J_0(x) + C. \qquad (4.108)$$

(See Exercise 4.4.1.) We get, by a substitution of $x = s_{0,j}\rho$,

$$\begin{aligned}
d_{0,j}(f) &= \frac{2}{s_{0,j}^2 J_1^2(s_{0,j}\ell)} \int_0^{s_{0,j}\ell} \left(x - \frac{x^3}{s_{0,j}^2 \ell^2}\right) J_0(x) \, dx \\
&= \frac{2}{s_{0,j}^2 J_1^2(s_{0,j}\ell)} \left[x J_1(x) - \frac{1}{s_{0,j}^2 \ell^2} \left((x^3 - 4x) J_1(x) + 2x^2 J_0(x)\right) \right]_0^{s_{0,j}\ell} \\
&= \frac{2}{s_{0,j}^2 J_1^2(s_{0,j}\ell)} \left[s_{0,j}\ell J_1(s_{0,j}\ell) - \frac{1}{s_{0,j}^2 \ell^2} (s_{0,j}^3 \ell^3 - 4 s_{0,j}\ell) J_1(s_{0,j}\ell) \right] \\
&= \frac{8}{s_{0,j}^3 \ell J_1(s_{0,j}\ell)}. \qquad (4.109)
\end{aligned}$$

For the next to the last step we used the fact that, by assumption, $J_0(s_{0,j}\ell) = 0$.

Exercises

4.4.1 Verify the formulas (4.107) and (4.108) by differentiating the righthand sides and checking that you get the associated integrands. Proposition 4.3.2 and Exercise 4.3.2 should be of help.

4.4.2 Show that

$$\rho^2 = \frac{\ell^2}{2} + 4 \sum_{j=1}^\infty \frac{J_0(s_{1,j}\rho)}{s_{1,j}^2 J_0(s_{1,j}\ell)} \quad (0 < \rho < \ell),$$

the $s_{1,j}$'s being the positive solutions s to $J_1(s\ell) = 0$.

4.4.3 Show that, for $h > 0$,

$$1 = 2\ell \sum_{j=1}^\infty \frac{s_{0,j} J_1(s_{0,j}\ell) J_0(s_{0,j}\rho)}{(s_{0,j}^2 \ell^2 + h^2) J_0^2(s_{0,j}\ell)} \quad (0 < \rho < \ell),$$

the $s_{0,j}$'s being the positive solutions s to $hJ_0(s\ell) + s\ell J_0'(s\ell) = 0$.

4.4.4 Show that, for $n \in \mathbb{Z}^+$,

$$\rho^n = 2\sum_{j=1}^{\infty} \frac{s_{n,j} J_{n+1}(s_{n,j}) J_n(s_{n,j}\rho)}{(s_{n,j}^2 - n^2) J_n^2(s_{n,j})} \quad (0 < \rho < 1),$$

the $s_{n,j}$'s being the positive zeroes of J_n'.

4.5 BESSEL FUNCTIONS III: BOUNDARY VALUE PROBLEMS

It's important to set a good example. Here are two.

Example 4.5.1 Complete the solution to the circular drum boundary value problem (2.266)–(2.269).

Solution. In Example 2.12.2 we arrived at separated solutions $z = z(\rho, \theta, t) = R(\rho)\Theta(\theta)T(t)$ to (2.266)–(2.268), where Θ is as in (2.275); R solves the singular Sturm-Liouville problem (2.276), which is really just (4.86), (4.87) in the case $\beta = \pi/2$; and T is as in (2.277). So Theorem 4.4.1 gives us, for each $n \in \mathbb{N}$ and $j \in \mathbb{Z}^+$, separated solutions

$$z_{n,j}(\rho, \theta, t) = J_n(s_{n,j}\rho)\bigl(\gamma e^{in\theta} + \delta e^{-in\theta}\bigr)\bigl(\sigma \cos s_{n,j}ct + \tau \sin s_{n,j}ct\bigr) \quad (4.110)$$

$(\gamma, \delta, \gamma, \tau \in \mathbb{C})$ to (2.266)–(2.268). Here the $s_{n,j}$'s are, for any given n, the positive zeroes of $J_n(s\ell)$.

It will be convenient to think of (4.110) a bit differently. Namely, by considering separately the "$e^{in\theta}$" and the "$e^{-in\theta}$" terms and replacing n by $-n$ in the the latter, we get, for *any* integer n and positive integer j, a separated solution

$$z_{n,j}(\rho, \theta, t) = J_{|n|}(s_{|n|,j}\rho)\bigl(D_{n,j} \cos s_{|n|,j}ct + E_{n,j} \sin s_{|n|,j}ct\bigr)e^{in\theta} \quad (4.111)$$

to equations (2.266)–(2.268). Since those equations are linear and homogeneous, we may, at least formally, also solve them with a series

$$z(\rho, \theta, t) = \sum_{n=-\infty}^{\infty} \sum_{j=1}^{\infty} J_{|n|}(s_{|n|,j}\rho)\bigl(D_{n,j} \cos s_{|n|,j}ct + E_{n,j} \sin s_{|n|,j}ct\bigr)e^{in\theta}.$$

(4.112)

Then, as usual, we plug in the inhomogeneous boundary conditions to try and determine the coefficients of this series. Let's see how this goes.

It goes like this: We first recall that the complex exponential functions in formula (4.112) constitute an orthogonal basis for $L^2(-\pi, \pi)$. And again, the set of Bessel functions in that formula is an orthogonal basis for $L_v^2(0, \ell)$, by Theorem 4.4.1(c). It follows that the set $\{X_{n,j}: n \in \mathbb{Z}, j \in \mathbb{Z}^+\}$, where

$$X_{n,j}(\rho, \theta) = J_{|n|}(s_{|n|,j}\rho)e^{in\theta}, \quad (4.113)$$

is an orthogonal basis for $L_V^2((0,\ell) \times (-\pi,\pi))$, where $V(\rho,\theta) = v(\rho) = \rho$. Note that, because the standard polar coordinate change of variables takes $\rho\,d\rho\,d\theta$ to $dx\,dy$, the space in question is really just the space $L^2(D)$ of square integrable functions on the disk D of radius ℓ and center $(0,0)$ in the Cartesian plane.

Actually, that (4.113) really *is* such a basis requires some proof. It does not follow *directly* from Proposition 3.8.1, because the set Λ there comprises product functions whose first factor is indexed by one integer and whose second is indexed by another. Not so in (4.113); the Bessel functions there depend on n and j. However, the *proof* of Proposition 3.8.1 adapts easily to give us our stated result concerning (4.113).

Now, we put the inhomogeneous condition $z(\rho,\theta,0) = f(\rho,\theta)$, from (2.269), into the series (4.112); we get

$$f(\rho,\theta) = \sum_{n=-\infty}^{\infty} \sum_{j=1}^{\infty} D_{n,j} J_{|n|}(s_{|n|,j}\rho) e^{in\theta}. \tag{4.114}$$

Then by Lemma 4.1.2 we have, for $n \in \mathbb{Z}$ and $j \in \mathbb{Z}^+$,

$$D_{n,j} = \frac{\langle f, X_{n,j}\rangle_V}{\|X_{n,j}\|_V^2}$$

$$= \frac{1}{\pi\ell^2 J_{|n|+1}^2(s_{|n|,j}\ell)} \int_{-\pi}^{\pi} \int_0^{\ell} f(\rho,\theta) J_{|n|}(s_{|n|,j}\rho) e^{-in\theta} \rho\,d\rho\,d\theta, \tag{4.115}$$

the last step because

$$\|X_{n,j}\|_V^2 = \int_{-\pi}^{\pi} \int_0^{\ell} \left|J_{|n|}(s_{|n|,j}\rho) e^{in\theta}\right|^2 \rho\,d\rho\,d\theta$$

$$= \left(\int_{-\pi}^{\pi} d\theta\right) \left(\int_0^{\ell} J_{|n|}^2(s_{|n|,j}\rho)\rho\,d\rho\right) = 2\pi \|R_{n,j}\|_v^2$$

$$= \pi\ell^2 J_{|n|+1}^2(s_{|n|,j}\ell), \tag{4.116}$$

by the definition (4.90) of $R_{n,j}$ and the formula (4.91) for the square of its norm (recall that, here, we're in the case $\beta = \pi/2$ of Theorem 4.4.1).

Finally, putting $z_t(\rho,\theta,0) = g(\rho,\theta)$ into (4.112) yields

$$g(\rho,\theta) = \sum_{n=-\infty}^{\infty} \sum_{j=1}^{\infty} E_{n,j} s_{|n|,j} c\, J_{|n|}(s_{|n|,j}\rho) e^{in\theta}, \tag{4.117}$$

so similar arguments give

$$E_{n,j} = \frac{1}{\pi\ell^2 s_{|n|,j} c\, J_{|n|+1}^2(s_{|n|,j}\ell)} \int_{-\pi}^{\pi} \int_0^{\ell} g(\rho,\theta) J_{|n|}(s_{|n|,j}\rho) e^{-in\theta} \rho\,d\rho\,d\theta \tag{4.118}$$

for $n \in \mathbb{Z}$ and $j \in \mathbb{Z}^+$.

In sum:

Proposition 4.5.1 *The boundary value problem (2.266)–(2.269) has formal solution (4.112), (4.115), (4.118), with the $s_{|n|,j}$'s the positive zeroes of $J_{|n|}(s\ell)$.*

We now consider a problem that's similar to Example 2.10.1 and that illustrates a general fact: Many of the techniques used in Chapter 2 to handle inhomogeneous behavior work just as well more generally, when the eigenfunctions involved are not necessarily sinusoids.

Example 4.5.2 Solve the boundary value problem corresponding to a circular drum as in the previous problem, but now with the acceleration G due to gravity taken into account.

Solution. Our boundary value problem here consists of the differential equation

$$c^2 \left(\frac{1}{\rho^2} z_{\theta\theta} + \frac{1}{\rho}[\rho z_\rho]_\rho \right) = z_{tt} - G \quad (0 < \rho < \ell,\ -\pi < \theta < \pi,\ t > 0) \quad (4.119)$$

together with the previous conditions (2.267)–(2.269). We proceed much as we did in Example 2.10.1. Namely, we first find a particular solution $z_i(\rho, \theta, t)$ to the differential equation (4.119) and to the homogeneous conditions (2.267), (2.268). We then add this to our general series solution (4.112) to the corresponding homogeneous problem (2.266)–(2.268); the result will, by the superposition principle II (Proposition 2.10.1), itself solve (4.119), (2.267), and (2.268), at least formally. We then determine what the coefficients of our series must be in order that conditions (2.269) be met.

In order to make our lives easier, we assume, as in Example 2.10.1, that our particular solution is t-independent. But while we're at it why don't we, in the present situation, assume θ-independence as well? Let's: so we write $z_i(\rho, \theta, t) = Z(\rho)$. Then the conditions (2.267) "in the θ direction" are automatically met. Also, that Z should solve the differential equation (4.119) means $c^2 \rho^{-1}[\rho Z'(\rho)]' = -G$, or $c^2[\rho Z'(\rho)]_\rho = -G\rho$, or $c^2 \rho Z'(\rho) = -G\rho^2/2 + H$, or $c^2 Z'(\rho) = -G\rho/2 + H/\rho$, or $c^2 Z(\rho) = -G\rho^2/4 + H/\ln \rho + K$, for constants K and H. We certainly don't want Z to blow up at $\rho = 0$, so we choose $H = 0$; the boundary condition (2.268) then gives $-G\ell^2/4 + K = 0$, so $K = G\ell^2/4$, so finally

$$Z(\rho) = \frac{G(\ell^2 - \rho^2)}{4c^2}. \quad (4.120)$$

Therefore, as just noted,

$$z(\rho, \theta, t) = \frac{G(\ell^2 - \rho^2)}{4c^2}$$
$$+ \sum_{n=-\infty}^{\infty} \sum_{j=1}^{\infty} J_{|n|}(s_{|n|,j}\rho)(D_{n,j} \cos s_{|n|,j}ct + E_{n,j} \sin s_{|n|,j}ct)e^{in\theta}$$

$$(4.121)$$

formally solves our boundary value problem, provided the $D_{n,j}$'s and $E_{n,j}$'s are chosen so that the boundary conditions (2.269) hold. We find such $D_{n,j}$'s and $E_{n,j}$'s as in our previous example, except that, now, the $D_{n,j}$'s must make the series in (4.114) equal to $f(\rho,\theta) - Z(\rho)$, instead of just $f(\rho,\theta)$. This means that, instead of the previous formula (4.115) for the $D_{n,j}$'s, we get

$$D_{n,j} = \frac{1}{\pi \ell^2 J_{|n|+1}^2(s_{|n|,j}\ell)} \int_{-\pi}^{\pi} \int_0^\ell f(\rho,\theta) J_{|n|}(s_{|n|,j}\rho) e^{-in\theta} \rho \, d\rho \, d\theta$$
$$- \frac{1}{\pi \ell^2 J_{|n|+1}^2(s_{|n|,j}\ell)} \int_{-\pi}^{\pi} \int_0^\ell \frac{G(\ell^2 - \rho^2)}{4c^2} J_{|n|}(s_{|n|,j}\rho) e^{-in\theta} \rho \, d\rho \, d\theta. \quad (4.122)$$

Note that, in the integral on the far right side, the only part of the integrand that's dependent on θ is $e^{-in\theta}$, whose integral over $[-\pi,\pi]$ is $2\pi\delta_{n,0}$, by Lemma 1.1.2. Therefore,

$$-\frac{1}{\pi \ell^2 J_{|n|+1}^2(s_{|n|,j}\ell)} \int_{-\pi}^{\pi} \int_0^\ell \frac{G(\ell^2 - \rho^2)}{4c^2} J_{|n|}(s_{|n|,j}\rho) e^{-in\theta} \rho \, d\rho \, d\theta$$
$$= -\frac{1}{\pi \ell^2 J_1^2(s_{0,j}\ell)} \cdot \frac{G}{4c^2} \cdot 2\pi \delta_{n,0} \int_0^\ell (\ell^2 - \rho^2) J_0(s_{0,j}\rho) \rho \, d\rho$$
$$= -\frac{2G}{c^2 s_{0,j}^3 \ell J_1(s_{0,j}\ell)} \delta_{n,0}, \quad (4.123)$$

the last step by the results of Example 4.4.1—see (4.109).

By the same reasoning as in our previous example, we get, here, the same $E_{n,j}$'s as we had there. Let's summarize.

Proposition 4.5.2 *The boundary value problem* (4.119), (2.267)–(2.269) *has formal solution given by* (4.121), *where*

$$D_{n,j} = \frac{1}{\pi \ell^2 J_{|n|+1}^2(s_{|n|,j}\ell)} \int_{-\pi}^{\pi} \int_0^\ell f(\rho,\theta) J_{|n|}(s_{|n|,j}\rho) e^{-in\theta} \rho \, d\rho \, d\theta$$
$$- \frac{2G}{c^2 s_{0,j}^3 \ell J_1(s_{0,j}\ell)} \delta_{n,0} \quad (4.124)$$

and $E_{n,j}$ is as in (4.118).

Exercises

4.5.1 Solve the two-dimensional heat problem $k \nabla_2^2 u = u_t$ on the disk D of radius ℓ and center $(0,0)$, assuming an initial temperature distribution $f(\rho)$ that's *independent* of θ and constant temperature 0 on the boundary. ANSWER:

$$u(\rho,t) = \frac{2}{\ell^2}\sum_{j=1}^{\infty}\frac{J_0(s_{0,j}\rho)}{J_1^2(s_{0,j}\ell)}\left[\int_0^\ell f(v)J_0(s_{0,j}v)\,v\,dv\right]e^{-s_{0,j}^2 kt},$$

the $s_{0,j}$'s being the positive solutions s to $J_0(s\ell) = 0$.

4.5.2 Solve the two-dimensional heat problem $k\nabla_2^2 u = u_t$, with initial temperature distribution $f(\rho,\theta)$, on the disk D of radius ℓ and center $(0,0)$, in the case where:

  a. the boundary of the disk is insulated;
  b. Newton's law of cooling applies on this boundary.

Hints: If you set things up properly, you should arrive at the boundary value problem

$$k\left(\frac{1}{\rho^2}u_{\theta\theta} + \frac{1}{\rho}[\rho u_\rho]_\rho\right) = u_t \quad (0 < \rho < \ell,\ -\pi < \theta < \pi,\ t > 0),$$
$$u(\rho,-\pi,t) = u(\rho,\pi,t), \quad u_\theta(\rho,-\pi,t) = u_\theta(\rho,\pi,t) \quad (0 < \rho < \ell,\ t > 0),$$
$$\sin\beta\, u(\ell,\theta,t) + \cos\beta\, \ell u_\rho(\ell,\theta,t) = 0 \quad (-\pi < \theta < \pi,\ t > 0),$$
$$u(\rho,\theta,0) = f(\rho,\theta) \quad (0 < \rho < \ell,\ -\pi < \theta < \pi),$$

where $\beta = 0$ in the case of part a; $\tan\beta = h\ell$ in the case of part b. (Explain.) Look first for solutions $u = R\Theta T$ to the homogeneous parts of the problem; then apply the initial conditions.

4.5.3 Our pal the disk D of radius ℓ and center $(0,0)$ has an internal heat source generating heat at a constant rate C. Find the temperature of the disk at time t if its perimeter has constant temperature zero and its initial temperature throughout (just before the heat source is switched on) is zero. That is, solve the boundary value problem

$$k\left(\frac{1}{\rho^2}u_{\theta\theta} + \frac{1}{\rho}[\rho u_\rho]_\rho\right) = u_t - C \quad (0 < \rho < \ell,\ -\pi < \theta < \pi,\ t > 0),$$
$$u(\rho,-\pi,t) = u(\rho,\pi,t), \quad u_\theta(\rho,-\pi,t) = u_\theta(\rho,\pi,t) \quad (0 < \rho < \ell,\ t > 0),$$
$$u(\ell,\theta,t) = 0 \quad (-\pi < \theta < \pi,\ t > 0),$$
$$u(\rho,\theta,0) = 0 \quad (0 < \rho < \ell,\ -\pi < \theta < \pi).$$

(Compare with Exercise 2.10.2.) Express your answer in terms of positive solutions $s_{0,j}$ to the equation $J_0(s\ell) = 0$.

4.5.4 Solve the following *damped* circular drum problem, with zero initial displacement and spatially constant initial velocity:

$$\frac{c^2}{\rho}[\rho z_\rho]_\rho = u_{tt} + 2dz_t \quad (0 < \rho < \ell,\ t > 0),$$
$$z(\ell,t) = 0 \quad (t > 0),$$
$$z(\rho,0) = 0, \quad z_t(\rho,0) = v_0 \quad (t > 0).$$

(Here $d > 0$.) Express your answer in terms of the positive zeroes of $J_0(s\ell)$. Assume for simplicity that the smallest of these zeroes is larger than $2d/c$. (Compare with Exercise 2.7.5.)

4.6 ORTHOGONAL POLYNOMIALS

Many Sturm-Liouville problems of interest have *polynomial* eigenfunctions. Here we present a theorem that is quite useful in the study of such problems, and consider a couple of examples.

Theorem 4.6.1 *Let $\{P_n : n \in \mathbb{N}\} \subset L_w^2(a,b)$ be an orthogonal set of polynomials; suppose w integrable on $[a,b]$, and P_n has degree n for each n.*

(a) *If Q is a polynomial of degree m on $[a,b]$, then $\langle Q, P_n \rangle_w = 0$ for $n > m$.*

(b) *$\{P_n : n \in \mathbb{N}\}$ is an orthogonal basis for $L_w^2(a,b)$.*

Proof. Let the P_n's and Q be as stipulated. We write Q as a linear combination of $P_1, P_2, \ldots, P_m$ as follows. Let a_m be the leading coefficient of Q and b_m that of P_m; then $Q - (a_m/b_m)P_m$ is a polynomial, call it Q_{m-1}, of degree at most $m-1$. (Our assumption on the degree of P_m assures $b_m \neq 0$.) But then, similarly, if a_{m-1} is the leading coefficient of Q_{m-1}, we find that $Q_{m-1} - (a_{m-1}/b_{m-1})P_{m-1}$ is a polynomial Q_{m-2} of degree at most $m-2$. And so on, until we arrive at the equation $Q_1 - (a_1/b_1)P_1 = Q_0$, where Q_0 is a constant. So

$$Q = \frac{a_m}{b_m}P_m + Q_{m-1} = \frac{a_m}{b_m}P_m + \frac{a_{m-1}}{b_{m-1}}P_{m-1} + Q_{m-2}$$

$$\cdots$$

$$= \frac{a_m}{b_m}P_m + \frac{a_{m-1}}{b_{m-1}}P_{m-1} + \cdots + \frac{a_1}{b_1}P_1 + \frac{Q_0}{P_0}P_0. \qquad (4.125)$$

Let's now define d_n to be the coefficient of P_n in (4.125) if $0 \le n \le m$ and 0 otherwise. Then

$$Q = \sum_{n=0}^{\infty} d_n P_n, \qquad (4.126)$$

so by the orthogonality of the P_n's,

$$\langle Q, P_n \rangle_w = \left\langle \sum_{k=0}^{\infty} d_k P_k, P_n \right\rangle_w = \sum_{k=0}^{\infty} d_k \langle P_k, P_n \rangle_w = \sum_{k=0}^{\infty} d_k \|P_k\|_w^2 \, \delta_{k,n}$$

$$= d_n \|P_n\|_w^2. \qquad (4.127)$$

This proves part (a).

To prove part (b), we need only, by Lemma 4.1.2, show that

$$f = \sum_{n=0}^{\infty} \frac{\langle f, P_n \rangle_w}{||P_n||_w^2} P_n \qquad (4.128)$$

for any $f \in L_w^2(a,b)$. We do so as follows: First we note that, by (4.126) and (4.127) and the surrounding discussions, (4.128) certainly holds for f a *polynomial* in $L_w^2(a,b)$. We then use the Stone-Weierstrass theorem (Theorem 5.8.1 in [33]), together with limiting arguments very much like those in the proof of Proposition 3.2.1(e), to get the general case of (4.128) from the polynomial case, and we're done. □

We consider some examples, whose significance we'll explain in due course. The first involves *Tchebyschev's equation*

$$[(1-x^2)^{1/2} T'(x)]' + \lambda (1-x^2)^{-1/2} T(x) = 0 \quad (-1 < x < 1). \qquad (4.129)$$

Note that this amounts to a Sturm-Liouville problem on $[-1, 1]$; in the language of Definition 4.1.1(a), we have $r(x) = (1-x^2)^{1/2}$, $q(x) = 0$, and $w(x) = (1-x^2)^{-1/2}$, and in the language of Definition 4.1.1(b)(i), we have $\alpha = \beta = 0$. This problem is singular because $r(\pm 1) = 0$ and $w(x)$ blows up at $x = \pm 1$.

We have:

Proposition 4.6.1 (a) *For each $n \in \mathbb{N}$, $\lambda_n = n^2$ is an eigenvalue of (4.129), with corresponding eigenfunction given by the nth Tchebyschev polynomial*

$$T_n(x) = \sum_{k=0}^{\lfloor n/2 \rfloor} \frac{n!}{(2k)!(n-2k)!} x^{n-2k} (x^2 - 1)^k. \qquad (4.130)$$

Here $\lfloor n/2 \rfloor$ denotes the greatest integer less than or equal to $n/2$; that is,

$$\left\lfloor \frac{n}{2} \right\rfloor = \begin{cases} \dfrac{n}{2} & \text{if } n \text{ is even,} \\ \dfrac{n-1}{2} & \text{if } n \text{ is odd.} \end{cases} \qquad (4.131)$$

(b) $T_n(\cos \theta) = \cos n\theta$ *for each n.*

(c) $\{T_n : n \in \mathbb{N}\}$ *is an orthogonal basis for $L_u^2(-1, 1)$, where $u(x) = (1-x^2)^{-1/2}$. Any eigenfunction of (4.129) is a constant multiple of one of the T_n's.*

(d) $||T_n||_u^2 = \pi/2$ *for $n \in \mathbb{Z}^+$ and $||T_0||_u^2 = \pi$.*

Proof. We prove part (b) first. We have

$$\begin{aligned}
T_n(\cos\theta) &= \sum_{k=0}^{\lfloor n/2 \rfloor} \frac{n!}{(2k)!(n-2k)!}(\cos\theta)^{n-2k}(\cos^2\theta - 1)^k \\
&= \sum_{k=0}^{\lfloor n/2 \rfloor} \frac{n!}{(2k)!(n-2k)!}(\cos\theta)^{n-2k}(i\sin\theta)^{2k} \\
&= \sum_{\substack{0 \le m \le n \\ m \text{ is even}}} \frac{n!}{m!(n-m)!}(\cos\theta)^{n-m}(i\sin\theta)^m \\
&= \operatorname{Re}\left(\sum_{m=0}^{n} \frac{n!}{m!(n-m)!}(\cos\theta)^{n-m}(i\sin\theta)^m\right) \\
&= \operatorname{Re}(\cos\theta + i\sin\theta)^n = \operatorname{Re} e^{in\theta} = \cos n\theta. \quad (4.132)
\end{aligned}$$

We explain: For the second equality, we wrote $\cos^2\theta - 1 = -\sin^2\theta = (i\sin\theta)^2$. For the third, we put $m = 2k$; as k ranges over integers in the interval $[0, \lfloor n/2 \rfloor]$, m ranges over *even* integers in $[0, 2\lfloor n/2 \rfloor]$; by (4.131), the latter are just the even integers in $[0,n]$. For the fourth equality, we noted that adding real numbers times odd powers of i to a complex quantity does not change the real part of that quantity. For the fifth, we used the binomial theorem

$$(a+b)^n = \sum_{m=0}^{n} \frac{n!}{m!(n-m)!} a^{n-m} b^m; \quad (4.133)$$

the remaining equalities are for familiar reasons.

So part (b) is proved. Part (a) then goes like this: We note that

$$\frac{d^2}{d\theta^2} T_n(\cos\theta) = \frac{d^2}{d\theta^2} \cos n\theta = -n^2 \cos n\theta = -n^2 T_n(\cos\theta), \quad (4.134)$$

so the substitution $x = \cos\theta$ and the chain rule give

$$\begin{aligned}
-n^2 T_n(x) &= \frac{d^2 T_n(x)}{d\theta^2} = \frac{d}{d\theta}\left[\frac{dT_n(x)}{d\theta}\right] = \frac{d}{d\theta}\left[\frac{dx}{d\theta}\frac{dT_n(x)}{dx}\right] \\
&= \frac{d^2 x}{d\theta^2}\frac{dT_n(x)}{dx} + \left(\frac{dx}{d\theta}\right)^2 \frac{d^2 T_n(x)}{dx^2} \\
&= -\cos\theta \frac{dT_n(x)}{dx} + \sin^2\theta \frac{d^2 T_n(x)}{dx^2} = -xT_n'(x) + (1-x^2)T_n''(x). \\
& \hspace{10cm} (4.135)
\end{aligned}$$

Dividing by $(1-x^2)^{1/2}$ gives

$$\begin{aligned}
-n^2(1-x^2)^{-1/2} T_n(x) &= -x(1-x^2)^{-1/2} T_n'(x) + (1-x^2)^{1/2} T_n''(x) \\
&= [(1-x^2)^{1/2} T_n'(x)]', \quad (4.136)
\end{aligned}$$

so $T = T_n$ satisfies (4.129), as required.

For part (c), we note first that the T_n's are, by Theorem 4.1.1(c), orthogonal. (The substitution $x = \cos\theta$ shows the integrability of $u(x) = (1-x^2)^{-1/2}$ on $(-1,1)$.) By its definition (4.130), T_n has degree n, so by Theorem 4.6.1(b), the T_n's form an orthogonal basis for $L_u^2(-1,1)$. So, to complete our proof of part (c), we need only show that any eigenfunction is a multiple of some T_n.

So suppose we have an eigenfunction T with eigenvalue λ. We write

$$T = \sum_{n=0}^{\infty} \frac{\langle T, T_n\rangle_u}{\|T_n\|_u^2} T_n \tag{4.137}$$

and note the following. First, λ must equal k^2 for some $k \in \mathbb{N}$: If not then, by Theorem 4.1.1(c), T is orthogonal to all T_n's, so by the expansion (4.137), T is zero, contradicting the fact that T is an eigenfunction. But then, by Theorem 4.1.1(c) again, $\langle T, T_n\rangle = 0$ for $n \neq k$. So by (4.137) again, T is a multiple of T_k, as required.

Part (d): By again substituting $x = \cos\theta$, we get, for $n \geq 1$,

$$\|T_n\|_u^2 = \int_{-1}^{1} \frac{T_n^2(x)}{(1-x^2)^{1/2}}\,dx = \int_0^\pi T_n^2(\theta)\,d\theta = \int_0^\pi \cos^2 n\theta\,d\theta$$
$$= \frac{1}{2}\int_0^\pi (1 + \cos 2n\theta)\,d\theta = \frac{1}{2}\left[\theta + \frac{\sin 2n\theta}{2n}\right]_0^\pi = \frac{\pi}{2}. \tag{4.138}$$

Since T_0 is identically 1, a similar calculation gives $\|T_0\|_u^2 = \pi$; this completes our proof. □

Tchebyschev polynomials are central to the theory of *uniform approximation*. Specifically, we have the following.

Proposition 4.6.2 *Let $n \in \mathbb{Z}^+$; let $\mathcal{P}_{n,1}([-1,1])$ denote the set of all polynomials on $[-1,1]$ having degree n and leading coefficient 1. Then $2^{1-n}T_n$ is an element of this set, and among all such elements has the smallest supremum.*

For a proof and for discussion of the importance of the result, see Chapters 43–45 in [30].

We turn, now, to another singular Sturm-Liouville problem—the one given by *Legendre's equation*

$$((1-x^2)P'(x))' + \lambda P(x) = 0 \quad (-1 < x < 1). \tag{4.139}$$

Here we have, in the language of Definition 4.1.1(a), $r(x) = 1 - x^2$, $q(x) = 0$, $w(x) = 1$, $\alpha = \beta = 0$. We'll see in the next section that Legendre's equation is relevant to certain boundary value problems of note. In the meantime, we have:

Proposition 4.6.3 (a) *For each $n \in \mathbb{N}$, $\lambda_n = n(n+1)$ is an eigenvalue of (4.139), with corresponding eigenfunction given by the nth Legendre polynomial*

$$P_n(x) = \frac{1}{2^n}\sum_{k=0}^{\lfloor n/2 \rfloor} \frac{(-1)^k(2n-2k)!}{k!(n-2k)!(n-k)!} x^{n-2k}. \tag{4.140}$$

(b) $\{P_n: n \in \mathbb{N}\}$ *is an orthogonal basis for* $L^2(-1,1)$. *Any eigenfunction of* (4.139) *is a constant multiple of one of the* P_n's.

Proof. Part (a): Let's fix $n \in \mathbb{N}$ and write

$$z_k = \frac{(-1)^k (2n-2k)!}{2^n k!(n-2k)!(n-k)!} \quad \left(0 \le k \le \left\lfloor \frac{n}{2} \right\rfloor\right). \tag{4.141}$$

Then by the definition (4.140) of P_n,

$$((1-x^2)P_n'(x))' + n(n+1)P_n(x)$$

$$= \sum_{k=0}^{\lfloor n/2 \rfloor} z_k \left[((1-x^2)(x^{n-2k})')' + n(n+1)x^{n-2k}\right]$$

$$= \sum_{k=0}^{\lfloor n/2 \rfloor} z_k \left((n-2k)(n-2k-1)x^{n-2k-2} + 2k(2n-2k+1)x^{n-2k}\right)$$

$$= \sum_{k=0}^{\lfloor n/2 \rfloor - 1} (n-2k)(n-2k-1)z_k\, x^{n-2k-2} + 2\sum_{k=1}^{\lfloor n/2 \rfloor} k(2n-2k+1)z_k\, x^{n-2k}. \tag{4.142}$$

Here, for the second equality, we performed the necessary differentiation and then collected coefficients of x^{n-2k}. For the last, we split our sum in two, and noted that the coefficient $(n-2k)(n-2k-1)$ of x^{n-2k-2} vanishes when $k = \lfloor n/2 \rfloor$, while the coefficient $2k(2n-2k+1)$ of x^{n-2k} does so when $k = 0$.

Now, in the first sum in (4.142), we replace k by $k-1$; we combine the resulting sums, to get

$$((1-x^2)P_n'(x))' + n(n+1)P_n(x) \tag{4.143}$$

$$= \sum_{k=1}^{\lfloor n/2 \rfloor} \left((n-2k+2)(n-2k+1)z_{k-1} + 2k(2n-2k+1)z_k\right) x^{n-2k}. \tag{4.144}$$

To finish with part (a), it will suffice to show that

$$(n-2k+2)(n-2k+1)z_{k-1} + 2k(2n-2k+1)z_k = 0 \tag{4.145}$$

for $1 \le k \le \lfloor n/2 \rfloor$. Let's show this. To do so we note that, by the definition (4.141) of z_k and the identity $m!/m = (m-1)!$,

$$\frac{z_k}{z_{k-1}} = \frac{(-1)^k(2n-2k)!}{2^n k!(n-2k)!(n-k)!} \cdot \frac{2^n(k-1)!(n-2k+2)!(n-k+1)!}{(-1)^{k-1}(2n-2k+2)!}$$

$$= -\frac{(n-2k+2)(n-2k+1)(n-k+1)}{k(2n-2k+2)(2n-2k+1)} = -\frac{(n-2k+2)(n-2k+1)}{2k(2n-2k+1)} \tag{4.146}$$

(the assumption $1 \leq k \leq \lfloor n/2 \rfloor$ assures a nonzero denominator), which *is* (4.145). So part (a) is proved.

Part (b) is deduced from Lemma 4.1.2 just as part (c) of Proposition 4.6.1 was. □

Exercises

4.6.1 Write down explicit formulas for the Tchebyschev polynomials T_0 through T_6.

4.6.2 For $1 \leq m \leq 6$, write $f_m(x) = x^m$ as a finite linear combination of Tchebyschev polynomials. Hint: You don't have to write down integrals to find the coefficients of the linear combination; just follow the line of reasoning commencing the proof of Theorem 4.6.1.

4.6.3 Use Proposition 4.6.1(b) and a trig identity to show that $T_n(\sin \theta)$ equals $(-1)^{n/2} \cos n\theta$ if n is even and equals $(-1)^{(n-1)/2} \sin n\theta$ if n is odd ($n \in \mathbb{N}$).

4.6.4 **a.** Use Proposition 4.6.1(b) and Exercises 4.6.1 and 4.6.3 to express $\cos 5\theta$ in terms of $\cos \theta$ and $\sin 5\theta$ in terms of $\sin \theta$.

b. Express $\sin 6\theta$ in terms of terms of $\cos \theta$ and $\sin \theta$. Hint: First expand $\sin(\theta + 5\theta)$.

4.6.5 **a.** Use simple trig identities to show that, for $m \in \mathbb{N}$,

$$\cos(m+2)\theta = 2\cos\theta \cos(m+1)\theta - \cos m\theta.$$

b. Deduce the recurrence relation

$$T_{m+2}(x) = 2xT_{m+1}(x) - T_m(x) \quad (m \in \mathbb{N}).$$

4.6.6 Write down explicit formulas for the Legendre polynomials P_0 through P_6.

4.6.7 For $1 \leq m \leq 6$, write $f_m(x) = x^m$ as a finite sum of Legendre polynomials. See the hint for Exercise 4.6.2.

4.7 MORE ON LEGENDRE POLYNOMIALS

We begin with a variety of important explicit results concerning the Legendre polynomials P_n of Proposition 4.6.3.

Proposition 4.7.1 *Let* $n \in \mathbb{N}$.

(a) P_{2n+1} *is an odd function; in particular,* $P_{2n+1}(0) = 0$. P_{2n} *is an even function, and*

$$P_{2n}(0) = \frac{(-1)^n (2n)!}{2^{2n}(n!)^2}. \tag{4.147}$$

(b) *We have* Rodrigues' formula

$$P_n(x) = \frac{1}{2^n n!} \frac{d^n}{dx^n} (x^2 - 1)^n. \tag{4.148}$$

(c) $P_n(-1) = (-1)^n$ *and* $P_n(1) = 1$.

(d)

$$(2n+1)xP_n(x) = (n+1)P_{n+1}(x) + nP_{n-1}(x), \tag{4.149}$$
$$(2n+1)P_n(x) = P'_{n+1}(x) - P'_{n-1}(x), \tag{4.150}$$

where (for the case $n = 0$) we define $P_{-1} = 0$.

(e) $||P_n||^2 = 2/(2n+1)$.

Proof. An odd polynomial in x is one containing only odd powers of x; similarly for even polynomials. Moreover, an odd function f satisfies $f(-0) = -f(0)$, which means $f(-0) = 0$.

Now by the definition (4.140), $P_{2n+1}(x)$ contains only the powers $x^{2n+1-2k}$ of x; the exponents are all odd. Similarly for P_{2n}. So we have proved all statements of part (a) except the one regarding $P_{2n}(0)$. For this, we note that all terms in the expression for P_{2n} given by (4.140) vanish at $x = 0$ *except* for the constant term, where $k = n$; that term equals

$$\frac{(-1)^n (2(2n) - 2n)!}{2^{2n} n! (2n - 2n)! (2n - n)!} = \frac{(-1)^n (2n)!}{2^{2n} (n!)^2}. \tag{4.151}$$

So part (a) is proved.

Next, by the binomial theorem (4.133),

$$(x^2 - 1)^n = n! \sum_{k=0}^{n} \frac{(-1)^k}{k!(n-k)!} x^{2n-2k}. \tag{4.152}$$

We differentiate both sides n times and note that, if $2n - 2k < n$, then the nth derivative of x^{2n-2k} is zero. Since $2n - 2k < n$ is the same as $k > \lfloor n/2 \rfloor$ (for $k, n \in \mathbb{Z}$), we get

$$\frac{1}{2^n n!} \frac{d^n}{dx^n} (x^2-1)^n = \frac{1}{2^n} \sum_{k=0}^{\lfloor n/2 \rfloor} \frac{(-1)^k}{k!(n-k)!} \frac{d^n}{dx^n} x^{2n-2k}$$

$$= \frac{1}{2^n} \sum_{k=0}^{\lfloor n/2 \rfloor} \frac{(-1)^k (2n-2k)(2n-2k-1)\cdots(n-2k+1)}{k!(n-k)!} x^{n-2k}$$

$$= \frac{1}{2^n} \sum_{k=0}^{\lfloor n/2 \rfloor} \frac{(-1)^k (2n-2k)!}{k!(n-2k)!(n-k)!} x^{n-2k} = P_n(x), \tag{4.153}$$

the next to last equality because

$$(2n-2k)(2n-2k-1)\cdots(n-2k+1) = \frac{(2n-2k)(2n-2k-1)\cdots 1}{(n-2k)(n-2k-1)\cdots 1}$$
$$= \frac{(2n-2k)!}{(n-2k)!}. \qquad (4.154)$$

This gives us part (b).

For part (c) we use *Leibniz'* rule

$$(fg)^{(n)} = \sum_{m=0}^{n} \frac{n!}{m!(n-m)!} f^{(m)} g^{(n-m)} \qquad (4.155)$$

(assuming all derivatives on the right exist), which is readily verified by mathematical induction. Since $(x^2-1)^n = (x+1)^n(x-1)^n$, this rule and Rodrigues' formula give

$$P_n(x) = \frac{1}{2^n} \sum_{m=0}^{n} \frac{1}{m!(n-m)!} \frac{d^m(x+1)^n}{dx^m} \frac{d^{n-m}(x-1)^n}{dx^{n-m}}$$
$$= \frac{1}{2^n} \sum_{m=0}^{n} \frac{(n!)^2}{(m!)^2((n-m)!)^2} (x+1)^{n-m}(x-1)^m \qquad (4.156)$$

(to simplify the factors coming from repeated differentiation, we've used formulas like (4.154)). Now if we put $x = 1$ into the above then, because of the factors of $(x-1)^m$, all summands vanish except the 0th. So we get

$$P_n(1) = \frac{1}{2^n} \cdot \frac{(n!)^2}{(0!)^2(n!)^2}(1+1)^n = 1, \qquad (4.157)$$

as claimed. Similarly, putting $x = -1$ into the above formula for $P_n(x)$ leaves only the nth term, so

$$P_n(-1) = \frac{1}{2^n} \cdot \frac{(n!)^2}{(n!)^2(0!)^2}(-1-1)^n = (-1)^n. \qquad (4.158)$$

So we're done with part (c).

On to part (d). We first check directly, using the definition (4.140) of P_n, that both of the equations stated in this part are true for $n = 0$, under the stated convention $P_{-1} = 0$. So now we assume $n > 0$. We let $k_n(x) = (x^2-1)^n$, so that by Rodrigues' formula,

$$2^n n! P_n(x) = k_n^{(n)}(x). \qquad (4.159)$$

We note that

$$k'_{n+1}(x) = 2(n+1)x(x^2-1)^n = 2(n+1)xk_n(x); \qquad (4.160)$$

consequently, by Leibniz' rule (4.155),

$$2^{n+1}(n+1)!P_{n+1}(x) = k_{n+1}^{(n+1)}(x) = (k'_{n+1}(x))^{(n)} = 2(n+1)(xk_n(x))^{(n)}$$
$$= 2(n+1) \sum_{m=0}^{n} \frac{n!}{m!(n-m)!} x^{(m)} k_n^{(n-m)}(x)$$
$$= 2(n+1)\left[xk_n^{(n)}(x) + nk_n^{(n-1)}(x)\right]$$
$$= 2^{n+1}(n+1)!xP_n(x) + 2n(n+1)k_n^{(n-1)}(x). \quad (4.161)$$

Here, for the next to the last equality, we used the fact that the mth derivative of x is zero for $m > 1$; for the last, we used (4.159) again.

On the other hand, (4.160) gives

$$k''_{n+1}(x) = [2(n+1)xk_n(x)]' = 2(n+1)[k_n(x) + xk'_n(x)]$$
$$= 2(n+1)[(x^2-1)^n + 2nx^2(x^2-1)^{n-1}]$$
$$= 2(n+1)[(x^2-1)^n + 2n([x^2-1]+1)(x^2-1)^{n-1}]$$
$$= 2(n+1)[(2n+1)(x^2-1)^n + 2n(x^2-1)^{n-1}]$$
$$= 2(n+1)[(2n+1)k_n(x) + 2nk_{n-1}(x)], \quad (4.162)$$

so again by (4.159),

$$2^{n+1}(n+1)!P_{n+1}(x) = k_{n+1}^{(n+1)}(x) = (k''_{n+1}(x))^{(n-1)}$$
$$= 2(n+1)[(2n+1)k_n^{(n-1)}(x) + 2nk_{n-1}^{(n-1)}(x)]$$
$$= 2(n+1)(2n+1)k_n^{(n-1)}(x) + 2^{n+1}(n+1)!P_{n-1}(x). \quad (4.163)$$

Multiplying (4.161) by $2n+1$, (4.163) by n, subtracting, and dividing by $2^{n+1}(n+1)!$, we arrive at (4.149).

Further, differentiating both sides of (4.163), applying (4.159), and again dividing by $2^{n+1}(n+1)!$ give (4.150). So part (d) of our proposition is proved.

Finally, part (e): Putting $n = 0$ into this formula gives $||P_0||^2 = 2$, which is readily verified by direct means, so we assume $n \geq 1$. We replace n by $n-1$ in (4.149) and multiply the result through by $(2n+1)P_n(x)$, to get

$$(2n+1)(2n-1)xP_{n-1}P_n(x) = (2n+1)nP_n^2(x)$$
$$+ (2n+1)(n-1)P_{n-2}(x)P_n(x). \quad (4.164)$$

We next multiply (4.149) itself by $(2n-1)P_{n-1}(x)$ and subtract the result from (4.164), yielding

$$(2n+1)nP_n^2(x) + (2n+1)(n-1)P_{n-2}(x)P_n(x)$$
$$= (2n-1)(n+1)P_{n-1}P_n(x) + (2n-1)nP_{n-1}^2(x). \quad (4.165)$$

Integrating both sides over $[-1, 1]$, dividing by n, and using the fact that $\langle P_n, P_k \rangle = 0$ for $n \neq k$, we get

$$(2n + 1)||P_n||^2 = (2n - 1)||P_{n-1}||^2. \tag{4.166}$$

Repeated application of this formula gives

$$(2n + 1)||P_n||^2 = (2n - 1)||P_{n-1}||^2 = (2(n-1) - 1)||P_{(n-1)-1}||^2$$
$$= (2(n-2) - 1)||P_{(n-2)-1}||^2 \ldots = (2(1) - 1)||P_{(1)-1}||^2 = 2, \tag{4.167}$$

which is part (e), and we're done. □

An important application of Legendre polynomials is to the solution of a cognate problem, which we now investigate.

Corollary 4.7.1 *The singular Sturm-Liouville problem*

$$[\sin \phi \, \Phi'(\phi)]' + \lambda \sin \phi \, \Phi(\phi) = 0 \quad (0 < \phi < \pi) \tag{4.168}$$

has, for each $n \in \mathbb{N}$, the following eigenvalue and corresponding eigenfunction:

$$\lambda_n = n(n+1), \quad \Phi_n(\phi) = P_n(\cos \phi). \tag{4.169}$$

These eigenfunctions form an orthogonal basis for $L_w^2(0, \pi)$, where $w(\phi) = \sin \phi$; moreover, for each n,

$$||\Phi_n||_w^2 = \frac{2}{2n+1}. \tag{4.170}$$

Proof. That the Φ_n's are orthogonal, with the stated norms, in $L_w^2(0, \pi)$ follows from the corresponding properties of the P_n's and the substitution $x = \cos \phi$, in this way:

$$\langle \Phi_n, \Phi_k \rangle_w = \int_0^\pi \Phi_n(\phi) \Phi_k(\phi) \sin \phi \, d\phi = \int_0^\pi P_n(\cos \phi) P_k(\cos \phi) \sin \phi \, d\phi$$
$$= \int_{-1}^1 P_n(x) P_k(x) \, dx = \langle P_n, P_k \rangle. \tag{4.171}$$

To show that the Φ_n's form a basis, let's suppose $f \in L_w^2(0, \pi)$. Then by arguments like those in (4.171), we find that $g(x) = f(\arccos x)$ defines a function $g \in L^2(-1, 1)$ and that

$$\lim_{N \to \infty} \left\| \sum_{n=0}^N \frac{\langle f, \Phi_n \rangle_w}{||\Phi_n||_w^2} - f \right\|_w = \lim_{N \to \infty} \left\| \sum_{n=0}^N \frac{\langle g, P_n \rangle}{||P_n||^2} - g \right\| = 0, \tag{4.172}$$

the last step because the P_n's constitute a basis for $L^2(-1, 1)$.

To finish our proof, we must show that (4.168) has precisely the stated eigenvalues and eigenfunctions. It will suffice, by Proposition 4.6.3, to show that (λ, P) solves Legendre's equation (4.139) if and only if (λ, Φ), where $\Phi(\phi) = P(\cos \phi)$, solves our present equation (4.168). Let's do this—actually, we'll prove only the "if" part, but our argument is entirely reversible.

We assume, then, that

$$\lambda \sin \phi \, P(\cos \phi) = -[\sin \phi \, (P(\cos \phi))']' \quad (0 < \phi < \pi). \tag{4.173}$$

A calculation gives

$$[\sin \phi \, (P(\cos \phi))']' = \sin \phi \, (\sin^2 \phi \, P''(\cos \phi) - 2\cos \phi \, P'(\cos \phi)); \tag{4.174}$$

consequently, (4.173) and the substitution $x = \cos \phi$ yield

$$\begin{aligned}\lambda(1-x^2)^{1/2} P(x) &= -(1-x^2)^{1/2} \left((1-x^2)P''(x) - 2xP'(x)\right) \\ &= -(1-x^2)^{1/2} \left((1-x^2)P'(x)\right)'. \end{aligned} \tag{4.175}$$

Dividing by $(1-x^2)^{1/2}$ gives precisely (4.139), and we're done. □

The following example illustrates the significance of the above corollary.

Example 4.7.1 Solve the Dirichlet problem for a solid ball of radius ℓ, assuming longitudinally independent boundary conditions.

Solution. We use the spherical coordinates (r, θ, ϕ) of Section 2.6; "longitudinally independent" means independent of θ. So by Proposition 2.6.1, we have the boundary value problem

$$\frac{1}{r}[ru]_{rr} + \frac{1}{r^2 \sin \phi}[\sin \phi \, u_\phi]_\phi = 0 \quad (0 < r < \ell, \ 0 < \phi < \pi), \tag{4.176}$$

$$u(\ell, \phi) = f(\phi) \quad (0 < \phi < \pi). \tag{4.177}$$

Putting $u(r, \phi) = R(r)\Phi(\phi)$ into the differential equation and multiplying through by r^2 gives

$$r[rR(r)]''\Phi(\phi) + R(r) \csc \phi [\sin \phi \, \Phi'(\phi)]' = 0. \tag{4.178}$$

This will be true if $\csc \phi [\sin \phi \, \Phi'(\phi)]' = -\lambda \Phi(\phi)$ and $r[rR(r)]'' = \lambda R(r)$, so to find separated solutions $u = R\Phi$ to the differential equation, it will suffice to solve the problem

$$[\sin \phi \, \Phi'(\phi)]' + \lambda \sin \phi \, \Phi(\phi) = 0 \quad (0 < \phi < \pi) \tag{4.179}$$

and then solve, for each eigenvalue of this problem, the equation

$$r[rR(r)]'' - \lambda R(r) = 0 \quad (0 < r < \ell). \tag{4.180}$$

Now the problem in Φ is just that of the above Corollary 4.7.1; the eigenfunctions of this problem are, again, the functions $\Phi_n(\phi) = P_n(\cos\phi)$. The corresponding eigenvalues are the numbers $\lambda_n = n(n+1)$ ($n \in \mathbb{N}$); so (4.180) becomes

$$r[rR(r)]'' - n(n+1)R(r) = 0. \tag{4.181}$$

This equation is very much like, and is handled very much like, (2.114). We get

$$R(r) = \gamma r^n + \delta r^{-1-n} \tag{4.182}$$

as our general solution. We put $\delta = 0$ because we want $R(0^+)$ to exist. So we get separated solutions $u_n(r,\phi) = r^n \Phi_n(\phi) = r^n P_n(\cos\phi)$ and formal series solutions

$$u(r,\phi) = \sum_{n=0}^{\infty} D_n r^n P_n(\cos\phi) \tag{4.183}$$

to the differential equation (4.176).

Now we plug the inhomogeneous condition (4.177) into this series, to get

$$f(\phi) = \sum_{n=0}^{\infty} D_n \ell^n P_n(\cos\phi), \tag{4.184}$$

so that by Corollary 4.7.1,

$$D_n = \frac{2n+1}{2\ell^n} \int_0^\pi f(\phi) P_n(\cos\phi) \sin\phi \, d\phi. \tag{4.185}$$

We summarize:

Proposition 4.7.2 *The boundary value problem (4.176), (4.177) has formal solution (4.183), with the D_n's as in (4.185).*

If one wants to remove the restriction of longitudinal independence, then one needs *associated Legendre functions* P_n^m ($m, n \in \mathbb{N}$). See, for example, Section 6.3 in [19].

Exercises

4.7.1 Show that the Dirichlet problem of Example 4.7.1, with constant boundary condition $f(\phi) = c$, has solution $u(r,\phi) = c$ throughout the ball. Hints: The P_n's are orthogonal and P_0 is a constant.

4.7.2 Solve the Dirichlet problem of Example 4.7.1, with boundary condition $f(\phi) = \cos^3\phi$. Hint: Use Exercise 4.6.7.

4.7.3 Show that the Dirichlet problem of Example 4.7.1, with boundary condition

$$f(\phi) = \begin{cases} 1 & \text{if } 0 < \phi < \pi/2, \\ 0 & \text{if } \pi/2 \le \phi < \pi, \end{cases}$$

has formal solution

$$u(r,\phi) = \frac{1}{2} + \frac{1}{2}\sum_{m=0}^{\infty}(P_{2m}(0) - P_{2m+2}(0))\left(\frac{r}{\ell}\right)^{2m+1} P_{2m+1}(\cos\phi)$$

$$= \frac{1}{2} + \frac{1}{2}\sum_{m=0}^{\infty}\frac{(-1)^m(2m)!(4m+3)}{2^{2m+1}m!(m+1)!}\left(\frac{r}{\ell}\right)^{2m+1} P_{2m+1}(\cos\phi).$$

Hint: To antidifferentiate, put $u = \cos\phi$ and use (4.150); to simplify, use various parts of Proposition 4.7.1.

4.7.4 Solve the Dirichlet problem of Example 4.7.1, with boundary condition

$$f(\phi) = \begin{cases} \cos\phi & \text{if } 0 < \phi < \pi/2, \\ 0 & \text{if } \pi/2 \leq \phi < \pi. \end{cases}$$

Hint: To antidifferentiate, put $u = \cos\phi$, integrate by parts, and use (4.150). Simplify as much as you can.

4.7.5 Use Proposition 4.6.3 and part of Proposition 4.7.1 to show the following.

a. The Sturm-Liouville problem

$$[\sin\phi\,\Phi'(\phi)]' + \lambda\sin\phi\,\Phi(\phi) = 0 \quad (0 < \phi < \pi/2), \quad \Phi(\pi/2) = 0$$

has, for each $m \in \mathbb{N}$, an eigenvalue $\lambda_m = (2m+1)(2m+2)$ and an eigenfuction $\Phi_{2m+1}(\phi) = P_{2m+1}(\cos\phi)$. Moreover, these eigenfunctions form an orthogonal basis for $L^2_w(0, \pi/2)$, where $w(\phi) = \sin\phi$; and for each m, $\|\Phi_{2m+1}\|^2_w = 1/(4m+3)$.

b. The Sturm-Liouville problem

$$[\sin\phi\,\Phi'(\phi)]' + \lambda\sin\phi\,\Phi(\phi) = 0 \quad (0 < \phi < \pi/2), \quad \Phi'(\pi/2) = 0$$

has, for each $m \in \mathbb{N}$, an eigenvalue $\lambda_m = 2m(2m+1)$ and an eigenfuction $\Phi_{2m}(\phi) = P_{2m}(\cos\phi)$. Moreover, these eigenfunctions form an orthogonal basis for $L^2_w(0, \pi/2)$, where $w(\phi) = \sin\phi$; and for each m, $\|\Phi_{2m}\|^2_w = 1/(4m+1)$.

Hint: This is kind of like the derivation of Fourier cosine and sine series, cf. Section 1.12, from standard Fourier series on $(-\ell, \ell)$.

4.7.6 a. Consider the following problem on the *upper half* of the ball of Example 4.7.1: The hemispherical part of the boundary has (longitudinally independent) temperature distribution $f(\phi)$, and the base of the boundary is insulated (so $u_\phi(r, \pi/2) = 0$ for $0 < r < \ell$; see Exercise 2.6.3). Use separation of variables and the previous exercise to derive the formal solution

$$u(r,\phi) = \sum_{m=0}^{\infty} D_m r^{2m} P_{2m}(\cos\phi),$$

where

$$D_m = \frac{4m+1}{\ell^{2m}} \int_0^{\pi/2} f(\phi) P_{2m}(\cos\phi) \sin\phi\, d\phi.$$

b. What do the formal solutions look like if the base is instead held at temperature zero?

4.7.7 **a.** Determine the formal solution if, in part a of the previous exercise, the condition $u(\ell, \phi) = f(\phi)$ is replaced by the constraint $u_r(1, \phi) = g(\phi)$ on the *flux* through the upper hemisphere and if the *total* flux through this hemisphere is zero: $\int_0^{\pi/2} g(\phi) \sin \phi \, d\phi = 0$. (This condition will be necessary to solve for your coefficients, after plugging in the condition $u_r(1, \phi) = g(\phi)$.) Remark: Your answer should involve an arbitrary constant. Why does this make physical sense? What physical quantity does this constant correspond to?

5

Convolution and the Delta Function: A Splat and a Spike

We wish, ultimately, to turn our attention to Fourier analysis of *aperiodic* functions, which are functions defined (almost everywhere) on $\mathbb{R}$ but not periodic there.

The situation here is more delicate than that of periodic functions or of functions on bounded intervals. This is essentially because complex exponentials, having absolute value 1, *are* integrable and square integrable over bounded intervals (and hence over tori), but for the same reason are *not* integrable or square integrable over unbounded intervals, such as $\mathbb{R}$.

So in the aperiodic setting, a correspondingly more delicate analysis is required. We begin that analysis in this chapter, by looking closely at some general concepts and constructs related to aperiodic functions. Specifically, we investigate the *convolution* operation, also referred to as "splat," and a related entity known as the delta "spike," also called the *Dirac delta function*. (This entity is not actually a function at all, though this won't much bother us; it didn't particularly concern Dirac.)

5.1 CONVOLUTION: WHAT IS IT?

Recall formula (1.82) for the partial sum S_N^f of the Fourier series for $f \in \text{PC}(\mathbb{T})$:

$$S_N^f(x) = \frac{1}{2\pi} \int_{\mathbb{T}} f(x-y) D_N(y)\, dy \tag{5.1}$$

(with D_N as in (1.81)). Note that this formula represents a method of *combining* two 2π-periodic functions (in this case, f and D_N) to obtain a new one (in this case, S_N^f). And the fundamental role played by this formula in our proof of Theorem 1.4.1 suggests that this method deserves a closer look.

So let's take a closer look, but from a slightly different perspective, namely, from an *aperiodic* perspective. From this point of view, the torus $\mathbb{T}$ of integration should be replaced by the real line $\mathbb{R}$. This leads us to the following definition, in which the functions f and g *are* often chosen to be aperiodic, although, as we'll see, somewhat greater generality is permitted.

Definition 5.1.1 If f and g are functions on $\mathbb{R}$, then the *convolution of f and g* is the function $f * g$ defined by

$$f * g(x) = \int_{-\infty}^{\infty} f(x-y)g(y)\,dy \tag{5.2}$$

for any $x \in \mathbb{R}$ such that the integral exists.

Note that, for a given real number x, the existence of $f * g(x)$ is equivalent to the condition that the function h_x defined by $h_x(y) = f(x-y)g(y)$ belong to $L^1(\mathbb{R})$.

Let's consider an example.

Example 5.1.1 Find $f * g$ if

$$f(x) = \sin x, \quad g(x) = e^{-|x|}. \tag{5.3}$$

Solution. For $x \in \mathbb{R}$, we have

$$f * g(x) = \int_{-\infty}^{\infty} f(x-y)g(y)\,dy = \int_{-\infty}^{\infty} \sin(x-y)\,e^{-|y|}\,dy$$

$$= \int_{-\infty}^{0} \sin(x-y)\,e^{y}\,dy + \int_{0}^{\infty} \sin(x-y)\,e^{-y}\,dy$$

$$= \int_{0}^{\infty} \sin(x+u)\,e^{-u}\,du + \int_{0}^{\infty} \sin(x-y)\,e^{-y}\,dy$$

$$= \int_{0}^{\infty} (\sin(x+y) + \sin(x-y))\,e^{-y}\,dy = 2\sin x \int_{0}^{\infty} \cos y\, e^{-y}\,dy$$

$$= 2\sin x \cdot \frac{1}{2} = \sin x. \tag{5.4}$$

The fourth equality is by the change of variable $u = -y$ into the first of two integrals; the fifth is by combining the resulting integrals; the sixth is by a trig identity (see Exercise A.0.16); the seventh is by way of the formula

$$\int_{0}^{\infty} \cos y\, e^{-by}\,dy = \frac{b}{1+b^2} \quad (b > 0). \tag{5.5}$$

It would be nice if there were a succinct way to say "the convolution of f and g" (which, we'll see soon enough, is the *same* as "the convolution of g and f"), much as the product of f and g is called "f times g" and the sum of f and g "f plus g." And there is a nice way: $f * g$ is often pronounced "f splat g." We're not making it up.

We stress, though, that convolution is quite different from addition or multiplication. Much more different than the latter two are from each other—different not in the way that 3 is from 4, but more in the way that 3 is from a strawberry tart, say. Different, in particular, in this way: While knowledge of the two *numbers* $f(x)$ and $g(x)$ suffices, for a given value of x, for determination of $(f+g)(x)$ and $fg(x)$ (the first is $f(x) + g(x)$ and the second $f(x)g(x)$), this is not so for $f * g(x)$. The latter is given by the integration of $f(x-y)g(y)$ over the real line and thus depends on the behavior of f and g across that whole line. (For these reasons, we strongly advocate *avoidance* of the notation $f(x) * g(x)$, occasionally employed in the literature to signify $f * g(x)$.)

We now present some fundamental properties of the convolution operation. Here f, g, h are functions and α, β are complex numbers.

Proposition 5.1.1 (a) **(the commutative property of convolution)** $f * g(x) = g * f(x)$, and in particular both sides exist, whenever either side does.

(b) **(the associative property of convolution)** $(f * g) * h(x) = f * (g * h)(x)$, and in particular both sides exist, whenever either side does.

(c) **(the distributive property of convolution)** $(\alpha f + \beta g) * h(x) = \alpha(f * h(x)) + \beta(g * h(x))$, and in particular the left side exists, provided the right side does.

Proof. Regarding part (a), we have

$$g * f(x) = \int_{-\infty}^{\infty} g(x-y)f(y)\,dy = \int_{-\infty}^{\infty} g(u)f(x-u)\,du = f * g(x), \quad (5.6)$$

the second equality coming from the substitution $u = x - y$. So the existence of either $g * f(x)$ or $f * g(x)$ assures that of the other, as well as the equality in question. For part (b), we argue as follows. If $f * (g * h)(x)$ exists, then

$$f * (g * h)(x) = \int_{-\infty}^{\infty} f(x-y)(g * h(y))\,dy$$

$$= \int_{-\infty}^{\infty} f(x-y) \int_{-\infty}^{\infty} g(y-u)h(u)\,du\,dy$$

$$= \int_{-\infty}^{\infty} \left(\int_{-\infty}^{\infty} f(x-y)g(y-u)\,dy \right) h(u)\,du$$

$$= \int_{-\infty}^{\infty} \left(\int_{-\infty}^{\infty} f(x-u-v)g(v)\,dv \right) h(u)\,du$$

$$= \int_{-\infty}^{\infty} (f * g(x-u))h(u)\,du = (f * g) * h(x). \quad (5.7)$$

The steps taken here merit some explanation: The first and second are by definition of convolution. The third is by Fubini's theorem (Proposition 3.5.3), the hypotheses of which do hold in the situation at hand. The fourth is by the substitution $v = y - u$ into the integral in y; the fifth and sixth are by definition of convolution, again.

Reversing the argument, one obtains the desired result where only $(f * g) * h(x)$ is assumed to exist.

Finally, for part (c), we have

$$\alpha(f * h(x)) + \beta(g * h(x)) = \alpha \int_{-\infty}^{\infty} f(x-y)h(y)\,dy + \beta \int_{-\infty}^{\infty} g(x-y)h(y)\,dy$$

$$= \int_{-\infty}^{\infty} (\alpha f(x-y) + \beta g(x-y))h(y)\,dy$$

$$= (\alpha f + \beta g) * h(x). \tag{5.8}$$

This completes our proof. □

The above proposition allows us to define the "triple convolution" $f * g * h(x)$ unambiguously. Namely, by this we simply mean either side of the equality in part (b) of the proposition. Similarly, by the quadruple convolution $f * g * h * k(x)$ we mean any of the convolutions

$$f * (g * (h * k))(x), \quad f * ((g * h) * k)(x), \quad (f * (g * h)) * k(x),$$
$$((f * g) * h) * k(x), \quad (f * g) * (h * k)(x), \tag{5.9}$$

all of which are the same provided any one exists, by a natural generalization of Proposition 5.1.1(b). And so on for general multiple convolutions. And of course, we can change not only the *grouping* but also, by part (a) of our proposition, the *ordering* of terms in such convolutions without changing their values.

We note that the above proposition remains true with "times" in place of "splat." Further, the first two parts of this proposition remain true with "plus" in place of "splat." So, notwithstanding the differences noted above, convolution *does* behave in a number of ways like certain more familar operations.

Exercises

In Exercises 5.1.1–5.1.8, evaluate $f * g$. (In some cases it might be easier to compute $g * f$, which is the same by Proposition 5.1.1(a).)

5.1.1 $f(x) = e^x, g(x) = \chi_{[0,\infty)}(x)$. (Recall the definition (3.96) of χ_R.)

5.1.2 $f(x) = e^{-|x|}, g(x) = \chi_{[0,\infty)}(x)$. Hint: Consider the case $x \leq 0$ first. Then, for $x > 0$, break the defining integral into two, one where $0 < y < x$ and one where $x < y < \infty$.

5.1.3 $f(x) = \cos x, g(x) = e^{-|x|}$.

5.1.4 $f(x) = x\chi_{[-2,1]}(x), g(x) = x^2 + 2$.

5.1.5 $f(x) = \sin 2\pi x\, \chi_{[1,3]}(x), g(x) = \cos \pi x$.

5.1.6 $f(x) = \sin 2\pi x\, \chi_{[1,3/2]}(x), g(x) = \cos \pi x$.

5.1.7 $f(x) = e^{-x}, g(x) = e^{-e^x}$.

5.1.8 $f(x) = 1/(1+x^2)^2, g(x) = x$. Hint: A substitution of $x = \tan\theta$ may help.

5.1.9 Show that $f * g$ is:
  a. even, provided f and g are both even or are both odd;
  b. odd, provided f is odd and g is even or vice versa.
Hint for both parts: Substitute $u = -y$ in the integral defining $f * g(-x)$.

5.2 CONVOLUTION: WHEN IS IT COMPACTLY SUPPORTED?

Another sense in which convolution is distinct from addition or multiplication is in that of *existence*. Indeed, while $(f+g)(x)$ and $fg(x)$ exist whenever $f(x)$ and $g(x)$ do, the same cannot be said for $f * g(x)$.

For a ridiculously simple example, let $f(x) = g(x) = 1$ for all x. Then

$$\int_{-\infty}^{\infty} f(x-y)g(y)\, dx = \int_{-\infty}^{\infty} 1\, dx \tag{5.10}$$

and the latter integral is infinite, so $f * g(x)$ fails to exist for even a single number x!

So now we ask: When *does* $f * g(x)$ exist? Also: What can we say about the *nature* of the function $f * g$, or of spaces in which it lives, given the natures of f and g? We'll answer these questions, over the course of this and the next several sections, with some quite general results, which we'll motivate and illustrate with some quite specific examples. The following proposition, which gives a useful formula for the convolution of compactly supported functions, will lead immediately to one of these results and will aid in the development of several of those examples.

Proposition 5.2.1 *Suppose $a, b, c, d \in \mathbb{R}, a < b, c < d$, and*

$$f(x) = \begin{cases} 0 & \text{if } x < a, \\ F(x) & \text{if } a < x < b, \\ 0 & \text{if } x > b, \end{cases} \qquad g(x) = \begin{cases} 0 & \text{if } x < c, \\ G(x) & \text{if } c < x < d, \\ 0 & \text{if } x > d. \end{cases} \tag{5.11}$$

(It does not matter how, or whether, $f(a), f(b), g(c),$ or $g(d)$ is defined.) Also assume that $d - c \leq b - a$. (If not, simply interchange the roles of f and g, and of F and G.) Then

$$f * g(x) = \begin{cases} 0 & \text{if } x \leq a + c, \\ \int_c^{x-a} F(x-y)G(y)\, dy & \text{if } a + c \leq x \leq a + d, \\ \int_c^d F(x-y)G(y)\, dy & \text{if } a + d \leq x \leq b + c, \\ \int_{x-b}^d F(x-y)G(y)\, dy & \text{if } b + c \leq x \leq b + d, \\ 0 & \text{if } x \geq b + d. \end{cases} \tag{5.12}$$

In all cases, $f * g(x)$ exists whenever the given expression for it does.

Proof. We assume $a < b, c < d$, and $d - c \leq b - a$. Since convolution is defined by integration, which does not recognize differences on negligible sets, the natures of $f(a), f(b), g(c)$, and $g(d)$ are indeed immaterial. Let's, for simplicity, assume that $f(a) = f(b) = g(c) = g(d) = 0$.

Now suppose $f(x - y)g(y) \neq 0$. Then $f(x - y) \neq 0$ *and* $g(y) \neq 0$. But if $g(y) \neq 0$, then $c < y < d$; if $f(x - y) \neq 0$, then $a < x - y < b$, or equivalently $x - b < y < x - a$. That is, if $f(x - y)g(y) \neq 0$, then

$$y \in I(x) = (c, d) \cap (x - b, x - a). \tag{5.13}$$

So by Definition 5.1.1,

$$f * g(x) = \int_{I(x)} f(x - y)g(y)\, dy = \int_{I(x)} F(x - y)G(y)\, dy, \tag{5.14}$$

the last step because $f(y) = F(y)$ for $y \in I(x) \subset (c, d)$ and $g(x - y) = G(x - y)$ for $y \in I(x) \subset (x - b, x - a)$.

If $I(x)$ is empty then, by (5.14), $f * g(x) = 0$. So let's determine for which values of x this interval *is* empty and, when it's not, what it looks like.

We claim that $I(x)$ is empty if $x \leq a + c$ or if $x \geq b + d$. To see this, note that, if $x \leq a + c$, then $x - a \leq c$, so that any number $y > c$ is larger than $x - a$, so that $(c, d) \cap (x - b, x - a) = I(x)$ contains no points, as asserted. Similarly observe that, if $x \geq b + d$, then $x - b \geq d$, so that any number $y < d$ is smaller than $x - b$, so that again $(c, d) \cap (x - b, x - a) = I(x)$ is empty.

So we need only consider the situation $a + c \leq x \leq b + d$, which we break into three subcases, the first given by $a + c \leq x \leq a + d$. Here $c \leq x - a \leq d$. Combining this with our assumption that $a - b \leq c - d$, we get $x - b = (x - a) + (a - b) \leq d + (c - d) = c$ as well. In sum, $x - b \leq c \leq x - a \leq d$, whence it's clear from (5.13) that $I(x) = (c, x - a)$. So by (5.14),

$$f * g(x) = \int_c^{x-a} F(x - y)G(y)\, dy \tag{5.15}$$

in our first subcase, as asserted.

The second subcase we consider is where $a + d \leq x \leq b + c$. Here we have $x - b \leq c < d \leq x - a$, so $(c, d) \subset (x - b, x - a)$, so $I(x) = (c, d)$ by (5.13). So by (5.14),

$$f * g(x) = \int_c^d F(x - y)G(y)\, dy \tag{5.16}$$

in this subcase, as promised.

The final subcase is where $b + c \leq x \leq b + d$. This gives $c \leq x - b \leq d$. But then because, again, $b - a \geq d - c$, we also have $x - a = (x - b) + (b - a) \geq c + (d - c) = d$. In sum, $c \leq x - b \leq d \leq x - a$, so by (5.13), $I(x) = (x - b, d)$. So by (5.14),

$$f * g(x) = \int_{x-b}^d F(x - y)G(y)\, dy \tag{5.17}$$

in this final subcase, and we're done. □

The above lemma has this useful consequence:

Corollary 5.2.1 *If f and g are compactly supported, then so is $f * g$.*

Proof. If f is supported on $[a, b]$ and g on $[c, d]$, then by Proposition 5.2.1, $f * g$ is supported on $[a + c, b + d]$. □

Example 5.2.1 Let $a < b$ and $c < d$; assume for simplicity that $d - c \leq b - a$. Find $\chi_{[a,b]} * \chi_{[c,d]}$, where χ_S, for $S \subset \mathbb{R}$, is as defined in (3.96).
Solution. We apply Proposition 5.2.1 with $f = \chi_{[a,b]}$, $g = \chi_{[c,d]}$, and $F = G = 1$. Then $F(x - y)G(y) = 1$ for all x and y. Since

$$\int_c^{x-a} dy = x - a - c, \quad \int_c^d dy = d - c, \quad \int_{x-b}^d dy = b + d - x, \qquad (5.18)$$

we conclude that

$$\chi_{[a,b]} * \chi_{[c,d]}(x) = \begin{cases} 0 & \text{if } x \leq a + c, \\ x - a - c & \text{if } a + c \leq x \leq a + d, \\ d - c & \text{if } a + d \leq x \leq b + c, \\ b + d - x & \text{if } b + c \leq x \leq b + d, \\ 0 & \text{if } x \geq b + d. \end{cases} \qquad (5.19)$$

For example, see Figure 5.1.

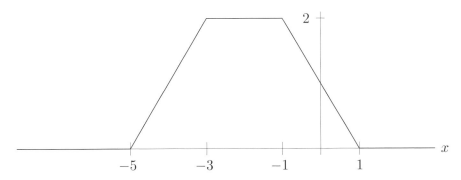

Fig. 5.1 The graph of $\chi_{[-4,0]} * \chi_{[-1,1]}$

We remark that Proposition 5.2.1 holds just as well, and is proved in much the same way, when one or more of the endpoints a, b, c, d is allowed to be *infinite*. Of course in such cases, some of the conditions on x in the piecewise formulas for f, g, and

$f * g$ become vacuous. For instance, here's what happens when we let $b = d = \infty$; we leave it to the reader to trace through the details of other, similar cases.

Proposition 5.2.2 *Suppose $a, c \in \mathbb{R}$, and*

$$f(x) = \begin{cases} 0 & \text{if } x < a, \\ F(x) & \text{if } x > a, \end{cases} \qquad g(x) = \begin{cases} 0 & \text{if } x < c, \\ G(x) & \text{if } x > c. \end{cases} \tag{5.20}$$

(It does not matter how, or whether, $f(a)$ or $g(c)$ is defined.) Then

$$f * g(x) = \begin{cases} 0 & \text{if } x \leq a + c, \\ \int_c^{x-a} F(x-y)G(y)\, dy & \text{if } x \geq a + c. \end{cases} \tag{5.21}$$

In the case $x \geq a + c$, $f * g(x)$ exists whenever the given expression for it does.

Example 5.2.2 Let $f = \chi_{[1,\infty)}$ and $g = \chi_{[-2,\infty)}$. Then, putting $F = G = 1$, we have

$$\int_{-2}^{x-1} F(x-y)G(y)\, dy = x - 1 - (-2) = x + 1, \tag{5.22}$$

so by Proposition 5.2.2,

$$f * g(x) = \begin{cases} 0 & \text{if } x \leq -1, \\ x + 1 & \text{if } x \geq -1. \end{cases} \tag{5.23}$$

Exercises

5.2.1 Show that, if f is supported on $[a, b]$ and g on $[c, d]$ (where $a, b, c,$ and d are finite), f is bounded by M and g by K, and f and g are measurable, then $f * g$ is bounded by $MK \min\{b - a, d - c\}$.

5.2.2 Show that, for $b > 0$, $\chi_{[-b,b]} * \chi_{[-b,b]}(x) = (2b - |x|)\chi_{[-2b,2b]}(x)$. Use Example 5.2.1.

In Exercises 5.2.3–5.2.11, evaluate $f * g$.

5.2.3 $f(x) = x\chi_{[-2,1]}(x), g(x) = \chi_{[0,2]}(x)$.

5.2.4 $f(x) = x\chi_{[-2,2]}(x), g(x) = x\chi_{[-3,3]}(x)$.

5.2.5 $f(x) = x\chi_{[-3,3]}(x), g(x) = x^2\chi_{[-2,2]}(x)$.

5.2.6 $f(x) = \sin x\, \chi_{[\pi/2,\pi]}(x), g(x) = \cos x\, \chi_{[\pi/2,\pi]}(x)$.

5.2.7 $f(x) = g(x) = \chi_{[0,\infty)}(x)/(1+x)$. Hint: $1/((1 + x - y)(1 + y)) = (2 + x)^{-1}(1/(1 + x - y) + 1/(1 + y))$.

5.2.8 $f(x) = x\chi_{[0,\infty)}(x), g(x) = x^2\chi_{[0,\infty)}(x)$.

5.2.9 $f(x) = x^2 \chi_{(-\infty,-1]}(x)$, $g(x) = (x-1)\chi_{(-\infty,2]}(x)$.

5.2.10 $f = \chi_{[-1,1]} * \chi_{[-1,1]}$, $g = \chi_{[-1,1]}$. Hint: Use Proposition 5.2.1 together with the result of Exercise 5.2.2. If you also invoke Exercise 5.1.9, you can reduce computation to the case $-3 \le x \le 0$. ANSWER:

$$f * g(x) = \begin{cases} 0 & \text{if } x \le -3, \\ \frac{1}{2}(3+x)^2 & \text{if } -3 \le x \le -1, \\ 3 - x^2 & \text{if } -1 \le x \le 1, \\ \frac{1}{2}(3-x)^2 & \text{if } 1 \le x \le 3, \\ 0 & \text{if } x \ge 3. \end{cases}$$

5.2.11 $f = g = \chi_{[-1,1]} * \chi_{[-1,1]}$.

5.2.12 Show that the space $C(\mathbb{R}^+)$ of continuous functions supported on $\mathbb{R}^+ = (0, \infty)$ is *closed under convolution*, meaning $f * g$ belongs to the space provided f and g do. Hint: Using Proposition 5.2.2, show that, for $f, g \in C(\mathbb{R}^+)$ and $x, x_0 \in \mathbb{R}^+$,

$$f * g(x) - f * g(x_0) = \int_0^{x_0} (f(x-y) - f(x_0 - y))g(y)\,dy + \int_{x_0}^x f(x-y)g(y)\,dy.$$

Apply the fact that a continuous function on a closed bounded interval is bounded there to show that each of the integrals on the right goes to zero as $x \to x_0$. (For the first of these integrals, the Lebesgue dominated convergence theorem, Proposition 3.5.1, will also be required.)

5.3 CONVOLUTION: WHEN IS IT BOUNDED AND CONTINUOUS?

Here and hereafter it will, from time to time (especially when several different norms are in play at once), be convenient to use the notation $||\cdot||_2$ in place of $||\cdot||$.

Let's look again at Example 5.2.1. Note that the convolution $\chi_{[a,b]} * \chi_{[c,d]}$ there is *continuous*, even though both $\chi_{[a,b]}$ and $\chi_{[c,d]}$ are discontinuous. This reflects a general principle, which says that *under some mild conditions, the convolution of two functions is at least as continuous as, and quite often more continuous than, either of the functions being convolved*. We wish to say something more specific about this principle (but less specific than Example 5.2.1); to do so, we'll first need:

Lemma 5.3.1 *Let $p = 1$ or 2. For $\delta \in \mathbb{R}$ and $g \in L^p(\mathbb{R})$, define a new function $L_\delta g$ by*

$$L_\delta g(x) = g(x + \delta). \tag{5.24}$$

Then $L_\delta g \in L^p(\mathbb{R})$, $||L_\delta g||_p = ||g||_p$, and

$$\lim_{\delta \to 0} ||L_\delta g - g||_p = 0. \tag{5.25}$$

Proof. For p, δ, and g as stipulated, we have

$$\int_{-\infty}^{\infty} |L_\delta\, g(x)|^p\, dx = \int_{-\infty}^{\infty} |g(x+\delta)|^p\, dx = \int_{-\infty}^{\infty} |g(y)|^p\, dy \qquad (5.26)$$

(we put $y = x + \delta$), so $L_\delta\, g$ belongs to $L^p(\mathbb{R})$ and has the same (mean or mean square) norm as g.

We prove the rest of the proposition first under the assumption that g is, in fact, in $C_c(\mathbb{R}) \subset L^p(\mathbb{R})$. Let's suppose that, specifically, $g(x) = 0$ for $|x| > R$. Let's also suppose that $|\delta| < 1$, which we may do without sacrificing generality, since we're concerned only with what happens as δ approaches zero. Then both $g(x)$ and $L_\delta\, g(x)$ are zero for $|x| > R + 1$.

So we have

$$\|L_\delta\, g - g\|_p = \left(\int_{-\infty}^{\infty} |L_\delta\, g(x) - g(x)|^p\, dx \right)^{1/p}$$

$$= \left(\int_{-R-1}^{R+1} |g(x+\delta) - g(x)|^p\, dx \right)^{1/p}. \qquad (5.27)$$

Now a continuous function on a closed, bounded interval attains a maximum there, so we have $|g(x)| \leq M$ on $[-R, R]$ for some $M \geq 0$. Then

$$|g(x+\delta) - g(x)|^p \leq (|g(x+\delta)| + |g(x)|)^p \leq (2M)^p \qquad (5.28)$$

on $[-R-1, R+1]$. That is, the integrand on the right side of (5.27) is bounded, independently of δ, by a constant and therefore integrable function on this interval. But for any x this integrand goes to zero as δ does, by the continuity of g. So the Lebesgue dominated convergence theorem (Proposition 3.5.1) may be applied to (5.27), yielding

$$\lim_{\delta \to 0} \|L_\delta\, g - g\|_p = \left(\int_{-R-1}^{R+1} \lim_{\delta \to 0} |g(x+\delta) - g(x)|^p\, dx \right)^{1/p} = 0^{1/p} = 0, \qquad (5.29)$$

as claimed.

We now consider the more general case: If $g \in L^p(\mathbb{R})$ then, since $C_c(\mathbb{R})$ is dense in this space, there is a sequence of functions $g_M \in C_c(\mathbb{R})$ converging to g in the $L^p(\mathbb{R})$ norm. Noting that

$$L_\delta\, g - g = (L_\delta\, g - L_\delta\, g_M) + (L_\delta\, g_M - g_M) + (g_M - g), \qquad (5.30)$$

we deduce from the triangle inequality on $L^p(\mathbb{R})$ that

$$\|L_\delta\, g - g\|_p \leq \|L_\delta\, g - L_\delta\, g_M\|_p + \|L_\delta\, g_M - g_M\|_p + \|g_M - g\|_p. \qquad (5.31)$$

But

$$\|L_\delta\, g - L_\delta\, g_M\|_p = \|L_\delta(g - g_M)\|_p = \|g - g_M\|_p \qquad (5.32)$$

(the second equality by results just proved), so (5.31) gives

$$||L_\delta g - g||_p \leq ||L_\delta g_M - g_M||_p + 2||g_M - g||_p. \qquad (5.33)$$

Since $g_M \in C_c(\mathbb{R})$, the first norm on the right goes to zero as $\delta \to 0$ for any fixed M, as just shown (with g in place of g_M); also, by assumption on the g_M's, the second norm on the right goes to zero as $M \to \infty$. So by (a variant, with "$s \to \infty$" replaced by "$\delta \to 0$," of) the squeeze law II(b) (Proposition 3.4.1(b)), the norm on the left approaches zero as $\delta \to 0$, as required. $\square$

The above lemma gives us, among other things:

Proposition 5.3.1 *If f is bounded and measurable and $g \in L^1(\mathbb{R})$, or $f \in L^1(\mathbb{R})$ and g is bounded and measurable, or $f, g \in L^2(\mathbb{R})$, then $f * g(x)$ exists for all $x \in \mathbb{R}$, and moreover $f * g \in \mathrm{BC}(\mathbb{R})$, where by definition*

$$\mathrm{BC}(\mathbb{R}) = \{\text{bounded, continuous functions on } \mathbb{R}\}. \qquad (5.34)$$

Moreover, suppose we define the "infinity norm" (or "sup norm") $||h||_\infty$ of a bounded function h by

$$||h||_\infty = \sup_{x \in \mathbb{R}} |h(x)|. \qquad (5.35)$$

Then in the first of our three cases we have $||f * g||_\infty \leq ||f||_\infty ||g||_1$; in the second we have $||f * g||_\infty \leq ||f||_1 ||g||_\infty$; in the third we have $||f * g||_\infty \leq ||f||_2 ||g||_2$.

Proof. We first assume f is bounded and measurable and $g \in L^1(\mathbb{R})$. Then for $x \in \mathbb{R}$,

$$|f * g(x)| = \left| \int_{-\infty}^{\infty} f(x-y)g(y)\,dy \right| \leq \int_{-\infty}^{\infty} |f(x-y)g(y)|\,dy$$

$$\leq ||f||_\infty \int_{-\infty}^{\infty} |g(y)|\,dy = ||f||_\infty ||g||_1 \qquad (5.36)$$

for all $x \in \mathbb{R}$. (We substituted $v = x - y$.) This not only guarantees the existence of $f * g(x)$ everywhere (what we're really doing here is applying Proposition 3.5.2) but also provides the stated bound on $||f * g||_\infty$.

To show continuity of $f * g$, in the case at hand, let $x, \delta \in \mathbb{R}$. Then

$$|f * g(x + \delta) - f * g(x)|$$

$$= \left| \int_{-\infty}^{\infty} f(x + \delta - y)g(y)\,dy - \int_{-\infty}^{\infty} f(x-y)g(y)\,dy \right|$$

$$= \left| \int_{-\infty}^{\infty} f(x-u)g(u+\delta)\,du - \int_{-\infty}^{\infty} f(x-y)g(y)\,dy \right|$$

$$= \left| \int_{-\infty}^{\infty} f(x-y)(g(y+\delta) - g(y))\,dy \right|$$

$$\leq ||f||_\infty \int_{-\infty}^{\infty} |g(y+\delta) - g(y)|\,dy = ||f||_\infty ||L_\delta g - g||_1. \qquad (5.37)$$

For the second equality, we substituted $u = y + \delta$; for the third we combined two integrals; for the inequality, we applied the bound on f; for the last equality, we used the definition of L_δ, cf. Lemma 5.3.1. By that lemma, the right side of (5.37) goes to zero as $\delta \to 0$. So the left side does too, by the squeeze law, and this is precisely what it means for $f * g$ to be continuous.

If g is bounded and measurable and $f \in L^1(\mathbb{R})$, then, because $f * g(x) = g * f(x)$ whenever either exists, we can apply the above arguments with the roles of f and g switched to get the desired results concerning $f * g$.

Finallly, we assume $f, g \in L^2(\mathbb{R})$. The Cauchy-Schwarz inequality (3.15) on $L^2(\mathbb{R})$ tells us

$$\left| \int_{-\infty}^{\infty} F(y) G(y)\, dy \right| \leq \left(\int_{-\infty}^{\infty} |F(y)|^2\, dy \right)^{1/2} \left(\int_{-\infty}^{\infty} |G(y)|^2\, dy \right)^{1/2} \quad (5.38)$$

for $F, G \in L^2(\mathbb{R})$; so

$$|f * g(x)| = \left| \int_{-\infty}^{\infty} f(x-y) g(y)\, dy \right|$$
$$\leq \left(\int_{-\infty}^{\infty} |f(x-y)|^2\, dy \right)^{1/2} \left(\int_{-\infty}^{\infty} |g(y)|^2\, dy \right)^{1/2} = ||f||_2\, ||g||_2, \quad (5.39)$$

so that (by Proposition 3.5.2) $f * g(x)$ is defined everywhere and has the appropriate bound, as in our previous cases. Similarly, to demonstrate continuity of $f * g$, we observe that

$$\left| \int_{-\infty}^{\infty} f(x-y)(g(y+\delta) - g(y))\, dy \right|$$
$$\leq \left(\int_{-\infty}^{\infty} |f(x-y)|^2\, dy \right)^{1/2} \left(\int_{-\infty}^{\infty} |g(y+\delta) - g(y)|^2\, dy \right)^{1/2}$$
$$= \left(\int_{-\infty}^{\infty} |f(y)|^2\, dy \right)^{1/2} \left(\int_{-\infty}^{\infty} |L_\delta g(y) - g(y)|^2\, dy \right)^{1/2} = ||f||_2\, ||L_\delta g - g||_2 \quad (5.40)$$

by (5.38), so that, by (the first three equalities in) (5.37),

$$|f * g(x + \delta) - f * g(x)| \leq ||f||_2\, ||L_\delta g - g||_2 \quad (5.41)$$

as well. Again by Lemma 5.3.1, the right side goes to zero as $\delta \to 0$, so again by the squeeze law, the left side does too, as required. □

Exercises

In Exercises 5.3.1–5.3.2, show that the given vector space is closed under convolution (for the meaning of which, see Exercise 5.2.12).

5.3.1 The space $C_c(\mathbb{R})$ of continuous, compactly supported functions on $\mathbb{R}$. Hint: Corollary 5.2.1 and Proposition 5.3.1.

5.3.2 The space A of compactly supported functions in $L^2(\mathbb{R})$.

5.3.3 Show that, if $f, g \in L^2(\mathbb{R})$, then

$$\lim_{x \to \pm\infty} f * g(x) = 0, \tag{$*$}$$

as follows.

  a. First show ($*$) holds for $f, g \in C_c(\mathbb{R})$.

  b. Now show ($*$) holds for $f \in C_c(\mathbb{R})$ and $g \in L^2(\mathbb{R})$, like this. Let g_N be a sequence in $C_c(\mathbb{R})$ converging in mean square norm to g. Show that

$$|f * g(x)| \leq |f * g_N(x)| + \|f\|_2 \|g - g_N\|_2$$

using the triangle inequality on $\mathbb{C}$ together with Proposition 5.3.1. Now deduce the desired result from part (a) of this exercise and Proposition 3.4.1(b).

  c. Now show ($*$) holds for $f, g \in L^2(\mathbb{R})$, like this. Let f_N be a sequence in $C_c(\mathbb{R})$ converging in mean square norm to f. Show that

$$|f * g(x)| \leq |f_N * g(x)| + \|f - f_N\|_2 \|g\|_2$$

using the triangle inequality on $\mathbb{C}$ together with Proposition 5.3.1. Now deduce the desired result from part (b) of this exercise and Proposition 3.4.1(b).

5.3.4 Does the result ($*$) of the previous exercise necessarily hold if $f \in L^1(\mathbb{R})$ and g is bounded and measurable on $\mathbb{R}$ (or vice versa)?

5.4 CONVOLUTION: WHEN IS IT DIFFERENTIABLE?

In the previous section we showed that, in many situations, convolution increases continuity. We now show that it has similar effects on differentiability.

To motivate our first general result along these lines, we recall Example 5.1.1. There, we had a function f differentiable on $\mathbb{R}$ convolved with a function g that's not (as $g'(0)$ fails to exist), resulting in a function $f * g$ that is. That is, convolution in this case preserved the nice properties, differentiably speaking, of the nicer of the two functions being convolved, but lost the less nice properties of the less nice one. That example illustrates Proposition 5.4.1 below, whose proof will require the following lemma, which like so many of our lemmas will serve more than one purpose. (See also the proofs of Proposition 5.6.2 and Proposition 5.7.1.)

Lemma 5.4.1 *Suppose h is measurable on $\mathbb{R}^2$. Then*

$$\left(\int_{-\infty}^{\infty} \left| \int_{-\infty}^{\infty} h(u,v)\, du \right|^2 dv \right)^{1/2} \leq \int_{-\infty}^{\infty} \left(\int_{-\infty}^{\infty} |h(u,v)|^2\, dv \right)^{1/2} du. \tag{5.42}$$

assuming existence of the integral on the right.

Proof. We use the handy device of writing a product of integrals as a double integral. Specifically, we note that

$$\left| \int_{-\infty}^{\infty} h(u,v)\, du \right|^2 = \left(\int_{-\infty}^{\infty} h(u,v)\, du \right) \overline{\left(\int_{-\infty}^{\infty} h(x,v)\, dx \right)}$$

$$= \left(\int_{-\infty}^{\infty} h(u,v)\, du \right) \left(\int_{-\infty}^{\infty} \overline{h(x,v)}\, dx \right)$$

$$= \int_{-\infty}^{\infty} \int_{-\infty}^{\infty} h(u,v)\overline{h(x,v)}\, du\, dx \qquad (5.43)$$

(the integral in u is just a constant with respect to x and therefore may be brought inside the integral in x). We integrate both sides with respect to v:

$$\int_{-\infty}^{\infty} \left| \int_{-\infty}^{\infty} h(u,v)\, du \right|^2 dv = \int_{-\infty}^{\infty} \int_{-\infty}^{\infty} \int_{-\infty}^{\infty} h(u,v)\overline{h(x,v)}\, du\, dx\, dv$$

$$= \int_{-\infty}^{\infty} \int_{-\infty}^{\infty} \left(\int_{-\infty}^{\infty} h(u,v)\overline{h(x,v)}\, dv \right) du\, dx$$

$$\leq \int_{-\infty}^{\infty} \int_{-\infty}^{\infty} \left(\int_{-\infty}^{\infty} \left| h(u,v)\overline{h(x,v)} \right| dv \right) du\, dx, \qquad (5.44)$$

the middle step by an application of Fubini's theorem. This application is justified provided the integral on the far right side of (5.44) exists; it does for the following reason. The Cauchy-Schwarz inequality (5.38) gives

$$\int_{-\infty}^{\infty} \int_{-\infty}^{\infty} \left(\int_{-\infty}^{\infty} \left| h(u,v)\overline{h(x,v)} \right| dv \right) du\, dx$$

$$\leq \int_{-\infty}^{\infty} \int_{-\infty}^{\infty} \left(\int_{-\infty}^{\infty} |h(u,v)|^2 dv \right)^{1/2} \left(\int_{-\infty}^{\infty} |h(x,y)|^2 dy \right)^{1/2} du\, dx$$

$$= \left(\int_{-\infty}^{\infty} \left(\int_{-\infty}^{\infty} |h(u,v)|^2 dv \right)^{1/2} du \right) \left(\int_{-\infty}^{\infty} \left(\int_{-\infty}^{\infty} |h(x,y)|^2 dy \right)^{1/2} dx \right)$$

$$= \left(\int_{-\infty}^{\infty} \left(\int_{-\infty}^{\infty} |h(u,v)|^2 dv \right)^{1/2} du \right)^2, \qquad (5.45)$$

and the integral on the right is assumed to exist.

Combining (5.44) with (5.45) gives

$$\int_{-\infty}^{\infty} \left| \int_{-\infty}^{\infty} h(u,v)\, du \right|^2 dv \leq \left(\int_{-\infty}^{\infty} \left(\int_{-\infty}^{\infty} |h(u,v)|^2 dv \right)^{1/2} du \right)^2; \qquad (5.46)$$

taking square roots of both sides yields the desired result. □

We now demonstrate that a convolution is differentiable if only one of its factors is (provided each of these factors is sufficiently reasonable in other respects).

Proposition 5.4.1 (a) *Suppose f is differentiable on $\mathbb{R}$. If f, f' are bounded and measurable and $g \in L^1(\mathbb{R})$, or $f, f', g \in L^2(\mathbb{R})$, then $f * g$ is differentiable on $\mathbb{R}$, $f * g, (f * g)' \in \mathrm{BC}(\mathbb{R})$, and*

$$(f * g)' = f' * g. \tag{5.47}$$

(b) *Suppose g is differentiable on $\mathbb{R}$. If $f \in L^1(\mathbb{R})$ and g, g' are bounded and measurable, or $f, g, g' \in L^2(\mathbb{R})$, then $f * g$ is differentiable on $\mathbb{R}$, $f * g, (f * g)' \in \mathrm{BC}(\mathbb{R})$, and*

$$(f * g)' = f * g'. \tag{5.48}$$

Proof. By symmetry (the fact that $f * g = g * f$), we need only consider part (a). The essential idea is the interchange of a derivative with an integral, as follows:

$$(f * g)'(x) = \frac{d}{dx} \int_{-\infty}^{\infty} f(x-y)g(y)\,dy = \int_{-\infty}^{\infty} \left(\frac{d}{dx} f(x-y)g(y)\right) dy$$
$$= \int_{-\infty}^{\infty} f'(x-y)g(y)\,dy = f' * g(x). \tag{5.49}$$

But such an interchange is not always valid. We must show that, under either of the given sets of assumptions in part (a), it is.

So let's assume, first, that f, f' are bounded and measurable and $g \in L^1(\mathbb{R})$. Then $f * g \in \mathrm{BC}(\mathbb{R})$ by Proposition 5.3.1. Moreover, if $x, \delta \in \mathbb{R}$, then

$$f * g(x + \delta) - f * g(x) = \int_{-\infty}^{\infty} (f(x+\delta-y) - f(x-y))g(y)\,dy, \tag{5.50}$$

so

$$(f * g)'(x) = \lim_{\delta \to 0} \frac{f * g(x+\delta) - f * g(x)}{\delta}$$
$$= \lim_{\delta \to 0} \int_{-\infty}^{\infty} \frac{f(x+\delta-y) - f(x-y)}{\delta} g(y)\,dy \tag{5.51}$$

provided the limit on the right exists.

To see that it does, and to evaluate it, we note first that

$$\lim_{\delta \to 0} \frac{f(x+\delta-y) - f(x-y)}{\delta} = f'(x-y), \tag{5.52}$$

by definition of the derivative. Moreover, the integrand in (5.51) is bounded by a constant times $|g(y)|$, as follows. By the mean value theorem,

$$\frac{f(x+\delta-y) - f(x-y)}{\delta} = f'(X) \tag{5.53}$$

for some number X between $x - y$ and $x + \delta - y$. But f' is bounded, say by M', so (5.53) implies

$$\left| \frac{f(x + \delta - y) - f(x - y)}{\delta} g(y) \right| = |f'(X)g(y)| \leq M'|g(y)|, \tag{5.54}$$

as claimed.

Since $g \in L^1(\mathbb{R})$, $M'|g|$ is integrable on $\mathbb{R}$, so we may apply the Lebesgue dominated convergence theorem (Proposition 3.5.1) to (5.51). We get

$$\begin{aligned}(f * g)'(x) &= \int_{-\infty}^{\infty} \lim_{\delta \to 0} \frac{f(x + \delta - y) - f(x - y)}{\delta} g(y)\, dy \\ &= \int_{-\infty}^{\infty} f'(x - y) g(y)\, dy = f' * g(x),\end{aligned} \tag{5.55}$$

as promised.

To now show $(f * g)' \in \mathrm{BC}(\mathbb{R})$ we observe that, since $f' \in L^1(\mathbb{R})$ and g is bounded and measurable, we have $f' * g \in \mathrm{BC}(\mathbb{R})$, by Proposition 5.3.1. But as we just saw, $f' * g$ equals $(f * g)'$.

Next, we suppose $f, f', g \in L^2(\mathbb{R})$; we observe that

$$\begin{aligned}&\frac{f * g(x + \delta) - f * g(x)}{\delta} - f' * g(x) \\ &= \int_{-\infty}^{\infty} \left(\frac{f(x + \delta - y) - f(x - y)}{\delta} - f'(x - y) \right) g(y)\, dy \\ &= \int_{-\infty}^{\infty} H_\delta(x - y) g(y)\, dy = H_\delta * g(x),\end{aligned} \tag{5.56}$$

where

$$H_\delta(v) = \frac{f(v + \delta) - f(v)}{\delta} - f'(v). \tag{5.57}$$

Note that, for any δ, H_δ is in $L^2(\mathbb{R})$ because f and f' are.

Now by Proposition 5.3.1, the right side of (5.56) is bounded, independently of x, by $\|H_\delta\|_2 \|g\|_2$. If we can show that

$$\lim_{\delta \to 0} \|H_\delta\|_2 = 0, \tag{5.58}$$

then the squeeze law will tell us that the limit as $\delta \to 0$ of the left side of (5.56) is also zero. But this limit *is* $(f * g)'(x) - f' * g(x)$, so we'll be done.

So let's demonstrate (5.58). The trick here is to note the following:

$$\int_0^1 (f'(v + \delta u) - f'(v)) \, du = \int_0^1 f'(v + \delta u) \, du - f'(v) \int_0^1 du$$
$$= \frac{1}{\delta} \int_0^\delta f'(v + x) \, dx - f'(v) \int_0^1 du$$
$$= \frac{1}{\delta} f(v + x) \Big|_0^\delta - f'(v)$$
$$= \frac{1}{\delta} \big(f(v + \delta) - f(v)\big) - f'(v) = H_\delta(v), \quad (5.59)$$

the second equality by substituting $x = \delta u$ and the last by definition of H_δ. Then

$$\|H_\delta\|_2 = \left(\int_{-\infty}^\infty |H_\delta(v)|^2 \, dv \right)^{1/2}$$
$$= \left(\int_{-\infty}^\infty \left| \int_0^1 (f'(v + \delta u) - f'(v)) \, du \right|^2 dv \right)^{1/2}. \quad (5.60)$$

By Lemma 5.4.1 we can conclude

$$\|H_\delta\|_2 \leq \int_0^1 \left(\int_{-\infty}^\infty |f'(v + \delta u) - f'(v)|^2 \, dv \right)^{1/2} du. \quad (5.61)$$

(Each interval of integration in that lemma is the entire real line, but replacing the function $h(u, v)$ there with $h(u, v) \chi_{[0,1]}(u)$, we get a result directly applicable the present situation.) Note that (5.61) reads

$$\|H_\delta\|_2 \leq \int_0^1 \|L_{\delta u} f' - f'\|_2 \, du, \quad (5.62)$$

where L_δ is as in Lemma 5.3.1. But that lemma says that the integrand in (5.62) converges, for any given u, to zero as $\delta \to 0$. Moreover, by the triangle inequality on $L^2(\mathbb{R})$,

$$\|L_{\delta u} f' - f'\|_2 \leq \|L_{\delta u} f'\|_2 + \|f'\|_2 = 2\|f'\|_2, \quad (5.63)$$

the last step by the same lemma again. That is, the integrand in question is bounded, independently of δ, by a number constant with respect to u. Since a constant is integrable over $[0, 1]$, the right side of (5.62) goes to zero with δ, by the Lebesgue dominated convergence theorem (Proposition 3.5.1). But then the left side does too, by the squeeze law, and we're done. □

The following result generalizes Proposition 5.4.1, saying essentially that whatever goes for first derivatives there also goes for higher ones.

278 CONVOLUTION AND THE DELTA FUNCTION: A SPLAT AND A SPIKE

Proposition 5.4.2 *Let n, ℓ be nonnegative integers; suppose f is n times differentiable on $\mathbb{R}$ and g is ℓ times differentiable on $\mathbb{R}$. Suppose further that $f^{(n)}, g \in L^1(\mathbb{R})$ and $f, f', f'', \ldots, f^{(n)}, g, g', g'', \ldots, g^{(\ell)}$ are bounded and measurable, or that all these derivatives (including the zeroth ones) are in $L^2(\mathbb{R})$. Then $(f * g)^{(n+\ell)} \in \mathrm{BC}(\mathbb{R})$, and*

$$(f * g)^{(n+\ell)} = f^{(n)} * g^{(\ell)}. \tag{5.64}$$

We omit the proof.

Essentially, what the proposition says is this: In hospitable circumstances, you can obtain the $(n + \ell)$th derivative of $f * g$ from the nth of f and the ℓth of g. In particular, since $n + \ell$ is larger than either n or ℓ if both are positive, $f * g$ is in many situations smoother than either f or g.

In fact, this latter claim applies even to certain situations not covered by the above proposition: A convolution may be differentiable even when *neither* of its factors is. See Exercise 5.4.2.

At any rate, the discussions of this and the previous section certainly bear out this general principle: *Convolution increases smoothness*.

Exercises

5.4.1 Show that the vector space

$$C_c^\infty(\mathbb{R}) = \{\text{compactly supported functions } f \colon \mathbb{R} \to \mathbb{C} \colon$$
$$f^{(k)}(x) \text{ exists for all } k \in \mathbb{N} \text{ and } x \in \mathbb{R}\}$$

is closed under convolution.

5.4.2 Use Exercise 5.2.10 to show that $f * g$ can be differentiable throughout $\mathbb{R}$ even when neither of its factors is. Hint: At most points x, the existence of $(f * g)'(x)$, for f and g as in that exercise, is clear. At the "joins" in the piecewise definition of $f * g$, existence of $(f * g)'$ may be checked using Exercise 1.3.17.

5.4.3   **a.** Let $f = \chi_{[-1,1]}$ and $g = \chi_{[-1,1]}$. Is it true that $(f * g)'(x) = f' * g(x)$ at all points x where the derivative on the left exists? (Define f' arbitrarily at points where f is not differentiable.) You may want to consult Exercise 5.2.2.

  b. Repeat the above with $f = \chi_{[-1,1]} * \chi_{[-1,1]}$ and $g = \chi_{[-1,1]}$. See Exercise 5.2.10.

  c. What does this make you think about possible generalizations of Proposition 5.4.1?

5.5 CONVOLUTION: AN EXAMPLE

The analyses of the previous two sections were rather general and abstract, and those of the next two sections will be much the same. Not that there's anything wrong

CONVOLUTION: AN EXAMPLE

with that. But here, in between, we develop a relatively concrete example, which will illustrate some of these earlier abstractions and anticipate some of those still to come.

We first need a definition and a lemma.

Definition 5.5.1 (a) The *gamma function* Γ is defined by

$$\Gamma(p) = \int_0^\infty x^{p-1} e^{-x}\, dx \tag{5.65}$$

for any $p \in \mathbb{R}$ such that the integral exists.

(b) The *beta function* B is defined by

$$\mathrm{B}(p, q) = \int_0^1 v^{p-1}(1-v)^{q-1}\, dv \tag{5.66}$$

for any $p, q \in \mathbb{R}$ such that the integral exists.

One may also consider complex values of p and q, but to keep the discussions simple, we'll take Γ and B to be functions of real numbers only. With this restriction in force, we have:

Lemma 5.5.1 (a) *The integral defining $\Gamma(p)$ exists if and only if $p > 0$.*

(b) *The integral defining $\mathrm{B}(p, q)$ exists if and only if $p, q > 0$.*

Proof. Concerning part (a), we pick $p \in \mathbb{R}$ and observe that, by l'Hôpital's rule,

$$\lim_{x \to \infty} \frac{(p-1)\ln x}{x} = (p-1) \lim_{x \to \infty} \frac{1/x}{1} = 0. \tag{5.67}$$

But certainly a quantity approaching zero is eventually less than $\frac{1}{2}$. So there's a number $K > 0$ such that, if $x \geq K$, then

$$\frac{(p-1)\ln x}{x} \leq \frac{1}{2}. \tag{5.68}$$

Multiplying both sides by x and then exponentiating both sides of the result, we get $x^{p-1} \leq e^{x/2}$ for $x \geq K$.

Let's write

$$\int_0^\infty x^{p-1} e^{-x}\, dx = \int_0^K x^{p-1} e^{-x}\, dx + \int_K^\infty x^{p-1} e^{-x}\, dx. \tag{5.69}$$

Since

$$\int_K^\infty x^{p-1} e^{-x}\, dx \leq \int_K^\infty e^{x/2} e^{-x}\, dx = \int_K^\infty e^{-x/2}\, dx$$

$$= -2e^{-x/2}\Big|_K^\infty = -2(0 - e^{-K/2}) < \infty, \tag{5.70}$$

the integral defining $\Gamma(p)$ exists if and only if the first one on the right side of (5.69) does.

But $e^{-K} \leq e^{-x} \leq 1$ for $0 \leq x \leq K$, so

$$e^{-K} \int_0^K x^{p-1}\, dx \leq \int_0^K x^{p-1} e^{-x}\, dx \leq \int_0^K x^{p-1}\, dx. \tag{5.71}$$

The integral on either side converges if and only if $p > 0$, so that, by Proposition 3.5.2, the same can be said about the one in the middle. Thus, part (a).

For part (b) we note that, if $v \in (0,1)$ and $p > 0$, then $v^{-p} > 1$, so $v^{1-p} > v$. But for $v \in (0,1)$ we also have $0 < 1-v < 1$, so if also $q > 0$, then $(1-v)^{1-q} > 1-v$. So, under these assumptions regarding v, p, q, we have

$$v^{1-p} + (1-v)^{1-q} > v + (1-v) = 1, \tag{5.72}$$

which gives, upon multiplication by $v^{p-1}(1-v)^{q-1}$,

$$(1-v)^{q-1} + v^{p-1} > v^{p-1}(1-v)^{q-1}. \tag{5.73}$$

So for $p, q > 0$,

$$\int_0^1 v^{p-1}(1-v)^{q-1}\, dv < \int_0^1 v^{p-1}\, dv + \int_0^1 (1-v)^{q-1}\, dv$$

$$= \frac{v^p}{p}\Big|_0^1 + \frac{-(1-v)^q}{q}\Big|_0^1 = \frac{1^p - 0^p}{p} + \frac{-0^q + 1^q}{q}$$

$$= \frac{1}{p} + \frac{1}{q} < \infty, \tag{5.74}$$

the second-to-last equality because $0^p = 0^q = 0$ for $p, q > 0$. This proves the "if" direction of part (b).

To get the other direction, suppose first that $p \leq 0$: then for $v \in (0,1)$, we have $v^{p-1} \geq v^{-1}$. We also check that, for any fixed $q \in \mathbb{R}$, the function $f(v) = (1-v)^{q-1}$ attains a nonzero minimum m on the interval $[0, \frac{1}{2}]$. (This minimum equals 1 if $q \leq 1$ and $(\frac{1}{2})^{q-1}$ otherwise). So $p \leq 0$ implies

$$\int_0^1 v^{p-1}(1-v)^{q-1}\, dv \geq m \int_0^1 v^{p-1}\, dv; \tag{5.75}$$

the integral on the right side equals $m \ln v\big|_0^1$ if $p = 0$ and equals $(m/p)v^p\big|_0^1$ if $p < 0$. In either case this integral diverges, so the one on the left does too.

The case $q \leq 0$ is handled similarly. We omit the details. □

In Exercise 5.6.3, we examine an interesting relationship between Γ and B. Now for our promised example, which we present in the form of a proposition.

Proposition 5.5.1 *For $p \in \mathbb{R}$, define*

$$h_p(x) = \begin{cases} 0 & \text{if } x \leq 0, \\ x^{p-1}e^{-x} & \text{if } x > 0 \end{cases} \qquad (5.76)$$

(see Fig. 5.2). Then:

(a) $h_p \in L^1(\mathbb{R})$ *if and only if* $p > 0$.

(b) $h_p \in L^2(\mathbb{R})$ *if and only if* $p > \frac{1}{2}$.

(c) h_p *is continuous if and only if* $p > 1$.

(d) h_p *is bounded if and only if* $p \geq 1$.

(e) $h_p * h_q$ *exists as a function on* $\mathbb{R}$ *if and only if* $p, q > 0$; *moreover, in this case*

$$h_p * h_q = \mathrm{B}(p,q) h_{p+q}. \qquad (5.77)$$

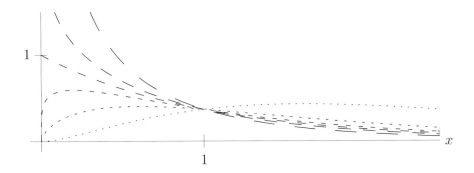

Fig. 5.2 The functions $h_{2/5}, h_{4/5}, h_1, h_{6/5}, h_{8/5}$, and $h_{13/5}$ (shorter dashes correspond to larger subscripts)

Proof. We have

$$\int_{-\infty}^{\infty} |h_p(x)|\, dx = \int_0^{\infty} x^{p-1}e^{-x}\, dx, \qquad (5.78)$$

so part (a) of our proposition follows from Lemma 5.5.1(a). Similarly,

$$\int_{-\infty}^{\infty} |h_p(x)|^2\, dx = \int_0^{\infty} x^{2p-2}e^{-2x}\, dx = 2^{1-2p}\int_0^{\infty} u^{2p-2}e^{-u}\, du, \qquad (5.79)$$

the last step by the substitution $u = 2x$. Lemma 5.5.1(a) says that the right side is defined if and only if $2p - 1 > 0$, whence part (b) of our proposition as well.

To get part (c), we need only note that $\lim_{x \to 0+} x^{p-1}e^{-x} = 0$ if and only if $p > 1$. For part (d), we observe that $\lim_{x \to \infty} x^{p-1}e^{-x} = 0$ for any p and that $\lim_{x \to 0} x^{p-1}e^{-x}$ is finite if and only if $p \geq 1$.

Finally, we consider part (e). By Proposition 5.2.2 with $F(x) = x^{p-1}e^{-x}$, $G(x) = x^{q-1}e^{-x}$, and $a = c = 0$, we have $h_p * h_q(x) = 0$ for $x \leq 0$, while for $x > 0$ we have

$$h_p * h_q(x) = \int_0^x F(x-y)G(y)\, dy = \int_0^x [(x-y)^{p-1}e^{-(x-y)}][y^{q-1}e^{-y}]\, dy$$

$$= e^{-x} \int_0^x (x-y)^{p-1} y^{q-1}\, dy$$

$$= e^{-x} \int_1^0 (xv)^{p-1}(x(1-v))^{q-1}\, (-x\, dv)$$

$$= x^{p+q-1}e^{-x} \int_0^1 v^{p-1}(1-v)^{q-1}\, dv, \qquad (5.80)$$

the next to the last step by the substitution $v = 1 - y/x$. But by Lemma 5.5.1(b), the rightmost integral is defined if and only if $p, q > 0$, in which case it equals $\mathrm{B}(p, q)$, so that

$$h_p * h_q(x) = \begin{cases} 0 & \text{if } x \leq 0, \\ \mathrm{B}(p,q) x^{p+q-1} e^{-x} & \text{if } x > 0 \end{cases} \qquad (5.81)$$

$$= \mathrm{B}(p,q) h_{p+q}(x), \qquad (5.82)$$

as claimed. This completes our proof. □

Notice how nicely the h_p's reflect Proposition 5.3.1. Indeed, to say that h_p is bounded and $h_q \in L^1(\mathbb{R})$, or that $h_p \in L^1(\mathbb{R})$ and h_q is bounded, or that $h_p, h_q \in L^2(\mathbb{R})$ is, by Proposition 5.5.1, to say $p > 1$ and $q > 0$, or $p > 0$ and $q > 1$, or $p, q > \frac{1}{2}$, respectively. In any of these cases, $p, q > 0$ and $p + q > 1$, which by Proposition 5.5.1 again is *precisely* what's required to have $h_p * h_q = \mathrm{B}(p,q) h_{p+q} \in \mathrm{BC}(\mathbb{R})$, as Proposition 5.3.1 says it will be.

At the same time, the h_p's demonstrate that the converse to Proposition 5.3.1 is false: $h_{1/3} * h_{5/6} \in \mathrm{BC}(\mathbb{R})$, since $\frac{1}{3} + \frac{5}{6} > 1$, though neither factor is bounded, nor is the first in $L^2(\mathbb{R})$.

The h_p's also beautifully exemplify Proposition 5.4.1. Namely, let's note that, by direct computation,

$$h_p' = (p-1) h_{p-1} - h_p \qquad (5.83)$$

for $p > 0$. So, to say that h_p, h_p' are bounded and $h_q \in L^1(\mathbb{R})$, or that $h_p, h_p', h_q \in L^2(\mathbb{R})$, is, by Proposition 5.5.1, to say $p > 2$ and $q > 0$, or $p > \frac{3}{2}$ and $q > \frac{1}{2}$, respectively. In either of these cases, $p, q > 0$ and $p + q > 2$, which *again* by Proposition 5.5.1 is just what's needed to ensure that both $h_p * h_q = \mathrm{B}(p,q) h_{p+q}$ and

$$(h_p * h_q)' = (\mathrm{B}(p,q) h_{p+q})' = \mathrm{B}(p,q)((p+q-1) h_{p+q-1} - h_{p+q}) \qquad (5.84)$$

are in $\mathrm{BC}(\mathbb{R})$, as Proposition 5.4.1(a) says they will be.

In the next section, we'll invoke the h_p's to suggest still other general properties of convolution.

Exercises

5.5.1 a. Show that, for $p > 0$, $\Gamma(p+1) = p\Gamma(p)$. Hint: integration by parts.
  b. Use part a and mathematical induction to show that $\Gamma(n+1) = n!$ for $n \in \mathbb{N}$.

5.5.2 Show that $B(p,q) = B(q,p)$ for $p, q > 0$.

5.5.3 Use Proposition 5.5.1(e) and the associative property of convolution (Proposition 5.1.1(b)) to show that, for $p, q, r > 0$, $B(p,q)B(p+q,r) = B(q,r)B(p,q+r)$. (In Exercise 5.6.4, we'll verify this in a different way.)

5.5.4 Note that Proposition 5.4.1, Proposition 5.5.1(e), and (5.83) yield

$$(h_p * h_q)' = h_p' * h_q = ((p-1)h_{p-1} - h_p) * h_q$$
$$= (p-1)B(p-1,q)h_{p+q-1} - B(p,q)h_{p+q}$$

for $p > 1$ and $q > 0$. Reconcile this with (5.84) by showing that, for such p and q, $(p-1)B(p-1,q) = (p+q-1)B(p,q)$. Hint: In the integral defining $B(p,q)$, write $v = 1 - (1-v)$ to break the integral into a difference of two such. In the second of these resulting integrals, integrate by parts to get back a constant times $B(p,q)$. (In Exercise 5.6.5, we'll prove this identity in a different way.)

5.5.5 a. Is a convolution of functions in $L^1(\mathbb{R})$ necessarily in $L^2(\mathbb{R})$?
  b. Is a convolution of a function in $L^1(\mathbb{R})$ with one in $L^2(\mathbb{R})$ necessarily in $BC(\mathbb{R})$?

5.6 CONVOLUTION: WHEN IS IT IN $L^1(\mathbb{T})$? IN $L^2(\mathbb{T})$?

Recall the h_p's of Proposition 5.5.1. Note that, if $p, q > 0$, so that $h_p, h_q \in L^1(\mathbb{R})$ by part (a) of that proposition, then certainly $p + q > 0$, so that $h_p * h_q \in L^1(\mathbb{R})$ as well, by parts (a) and (e) of that proposition.

Certainly this doesn't *prove* that $L^1(\mathbb{R})$ is closed under convolution. But it does suggest it, and it's a pretty good suggestion, we think. Because:

Proposition 5.6.1 *Suppose $f, g \in L^1(\mathbb{R})$. Then $f * g(x)$ exists for almost all x, $f * g \in L^1(\mathbb{R})$,*

$$\int_{-\infty}^{\infty} f * g(x)\, dx = \left(\int_{-\infty}^{\infty} f(v)\, dv\right)\left(\int_{-\infty}^{\infty} g(y)\, dy\right), \quad (5.85)$$

and $\|f * g\|_1 \leq \|f\|_1 \|g\|_1$.

Proof. For such f and g, we have

$$\begin{aligned}\int_{-\infty}^{\infty} f*g(x)\,dx &= \int_{-\infty}^{\infty}\left(\int_{-\infty}^{\infty} f(x-y)g(y)\,dy\right)dx \\ &= \int_{-\infty}^{\infty}\int_{-\infty}^{\infty} f(x-y)g(y)\,dx\,dy \\ &= \int_{-\infty}^{\infty} g(y)\left(\int_{-\infty}^{\infty} f(x-y)\,dx\right)dy \\ &= \int_{-\infty}^{\infty} g(y)\left(\int_{-\infty}^{\infty} f(v)\,dv\right)dy = \int_{-\infty}^{\infty} f(v)\,dv \int_{-\infty}^{\infty} g(y)\,dy,\end{aligned}$$
(5.86)

which is (5.85). We explain our steps here: The first is by definition of convolution; the second by Fubini's theorem (Proposition 3.5.3); the third because $g(y)$, being constant in x, can be pulled out of the integral in x; the fourth by substituting $v = x - y$ into that integral; the fifth by pulling the integral in v outside the integral in y. The application of Fubini's theorem is valid for the following reason: If we repeat the third through fifth steps of (5.86) with $|f|$ and $|g|$ in place of f and g, we get

$$\int_{-\infty}^{\infty}\left(\int_{-\infty}^{\infty} |f(x-y)g(y)|\,dx\right)dy = \int_{-\infty}^{\infty} |g(v)|\,dv \int_{-\infty}^{\infty} |f(y)|\,dy = \|f\|_1\|g\|_1;$$
(5.87)

since $f, g \in L^1(\mathbb{R})$, the norms on the right are finite, whence so is the double integral on the left, which is just what's required.

Now (5.86) tells us that $f*g \in L^1(\mathbb{R})$, so in particular $f*g(x)$ exists almost everywhere, as values of functions in $L^1(\mathbb{R})$ do. Finally, our desired bound on $\|f*g\|_1$ follows from (5.87) and the fact that

$$\begin{aligned}\int_{-\infty}^{\infty}\left(\int_{-\infty}^{\infty} |f(x-y)g(y)|\,dx\right)dy &= \int_{-\infty}^{\infty}\left(\int_{-\infty}^{\infty} |f(x-y)g(y)|\,dy\right)dx \\ &\geq \int_{-\infty}^{\infty}\left|\int_{-\infty}^{\infty} f(x-y)g(y)\,dy\right|dx \\ &= \int_{-\infty}^{\infty} |f*g(x)|\,dx = \|f*g\|_1.\end{aligned}$$
(5.88)

$\square$

The above proposition will be relevant to some investigations in the next section and some others in the next chapter. These investigations will also require that we know what happens when an $f \in L^2(\mathbb{R})$ is convolved with a $g \in L^1(\mathbb{R})$. Here, again, we use the h_p's as our guide. If $p > \frac{1}{2}$ and $q > 0$, so that $h_p \in L^2(\mathbb{R})$ and $h_q \in L^1(\mathbb{R})$, by parts (b) and (a), respectively, of Proposition 5.5.1, then $p+q > \frac{1}{2}$ so that, by parts (b) and (e) of that proposition, $h_p * h_q \in L^2(\mathbb{R})$. All of this is reflective of the following general fact.

Proposition 5.6.2 *Suppose $f \in L^2(\mathbb{R})$ and $g \in L^1(\mathbb{R})$. Then $f * g(x)$ exists for almost all x, $f * g \in L^2(\mathbb{R})$, and $||f * g||_2 \leq ||f||_2 \, ||g||_1$.*

Proof. We have

$$\left(\int_{-\infty}^{\infty} |f * g(x)|^2 \, dx\right)^{1/2} = \left(\int_{-\infty}^{\infty} \left|\int_{-\infty}^{\infty} f(x-y)g(y) \, dy\right|^2 dx\right)^{1/2}$$

$$\leq \int_{-\infty}^{\infty} \left(\int_{-\infty}^{\infty} |f(x-y)g(y)|^2 \, dx\right)^{1/2} dy$$

$$= \int_{-\infty}^{\infty} |g(y)| \left(\int_{-\infty}^{\infty} |f(x-y)|^2 \, dx\right)^{1/2} dy$$

$$= \int_{-\infty}^{\infty} |g(y)| \left(\int_{-\infty}^{\infty} |f(v)|^2 \, dv\right)^{1/2} dy$$

$$= \int_{-\infty}^{\infty} |g(y)| \, ||f||_2 \, dy = ||f||_2 \int_{-\infty}^{\infty} |g(y)| \, dy$$

$$= ||f||_2 \, ||g||_1. \tag{5.89}$$

For the inequality we used Lemma 5.4.1; the other steps are similar to those employed in the proof of Proposition 5.6.1. The left side of (5.89) is $||f * g||_2$, so as in that earlier proof, we're done. □

The parallel between the above two propositions can hardly be missed. Nor can one ignore the similarities between these results and Proposition 5.3.1. Actually, each of these three propositions is an instance of the following more general one.

Proposition 5.6.3 *For $1 \leq r < \infty$, define*

$$L^r(\mathbb{R}) = \left\{ \text{measurable functions } f \text{ on } \mathbb{R} \colon \int_{-\infty}^{\infty} |f(x)|^r \, dx < \infty \right\}, \tag{5.90}$$

and for $f \in L^r(\mathbb{R})$, put

$$||f||_r = \left(\int_{-\infty}^{\infty} |f(x)|^r \, dx\right)^{1/r}. \tag{5.91}$$

Also write $L^\infty(\mathbb{R})$ for the space of bounded and measurable functions on $\mathbb{R}$, and for h an element of this space, define $||h||_\infty$ as in (5.35).

Suppose $f \in L^r(\mathbb{R})$ and $g \in L^s(\mathbb{R})$, where r and s are between 1 and ∞ inclusive, and $r^{-1} + s^{-1} \geq 1$. Then $f * g(x)$ exists for almost all x, $f * g \in L^t(\mathbb{R})$, where

$$t = \frac{1}{r^{-1} + s^{-1} - 1}, \tag{5.92}$$

and $||f * g||_t \leq ||f||_r \, ||g||_s$. (Here, by $1/\infty$ we mean zero and by $1/0$ we mean ∞.)

Finally, if $t = \infty$ then $f * g(x)$ exists for all x, and $f * g$ is continuous on $\mathbb{R}$ (so that $f * g \in \mathrm{BC}(\mathbb{R})$).

We omit the proof; see Section 6.3 in [43].

The reader should check the following: Putting $(r, s) = (2, 1)$ into Proposition 5.6.3 yields Proposition 5.6.2; putting $(r, s) = (1, 1)$ yields Proposition 5.6.1; putting $(r, s) = (\infty, 1), (1, \infty)$, and $(2, 2)$ yields, in order, the three cases of Proposition 5.3.1.

Exercises

5.6.1 Show that the space $BL^1(\mathbb{R})$ of bounded integrable functions on $\mathbb{R}$ is closed under convolution. Hint: Corollary 3.5.1(b) and Proposition 5.6.1.

5.6.2 Show that $L^1(\mathbb{R}) \cap L^2(\mathbb{R})$ is closed under convolution.

5.6.3 Let $p, q > 0$. Show that

$$\mathrm{B}(p, q) = \frac{\Gamma(p)\Gamma(q)}{\Gamma(p+q)}.$$

(This is an incredibly wonderful formula connecting the beta and gamma functions, and has all sorts of very nice applications and consequences, a few of which are given in the following exercises.) Hint: Integrate both sides of (5.77); use (5.85) to simplify the resulting integral on the left.

5.6.4 Use the previous exercise to give another proof of the fact (cf. Exercise 5.5.3) that, for $p, q, r > 0$, $\mathrm{B}(p, q)\mathrm{B}(p+q, r) = \mathrm{B}(q, r)\mathrm{B}(p, q+r)$.

5.6.5 Use Exercises 5.5.1 and 5.6.3 to give another proof of the fact (cf. Exercise 5.5.4) that, for $p > 1$ and $q > 0$, $(p-1)\mathrm{B}(p-1, q) = (p+q-1)\mathrm{B}(p, q)$.

5.6.6   **a.** Use a change of variable to show that, for $p, q > 0$,

$$\mathrm{B}(p, q) = 2 \int_0^1 u^{2p-1}(1-u^2)^{q-1}\, du.$$

b. Use the above result and Exercise 5.6.3 to show that $\Gamma(\tfrac{1}{2}) = \sqrt{\pi}$. Hint: Consider $\mathrm{B}(\tfrac{1}{2}, \tfrac{1}{2})$; also, an antiderivative of $1/\sqrt{1-u^2}$ is $\arcsin u$.

c. Show that, for $q > 0$, $\Gamma(q)\Gamma(q + \tfrac{1}{2}) = \sqrt{\pi}\, 2^{1-2q}\Gamma(2q)$ (this is the so-called duplication formula for the gamma function). Hint: put $p = \tfrac{1}{2}$ into the formula from part a. Using evenness of the resulting integrand, write the resulting integral as an integral over $(-1, 1)$. Put $w = (u+1)/2$ into this result; then use Exercise 5.6.3.

5.6.7 A family $\{g_p : p > 0\}$ of functions on $\mathbb{R}$ is called a *convolution semigroup* if

$$g_p * g_q = g_{p+q}$$

for all $p, q > 0$. Show that, if h_p is as in (5.76), then $g_p = h_p/\Gamma(p)$ defines a convolution semigroup.

In the next two exercises, we fill in some gaps in our earlier study of Bessel functions.

5.6.8 Let $\nu \geq 0$. The *Bessel function J_ν of the first kind, of order ν*, is defined by

$$J_\nu(x) = \sum_{m=0}^{\infty} \frac{(-1)^m}{m!\,\Gamma(m+\nu+1)} \left(\frac{x}{2}\right)^{2m+\nu}.$$

a. Show that this definition of J_ν coincides with:
  (i) the definition (4.80), in the case where ν is a nonnegative integer n;
  (ii) the definition from Exercise 4.3.4, in the case where $\nu = k + \frac{1}{2}$ for some $k \in \mathbb{N}$.
Hints: $\Gamma(k+1) = k!$ for $k \in \mathbb{N}$. Also, for part (ii), use the multiplication formula of Exercise 5.6.6.

b. Show that J_ν satisfies:
  (i) *Bessel's equation of order ν*, meaning the differential equation (4.66) with ν in place of n;
  (ii) the recurrence relation (4.81) with ν in place of n.
Hint for both parts: Differentiate term by term.

5.6.9 **a.** Show that, for $\nu \geq 0$,

$$J_\nu(x) = \frac{x^\nu}{2^{\nu-1}\sqrt{\pi}\,\Gamma(\nu+\frac{1}{2})} \int_0^1 (1-t^2)^{\nu-1/2} \cos xt\, dt.$$

Hints: Replace the cosine factor by its Maclaurin series, cf. (A.21); bring the sum outside the integral (this step may readily be justified), evaluate the resulting integrals using Exercise 5.6.6.

b. Deduce from part a of this exercise the formula for $J_1(r)$ given in Exercise 1.8.9.

5.7 APPROXIMATE IDENTITIES AND THE DIRAC DELTA "FUNCTION"

By way of introducing and motivating our discussion of approximate identities, we go way back to Example 5.1.1. Note that, for the functions f and g there, we had $f * g = f$. This brings to mind the following question: Is there any function g such that $f * g = f$ for *all*, or at least all sufficiently reasonable, functions f? Such a g would be an "identity function for convolution," much as the constant function 0 is the identity function for addition (that is, $f + 0 = f$ for all f) and the constant function 1 the identity function for multiplication (that is, $f \cdot 1 = f$ for all f).

The answer is: NO, there is no such function g. This is due, essentially, to convolution's tendency to increase smoothness. The basic idea is that $f * g$ can't equal f if it's smoother than f.

For example, is there a $g \in L^1(\mathbb{R})$ such that $f * g = f$ for all $f \in L^1(\mathbb{R})$? No: If $f, g \in L^1(\mathbb{R})$, and f is bounded and measurable but *not* continuous (for example, let

f be the characteristic function of a bounded interval), then $f * g$, being in $BC(\mathbb{R})$ by Proposition 5.3.1, cannot equal f.

Similar arguments apply in other situations. So while the equation $f * g = f$ does admit solutions in specialized circumstances, the bottom line is that no identity function for convolution exists for any sufficiently reasonable class of sufficiently reasonable functions.

However, the nonexistence of such an identity function should not dissuade us from our quest for one! As it is with many unattainable goals, the pursuit of this one will take us to many fascinating places that we might never have visited had we stayed at home watching Iron Chef.

The first of these places is a land of functions that, when convolved with any sufficiently reasonable function f, will get us *as close as we want* to f. Specifically, we have:

Proposition 5.7.1 *Suppose "g has mass 1," meaning*

$$\int_{-\infty}^{\infty} g(v)\, dv = 1, \tag{5.93}$$

so that in particular $g \in L^1(\mathbb{R})$. For any $\varepsilon > 0$, let $g_{[\varepsilon]}$ be the function defined by

$$g_{[\varepsilon]}(y) = \varepsilon^{-1} g(\varepsilon^{-1} y). \tag{5.94}$$

We have the following:

(a) *Suppose f is bounded and measurable. Then for any $\varepsilon > 0$, $f * g_{[\varepsilon]}(x)$ exists for all $x \in \mathbb{R}$ and $f * g_{[\varepsilon]} \in BC(\mathbb{R})$; moreover*

$$\lim_{\varepsilon \to 0^+} f * g_{[\varepsilon]}(x) = f(x) \tag{5.95}$$

for any point x at which f is continuous.

(b) *Let $p = 1$ or 2. If $f \in L^p(\mathbb{R})$, then so is $f * g_{[\varepsilon]}$ for all $\varepsilon > 0$, and $f * g_{[\varepsilon]}$ converges in the $L^p(\mathbb{R})$ norm to f: that is, $\lim_{\varepsilon \to 0^+} \|f * g_{[\varepsilon]} - f\|_p = 0$.*

Proof. We begin with part (a). Let $g, g_{[\varepsilon]}$, and f be as stipulated there. We have

$$\int_{-\infty}^{\infty} |g_{[\varepsilon]}(x)|\, dx = \varepsilon^{-1} \int_{-\infty}^{\infty} |g(\varepsilon^{-1} x)|\, dx = \int_{-\infty}^{\infty} |g(v)|\, dv \tag{5.96}$$

(we substituted $v = \varepsilon^{-1} x$), so g being in $L^1(\mathbb{R})$ implies $g_{[\varepsilon]}$ is. But then f being bounded and measurable, together with Proposition 5.3.1, imply $f * g_{[\varepsilon]} \in BC(\mathbb{R})$.

Moreover, we compute

$$f * g_{[\varepsilon]}(x) - f(x) = \int_{-\infty}^{\infty} f(x-y)g_{[\varepsilon]}(y)\,dy - f(x)\int_{-\infty}^{\infty} g(v)\,dv$$

$$= \varepsilon^{-1}\int_{-\infty}^{\infty} f(x-y)g(\varepsilon^{-1}y)\,dy - \int_{-\infty}^{\infty} f(x)g(v)\,dv$$

$$= \int_{-\infty}^{\infty} f(x-\varepsilon v)g(v)\,dv - \int_{-\infty}^{\infty} f(x)g(v)\,dv$$

$$= \int_{-\infty}^{\infty} (L_{-\varepsilon v}f(x) - f(x))g(v)\,dv, \tag{5.97}$$

with $L_{-\varepsilon v}$ as in Lemma 5.3.1. Here, the first equality is by definition of convolution and the fact that g has mass 1, the second is by (5.94), the third is by a substitution of $v = \varepsilon^{-1}y$. We now note two things: First,

$$\big|(L_{-\varepsilon v}f(x) - f(x))g(v)\big| = \big|f(x-\varepsilon v) - f(x)\big|\,|g(v)|$$
$$\leq (|f(x-\varepsilon v)| + |f(x)|)\,|g(v)| \leq 2||f||_\infty |g(v)| \tag{5.98}$$

for all v, x, and ε. That is, the rightmost integrand in (5.97) is bounded, independently of x and ε, by an integrable function of v. Second, if f is continuous at x, then

$$\lim_{\varepsilon \to 0^+} L_{-\varepsilon v}f(x) = f(x). \tag{5.99}$$

So for such x, the Lebesgue dominated convergence theorem (Proposition 3.5.1) applied to (5.97) yields

$$\lim_{\varepsilon \to 0^+}(f * g_{[\varepsilon]}(x) - f(x)) = \int_{-\infty}^{\infty}\lim_{\varepsilon \to 0^+}(L_{-\varepsilon v}f(x) - f(x))g(v)\,dv$$
$$= \int_{-\infty}^{\infty}(f(x) - f(x))g(v)\,dv = 0, \tag{5.100}$$

as required.

Now for part (b). The fact that $f * g_{[\varepsilon]} \in L^p(\mathbb{R})$ whenever $f \in L^p(\mathbb{R})$ and $g \in L^1(\mathbb{R})$ follows from Propositions 5.6.1 (which gives the case $p = 1$) and 5.6.2 (which takes care of the $p = 2$ situation). We note that

$$||f * g_{[\varepsilon]} - f||_p = \left(\int_{-\infty}^{\infty} |f * g_{[\varepsilon]}(x) - f(x)|^p\,dx\right)^{1/p}$$
$$= \left(\int_{-\infty}^{\infty}\left|\int_{-\infty}^{\infty}(L_{-\varepsilon v}f(x) - f(x))g(v)\,dv\right|^p\,dx\right)^{1/p}, \tag{5.101}$$

the last step by (5.97). In case $p = 1$, the right side of (5.101) is

$$\leq \int_{-\infty}^{\infty}\int_{-\infty}^{\infty} \big|(L_{-\varepsilon v}f(x) - f(x))g(v)\big|\,dv\,dx$$
$$= \int_{-\infty}^{\infty} |g(v)|\left(\int_{-\infty}^{\infty}\big|L_{-\varepsilon v}f(x) - f(x)\big|\,dx\right)dv, \tag{5.102}$$

by Fubini's Theorem (Proposition 3.5.3). In case $p = 2$, it's

$$\leq \left(\int_{-\infty}^{\infty} \int_{-\infty}^{\infty} |(L_{-\varepsilon v} f(x) - f(x)) g(v)|^2 \, dx \right)^{1/2} dv$$

$$= \int_{-\infty}^{\infty} |g(v)| \left(\int_{-\infty}^{\infty} |L_{-\varepsilon v} f(x) - f(x)|^2 \, dx \right)^{1/2} dv, \qquad (5.103)$$

by Lemma 5.4.1. So in either case, we can conclude that

$$\|f * g_{[\varepsilon]} - f\|_p \leq \int_{-\infty}^{\infty} |g(v)| \, \|L_{-\varepsilon v} f - f\|_p \, dv. \qquad (5.104)$$

The integrand on the right is, by the triangle inequality on $L^p(\mathbb{R})$, less than or equal to

$$|g(v)| (\|L_{-\varepsilon v} f\|_p + \|f\|_p) = 2|g(v)| \, \|f\|_p, \qquad (5.105)$$

which is independent of ε, and integrable on $\mathbb{R}$ since g is. Moreover, for any v this integrand goes to zero as $\varepsilon \to 0^+$, by Lemma 5.3.1. Then by the Lebesgue dominated convergence theorem (Proposition 3.5.1), the right side of (5.104) goes to zero as $\varepsilon \to 0^+$. So by the squeeze law, $\|f * g_{[\varepsilon]} - f\|_p$ does too, as required. □

We think of Proposition 5.7.1 as saying that, generally,

$$f * g_{[\varepsilon]} \approx f \qquad (5.106)$$

for ε small. This motivates:

Definition 5.7.1 If g is in $L^1(\mathbb{R})$ and has mass 1 and $g_{[\varepsilon]}$ is as in (5.94), then the collection

$$\{g_{[\varepsilon]} : \varepsilon > 0\} \qquad (5.107)$$

of functions is called *the approximate identity for convolution*—or simply *the approximate identity*—generated by g.

Figure 5.3 illustrates the kinds of things that approximate identities do.

To make all this more concrete, let's look at some specific examples of approximate identities. Perhaps the simplest are those generated by appropriately scaled characteristic functions of bounded intervals: $\chi_{[-1/2,1/2]}$; $\chi_{[-1,1]}/2$; in general $\chi_{[a,b]}/(b-a)$. (The chosen scaling gives these functions mass 1.) Such choices for g have the advantage that the corresponding convolutions $f * g_{[\varepsilon]}$ are often, or at least more often than for less elementary choices, explicitly computable.

They have the disadvantage, though, of not being infinitely differentiable, or even once differentiable, or even continuous on all of $\mathbb{R}$. A function g that's all of these things is given by

$$g(v) = \frac{1}{\pi(1+v^2)}. \qquad (5.108)$$

APPROXIMATE IDENTITIES AND THE DIRAC DELTA "FUNCTION" 291

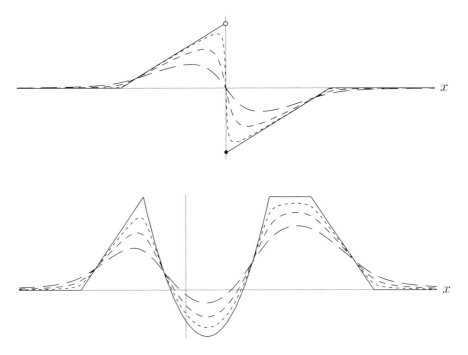

Fig. 5.3 A discontinuous (top) and a continuous (bottom) compactly supported, piecewise smooth function $f \in L^2(\mathbb{R})$ (solid), and $f * g_{[\varepsilon]}$ (dashed) for g as in (5.107), and for various values of ε. Shorter dashes connote smaller ε

We compute that

$$\int_{-\infty}^{\infty} g(v)\,dv = \frac{1}{\pi}\int_{-\infty}^{\infty}\frac{dv}{1+v^2} = \frac{1}{\pi}\arctan v\Big|_{-\infty}^{\infty} = \frac{1}{\pi}\left(\frac{\pi}{2}-\left(-\frac{\pi}{2}\right)\right) = 1, \tag{5.109}$$

so that g has mass 1. So $\{g_{[\varepsilon]} : \varepsilon > 0\}$ is indeed an approximate identity. And because g is a simple rational function and is readily antidifferentiated, there's still a decent chance that, for f also not too complicated, $f * g_{[\varepsilon]}$ will be explicitly evaluable.

If one wants $g_{[\varepsilon]}$'s in $C_c(\mathbb{R})$ or, even better, compactly supported and *infinitely* differentiable, one needs a g that's the same. Here's one:

$$g(v) = \begin{cases} C^{-1}e^{-1/(1-v^2)} & \text{if } -1 < v < 1, \\ 0 & \text{if not,} \end{cases} \tag{5.110}$$

where

$$C = \int_{-1}^{1} e^{-1/(1-u^2)}\,dy. \tag{5.111}$$

(The factor C^{-1} ensures that h has mass 1.) One checks that g is, indeed, in the space $C_c^\infty(\mathbb{R})$ of infinitely differentiable, compactly supported functions on $\mathbb{R}$ (cf. Exercise 5.7.5). For this reason g is quite useful as a generator of an approximate identity, even if the convolution integrals to which it leads are rarely easy to evaluate symbolically.

We digress for a moment, to remark on a rather vast realm of "real-life" applicability of approximate identities. Let's recall our discussions surrounding Propositions 5.3.1, 5.4.1, and 5.4.2. These discussions imply that $f * g_{[\varepsilon]}$ will, generally, be (at least) as smooth as $g_{[\varepsilon]}$, even when f is considerably less smooth. But again, Proposition 5.7.1 says that $f * g_{[\varepsilon]}$ will look quite a bit like f, in some reasonable sense or another, for ε small. Conclusion: For such ε and $g_{[\varepsilon]}$, we can think of $f * g_{[\varepsilon]}$ as "f, smoothed out."

And there are many situations where smoothing of a function is desired. For one of these, consider information that's been digitally encoded. Such encoding, because it relegates data to a discrete rather than a continuous set of values, by its very nature introduces nonsmoothness. But the latter may be mitigated by convolution of the digitized information with an appropriate approximate identity. (More precisely, the function to be smoothed is convolved with a $g_{[\varepsilon]}$ for some *particular*, sufficiently small, number ε).

In a variety of other circumstances, convolution with approximate identities is employed as a smoothing technique. For example, graphics and audio editing software makes of use of it in this way. Of course, should the information to be smoothed be two dimensional in nature, as for instance a graphic image is, then a two-dimensional version of convolution will be required. Similarly for three or more dimensions. We'll say more about multidimensional situations in Section 6.8, but for now we note simply that both the specifics and generalities of the present chapter generalize to these situations in natural and expected ways.

We also note that there are many situations where one wants to perform *deconvolution*. By this we mean the process that's reverse to, or *undoes*, convolution, much as division undoes multiplication and subtraction undoes addition. Such a process is, in particular, valuable when one seeks results opposite those effected by convolution with an approximate identity—in other words when, instead of smoothing, one wants to *sharpen*. As one does in refining graphic images, restoring degraded audio files, and so on.

Because convolution is, as discussed earlier, a more subtle and difficult creature than is multiplication or addition, deconvolution is intrinsically more difficult and subtle than is division or subtraction. In fact, our studies so far give us no means whatsoever of undoing a convolution. But here's the amazing thing: We can do so (or, more exactly, "undo" so) using the *Fourier transform!* See Section 7.8. What's amazing about this is (among other things) the fact that we have introduced convolution in the first place with an eye toward uncovering the mysteries of aperiodic Fourier analysis. Yet, as just noted, the latter will ultimately allow us to unlock, or unfold, the former as well.

But now, let's briefly get back on the road to nowhere—nowhere being, as mentioned at the start of this section, where the identity function for convolution resides.

APPROXIMATE IDENTITIES AND THE DIRAC DELTA "FUNCTION"

Let's allow Proposition 5.7.1 to take us one step further along this road and at the same time to take us to a place that is very much somewhere. As follows: We encapsulate that proposition not with the approximate equality (5.106), but this time with the equation

$$\lim_{\varepsilon \to 0^+} (f * g_{[\varepsilon]}) = f. \qquad (5.112)$$

(The limit is to be understood in an appropriate sense: The proposition gives us several such senses, according to the natures of f and g.) And we imagine, just for the moment, that we can switch the limit with the convolution, to get

$$f * \left(\lim_{\varepsilon \to 0^+} g_{[\varepsilon]} \right) = f. \qquad (5.113)$$

Let's rewrite this for a better look at what it tells us. If we define

$$\delta = \lim_{\varepsilon \to 0^+} g_{[\varepsilon]}, \qquad (5.114)$$

then what (5.113) says is that

$$f * \delta = f. \qquad (5.115)$$

And since f was arbitrary (within reason), we find that δ IS an identity function for convolution!

But again, no such function exists. So there must be something wrong with our above argument. And there is: The limit in (5.114) does not, itself, exist—at least not in any sense we've discussed so far.

Why not? Well, consider the following. First of all, replacing y by $\varepsilon^{-1}y$ in the argument of a function g causes a compression of the graph of g in the horizontal direction, by a factor of ε^{-1}. Second, multiplying $g(\varepsilon^{-1}y)$ by ε^{-1} elongates the graph of $g(\varepsilon^{-1}y)$ in the vertical direction, also by a factor of ε^{-1}. CONCLUSION: By (5.94), $g_{[\varepsilon]}$ is "g, stretched by a factor of ε^{-1} vertically and squeezed by a factor of ε^{-1} horizontally." See Figure 5.4.

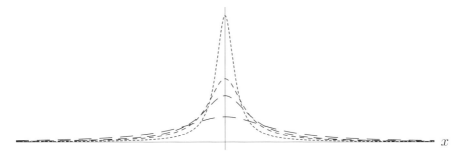

Fig. 5.4 The approximate identity $\{g_{[\varepsilon]} : \varepsilon > 0\}$ generated by the function g of (5.107). Shorter dashes connote smaller ε

So were the limit δ, given by (5.114), to exist as a function, then intuitively it would have to be a "spike" of infinite height and infinitesimal width (concentrated at $x = 0$). Moreover, it would have to have mass

$$\int_{-\infty}^{\infty} \delta(v)\,dv = \lim_{\varepsilon \to 0+} \int_{-\infty}^{\infty} g_{[\varepsilon]}(v)\,dv = \lim_{\varepsilon \to 0+} \varepsilon^{-1} \int_{-\infty}^{\infty} g\left(\varepsilon^{-1} v\right) dv$$
$$= \lim_{\varepsilon \to 0+} \int_{-\infty}^{\infty} g(y)\,dy = 1. \qquad (5.116)$$

(We substituted $y = \varepsilon^{-1} v$. And we switched an integral with a limit, the validity of which we gloss over here because we're only arguing intuitively, anyway.) Clearly, there can be no such function δ.

Still, we sometimes pretend there *is*, and that it *is* obtained as in (5.114), where again $\{g_{[\varepsilon]} : \varepsilon > 0\}$ is any approximate identity. And we imagine this function to be just as we argued it should be, were it at all: infinitely tall, infinitesimally thin, and of mass 1. Of course such an object is a bit hard to draw, so we sometimes depict δ using a bold vertical arrow, or spike, of height one (Fig. 5.5).

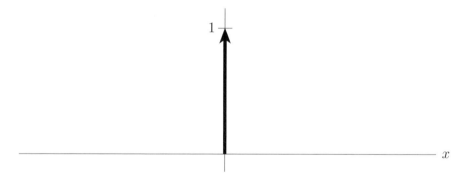

Fig. 5.5 δ, in spike form

And we call δ the "Dirac delta function," not surprisingly after Dirac, who used it freely without worrying especially about its nonexistence.

Should it make us queasy to refer to it thus, when in fact it isn't a function, we needn't do so. We can call it the delta spike, or arrow, or whatever. And we note that, actually, this arrow or spike or whatever does exist *somewhere*, namely, in the world of *tempered distributions*. In this world, there's yet another sense of convergence, known as *temperate* convergence, under which the limit of elements of an approximate identity is defined. But this limit turns out to be a tempered distribution instead of a function.

We'll say more about this strange new world in Section 6.9. But for the most part, we'll consider it enough simply to understand δ as shorthand for a certain limiting process. That is, whenever we see $f * \delta$, we can think "$\lim_{\varepsilon \to 0+} f * g_{[\varepsilon]}$." (The sense in which the limit is to be taken will depend on the context.) Whenever we see the Fourier transform of δ, as we will in the next chapter, we can think of some kind of limit of Fourier transforms of $g_{[\varepsilon]}$'s. And so on. At any rate, we *will* believe in δ, and will be rewarded for our faith.

Exercises

5.7.1 Let f and g be in $L^2(\mathbb{R})$; let $f_{[\varepsilon]}$ and $g_{[\varepsilon]}$ be as described by (5.94). Show that $f_{[\varepsilon]} * g_{[\varepsilon]}(x) = (f * g)_{[\varepsilon]}(x)$ for all x.

5.7.2 Let $g = \chi_{[-1,1]}/2$. Note that g has mass 1.
  a. Let $f(x) = x^3 - x$. Evaluate $f * g_{[\varepsilon]}(x)$, where $g_{[\varepsilon]}$ is as in (5.94), *explicitly* as a function of x and ε.
  b. Show directly, by taking limits of your result from part a, that
$$\lim_{[\varepsilon]\to 0^+} f * g_{[\varepsilon]}(x) = f(x)$$
for any $x \in \mathbb{R}$.

5.7.3 Repeat Exercise 5.7.2, but this time with $f(x) = x^2 - 1$ and $g(x) = 1/(\pi(1+x^2))$. (We already showed in (5.109) that g has mass 1.) An integration by parts with $u = y$ and $dv = y\,dy/(1 + (\varepsilon^{-1}y)^2)$ may help.

5.7.4 Repeat Exercise 5.7.2, but this time with $f(x) = \cos x$ and $g(x) = e^{-2|x|}$. (That g has mass 1 is an easy computation.) You might want to follow the lead of Example 5.1.1.

5.7.5 Let g be as in (5.110). Show that $g \in C_c^\infty(\mathbb{R})$. Hint: Show, by induction on k, that
$$g^{(k)}(x) = \begin{cases} q_k(x)(1-x^2)^{-k}e^{-1/(1-x^2)} & \text{if } |x| < 1, \\ 0 & \text{if } |x| \geq 1, \end{cases}$$
where q_k is a polynomial. Be *careful* about the derivatives at $x = \pm 1$. Exercise 1.3.17 may help here.

5.7.6 Show that, given any positive number λ, there exists a function $h \in C_c^\infty(\mathbb{R})$ such that $0 \leq h(x) \leq 1$ for all x, $h(x) = 1$ for $x \in [-1,1]$, and $h(x) = 0$ for $|x| > 1 + \lambda$. Hint: Consider $f * g_\varepsilon$, where g is as in the previous exercise and $f = \chi_{[-1-\lambda/2, 1+\lambda/2]}$. Choose ε small enough; use Propositions 5.2.1 and 5.4.2 and Exercise 5.2.1.

6

Fourier Transforms and Fourier Integrals

6.1 THE FOURIER TRANSFORM ON $L^1(\mathbb{R})$: BASICS

Perhaps the most striking difference between frequency decompositions of periodic functions, or functions on bounded intervals, and those of aperiodic functions is this: Functions of the latter type have *continuous* frequency domains. That is, synthesis from frequency components of an aperiodic function requires a *continuous superposition* of these components. Let's take a very brief, intuitive look at why this should be so, as a prelude to our rigorous analyses of exactly *in what way* it is so and our demonstrations that it *is* so.

We revisit the frequency decomposition, cf. (1.4) and (1.18), of a reasonable P-periodic function f:

$$f(x) = \sum_{n=-\infty}^{\infty} \left[\frac{1}{P} \int_{-P/2}^{P/2} f(x) e^{-2\pi i n x / P} \, dx \right] e^{2\pi i n x / P}. \tag{6.1}$$

(Theorem 1.4.1 and Corollary 1.4.1, and Theorems 1.7.1, 3.4.1, and 3.6.2, give precise conditions under which (6.1) is true, for corresponding precise interpretations of "true." But for now we argue imprecisely, as we did in Section 1.1.) It will be illuminating to write (6.1) in the form

$$f(x) = \sum_{s \in \mathbb{Z}/P} \left[\int_{-P/2}^{P/2} f(x) e^{-2\pi i s x} \, dx \right] e^{2\pi i s x} \, \Delta s, \tag{6.2}$$

where $\mathbb{Z}/P$ denotes the set $\{n/P \colon n \in \mathbb{Z}\}$ and $\Delta s = 1/P$ is just the spacing between elements of this set.

297

Let's now consider, quite informally for the moment, what happens to (6.2) in the limit as $P \to \infty$. What does happen? Well first of all, the function f on the left side takes longer and longer to repeat itself, meaning it becomes, in the limit, "∞-periodic," meaning it becomes *aperiodic*. On the right side, several things transpire: The increment Δs becomes *infinitesimal*; let's say it becomes ds. The set $\mathbb{Z}/P$, whose elements extend from $-\infty$ to ∞ and are Δs units apart, becomes a set of numbers that have the same extent but are now infinitesimally close together; that is, $\mathbb{Z}/P$ becomes the *real line* $\mathbb{R}$. Consequently, the sum on the right side of (6.2) becomes an *integral* over $\mathbb{R}$. Finally, the limits $\pm P/2$ of integration become $\pm \infty$ respectively. SO: Letting $P \to \infty$ in (6.2) gives, intuitively, the following frequency decomposition for a reasonable aperiodic function f:

$$f(x) = \int_{-\infty}^{\infty} \widehat{f}(s) e^{2\pi i s x} \, ds \qquad (x \in \mathbb{R}), \tag{6.3}$$

where by definition

$$\widehat{f}(s) = \int_{-\infty}^{\infty} f(x) e^{-2\pi i s x} \, dx \qquad (s \in \mathbb{R}). \tag{6.4}$$

The quantity $\widehat{f}(s)$ is an aperiodic analog of a Fourier coefficient; the representation (6.3) is an aperiodic analog of a Fourier series. It's edifying to reformulate these aperiodic phenomena in terms of the inner product, as we did the corresponding periodic ones in Section 3.6. We do so as follows: We define

$$e_s(x) = e^{2\pi i s x} \tag{6.5}$$

for $s, x \in \mathbb{R}$, so that (6.4) yields

$$\widehat{f}(s) = \int_{-\infty}^{\infty} f(x) \overline{e_s(x)} \, dx = \langle f, e_s \rangle, \tag{6.6}$$

and consequently (6.3) reads

$$f(x) = \int_{-\infty}^{\infty} \langle f, e_s \rangle \, e_s(x) \, ds. \tag{6.7}$$

(Previously, we've used the notation $\langle f, g \rangle$ only for f and g both *square integrable*; since $e_s \notin L^2(\mathbb{R})$ for any s, our above application of this notation is somewhat more liberal. But we'll allow it, here and henceforth, because of its convenience and suggestiveness.)

Often, one abbreviates (6.7) by dropping the argument x on either side; thus

$$f = \int_{-\infty}^{\infty} \langle f, e_s \rangle \, e_s \, ds. \tag{6.8}$$

Note that this last representation is reminiscent not only of a Fourier series, but also of the general orthonormal decompositions

$$f = \sum_{n \in C} \langle f, \sigma_n \rangle \, \sigma_n, \tag{6.9}$$

and the still more general orthogonal decompositions

$$f = \sum_{n \in C} \frac{\langle f, \rho_n \rangle}{||\rho_n||^2} \rho_n, \qquad (6.10)$$

of Section 3.6. (See Proposition 3.6.1 and Corollary 3.6.2.) Of course, (6.8) is *not* an orthogonal decomposition—it's of an integral, rather than a series, form, and the e_s's, being not even square integrable, are certainly not orthogonal. Still, the parallels are striking and merit some contemplation. In particular, these parallels, together with the remarks concluding Section 3.7, encourage us to think of $\widehat{f}(s) e_s = \langle f, e_s \rangle e_s$ as "that part of f that looks like e_s," and of $\widehat{f}(s) = \langle f, e_s \rangle$ itself as measuring "the extent to which e_s contributes to the makeup of f."

Now, again, the above discussions are rather nonrigorous. But they do lead us to a good place to start, should we wish to *make* things rigorous. To be specific, since $|f(x) e^{-2\pi i s x}| \, dx = |f(x)|$, Proposition 3.5.2 implies that the integral in (6.4) exists, for any $s \in \mathbb{R}$, *precisely* when $f \in L^1(\mathbb{R})$. We are therefore led to:

Definition 6.1.1 If $f \in L^1(\mathbb{R})$, then *the Fourier transform* $\widehat{f}$, also denoted $F[f(x)]$, of f is the function defined by (6.4) (or, equivalently, by (6.6)).

It's immediate that

$$F[\alpha f(x) + \beta g(x)] = \alpha F[f(x)] + \beta F[g(x)] \qquad (6.11)$$

for $\alpha, \beta \in \mathbb{C}$ and $f, g \in L^1(\mathbb{R})$; that is, the Fourier transform is linear. What else is it? For the purposes of examining this question, a few examples might be in order. We start with Example 6.1.1, so as not to be out of order.

Example 6.1.1 Find $\widehat{\chi_{[-b,b]}}$, for $b > 0$.

Solution. It's a straightforward computation:

$$\widehat{\chi_{[-b,b]}}(s) = \int_{-\infty}^{\infty} \chi_{[-b,b]}(x) e^{-2\pi i s x} \, dx = \int_{-b}^{b} e^{-2\pi i s x} \, dx = \left[\frac{e^{-2\pi i s x}}{-2\pi i s} \right]_{-b}^{b}$$

$$= \frac{e^{-2\pi i s b} - e^{2\pi i s b}}{-2\pi i s} = \frac{1}{\pi s} \cdot \frac{e^{2\pi i s b} - e^{-2\pi i s b}}{2i} = \frac{\sin 2\pi b s}{\pi s}, \qquad (6.12)$$

all of which is valid, strictly speaking, when $s \neq 0$. Also, $\widehat{\chi_{[-b,b]}}(0) = \int_{-b}^{b} dx = 2b$, which is not surprising because

$$\lim_{s \to 0} \widehat{\chi_{[-b,b]}}(s) = \lim_{s \to 0} \frac{\sin 2\pi b s}{\pi s} = 2b, \qquad (6.13)$$

by l'Hôpital's rule. Let's summarize: Defining

$$\operatorname{sinc} u = \begin{cases} \dfrac{\sin u}{u} & \text{if } u \neq 0, \\ 1 & \text{if } u = 0, \end{cases} \qquad (6.14)$$

we have
$$\widehat{\chi_{[-b,b]}}(s) = 2b\operatorname{sinc} 2\pi bs. \tag{6.15}$$

We observe that $\widehat{\chi_{[-b,b]}}$ is, among other things, bounded and continuous (continuity at zero is by (6.13)). So our first Fourier transform example illustrates, among other things:

Proposition 6.1.1 *If $f \in L^1(\mathbb{R})$, then $\widehat{f} \in \mathrm{BC}(\mathbb{R})$ and, more specifically, $\|\widehat{f}\|_\infty$ is less than or equal to $\|f\|_1$.*

Proof. Assume $f \in L^1(\mathbb{R})$ and $s \in \mathbb{R}$. Then

$$\|\widehat{f}\|_\infty = \sup_{s \in \mathbb{R}} |\widehat{f}(s)| = \sup_{s \in \mathbb{R}} \left| \int_{-\infty}^{\infty} f(x) e^{-2\pi i s x} \, dx \right| \leq \sup_{s \in \mathbb{R}} \int_{-\infty}^{\infty} \left| f(x) e^{-2\pi i s x} \right| dx$$

$$= \sup_{s \in \mathbb{R}} \int_{-\infty}^{\infty} |f(x)| \, dx = \|f\|_1. \tag{6.16}$$

Moreover, suppose $s_0 \in \mathbb{R}$. Since

$$\lim_{s \to s_0} f(x) e^{-2\pi i s x} \, dx = f(x) e^{-2\pi i s_0 x} \, dx \tag{6.17}$$

for any $x \in \mathbb{R}$, and since again $\left| f(x) e^{-2\pi i s x} \right| = |f(x)|$, which is integrable on $\mathbb{R}$ and independent of s, we may apply the Lebesgue dominated convergence theorem (Proposition 3.5.1) as follows:

$$\lim_{s \to s_0} \widehat{f}(s) = \lim_{s \to s_0} \int_{-\infty}^{\infty} f(x) e^{-2\pi i s x} \, dx = \int_{-\infty}^{\infty} \left(\lim_{s \to s_0} f(x) e^{-2\pi i s x} \right) dx$$

$$= \int_{-\infty}^{\infty} f(x) e^{-2\pi i s_0 x} \, dx = \widehat{f}(s_0), \tag{6.18}$$

which is exactly what's required for $\widehat{f}$ to be continuous.
So $\widehat{f} \in \mathrm{BC}(\mathbb{R})$, as claimed. □

We work through another example, in order to produce not just another example, but one that compares in suggestive ways with the previous one.

Example 6.1.2 Find $\widehat{f}$ if

$$f(x) = e^{-2\pi |x|}. \tag{6.19}$$

Solution. We compute

$$\widehat{f}(s) = \int_{-\infty}^{\infty} e^{-2\pi|x|} e^{-2\pi i s x} \, dx = \int_{-\infty}^{0} e^{2\pi x} e^{-2\pi i s x} \, dx + \int_{0}^{\infty} e^{-2\pi x} e^{-2\pi i s x} \, dx$$

$$= \int_{-\infty}^{0} e^{2\pi x(1-is)} \, dx + \int_{0}^{\infty} e^{-2\pi x(1+is)} \, dx$$

$$= \frac{e^{2\pi x(1-is)}}{2\pi(1-is)} \bigg|_{-\infty}^{0} + \frac{e^{-2\pi x(1+is)}}{-2\pi(1+is)} \bigg|_{0}^{\infty} = \frac{1}{2\pi} \left[\frac{1}{1-is} + \frac{1}{1+is} \right]$$

$$= \frac{1}{2\pi} \left[\frac{1+is+1-is}{(1-is)(1+is)} \right] = \frac{1}{\pi(1+s^2)}. \tag{6.20}$$

(We used the facts that

$$\lim_{x \to -\infty} e^{2\pi x(1-is)} = \lim_{x \to \infty} e^{-2\pi x(1+is)} = 0 \tag{6.21}$$

for any s. These follow from the squeeze law and the fact that $e^{-2\pi i s x}$ is bounded in x.)

Note that the function f of the above example is *continuous*, unlike the function $\chi_{[-b,b]}$ of Example 6.1.1. Also note that $\widehat{f}$, being a constant divided by a degree-2 polynomial in s, *decays* roughly like $1/s^2$ as $s \to \pm\infty$, while $\widehat{\chi_{[-b,b]}}$, being a bounded, nondecaying function divided by s, does so only about like $1/s$. See Figure 6.1.

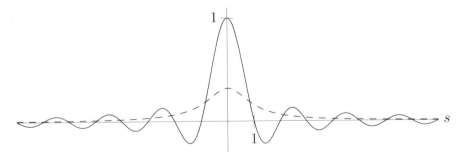

Fig. 6.1 $F[\chi_{[-1/2,1/2]}(x)]$ (solid) and $F[e^{-2\pi|x|}]$ (dashed)

Perhaps there's something going on here. That is, perhaps there's some general correspondence between the smoothness of a function $f \in L^1(\mathbb{R})$ and the rate of decay of $\widehat{f}$. Such a correspondence would mirror the one in the periodic situation: There, the rate of decay of the Fourier *coefficients* $c_n(f)$, as $n \to \pm\infty$, was seen to increase with the degree of smoothness of f. (See the discussions at the ends of Section 1.3, and also Corollary 1.8.1 and the comments below it.) And it would provide another manifestation of the local/global principle. (Again see the discussion following Corollary 1.8.1.)

That there *is* something going on here will follow ultimately from part (c)(i) of Proposition 6.2.1, below. But for the present, let's consider another interesting consequence of the slow decay of $\widehat{\chi_{[-b,b]}}$. Namely, *this Fourier transform does not belong to* $L^1(\mathbb{R})$. See Exercise 6.1.6: The essential difficulty is that $1/s$ is not "integrable out to ∞ (or $-\infty$)." (Antidifferentiating $1/s$ gives $\ln|s|$, which blows up as $|s| \to \infty$.) The upshot is that there *are* functions in $L^1(\mathbb{R})$ for which (6.4) fails miserably, in that its right side is not even defined.

Of course a simple way around this difficulty is to restrict our attention to the space

$$FL^1(\mathbb{R}) = \{f \in L^1(\mathbb{R}) : \widehat{f} \in L^1(\mathbb{R})\}. \tag{6.22}$$

(For example, the function f of Example 6.1.2 is in this space—that $\widehat{f} \in L^1(\mathbb{R})$ here follows from (5.109).) For $f \in FL^1(\mathbb{R})$ not only is $\widehat{f}(s)$ defined for all s, but so is the integral in equation (6.3) for all x. So in this case, at least it makes sense to consider the truth of that equation.

In Theorem 6.4.1, we *will* restrict our attention to $FL^1(\mathbb{R})$, and will see that (6.3) *is* true (almost everywhere) for f in this space. Still, there's a certain inelegance to this solution: $FL^1(\mathbb{R})$ is not all that well understood. (Of course, it's perfectly unmysterious in one sense: It's just the space given by (6.22). What we mean is that there's no simple characterization of it that does not directly involve the Fourier transform.)

If one considers Fourier transforms of *square* integrable, instead of integrable, aperiodic functions, one arrives at a more elegant and in many ways more powerful theory. In light of our discussions in the periodic context, this should not be surprising. But this theory is also more subtle and less direct; in particular, it requires modification of the very *definition* of Fourier transform, since the integral in (6.4) does not make sense if $f \notin L^1(\mathbb{R})$. (And again, $L^2(\mathbb{R}) \not\subset L^1(\mathbb{R})$.)

The moral is that the $L^1(\mathbb{R})$ and the $L^2(\mathbb{R})$ theories both have their advantages. We'll continue for a while with the former and will, ultimately, use it to sneak up on the latter.

Exercises

In Exercises 6.1.1–6.1.5, find the Fourier transform $\widehat{f}$ of the given function f.

6.1.1 $f = \chi_{[-2,0]} + \chi_{[0,2]}$.

6.1.2 $f(x) = \operatorname{sgn} x \, e^{-2\pi|x|}$, where $\operatorname{sgn} x$ is as in (A.12).

6.1.3 $f(x) = x\chi_{[-1,1]}(x)$.

6.1.4 $f(x) = (2b-|x|)\chi_{[-2b,2b]}(x) \, (b > 0)$. (For the time being the direct approach is probably best; in the exercises for the next section, we'll see a better way.)

6.1.5 $f(x) = \cos 2\pi ax \, e^{-2\pi b|x|}$, $a, b > 0$.

6.1.6 Let $g(x) = \operatorname{sinc} x$.
   a. Show that $g \in L^2(\mathbb{R})$.

b. Show that $g \notin L^1(\mathbb{R})$. Hint:

$$\int_\pi^\infty |g(x)|\, dx = \sum_{K=1}^\infty \int_{\pi K}^{\pi(K+1)} \frac{|\sin x|}{x}\, dx.$$

Show that the right side is larger than a constant times $\sum_{K=1}^\infty (K+1)^{-1}$ (hint: consider $|\sin x|$ for $x \in [\pi K + \pi/4, \pi K + 3\pi/4]$); the latter series diverges.

c. Show that the improper Riemann integral

$$\int_{-\infty}^\infty g(x)\, dx$$

converges. Hint: Write the integral as

$$2 \sum_{K=0}^\infty \int_{\pi K}^{\pi(K+1)} \frac{\sin x}{x}\, dx;$$

use the alternating series test to show the series converges.

6.2 MORE ON THE FOURIER TRANSFORM ON $L^1(\mathbb{R})$

The Fourier transform behaves nicely when you *compose* it with certain other operations, like modulation (multiplication by a complex exponential), translation, dilation (rescaling of the independent variable), differentiation, multiplication by the independent variable, and convolution. Specifically, we have:

Proposition 6.2.1 *Let $f, g \in L^1(\mathbb{R})$ and $s \in \mathbb{R}$.*

(a) *For any $a \in \mathbb{R}$,*
  (i) $F\left[e^{2\pi i a x} f(x)\right](s) = \widehat{f}(s-a)$ and
  (ii) $F[f(x-a)](s) = e^{-2\pi i s a} \widehat{f}(s)$.

(b) (i) *For any real, nonzero b, $F[f(bx)](s) = |b|^{-1} \widehat{f}(b^{-1} s)$.*
  (ii) $F[f(-x)](s) = \widehat{f}(-s)$.
  (iii) $F\left[\overline{f(x)}\right](s) = \overline{\widehat{f}(-s)}$.

(c) (i) $F[f'(x)](s) = 2\pi i s \widehat{f}(s)$, provided $f \in \mathrm{PSC}(\mathbb{R})$ and $f' \in L^1(\mathbb{R})$;
  (ii) $F[xf(x)](s) = (-2\pi i)^{-1} (\widehat{f})'(s)$, provided $xf(x)$ defines a function in $L^1(\mathbb{R})$.

(d) **(the composition theorem)** *For $x \in \mathbb{R}$,*

$$\int_{-\infty}^\infty \widehat{f}(s) g(s) e^{2\pi i s x}\, ds = \int_{-\infty}^\infty f(v) \widehat{g}(v-x)\, dv. \qquad (6.23)$$

In particular,

$$\int_{-\infty}^{\infty} \widehat{f}(s)g(s)\,ds = \int_{-\infty}^{\infty} f(v)\widehat{g}(v)\,dv. \tag{6.24}$$

(e) (the convolution theorem)

$$\widehat{f*g}(s) = \widehat{f}(s)\widehat{g}(s). \tag{6.25}$$

Proof. Part (a)(i) goes like this:

$$F\left[e^{2\pi iax}f(x)\right](s) = \int_{-\infty}^{\infty} e^{2\pi iax}f(x)e^{-2\pi isx}\,dx = \int_{-\infty}^{\infty} f(x)e^{-2\pi i(s-a)x}\,dx$$
$$= \widehat{f}(s-a). \tag{6.26}$$

Part (a)(ii) is left to the exercises.

Part (b)(i) amounts to a substitution of $u = bx$, like this:

$$F[f(bx)](s) = \int_{-\infty}^{\infty} f(bx)e^{-2\pi isx}\,dx = |b|^{-1}\int_{-\infty}^{\infty} f(u)e^{-2\pi i(b^{-1}s)u}\,du$$
$$= |b|^{-1}\widehat{f}(b^{-1}s). \tag{6.27}$$

(The absolute values come in because, if $b < 0$, then the indicated substitution flips the limits of integration; replacing the factor of b in front of the integral by $-b = |b|$ then flips them back.) Part (b)(ii) is part (b)(i) with $b = -1$; we leave part (b)(iii) to the exercises.

For part (c)(i), we integrate by parts with $u = e^{-2\pi isx}$ and $dv = f'(x)\,dx$, whereby $du = -2\pi is\,e^{-2\pi isx}\,dx$ and $v = f(x)$, to get the following:

$$F[f'(x)](s) = \int_{-\infty}^{\infty} f'(x)e^{-2\pi isx}\,dx$$
$$= f(x)e^{-2\pi isx}\Big|_{-\infty}^{\infty} + 2\pi is\int_{-\infty}^{\infty} f(x)e^{-2\pi isx}\,dx = 2\pi is\widehat{f}(s). \tag{6.28}$$

We used the fact that

$$\lim_{x\to\pm\infty} f(x)e^{-2\pi isx} = 0; \tag{6.29}$$

this is so for the following reason. For any real number x, we have

$$f(x) - f(0) = \int_0^x f'(t)\,dt \tag{6.30}$$

by continuity of f and the fundamental theorem of calculus. The integral on the right exists for any $x \in \mathbb{R}$ and has finite limit as $x \to \pm\infty$, because $f' \in L^1(\mathbb{R})$; consequently, the limits $\lim_{x \to \pm\infty} f(x)$ exist too. But since $f \in L^1(\mathbb{R})$, the area under the graph of $|f|$ must be finite, so these limits must be zero. The squeeze law then gives (6.29). (The continuity of f is essential to the validity of our integration-by-parts argument. In the next section, though, we'll consider a useful generalization—in which continuity is *not* required—of the result just proved.)

Next, we prove part (c)(ii) by a differentiation-under-the-integral-sign argument:

$$(-2\pi i)^{-1}(\widehat{f})'(s) = (-2\pi i)^{-1} \frac{d}{ds} \int_{-\infty}^{\infty} f(x)e^{-2\pi i s x}\, dx$$

$$= (-2\pi i)^{-1} \int_{-\infty}^{\infty} f(x)\left(\frac{d}{ds}e^{-2\pi i s x}\right) dx$$

$$= (-2\pi i)^{-1} \int_{-\infty}^{\infty} -2\pi i x f(x)e^{-2\pi i s x}\, dx$$

$$= \int_{-\infty}^{\infty} x f(x)e^{-2\pi i s x}\, dx = F[x f(x)](s). \qquad (6.31)$$

The validation of such an argument in this case is much as it was in the proof of Proposition 5.4.1. We omit the details.

The first statement in part (d) is because

$$\int_{-\infty}^{\infty} g(s)\widehat{f}(s)e^{2\pi i s x}\, ds = \int_{-\infty}^{\infty} g(s)\left(\int_{-\infty}^{\infty} f(v)e^{-2\pi i s v}\, dv\right) e^{2\pi i s x}\, ds$$

$$= \int_{-\infty}^{\infty} f(v)\left(\int_{-\infty}^{\infty} g(s)e^{-2\pi i (v-x)s}\, ds\right) dv$$

$$= \int_{-\infty}^{\infty} f(v)\widehat{g}(v - x)\, dy. \qquad (6.32)$$

(The interchange of the integrals here is justified by Fubini's theorem (Proposition 3.5.3).) The second statement is the first one with $x = 0$.

Finally, for part (e) we have, for $f, g \in L^1(\mathbb{R})$ and $s \in \mathbb{R}$,

$$\widehat{f * g}(s) = \int_{-\infty}^{\infty} f * g(x)e^{-2\pi i s x}\, dx = \int_{-\infty}^{\infty}\left(\int_{-\infty}^{\infty} f(x-y)g(y)\, dy\right) e^{-2\pi i s x}\, dx$$

$$= \int_{-\infty}^{\infty} g(y)\left(\int_{-\infty}^{\infty} f(x-y)e^{-2\pi i s x}\, dx\right) dy$$

$$= \int_{-\infty}^{\infty} g(y) F[f(x-y)](s)\, dy = \int_{-\infty}^{\infty} g(y)e^{-2\pi i s y}\widehat{f}(s)\, dy$$

$$= \widehat{f}(s) \int_{-\infty}^{\infty} g(y)e^{-2\pi i s y}\, dy = \widehat{f}(s)\widehat{g}(s). \qquad (6.33)$$

(Again we've used Fubini's theorem; we've also applied part (a)(ii) of the present proposition.) We're done. □

The above proposition may be summarized quite succinctly with the aid of some new notation. Specifically, let's define, for any function f and real numbers a and b, with $b \neq 0$, new functions $f_{(a)}, f_{\{b\}}, f^-$, and $f^\times$ by

$$f_{(a)}(x) = f(x-a), \quad f_{\{b\}}(x) = f(bx), \quad f^-(x) = f(-x), \quad f^\times(x) = xf(x). \tag{6.34}$$

(Note that, for $f_{[\varepsilon]}(x) = \varepsilon^{-1} f(\varepsilon^{-1} x)$ as in (5.94), we have $f_{[\varepsilon]} = \varepsilon^{-1} f_{\{\varepsilon^{-1}\}}$.) Also let $e_s(x) = e^{2\pi i s x}$, as in (6.5), and let $\operatorname{sgn} b$ denote the sign function, as defined in (A.12). Then the proposition tells us that, for $f \in L^1(\mathbb{R})$, all of the following are true:

$$\widehat{e_a f} = (\widehat{f})_{(a)}, \quad \widehat{f_{(a)}} = e_{-a}\widehat{f}, \quad \widehat{f_{\{b\}}} = \operatorname{sgn} b\, (\widehat{f})_{[b]}, \quad \widehat{f^-} = (\widehat{f})^-, \quad \widehat{\overline{f}} = \overline{(\widehat{f})^-}, \tag{6.35}$$

$$f \in \mathrm{PSC}(\mathbb{R}), f' \in L^1(\mathbb{R}) \Rightarrow \widehat{f'} = 2\pi i (\widehat{f})^\times, \quad f^\times \in L^1(\mathbb{R}) \Rightarrow \widehat{f^\times} = \frac{-(\widehat{f})'}{2\pi i}, \tag{6.36}$$

$$g \in L^1(\mathbb{R}) \Rightarrow (\widehat{fg})^- = \widehat{f} * \widehat{g}^- \text{ and } \widehat{f * g} = \widehat{f}\,\widehat{g}. \tag{6.37}$$

The disadvantage of all this compactness is that it means keeping track of quite a few "augmentation symbols" (subscripts, superscripts, and so on, used to denote operations on functions). We'll endeavor to strike the ever-elusive balance between lengthiness of exposition and overwhelmingness of notation.

Let's now turn to some examples, each of which employs different parts of the above proposition.

Example 6.2.1 Find $\widehat{k}$ if

$$k(x) = e^{4\pi i x} e^{-2\pi |x - 1/8|}. \tag{6.38}$$

Solution. We write

$$k(x) = e^{4\pi i x} f\left(x - \frac{1}{8}\right) = e_2(x) f_{(1/8)}(x), \tag{6.39}$$

where $f(x) = e^{-2\pi |x|}$ as in Example 6.1.2 above. By (6.35) and the result of that example, then,

$$\widehat{k}(s) = F[e_2(x) f_{(1/8)}(x)](s) = \widehat{f_{(1/8)}}(s - 2) = e_{-1/8}(s-2)\widehat{f}(s-2)$$

$$= e^{-2\pi i (s-2)/8} \widehat{f}(s-2) = \frac{ie^{-\pi i s/4}}{\pi(1 + (s-2)^2)}, \tag{6.40}$$

since $e^{\pi i/2} = i$.

We now apply the convolution theorem to a somewhat "bogus" example. We undertake this example in the spirit of fun, and because it will be of use to us despite its dubiousness.

Example 6.2.2 Let $\{g_{[\varepsilon]} : \varepsilon > 0\}$ be an approximate identity; let

$$\delta = \lim_{\varepsilon \to 0^+} g_{[\varepsilon]}. \tag{6.41}$$

See Section 5.7; again, δ doesn't really exist as a function, but if it did, what would its Fourier transform be?

Solution. The answer lies in equation (5.115):

$$f * \delta = f. \tag{6.42}$$

Assume we can take the Fourier transform of both sides; then

$$\widehat{f * \delta} = \widehat{f}, \tag{6.43}$$

or by Proposition 6.2.1(e) and the pretense that it applies in the present situation,

$$\widehat{f}\,\widehat{\delta} = \widehat{f}. \tag{6.44}$$

Now divide by $\widehat{f}$ to get

$$\widehat{\delta} = 1. \tag{6.45}$$

So, the Fourier transform of the delta "function" is the constant function 1.

One could have a field day finding holes in the arguments of the above example. But again, it's all in fun.

Our next example will be instrumental in many places: In the presentation of the one following it; in the proof of Theorem 6.4.1 below; in our study of local frequency analysis, later in Chapter 8; and so on. Its exposition will require the following lemma.

Lemma 6.2.1

$$\int_{-\infty}^{\infty} e^{-\pi x^2}\, dx = 1. \tag{6.46}$$

Proof. To evaluate the integral, we multiply it by itself! Specifically, let's denote the integral by I. Then

$$I^2 = I \cdot I = I \int_{-\infty}^{\infty} e^{-\pi y^2}\, dy = \int_{-\infty}^{\infty} I e^{-\pi y^2}\, dy$$

$$= \int_{-\infty}^{\infty} \left(\int_{-\infty}^{\infty} e^{-\pi x^2}\, dx \right) e^{-\pi y^2}\, dy = \int_{-\infty}^{\infty} \int_{-\infty}^{\infty} e^{-\pi(x^2+y^2)}\, dx\, dy$$

$$= \int_0^{\infty} \int_{-\pi}^{\pi} e^{-\pi \rho^2} \rho\, d\rho\, d\theta = \int_{-\pi}^{\pi} d\theta \int_0^{\infty} e^{-\pi \rho^2} \rho\, d\rho$$

$$= 2\pi \int_0^{\infty} e^{-\pi \rho^2} \rho\, d\rho = \int_0^{\infty} e^{-u}\, du = e^{-u}\Big|_0^{\infty} = 1. \tag{6.47}$$

Here we've used Fubini's theorem (Proposition 3.5.3), valid because all integrals in sight converge spectacularly (because of the rapid decay of $e^{-\pi x^2}$ in x). We've also used the polar coordinate change of variables

$$x = \rho \cos \theta, \quad y = \rho \sin \theta, \tag{6.48}$$

which turns $x^2 + y^2$ into $\rho^2(\cos^2 \theta + \sin^2 \theta) = \rho^2$, turns $dx\, dy$ into $\rho\, d\rho\, d\theta$, and turns the domain $\{(x, y): -\infty < x, y < \infty\}$ of integration into $\{(\rho, \theta): -\pi \leq \theta < \pi, 0 \leq \rho < \infty\}$. (Both sets sweep out the entire xy plane, the first in rectangular and the second in polar coordinates.) Finally, to evaluate the integral in ρ, we substituted $u = \pi \rho^2$. (While our original integral I cannot be done by a substitution, the integral in ρ can, because changing to polar coordinates introduces an extra factor of ρ, which becomes part of the "du" when we put $u = \pi \rho^2$!)

Now if $I^2 = 1$, then $I = \pm 1$; but I, being the integral of a positive function, must be positive, so $I = 1$. □

(For an alternative proof of this lemma, see Exercise 6.2.6.)
We now present our promised multipurpose example.

Example 6.2.3 Find $\widehat{G}$ if

$$G(x) = e^{-\pi x^2} \tag{6.49}$$

(Fig. 6.2).

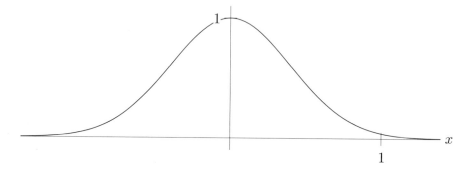

Fig. 6.2 The function G

Solution. We note that

$$G'(x) = \frac{d}{dx} e^{-\pi x^2} = -2\pi x\, e^{-\pi x^2} = -2\pi x G(x). \tag{6.50}$$

Taking the Fourier transform of the left and right sides yields

$$F[G'(x)](s) = -2\pi F[xG(x)](s). \tag{6.51}$$

Applying part (c)(i) of Proposition 6.2.1 to the left side of this and part (c)(ii) to the right (it's readily checked that the conditions for such application are met here), we get

$$2\pi i s \widehat{G}(s) = -2\pi(-2\pi i)^{-1}(\widehat{G})'(s) \tag{6.52}$$

or, rearranging and simplifying,

$$(\widehat{G})'(s) = -2\pi s \widehat{G}(s). \tag{6.53}$$

Let's take a look at what we have. We see that (6.53) is just (6.50) with x replaced by s and G by $\widehat{G}$. So, $\widehat{G}$ satisfies the *same* first-order differential equation as does G. Invoking a basic result concerning uniqueness of solutions to differential equations (or solving (6.50)/(6.53) directly—see Exercise 6.2.7), we conclude that $\widehat{G}$ must, in fact, be a constant *multiple* of G. That is,

$$\widehat{G}(s) = Ke^{-\pi s^2} \tag{6.54}$$

for some constant K.

Now let's determine K. We do so by plugging $s = 0$ into (6.54). We get

$$K = \widehat{G}(0) = \int_{-\infty}^{\infty} G(x)e^{-2\pi i(0)x}\,dx = \int_{-\infty}^{\infty} e^{-\pi x^2}\,dx = 1, \tag{6.55}$$

by Lemma 6.2.1. Putting this value of K back into (6.54) gives

$$\widehat{G}(s) = e^{-\pi s^2}. \tag{6.56}$$

Note that the right side is just $G(s)$. So in short,

$$\widehat{G} = G: \tag{6.57}$$

G equals its Fourier transform! This is a very cool property of G.

Any function g of the form $g(x) = aG(bx)$, with G as above, a complex and nonzero, and $b > 0$, is called a *Gaussian*. From the above example and Proposition 6.2.1(b)(i), it follows that the Fourier transform of a Gaussian *is* a Gaussian.

Remember the fun we had with Example 6.2.2. Let's now have even more, by turning that example around. That is, let's ask: What's the Fourier transform of the function on the *right* side of (6.45)? In answering this question, we'll accomplish five things. First, we'll answer this question. Second, we'll presage another result (Example 6.2.5) that nicely encapsulates the Fourier transform BIG idea, sinusoidally speaking. Third, we'll exemplify Theorem 6.4.1 below (okay, not really—that theorem regards functions in $L^1(\mathbb{R})$, which the function transformed in the following example is not, but still this example illustrates the *gestalt* of that theorem); fourth, we'll establish the framework for a bogus but suggestive proof of that theorem; and fifth, we'll see some of the ideas necessary to legitimize that proof.

Example 6.2.4 What is $\hat{1}$?

Solution. Actually, $\hat{1}$ doesn't exist, since $1 \notin L^1(\mathbb{R})$. But again, we shouldn't let that spoil our festivities.

To "find" $\hat{1}$, we begin with Example 6.2.3, which says

$$F\left[e^{-\pi x^2}\right](s) = e^{-\pi s^2}. \tag{6.58}$$

We note that this and Proposition 6.2.1(b)(i) give

$$F\left[e^{-\pi(\varepsilon x)^2}\right](s) = \varepsilon^{-1} e^{-\pi(s/\varepsilon)^2} = G_{[\varepsilon]}(s) \tag{6.59}$$

for $\varepsilon > 0$, where again $G(s) = e^{-\pi s^2}$.

Now G is clearly bounded (by 1); it's in $L^1(\mathbb{R})$ and has mass 1 by Lemma 6.2.1. So $\{G_{[\varepsilon]} : \varepsilon > 0\}$ *is* an approximate identity, whence

$$\delta = \lim_{\varepsilon \to \infty} G_{[\varepsilon]}. \tag{6.60}$$

So (6.59) implies

$$\lim_{\varepsilon \to 0^+} F\left[e^{-\pi(\varepsilon x)^2}\right] = \delta. \tag{6.61}$$

(We've dropped the (s) from either side for clarity.)

On the other hand, certainly

$$\lim_{\varepsilon \to 0^+} e^{-\pi(\varepsilon x)^2} = e^0 = 1 \tag{6.62}$$

for any x. So, assuming—which we do because we're only pretending, so we might as well make things as easy as possible—that the limit and the Fourier transform in equation (6.61) can be interchanged (and that the resulting limit may be taken pointwise), that equation gives us

$$F[1(x)] = \delta : \tag{6.63}$$

the Fourier transform of the constant function 1 is δ!

Finally, if we believe in $\hat{f}(s)$ as an indicator of frequency content or, more specifically, as a gauge of the extent to which f comprises the complex exponential $e_s(x) = e^{2\pi i s x}$ (cf. the arguments preceding Definition 6.1.1), then we should expect the Fourier transform of the complex exponential e_{s_0} to be concentrated at the single point $s = s_0$. (It stands to reason that e_{s_0} should be the *only* complex exponential contributing to the makeup of e_{s_0}.) The following example bears out this expectation and thereby supports this belief.

Example 6.2.5 What is $\widehat{e_{s_0}}$?

Solution. We have $e_{s_0}(x) = e^{2\pi i s_0 x} 1(x)$. So by Proposition 6.2.1(a)(i), $\widehat{e_{s_0}}(s) = \hat{1}(s - s_0)$, so by the result of the previous example,

$$\widehat{e_{s_0}}(s) = \delta(s - s_0). \tag{6.64}$$

Since δ has all of its "bulk" at $s = 0$, (6.64) does indeed tell us that $\widehat{e_{s_0}}$ is perfectly concentrated at $s = s_0$.

Exercises

In Exercises 6.2.1–6.2.4, find the Fourier transform $\widehat{f}$ of the given function f, *using the indicated method.*

6.2.1 $f(x) = \cos(\pi x/2)e^{-6\pi|x|}$. Write the cosine factor as a linear combination of complex exponentials; then apply parts of Proposition 6.2.1 to the result of Example 6.1.2.

6.2.2 $f(x) = x\chi_{[-1,1]}(x)$. Apply Proposition 6.2.1(c)(ii) to the result of Example 6.1.1.

6.2.3 $f(x) = x^2\chi_{[-3,3]}(x)$. Apply Proposition 6.2.1(c)(ii) twice to the result of Example 6.1.1.

6.2.4 $f(x) = (2b - |x|)\chi_{[-2b,2b]}(x)$ $(b > 0)$. Using Exercise 5.2.2, express f as $g * g$ for a certain g. Then use the convolution theorem (Proposition 6.2.1(e)).

6.2.5 Prove parts (a)(ii) and (b)(iii) of Proposition 6.2.1.

6.2.6 Prove Lemma 6.2.1 using the fact that $\Gamma(\frac{1}{2}) = \sqrt{\pi}$ (cf. Exercise 5.6.6), and the change of variable $u = \sqrt{x/\pi}$ in the integral defining $\Gamma(\frac{1}{2})$.

6.2.7 Solve (6.50) for G directly, using single-variable separation arguments. That is, put $y = G$, so that (6.50) reads $dy/dx = -2\pi xy$; get all y's on one side and x's on the other; antidifferentiate; solve. Check your answer against (6.54).

6.3 LOW-IMPACT FOURIER TRANSFORMS (INTEGRATION BY DIFFERENTIATION)

Proposition 6.2.1(c)(i) expresses the general principle (see also Lemma 1.7.1) that Fourier transformation turns differentiation into multiplication by a constant times the independent variable. And multiplication is, as a rule, much easier to undo than is differentiation. This explains, or at any rate is one way of understanding, the extraordinary power of Fourier analysis as a tool in the study of differential equations. See Chapter 2 and Sections 7.1–7.3.

Here, we wish to show that the flow of information can also travel in the opposite direction. That is, one can often apply the theory of differential equations to facilitate certain calculations in Fourier analysis—in particular, the evaluation of certain Fourier transforms. Actually, we've already seen this: In Example 6.2.3, we used Proposition 6.2.1(c)(i,ii) to deduce, from the differential equation (6.50) for G, the (same) equation (6.53) for $\widehat{G}$. We'd like to make further use of this same idea. To do so, we will first ease a rather severe restriction, namely the requirement, in Proposition 6.2.1(c)(i), that f be continuous.

Proposition 6.3.1 Let $f \in \mathrm{PS}(\mathbb{R})$; suppose also that $f, f' \in L^1(\mathbb{R})$. Assume there are only finitely many points $x_1, x_2, \ldots, x_N \in \mathbb{R}$ where f' fails to exist. Then

$$F[f'(x)](s) = 2\pi i s \widehat{f}(s) - \sum_{n=1}^{N} J_f(x_n) e^{-2\pi i s x_n}, \qquad (6.65)$$

where

$$J_f(x_n) = f(x_n^+) - f(x_n^-) \qquad (6.66)$$

($=$ the jump in f at x_n).

Proof. We'll supply a proof valid only for the case $N = 1$—that is, the case where f' fails to exist at exactly point x_1. From this case one readily deduces the general one.

We proceed in two steps. The first is to prove the desired result in the case $s = 0$. So we want to prove

$$F[f'(x)](0) = 0 \cdot \widehat{f}(0) - J_f(x_1) = f(x_1^-) - f(x_1^+). \qquad (6.67)$$

We do so as follows:

$$\begin{aligned} F[f'(x)](0) &= \int_{-\infty}^{\infty} f'(x) e^{-2\pi i (0) x} \, dx = \int_{-\infty}^{x_1} f'(x) \, dx + \int_{x_1}^{\infty} f'(x) \, dx \\ &= f(x) \Big|_{-\infty}^{x_1} + f(x) \Big|_{x_1}^{\infty} = f(x_1^-) - f(x_1^+) \end{aligned} \qquad (6.68)$$

by the fundamental theorem of calculus (that $\lim_{x \to \pm\infty} f(x) = 0$ follows from the assumptions on f; compare with the proof of Proposition 6.2.1(c)(i)). So we're done with this step.

For our second step, we let $s \in \mathbb{R}$ be arbitrary. We also write $h(x) = e^{-2\pi i s x} f(x)$; then by (6.67) with h in place of f and by the product rule,

$$\begin{aligned} h(x_1^-) - h(x_1^+) = F[h'(x)](0) &= F[-2\pi i s \, e^{-2\pi i s x} f(x) + e^{-2\pi i s x} f'(x)](0) \\ &= -2\pi i s F[e^{-2\pi i s x} f(x)](0) + F[e^{-2\pi i s x} f'(x)](0) \\ &= -2\pi i s \widehat{f}(0 - (-s)) + F[f'(x)](0 - (-s)) \\ &= -2\pi i s \widehat{f}(s) + F[f'(x)](s). \end{aligned} \qquad (6.69)$$

For the second equality we applied the product rule to the differentiation of h; for the third we used linearity of the Fourier transform; for the fourth we used Proposition 6.2.1(a)(ii); for the last we just cleaned up. But by definition of h and the continuity of complex exponentials,

$$h(x_1^-) - h(x_1^+) = (f(x_1^-) - f(x_1^+)) e^{-2\pi i s x_1}, \qquad (6.70)$$

so (6.69) gives

$$F[f'(x)](s) = 2\pi i s \widehat{f}(s) - (f(x_1^+) - f(x_1^-)) e^{-2\pi i s x_1}, \qquad (6.71)$$

which is the desired result in the case $N = 1$. □

We now apply Proposition 6.3.1 to the evaluation of Fourier transforms of some piecewise smooth, discontinuous functions f. The general strategy in each case will entail, first, finding a differential equation satisfied by f almost everywhere; second, taking Fourier transforms on both sides of that equation; third, using the proposition to derive an *algebraic* equation for $\hat{f}$; and fourth, solving that equation. This procedure will allow us to find the desired Fourier transforms *without ever having to integrate!!* (And there was much rejoicing.)

Example 6.3.1 Find $\hat{f}$ (again) if $f = \chi_{[-b,b]}$ ($b > 0$).

Solution. The only bad points are at $x = \pm b$; moreover $J_f(-b) = 1 - 0 = 1$ and $J_f(b) = 0 - 1 = -1$. Since f is piecewise constant, $f'(x) = 0$ almost everywhere. The Fourier transform of f' is therefore identically zero, so by Proposition 6.3.1,

$$0 = F[f'(x)](s) = 2\pi i s \hat{f}(s) - \left(1 \cdot e^{-2\pi i s(-b)} + (-1) \cdot e^{-2\pi i s(b)}\right). \quad (6.72)$$

Then

$$\hat{f}(s) = \frac{e^{2\pi i b s} - e^{-2\pi i b s}}{2\pi i s} = \frac{\sin 2\pi b s}{\pi s} \quad (6.73)$$

for $s \neq 0$; $\hat{f}(0)$ may be obtained by letting $s \to 0$. Thus we get the same result as in Example 6.1.1.

As our earlier direct computation of $F[\chi_{[-b,b]}(x)]$ was not particularly difficult or nasty, one could argue that our new, indirect method gains us little in this case. Still, the above example serves well to illustrate this method. And we now present some examples where the gain is greater.

Example 6.3.2 Find $\hat{f}$ if

$$f(x) = \begin{cases} e^x(\cos\sqrt{3}\,x + \sin\sqrt{3}\,x) & \text{if } x \leq 0, \\ e^{-2x} & \text{if } x \geq 0. \end{cases} \quad (6.74)$$

Solution. The trick is to differentiate f enough times that we get back a constant times f (almost everywhere) on the right side. We compute

$$f'(x) = \begin{cases} e^x\left((1+\sqrt{3})\cos\sqrt{3}\,x + (1-\sqrt{3})\sin\sqrt{3}\,x\right) & \text{if } x < 0, \\ -2e^{-2x} & \text{if } x > 0, \end{cases} \quad (6.75)$$

$$f''(x) = \frac{d}{dx}f'(x) = \begin{cases} 2e^x\left((-1+\sqrt{3})\cos\sqrt{3}\,x - (1+\sqrt{3})\sin\sqrt{3}\,x\right) & \text{if } x < 0, \\ 4e^{-2x} & \text{if } x > 0, \end{cases} \quad (6.76)$$

314 FOURIER TRANSFORMS AND FOURIER INTEGRALS

$$f'''(x) = \frac{d}{dx}f''(x) = \begin{cases} -8e^x\left(\cos\sqrt{3}\,x + \sin\sqrt{3}\,x\right) & \text{if } x < 0, \\ -8e^{-2x} & \text{if } x > 0; \end{cases} \quad (6.77)$$

that is,

$$f'''(x) = -8f(x) \quad (6.78)$$

(except at $x = 0$).

Taking the Fourier transform on both sides of this and applying Proposition 6.3.1 repeatedly give (since $e^{-2\pi i s(0)} = 1$)

$$\begin{aligned}
-8\widehat{f}(s) &= F[f'''(x)](s) = 2\pi i s F[f''(x)](s) - J_{f''}(0) \\
&= 2\pi i s(2\pi i s F[f'(x)](s) - J_{f'}(0)) - J_{f''}(0) \\
&= 2\pi i s(2\pi i s(2\pi i s \widehat{f}(s) - J_f(0)) - J_{f'}(0)) - J_{f''}(0) \\
&= (2\pi i s)^3 \widehat{f}(s) - (2\pi i s)^2 J_f(0) - 2\pi i s J_{f'}(0) - J_{f''}(0). \quad (6.79)
\end{aligned}$$

We see from our formulas for f, f', f'' that $J_f(0) = 1 - 1 = 0$, $J_{f'}(0) = -2 - (1+\sqrt{3}) = -3 - \sqrt{3}$, $J_{f''}(0) = 4 - 2(-1+\sqrt{3}) = 6 - 2\sqrt{3}$. So (6.79) gives

$$\begin{aligned}
\widehat{f}(s) &= \frac{(2\pi i s)^2(0) + 2\pi i s(-3-\sqrt{3}) + (6-2\sqrt{3})}{8 + (2\pi i s)^3} \\
&= \frac{3(1-i\pi s) - \sqrt{3}(1+i\pi s)}{4(1-i\pi^3 s^3)}. \quad (6.80)
\end{aligned}$$

Our low-impact technique is not omnipotent: It won't work unless $\widehat{f}$ satisfies some reasonable differential equation. In the previous example, this happened by sheer coincidence. (!!) On the other hand, if $f \in L^1(\mathbb{R})$ is *piecewise polynomial*, then for some $m \in \mathbb{Z}^+$, the differential equation $f^{(m)}(x) = 0$ will obtain for almost all x. Thus, our method will apply to *any* such f, as it did in Example 6.3.1 and as it does again in our next one.

Example 6.3.3 Find $\widehat{f}$ if

$$f(x) = \begin{cases} 0 & \text{if } x < -1, \\ 3 - (1+x)^3 & \text{if } -1 \leq x \leq 0, \\ 3 - (1-x)^3 & \text{if } 0 \leq x \leq 1, \\ 0 & \text{if } x > 1. \end{cases} \quad (6.81)$$

Solution. Since f is piecewise polynomial, with each "piece" of degree at most 3, $f^{(4)}(x)$ equals zero wherever it's defined. Moreover, from direct computation of the first three derivatives of f, we find readily that

$$\begin{aligned}
J_{f'''}(-1) &= -6 - 0 = -6, & J_{f'''}(0) &= 6 - (-6) = 12, & J_{f'''}(1) &= 0 - 6 = -6, \\
J_{f''}(-1) &= 0 - 0 = 0, & J_{f''}(0) &= -6 - (-6) = 0, & J_{f''}(1) &= 0 - 0 = 0, \\
J_{f'}(-1) &= 0 - 0 = 0, & J_{f'}(0) &= 3 - (-3) = 6, & J_{f'}(1) &= 0 - 0 = 0, \\
J_f(-1) &= 3 - 0 = 3, & J_f(0) &= 2 - 2 = 0, & J_f(1) &= 0 - 3 = -3. \quad (6.82)
\end{aligned}$$

(We leave the details to the reader.) So by Proposition 6.3.1,

$$\begin{aligned}
0 &= F[f^{(4)}(x)](s) = 2\pi is F[f'''(x)](s) + 6(e^{2\pi is} + e^{-2\pi is}) - 12 \\
&= 2\pi is(2\pi is F[f''(x)](s)) + 6(e^{2\pi is} + e^{-2\pi is}) - 12 \\
&= -4\pi^2 s^2 F[f''(x)](s) + 6(e^{2\pi is} + e^{-2\pi is}) - 12 \\
&= -4\pi^2 s^2 (2\pi is F[f'(x)](s) - 6) + 6(e^{2\pi is} + e^{-2\pi is}) - 12 \\
&= -8i\pi^3 s^3 F[f'(x)](s) + 24\pi^2 s^2 + 6(e^{2\pi is} + e^{-2\pi is}) - 12 \\
&= -8i\pi^3 s^3 (2\pi is \widehat{f}(s) - 3(e^{2\pi is} - e^{-2\pi is})) + 24\pi^2 s^2 \\
&\quad + 6(e^{2\pi is} + e^{-2\pi is}) - 12;
\end{aligned}$$ (6.83)

solving (DIY) yields

$$\widehat{f}(s) = \frac{12\pi^3 s^3 \sin 2\pi s - 6\pi^2 s^2 + 3(1 - \cos 2\pi s)}{4\pi^4 s^4}.$$ (6.84)

See the exercises below for more fun examples. (More examples, that is, and they're fun; not that they're necessarily more fun. Though certainly they're no less fun.)

Exercises

In each of the following exercises, use the methods of this section to compute the Fourier transform of the given function f.

6.3.1
$$f(x) = \begin{cases} 0 & \text{if } -\infty < x \leq -1, \\ x + 1 & \text{if } -1 < x \leq 0, \\ x^2 - 2x + 1 & \text{if } 0 < x \leq 1, \\ 0 & \text{if } 1 < x < \infty. \end{cases}$$

6.3.2
$$f(x) = \begin{cases} 0 & \text{if } -\infty < x \leq -2, \\ x^2 + 4x + 4 & \text{if } -2 < x \leq -1, \\ 2 - x^2 & \text{if } -1 < x \leq 1, \\ x^2 - 4x + 4 & \text{if } 1 < x \leq 2, \\ 0 & \text{if } 2 < x < \infty. \end{cases}$$

6.3.3 $f(x) = (2b - |x|)\chi_{[-b,b]}(x)$ ($b > 0$) (again).

6.3.4 $f(x) = (1 - x^2)^3 \chi_{[-1,1]}(x)$.

6.3.5 $f(x) = e^{-2\pi|x|}$ (again).

6.3.6 $f(x) = e^{-2\pi|x|}(\cos 2\pi|x| + \sin 2\pi|x|)$. ANSWER: $\widehat{f}(s) = 4/(\pi(4 + s^4))$.

6.4 FOURIER INVERSION ON $FL^1(\mathbb{R})$

We now present our first major theorem concerning frequency decompositions of aperiodic functions.

Theorem 6.4.1 (Fourier inversion on $FL^1(\mathbb{R})$)

(a) *Suppose $f \in L^1(\mathbb{R})$ and $\hat{f}$ is bounded. If f is continuous at x, then*

$$f(x) = \lim_{\varepsilon \to 0^+} \int_{-\infty}^{\infty} \hat{f}(s) e^{-\pi(\varepsilon s)^2} e^{2\pi i s x}\, ds. \tag{6.85}$$

(b) *Suppose $f \in FL^1(\mathbb{R})$ (recall (6.22)). Then the frequency decomposition (6.3) holds for almost all $x \in \mathbb{R}$. The integral on the right side is uniformly convergent; that is,*

$$I_b(x) = \int_{-b}^{b} \hat{f}(s) e^{2\pi i s x}\, ds \tag{6.86}$$

converges uniformly on $\mathbb{R}$ as $b \to \infty$.

(c) *If $f \in C(\mathbb{R}) \cap FL^1(\mathbb{R})$, then (6.3) is true for all $x \in \mathbb{R}$ (and the integral there converges uniformly to f on $\mathbb{R}$).*

Proof. Since $\hat{f}(s) = \hat{f}(s)1(s)$, $\hat{1} = \delta$, 1 is an even function, and $f * \delta = f$, the composition theorem (Proposition 6.2.1(d)) gives

$$\int_{-\infty}^{\infty} \hat{f}(s) e^{2\pi i s x}\, ds = \int_{-\infty}^{\infty} \hat{f}(s) 1(s) e^{2\pi i s x}\, ds = \int_{-\infty}^{\infty} f(v) \hat{1}(v-x)\, dv$$

$$= \int_{-\infty}^{\infty} f(v) \delta(x-v)\, dv = f * \delta(x) = f(x). \tag{6.87}$$

So part (c) is proved.

Well no, not really. Our "proof" isn't a proof; it involves (i) application of a theorem (the composition theorem) whose hypotheses aren't met, (ii) the Fourier transform of a function (the constant function 1) that's not in $L^1(\mathbb{R})$, and (iii) a function (δ) that doesn't exist. Still, the "proof" does communicate quite well the big ideas, which we now formalize.

We consider first part (a). So let $f \in L^1(\mathbb{R})$; then by the composition theorem (see also (6.37)),

$$\int_{-\infty}^{\infty} \hat{f}(s) e^{-\pi(\varepsilon s)^2} e^{2\pi i s x}\, ds = \int_{-\infty}^{\infty} f(v) F[e^{-\pi(\varepsilon s)^2}](v-x)\, dv = f * \widehat{g^-}(x), \tag{6.88}$$

where $g(s) = e^{-\pi(\varepsilon s)^2} = G(\varepsilon s)$, G being as always the Gaussian $G(x) = e^{-\pi x^2}$. Now it follows from Example 6.2.3, from the evenness of g, and from Proposition

6.2.1(b)(i) with $b = \varepsilon$ that $\widehat{g^-} = \widehat{g} = G_{[\varepsilon]}$. So (6.88) yields

$$\lim_{\varepsilon \to 0^+} \int_{-\infty}^{\infty} \widehat{f}(s) e^{-\pi(\varepsilon s)^2} e^{2\pi i s x} \, ds = \lim_{\varepsilon \to 0^+} f * G_{[\varepsilon]}(x), \qquad (6.89)$$

provided the limit on the right exists. But if f is bounded and is continuous at x, then this limit not only exists but equals $f(x)$, by Proposition 5.7.1(a) and the fact, noted in the course of Example 6.2.4, that $\{G_{[\varepsilon]} : \varepsilon > 0\}$ is an approximate identity. Thus, part (a).

For part (b), let's suppose that $f \in FL^1(\mathbb{R})$. Then $\widehat{f} \in L^1(\mathbb{R})$, so

$$\int_{-\infty}^{\infty} |\widehat{f}(s)| \, ds < \infty. \qquad (6.90)$$

From this, uniform convergence of the integral in (6.3) follows: The argument here is essentially the one we used earlier, to deduce uniform convergence of S_N^f from absolute convergence of the sum of the $|c_n(f)|$'s. (See the proof of Theorem 1.7.1.)

Further, note that $f * G_{[\varepsilon]}$ converges to f in the $L^1(\mathbb{R})$ norm by Proposition 5.7.1(b). So suppose we can show that $f * G_{[\varepsilon]}(x)$ converges *pointwise* to the integral, call it $f^*(x)$, in (6.3). Then we'll be done for the following reason. From an analog of Proposition 3.2.1(h), it will follow that $f^*(x) = f(x)$ almost everywhere. Moreover, suppose f is continuous. Since, by Proposition 6.1.1, f^* is too, $f^* - f$ will be continuous and zero almost everywhere. But then $f^* - f$ must be zero *everywhere* (by an analog of Proposition 3.1.1), meaning (6.3) holds everywhere, as claimed for such f.

To show what we need to show, we recall (6.89). Since the integrand of the integral on the left there is bounded in absolute value by $|\widehat{f}(s)|$ and since $\widehat{f} \in L^1(\mathbb{R})$, we may apply the Lebesgue dominated convergence theorem (Proposition 3.5.1) to conclude that, for any x, this integral converges to

$$\int_{-\infty}^{\infty} \left(\lim_{\varepsilon \to 0^+} \widehat{f}(s) e^{-\pi(\varepsilon s)^2} e^{2\pi i s x} \right) ds = \int_{-\infty}^{\infty} \widehat{f}(s) e^{2\pi i s x} \, ds \qquad (6.91)$$

as $\varepsilon \to 0^+$. So by (6.89), $f * G_{[\varepsilon]}(x)$ does the same, so we *are* done. □

Parts (b) and (c) of the above theorem admit the following very useful reformulation.

Theorem 6.4.2 (Fourier inversion on $FL^1(\mathbb{R})$, reprise) We have

$$\widehat{\widehat{f}}(-x) = f(x) \qquad (6.92)$$

($\widehat{\widehat{f}}$ denotes the Fourier transform of the Fourier transform of f) for almost all x if $f \in FL^1(\mathbb{R})$, and for all x if $f \in C(\mathbb{R}) \cap FL^1(\mathbb{R})$.

Moreover, let $f, g \in L^1(\mathbb{R})$. Then $g(x) = \widehat{f}(x)$ almost everywhere if and only if $\widehat{g}(s) = f(-s)$ almost everywhere. If f is continuous, then the former condition implies that the latter holds everywhere; if g is continuous then the latter condition implies that the former holds everywhere.

Proof. The right side of (6.3) equals

$$\int_{-\infty}^{\infty} \widehat{f}(s) e^{-2\pi i s(-x)} \, ds = F\bigl[\widehat{f}(s)\bigr](-x) = \widehat{\widehat{f}}(-x), \tag{6.93}$$

so parts (b) and (c) of Theorem 6.4.1 yield the stated results concerning $\widehat{\widehat{f}}$.

To get the remaining results, we argue as follows. If $g(x) = \widehat{f}(x)$ almost everywhere, then

$$\widehat{g}(s) = \widehat{\widehat{f}}(s) \tag{6.94}$$

everywhere. The right side equals $f(-s)$ almost everywhere, and everywhere if f is continuous, by what we just proved. This gives us the penultimate claim of our theorem.

Finally, if $\widehat{g}(s) = f(-s)$ almost everywhere, then $f(s) = \widehat{g}(-s)$ almost everywhere, so

$$\widehat{f}(x) = F\bigl[\widehat{g}(-s)\bigr](x) = \widehat{\widehat{g}}(-x) \tag{6.95}$$

everywhere (the last step is by Proposition 6.2.1(b)(ii)). The right side equals $g(x)$ almost everywhere, and everywhere if f is continuous, again by what we just proved, so we're done. $\square$

The above theorem tells us that the association $f \to (\widehat{f})^-$ undoes the association $f \to \widehat{f}$ under suitable circumstances. In light of this, we call $(\widehat{f})^-$ the *inverse Fourier transform* of f. Note by Proposition 6.2.1(b)(ii) that $(\widehat{f})^-$ is the same as $\widehat{f^-}$.

If $f, g \in L^1(\mathbb{R})$ are related as in the theorem, then the pair (f, g) is called a *Fourier transform pair*. We've already encountered examples of these. For instance, in Examples 6.2.2 and 6.2.4 we saw that $\widehat{1} = \delta$ and $\widehat{\delta} = 1^-$ (1^- being the same as 1). (Again, they're not "real" examples, but they'll do.) So $(\delta, 1)$ is a Fourier transform pair. Also remember Example 6.2.1; there we had $\widehat{G} = G = G^-$. That is, (G, G) is a Fourier transform pair.

A practical consequence of the theorem is this: *If you know what g is the Fourier transform of, then you know what the Fourier transform of g is* (for nice enough g). Here's an illustration of this.

Example 6.4.1 Find $\widehat{h}$ if

$$h(x) = \frac{e^{-\pi i x/4}}{1 + (x-2)^2}. \tag{6.96}$$

Solution. Direct evaluation of the integral

$$\int_{-\infty}^{\infty} \frac{e^{-\pi i x/4}}{1+(x-2)^2} e^{-2\pi i s x}\, dx \tag{6.97}$$

is a bit painful. On the other hand, let's recall that we've already seen h in a Fourier transform context. Namely, in Example 6.2.1 we saw that, if $k(x) = e^{4\pi i x} e^{-2\pi|x-1/8|}$, then

$$\widehat{k} = i\pi^{-1} h. \tag{6.98}$$

Now $h \in L^1(\mathbb{R})$ ($|h(x)| = 1/(1+(x-2)^2)$, which has integral $\arctan(x-2)|_{-\infty}^{\infty} = \pi < \infty$ over $\mathbb{R}$), and k is continuous, so by Theorem 6.4.2 and (6.98),

$$\widehat{i\pi^{-1} h}(s) = k(-s) \tag{6.99}$$

for all s. That is,

$$\widehat{h}(s) = -i\pi\, k(-s) = -i\pi\, e^{4\pi i(-s)} e^{-2\pi|-s-1/8|} = -i\pi\, e^{-4\pi i s} e^{-2\pi|s+1/8|} \tag{6.100}$$

since $i^{-1} = -i$ and $|-u| = |u|$.

Similarly (but more simply) we have the following, which results (DIY) from Fourier inversion applied to Example 6.1.2.

Example 6.4.2

$$F\left[\frac{1}{1+x^2}\right](s) = \pi e^{-2\pi|s|}. \tag{6.101}$$

We could also have deduced (6.101) from (6.100) (or vice versa) using Proposition 6.2.1(a) and the relationship between $e^{-\pi i x/4}/(1+(x-2)^2)$ and $(1+x^2)^{-1}$. Doing so would really just be the flip side, meaning the side to which Fourier inversion takes us, of our arguments relating Example 6.1.2 to Example 6.2.1. The degrees of symmetry and duality here are staggering.

The above two examples demonstrate quite dramatically the power of Fourier inversion. See also Example 6.3.2: The function on the right side of (6.80) is seen to be in $L^1(\mathbb{R})$, so we can use Theorem 6.4.2 to obtain the Fourier transform of that function.

On the other hand, recall Example 6.1.1: If we put $g = \widehat{\chi_{[-b,b]}}$, then, as we've noted, $g \notin L^1(\mathbb{R})$. So we can't even *make sense* of $\widehat{g}$. (Similar problems arise in the context of Example 6.3.3: The function on the right side of (6.84) is not in $L^1(\mathbb{R})$, because it only decays about like $|s^3/s^4| = 1/|s|$ as $|s| \to \infty$.)

Or can we? Maybe we can, if we can find some other place, beyond $L^1(\mathbb{R})$, in which to do Fourier transforms. The following will, shortly, help lead us to such a place.

Proposition 6.4.1 *Suppose* $f, g \in FL^1(\mathbb{R})$. *Then* $f, g, \widehat{f}, \widehat{g} \in L^2(\mathbb{R})$, *and:*

(a) **(the convolution-inversion theorem)** $f * g(x) = \widehat{(\widehat{f}\widehat{g})}(-x)$ for all $x \in \mathbb{R}$;

(b) **(the product theorem)** $\widehat{fg}(s) = \widehat{f} * \widehat{g}(s)$ for all $s \in \mathbb{R}$;

(c) **(the Plancherel theorem)** $\langle f, g \rangle = \langle \widehat{f}, \widehat{g} \rangle$. In particular, $||f|| = ||\widehat{f}||$.

Proof. Let f and g be as stipulated. Then certainly $f, g, \widehat{f}, \widehat{g} \in L^1(\mathbb{R})$. Moreover, $\widehat{f}$ and $\widehat{g}$ are, by Proposition 6.1.1, bounded; consequently, by that same proposition in conjunction with Theorem 6.4.1(b), f and g are equal almost everywhere to bounded functions. So $f, g, \widehat{f}, \widehat{g} \in L^2(\mathbb{R})$ by Corollary 3.5.1(b).

So $f * g$ is a convolution of functions in $L^1(\mathbb{R}) \cap L^2(\mathbb{R})$, and is therefore in $C(\mathbb{R}) \cap L^1(\mathbb{R})$, by Propositions 5.3.1 and 5.6.1. Further, $\widehat{f * g}$ is in $L^1(\mathbb{R})$ since, by the convolution theorem (Proposition 6.2.1(e)), it equals the product $\widehat{f}\widehat{g}$ of two functions in $L^2(\mathbb{R})$. So, by Fourier inversion (Theorem 6.4.2) and the convolution theorem again, we have

$$f * g(x) = F[\widehat{f * g}(s)](-x) = F[\widehat{f}(s)\widehat{g}(s)](-x), \qquad (6.102)$$

for $x \in \mathbb{R}$. Thus part (a) is proved.

Part (b) results from part (a), with f replaced by $\widehat{f}$ and g by $\widehat{g}$. To get the first statement of part (c) we observe that, by part (b),

$$\langle f, g \rangle = \int_{-\infty}^{\infty} f(x)\overline{g(x)}\,dx = \widehat{(f\overline{g})}(0) = \widehat{f} * \widehat{\overline{g}}(0) = \int_{-\infty}^{\infty} \widehat{f}(y)\widehat{\overline{g}}(0-y)\,dy$$

$$= \int_{-\infty}^{\infty} \widehat{f}(y)\overline{\widehat{g}(y)}\,dy = \langle \widehat{f}, \widehat{g} \rangle \qquad (6.103)$$

(the penultimate step by Proposition 6.2.1(b)(iii)).

Putting $g = f$ into the first statement of part (c) gives the second, and we're done. □

The full power of the above proposition will not be realized until we generalize it to $L^2(\mathbb{R})$, as we will in the next section.

Exercises

6.4.1 Find $\widehat{f}$ if $f(x) = 1/(1 + (ax+b)^2)$ $(a, b \in \mathbb{R}, a \neq 0)$.

6.4.2 Find $\widehat{f}$ if $f(x) = x/(1+x^2)^2$. Hint: First find an antiderivative of f.

6.4.3 Use Fourier inversion and Exercise 6.3.6 to find $\widehat{g}$ if

$$g(x) = \frac{1}{1+x^4}.$$

Hint: $1/(1+x^4) = 4/(4 + (\sqrt{2}\,x)^4)$. Now use Proposition 6.2.1(b)(i).

6.4.4 Show that, if $f, g \in FL^1(\mathbb{R})$ and $\widehat{f}(s) = \widehat{g}(s)$ for all s, then $f(x) = g(x)$ for almost all x, and for all x if f and g are continuous. Hint: Theorem 6.4.1.

6.4.5 Show that $FL^1(\mathbb{R})$ is closed under Fourier transformation, under products, and under convolution. Hint: A product of functions in $L^2(\mathbb{R})$ is in $L^1(\mathbb{R})$. Also use the convolution theorem and the product theorem from this section.

6.4.6 Let G be as in Example 6.2.3; define $G_b(x) = b^{-1/2}G(b^{-1/2}x)$ for $b > 0$. Show that $\{G_b : b > 0\}$ is a convolution semigroup; that is, show that $G_b * G_c = G_{b+c}$ for all b and c. To do so, use the result of Exercise 6.4.4 together with parts of Proposition 6.2.1.

6.4.7 Let
$$f(x) = \frac{1}{\pi(1+x^2)}$$
and, for $b > 0$, $f_{[b]}(x) = b^{-1}f(b^{-1}x)$ as always.
   a. Find $\widehat{f_{[b]}}$.
   b. Find the Fourier transform of $f_{[b]} * f_{[c]}$ for $b, c > 0$.
   c. Use Exercise 6.4.4 to show that $\{f_{[b]} : b > 0\}$ is a convolution semigroup, meaning $f_{[b]} * f_{[c]} = f_{[b+c]}$ for all positive b, c.

6.4.8 Let $f, g, h, k \in C(\mathbb{R}) \cap FL^1(\mathbb{R})$. Suppose
$$\widehat{f}(s) = (1+s^2)e^{-s^2}, \quad \widehat{g}(s) = \frac{1}{1+s^2}e^{-s^2}, \quad \widehat{h}(s) = \frac{1}{1+s^2}, \quad \widehat{k}(s) = s^3 e^{-s^2}.$$
Without actually computing f, g, h, or k (although you can do so to check your work, if you'd like), show that:
   a. $f * h * h(x) = g(x)$ for all x.
   b. $\pi i x f(x) = k(x)$ for all x.
   c. $\langle f, g \rangle = \sqrt{\pi/2}$.
Hints: for part c, use the Plancherel theorem. The other parts can be done using Exercise 6.4.4.

6.5 THE FOURIER TRANSFORM AND FOURIER INVERSION ON $L^2(\mathbb{R})$

If we are to have any hope of establishing a Fourier inversion result on $L^2(\mathbb{R})$, then of course we'll first need to define a *Fourier transform* there. This, as mentioned before (but it bears repeating), is problematic, because $L^2(\mathbb{R}) \not\subset L^1(\mathbb{R})$. Fortunately, these difficulties may be resolved by means of a certain limiting process. We need:

Proposition 6.5.1 *Let $p = 1$ or 2. Given $f \in L^p(\mathbb{R})$ and $N \in \mathbb{Z}^+$, define*
$$^N f = (f\chi_{[-N,N]}) * G_{[1/N]}, \tag{6.104}$$
where G is the Gaussian of Example 6.2.3 and $G_{[1/N]}$ is as described by (5.94). Then both $^N f$ and $\widehat{^N f}$ belong to $L^1(\mathbb{R}) \cap L^2(\mathbb{R})$ for any N. Moreover, the $^N f$'s converge to

f in the $L^p(\mathbb{R})$ norm, and the $\widehat{{}^N f}$'s form a Cauchy sequence in $L^2(\mathbb{R})$ with respect to the mean square norm.

Proof. Let p, f, and N be as stated. Because f is in $L^p(\mathbb{R})$, so is $f\chi_{[-N,N]}$, by Proposition 3.5.2. In fact, because it has compact support, $f\chi_{[-N,N]}$ is in $L^1(\mathbb{R})$ always (whether $f \in L^1(\mathbb{R})$ or $f \in L^2(\mathbb{R})$), by Corollary 3.5.1(c). Now G is in $L^1(\mathbb{R}) \cap L^2(\mathbb{R})$ and therefore so is $G_{[1/N]}$ for any N; but then ${}^N f$, being a convolution of a function in $L^1(\mathbb{R})$ with one in $L^1(\mathbb{R}) \cap L^2(\mathbb{R})$, is itself in $L^1(\mathbb{R}) \cap L^2(\mathbb{R})$, by Propositions 5.6.1 and 5.6.2.

To show $\widehat{{}^N f}$ belongs to $L^1(\mathbb{R}) \cap L^2(\mathbb{R})$, we observe that, by Propositions 6.1.1 and 6.2.1(e),

$$\left|\widehat{{}^N f}(s)\right| = \left|F[f(x)\chi_{[-N,N]}(x)](s)\, F[G_{[1/N]}(x)](s)\right|$$
$$\leq \|f\chi_{[-N,N]}\|_1 \, |F[G_{[1/N]}(x)](s)|. \tag{6.105}$$

But the Fourier transform of a Gaussian is a Gaussian; moreover, Gaussians are integrable and square integrable. So $\widehat{{}^N f}$ is too, by Proposition 3.5.2.

We now show that $\lim_{N \to \infty} \|{}^N f - f\|_p = 0$. We write

$$\begin{aligned}{}^N f - f &= ({}^N f - f * G_{[1/N]}) + (f * G_{[1/N]} - f) \\ &= (f\chi_{[-N,N]} - f) * G_{[1/N]} + (f * G_{[1/N]} - f). \end{aligned} \tag{6.106}$$

By the triangle inequality on $L^p(\mathbb{R})$, then,

$$\|{}^N f - f\|_p \leq \|(f\chi_{[-N,N]} - f) * G_{[1/N]}\|_p + \|f * G_{[1/N]} - f\|_p. \tag{6.107}$$

If we can show that each of the norms on the right approaches zero as $N \to \infty$, then we'll be done by the squeeze law.

Regarding the second one, we recall again that G has mass 1, so by Proposition 5.7.1(b), $f * G_{[1/N]}$ converges to f in L^p norm, as desired. For the first one, we argue as follows. We have $f\chi_{[-N,N]} - f \in L^p(\mathbb{R})$ and $G_{[1/N]} \in L^1(\mathbb{R})$, so by Propositions 5.6.1 and 5.6.2, $(f\chi_{[-N,N]} - f) * G_{[1/N]} \in L^p(\mathbb{R})$, and

$$\begin{aligned}\|(f\chi_{[-N,N]} - f) * G_{[1/N]}\|_p &\leq \|f\chi_{[-N,N]} - f\|_p \|G_{[1/N]}\|_1 \\ &= \|f\chi_{[-N,N]} - f\|_p \|G\|_1, \end{aligned} \tag{6.108}$$

the last step by (5.96). Now

$$\left(\|f\chi_{[-N,N]} - f\|_p\right)^p = \int_{|x|>N} |f(x)|^p \, dx. \tag{6.109}$$

Since $f \in L^p(\mathbb{R})$, the integral of $|f|^p$ over $\mathbb{R}$ is finite, so the right side of (6.109) must approach zero as $N \to \infty$. Then the right, and therefore also the left, side of (6.108) does too, as required.

Finally, we show that the sequence of $\widehat{{}^N f}$'s is Cauchy in $L^2(\mathbb{R})$. To do so, we use the linearity of the Fourier transform and the Plancherel theorem (Proposition 6.4.1(c)) to conclude that

$$||\widehat{{}^N f} - \widehat{{}^M f}|| = ||{}^N f - {}^M f||. \tag{6.110}$$

We now take the limit as $N \to \infty$ of the sup over $M > N$ of either side:

$$\lim_{N \to \infty} \sup_{M > N} ||\widehat{{}^N f} - \widehat{{}^M f}|| = \lim_{N \to \infty} \sup_{M > N} ||{}^N f - {}^M f||. \tag{6.111}$$

Any convergent sequence is Cauchy, so by Proposition 6.5.1, the right side of (6.111) goes to zero as $N \to \infty$. So the left side does too, whence our desired result. □

Let's, for the moment, restrict our attention to the case $p = 2$. Then the above ${}^N f$'s converge, in mean square norm, to $f \in L^2(\mathbb{R})$, and moreover the $\widehat{{}^N f}$'s converge in mean square norm to *something* in $L^2(\mathbb{R})$ (by completeness of this space). This something is unique by ($L^2(\mathbb{R})$ analogs of) the next to the next to the last paragraph of Section 3.3. It makes perfect sense, then, to *call* this something *the Fourier transform $\hat{f}$ of f*. Doesn't it?

Well yes, but an important issue must first be addressed. If f is also in $L^1(\mathbb{R})$, then, because of Definition 6.1.1, we have two at least superficially different candidates for $\hat{f}$. This is troubling.

But actually it's not so troubling, in light of the following proposition, which admonishes us not to be so superficial.

Proposition 6.5.2 *Let $f \in L^1(\mathbb{R}) \cap L^2(\mathbb{R})$. Then the function $\hat{f}$ of Definition 6.1.1 and the norm limit F of the above $\widehat{{}^N f}$'s are equal in $L^2(\mathbb{R})$.*

Proof. Let $f \in L^2(\mathbb{R})$; by definition of F, the $\widehat{{}^N f}$'s converge to F in the $L^2(\mathbb{R})$ norm. On the other hand note that, if $f \in L^1(\mathbb{R})$, then the $\widehat{{}^N f}$'s converge pointwise (uniformly, in fact) to $\hat{f}$; this follows from Proposition 6.1.1 and the fact that the ${}^N f$'s converge to f in the $L^1(\mathbb{R})$ norm.

So by the $L^2(\mathbb{R})$ analog of Proposition 3.2.1(h), $\hat{f} = F$ in $L^2(\mathbb{R})$, as claimed. □

NOW we may safely define the Fourier transform on $L^2(\mathbb{R})$.

Definition 6.5.1 *Given $f \in L^2(\mathbb{R})$, the Fourier transform $\hat{f} \in L^2(\mathbb{R})$ of f is the norm limit of the $\widehat{{}^N f}$'s.*

We summarize: To an arbitrary $f \in L^2(\mathbb{R})$, Definition 6.1.1 may not apply. But at any rate there will be a sequence of elements ${}^N f \in L^2(\mathbb{R})$ that converge in the $L^2(\mathbb{R})$ norm to f and to which the definition *does* apply. But then the $\widehat{{}^N f}$'s define a Cauchy sequence in $L^2(\mathbb{R})$, which necessarily converges to a unique element of $L^2(\mathbb{R})$, which we *define* to be $\widehat{f}$; this definition is consistent with Definition 6.1.1.

Of course, it would be nice for some kind of Fourier inversion theorem to exist on $f \in L^2(\mathbb{R})$. And it does. To see that this is so, we first need:

Proposition 6.5.3 *The conclusions of Proposition 6.4.1 — the convolution-inversion theorem, the product theorem, and the Plancherel theorem — hold for all $f, g \in L^2(\mathbb{R})$.*

Proof. This follows readily from Proposition 6.4.1 itself, together with the fact that the $^N f$'s of Proposition 6.5.1 are in $FL^1(\mathbb{R})$.

For example, the proof of the Plancherel theorem on $L^2(\mathbb{R})$ goes like this: We have

$$\langle ^N f, ^N g \rangle = \langle \widehat{^N f}, \widehat{^N g} \rangle \qquad (6.112)$$

by Proposition 6.5.1. The left side converges to $\langle f, g \rangle$, and the right side to $\langle \widehat{f}, \widehat{g} \rangle$, as $N \to \infty$ (cf. the definition (3.68) of the inner product on $L^2(\mathbb{R})$), yielding the desired result.

The proofs of the other parts of this proposition are similar; we omit the details. □

Finally, our main result.

Theorem 6.5.1 (Fourier inversion on $L^2(\mathbb{R})$) *For $f \in L^2(\mathbb{R})$,*

$$\widehat{\widehat{f}} = f^-. \qquad (6.113)$$

Moreover, let $f, g \in L^2(\mathbb{R})$. Then $g = \widehat{f}$ if and only if $\widehat{g} = f^-$.

Proof. Let $f, g \in L^2(\mathbb{R})$. We'll show that

$$g = \widehat{f} \Rightarrow \|\widehat{g} - f^-\| = 0. \qquad (6.114)$$

This will suffice for the following reasons: First, (6.114) certainly yields

$$\|\widehat{\widehat{f}} - f^-\| = 0, \qquad (6.115)$$

which is (6.113). Second, (6.115) and the assumption $\widehat{g} = f^-$ imply

$$\|\widehat{\widehat{f}} - \widehat{g}\| = 0; \qquad (6.116)$$

Proposition 6.5.3(c) then gives

$$\|\widehat{f} - g\| = 0, \qquad (6.117)$$

whence the remaining assertion of our theorem.

To demonstrate (6.114), let $^N g$ be as defined in Proposition 6.5.1. Then by the triangle inequality on $L^2(\mathbb{R})$,

$$\|\widehat{g} - f^-\| \leq \|\widehat{g} - \widehat{^N g}\| + \|\widehat{^N g} - f^-\|. \tag{6.118}$$

The first norm on the right goes to zero as $N \to \infty$, by definition of $\widehat{g}$ as the norm limit of the $\widehat{^N g}$'s. So we'll be done, according to the squeeze law, if we can show

$$g = \widehat{f} \Rightarrow \lim_{N \to \infty} \|\widehat{^N g} - f^-\| = 0. \tag{6.119}$$

To do so, we note that

$$\|\widehat{^N g} - f^-\| = \|\widehat{^N g} - \widehat{f^-}\| = \|^N g^- - f^-\|_2. \tag{6.120}$$

The first step here is by Proposition 6.5.3 and the linearity property of the Fourier transform (by limiting arguments, one checks that this property applies on $L^2(\mathbb{R})$ as well as on $L^1(\mathbb{R})$). The second step is because $^N g$, by Proposition 6.5.1, satisfies the hypotheses and therefore the conclusion of Theorem 6.4.2.

Now suppose $\widehat{f} = g$; then $\widehat{f^-} = g^-$, by a limiting argument applied to Proposition 6.2.1(b)(ii). So (6.120) gives

$$\|\widehat{^N g} - f^-\| = \|^N g^- - g^-\|. \tag{6.121}$$

In general, by a simple substitution, $\|k^-\| = \|k\|$ for $k \in L^2(\mathbb{R})$, so (6.121) yields

$$\|\widehat{^N g} - f^-\| = \|^N g - g\|. \tag{6.122}$$

The right side goes to zero as $N \to \infty$, by Proposition 6.5.1, so we're done. $\square$

Here's a nice application of the above theorem to the evaluation of the Fourier transform of a function *not* in $L^1(\mathbb{R})$.

Example 6.5.1 Find the Fourier transform of $\operatorname{sinc} 2\pi bx$, with the sinc function as in (6.14) and $b > 0$.

Solution. By Example 6.1.1, $F[\chi_{[-b,b]}(x)](s) = 2b \operatorname{sinc} 2\pi bs$, so by Theorem 6.5.1 and the fact that $\chi_{[-b,b]}$ is even and in $L^2(\mathbb{R})$,

$$F[\operatorname{sinc} 2\pi bx] = (2b)^{-1}(\chi_{[-b,b]})^- = (2b)^{-1}\chi_{[-b,b]}. \tag{6.123}$$

We make a few more useful observations concerning the Fourier transform and Fourier inversion on $L^2(\mathbb{R})$. We begin with:

Proposition 6.5.4 *If $f \in L^2(\mathbb{R})$ and f_N is any sequence of elements of $L^2(\mathbb{R})$ converging to f in $L^2(\mathbb{R})$ norm, then $\widehat{f_N}$ converges to $\widehat{f}$ in this norm.*

Proof. By linearity of the Fourier transform and by Proposition 6.5.3(c),

$$\lim_{N \to \infty} \|\widehat{f_N} - \widehat{f}\| = \lim_{N \to \infty} \|f_N - f\| = 0 \tag{6.124}$$

for f and f_N as stipulated. □

Some particularly pleasant consequences of the above proposition are the following equivalent—but on the surface simpler—formulations of the Fourier transform and Fourier inversion on $L^2(\mathbb{R})$.

Theorem 6.5.2 (Fourier inversion on $L^2(\mathbb{R})$ reprise) *Let $f \in L^2(\mathbb{R})$. Then with all notation as above, we have*

$$\widehat{f}(s) = \int_{-\infty}^{\infty} f(x)e^{-2\pi i s x}\, dx \tag{6.125}$$

and

$$f(x) = \int_{-\infty}^{\infty} \widehat{f}(s)e^{2\pi i s x}\, ds \tag{6.126}$$

provided, in either case, the integral on the right side is understood to denote the mean square norm limit *of the corresponding integrals over $[-N, N]$, and the equality is understood to mean that this limit is* the element of $L^2(\mathbb{R})$ represented on the left.

Proof. Because of the symmetry between f and $\widehat{f}$ manifested by Theorem 6.5.1, it suffices to prove that (6.125) is true in the norm limit sense just described. That is, it suffices to show that, if $f_N = f\chi_{[-N,N]}$, then $\widehat{f_N}$ converges to $\widehat{f}$ in norm. Let's do that.

By Proposition 6.5.4, it suffices in turn to show that $f_N \in L^2(\mathbb{R})$ and that f_N converges to f in norm. But actually, we've already verified both of these assertions in the proof of Proposition 6.5.1. (See the first paragraph of that proof for the first result and (6.109) for the second.) We're done. □

It's well worth recapitulating, so let's. The ^Nf's of Proposition 6.5.1 were seen to satisfy the hypotheses and therefore the conclusions of Theorem 6.4.1(c). Consequently we were able, using limiting arguments, to deduce both Fourier inversion and the Plancherel theorem on $L^2(\mathbb{R})$. The latter result gave us Proposition 6.5.4, which in turn yielded Theorem 6.5.2, which says that Fourier transforms and inversion on $L^2(\mathbb{R})$ are, in some sense, simpler than was initially indicated. That is, they may be phrased in straightforward terms of norm limits of "truncated" Fourier transforms, rather than the more complicated ^Nf's.

It should finally be noted that, by its very construction, the Fourier transform of an $f \in L^2(\mathbb{R})$ is an *equivalence class* of square integrable functions. On the other hand, recall from Proposition 6.1.1 that, if $f \in L^1(\mathbb{R})$, then $\widehat{f}$ is continuous. As noted earlier (see Exercise 3.3.3, whose result applies on any torus or interval just as it does on $\mathbb{T}$), there exists at most one continuous function per equivalence class. So if $f \in L^1(\mathbb{R}) \cap L^2(\mathbb{R})$, then its Fourier transform *is* uniquely defined, as a bona fide function rather than an equivalence class thereof.

Exercises

6.5.1 Find $\widehat{h}$ if
$$h(x) = (\operatorname{sinc} \pi x)^2.$$
Hints: Use the product theorem (see Propositions 6.4.1 and 6.5.3) together with the results of Example 6.5.1 and Exercise 5.2.2.

6.5.2 Use the previous exercise to find $\widehat{k}$ if
$$k(x) = \frac{\pi x \sin \pi x \cos \pi x - \sin^2 \pi x}{x^3}.$$
Hint: How is k related to h above?

6.5.3 Show that, if $f, g \in C(\mathbb{R}) \cap L^2(\mathbb{R})$ and $\widehat{f}(s) = \widehat{g}(s)$ for almost all s, then $f(x) = g(x)$ for all x.

6.5.4 Using Exercise 6.5.3, verify the identity $\sin 2\pi x = 2 \cos \pi x \sin \pi x$. Hint: Divide both sides by πx; write $\cos \pi x = (e^{\pi i x} + e^{-\pi i x})/2$; take Fourier transforms; use (6.35) (the identities there, originally posited for functions in $L^1(\mathbb{R})$, hold almost everywhere in $L^2(\mathbb{R})$).

6.5.5 Using the Plancherel theorem, find $\langle f, g \rangle$ if $f(x) = 1/(1+x^2)$ and $g(x) = e^{-4\pi i x} \operatorname{sinc} 6\pi x$. Parts of Proposition 6.2.1 should also be of help here.

6.5.6 Let $f(x) = \operatorname{sinc} 2\pi c x$ and $g(x) = \operatorname{sinc} 2\pi d x$, with $c, d > 0$.

  a. Using the convolution-inversion theorem (cf. Propositions 6.4.1 and 6.5.3), show that $f * g(x) = (2M)^{-1} \operatorname{sinc} 2\pi m x$, where m is the minimum and M the maximum of c and d.

  b. Evaluate (that is, express as *numbers*)
$$\int_{-\infty}^{\infty} \frac{\sin 4\pi y \sin 2\pi(1/2 - y)}{y(1/2 - y)} \, dy, \quad \int_{-\infty}^{\infty} \frac{\sin(\pi y/3) \sin \pi(1-y)}{y(1-y)} \, dy.$$

6.6 FOURIER INVERSION OF PIECEWISE SMOOTH, INTEGRABLE FUNCTIONS

In this section, we obtain aperiodic analogs of some of our Fourier series convergence theorems from Chapter 1.

We begin with the following, which gives us a large, *explicitly* characterized collection of functions such that the frequency decomposition (6.3) holds for all x. (By "explicitly" here, we mean "without direct reference to the Fourier transform.")

Theorem 6.6.1 Suppose $f \in \operatorname{PSC}(\mathbb{R}) \cap L^1(\mathbb{R})$ and $f' \in L^1(\mathbb{R}) \cap L^2(\mathbb{R})$. Then f satisfies the hypotheses and therefore the conclusions of Theorem 6.4.1(c).

Proof. Let f be as stated. Then certainly $f \in C(\mathbb{R}) \cap L^1(\mathbb{R})$, so we need only show that $\widehat{f} \in L^1(\mathbb{R})$. Since this Fourier transform is bounded (by Proposition 6.1.1), it's certainly integrable over $[-1, 1]$, say, so we need only show it's integrable over $(-\infty, -1) \cup (1, \infty)$. To do so, we write $\widehat{f}(s) = 2\pi i s \widehat{f}(s)/(2\pi i s)$, so that by (1.142),

$$|\widehat{f}(s)| \leq \frac{1}{2}\left(|2\pi i s \widehat{f}(s)|^2 + \frac{1}{|2\pi i s|^2}\right) = \frac{1}{2}\left(|\widehat{(f')}(s)|^2 + \frac{1}{4\pi^2 s^2}\right), \quad (6.127)$$

the last step by Proposition 6.2.1(c)(i). Since f' is by assumption in $L^2(\mathbb{R})$, so is $\widehat{(f')}$, by Proposition 6.5.2. But $|s|^{-2}$ is clearly integrable over $(-\infty, -1) \cup (1, \infty)$, so the right side of (6.127) is, and so $\widehat{f}(s)$ is, by Proposition 3.5.2, and we're done. □

We now wish to obtain a Fourier inversion result for piecewise smooth, integrable, but *not* necessarily continuous functions. For this we will need the following, which is an aperiodic analog of Lemma 1.4.1.

Proposition 6.6.1 (the Riemann-Lebesgue lemma for $L^1(\mathbb{R})$) *If $f \in L^1(\mathbb{R})$, then*

$$\lim_{s \to \pm\infty} \widehat{f}(s) = 0. \quad (6.128)$$

Proof. Let $f \in L^1(\mathbb{R})$. For any N and s we have

$$\widehat{f}(s) = \widehat{{}^N f}(s) + [\widehat{f}(s) - \widehat{{}^N f}(s)], \quad (6.129)$$

with ${}^N f$ as in (6.104). So by the triangle inequality on $\mathbb{C}$,

$$|\widehat{f}(s)| \leq |\widehat{{}^N f}(s)| + |\widehat{f}(s) - \widehat{{}^N f}(s)|. \quad (6.130)$$

But by the linearity of the Fourier transform and by Proposition 6.1.1,

$$|\widehat{f}(s) - \widehat{{}^N f}(s)| \leq \|f - {}^N f\|_1, \quad (6.131)$$

so (6.130) gives

$$|\widehat{f}(s)| \leq |\widehat{{}^N f}(s)| + \|{}^N f - f\|_1. \quad (6.132)$$

The norm on the right approaches zero as $N \to \infty$, by Proposition 6.5.1. So to prove our proposition it will suffice, by the squeeze law II(b) (Proposition 3.4.1(b)), to show that

$$\lim_{s \to \pm\infty} |\widehat{{}^N f}(s)| = 0 \quad (6.133)$$

for any *fixed* N.

Let's. We have, by (6.105),

$$|\widehat{{}^N f}(s)| \leq \|f\chi_{[-N,N]}\|_1 \, |F[G_{[1/N]}(x)](s)| = \|f\chi_{[-N,N]}\|_1 \, |NF[G(Nx)](s)|$$
$$= \|f\chi_{[-N,N]}\|_1 \, |N(N^{-1}\widehat{G}(N^{-1}s))| = \|f\chi_{[-N,N]}\|_1 \, \left|e^{-\pi(s/N)^2}\right|, \quad (6.134)$$

FOURIER INVERSION OF PIECEWISE SMOOTH, INTEGRABLE FUNCTIONS

the first equality by (5.94), the next by Proposition 6.2.1(b)(i), and the last by Example 6.2.3. The right side of (6.134) approaches 0 as $s \to \pm\infty$ (for fixed N), so the left side does too, as required. □

From the above proposition and Proposition 6.2.1(c)(i) (applied repeatedly), we deduce the following aperiodic analog of Corollary 1.8.1.

Proposition 6.6.2 *Let $m \in \mathbb{Z}^+$. If $f^{(m-1)} \in \mathrm{PSC}(\mathbb{R})$ and $f, f', f'', \ldots, f^{(m-1)} \in L^1(\mathbb{R})$, then*

$$\lim_{s \to \pm\infty} \frac{\widehat{f}(s)}{s^{-m}} = 0. \tag{6.135}$$

The above corollary is an aperiodic reflection of the local/global principle. (Recall the discussions following Corollary 1.8.1.)

We next need the following.

Lemma 6.6.1

$$\lim_{c \to \infty} \int_{-c}^{0} \operatorname{sinc} v \, dv = \frac{\pi}{2} = \lim_{c \to \infty} \int_{0}^{c} \operatorname{sinc} v \, dv. \tag{6.136}$$

Proof. The integrand is even, so it suffices to consider the second limit.

Since $\operatorname{sinc} v$ is bounded, it's integrable over $(0, c)$. Moreover, integrating by parts with $u = 1/v$ and $dv = \sin v \, dv$ gives

$$\int_{\pi/2}^{c} \operatorname{sinc} v \, dv = -\frac{\cos c}{c} - \int_{\pi/2}^{c} \frac{\cos v}{v^2} \, dv; \tag{6.137}$$

the limit on the right exists because $c^{-1}|\cos c| \le c^{-1} \to 0$ as $c \to \infty$ and because $|v^{-2} \cos v| \le v^{-2}$, the latter being integrable on $[\pi/2, \infty)$. We conclude that the second limit in (6.136) exists. So we may evaluate this limit by letting $c \to \infty$ through any *sequence* c_N.

In particular,

$$\lim_{c \to \infty} \int_{0}^{c} \operatorname{sinc} v \, dv = \lim_{N \to \infty} \int_{0}^{(N+1/2)\pi} \operatorname{sinc} v \, dv = \lim_{N \to \infty} \int_{0}^{\pi} \frac{\sin(N+1/2)y}{y} \, dy$$

$$= \frac{1}{2} \lim_{N \to \infty} \int_{0}^{\pi} \frac{\sin(N+1/2)y}{\sin(y/2)} \, dy$$

$$+ \lim_{N \to \infty} \int_{0}^{\pi} \left[\frac{1}{y} - \frac{1}{2\sin(y/2)}\right] \sin(N+1/2)y \, dy, \tag{6.138}$$

the second step by the substitution $y = v/(N+\frac{1}{2})$ and the third by some algebra. The first integrand on the right is the Dirichlet kernel $D_N(y)$, by (1.122); so the first integral on the right equals π, by (1.83). So (6.138) gives

$$\lim_{c \to \infty} \int_{0}^{c} \operatorname{sinc} v \, dv = \frac{\pi}{2} + \lim_{N \to \infty} \int_{0}^{\pi} g(y) \sin(N+1/2)y \, dy, \tag{6.139}$$

where

$$g(y) = \left[\frac{1}{y} - \frac{1}{2\sin(y/2)}\right]. \qquad (6.140)$$

But

$$\int_0^\pi g(y) \sin(N+1/2) y \, dy$$
$$= \int_0^\pi g(y)(\sin(Ny)\cos(y/2) + \cos(Ny)\sin(y/2)) \, dy$$
$$= \int_0^\pi g(y) \cos(y/2) \sin(Ny) \, dy + \int_0^\pi g(y) \sin(y/2) \cos(Ny) \, dy$$
$$= \frac{\pi}{2}(b_N(g_1) + a_N(g_2)), \qquad (6.141)$$

where g_1 is odd, 2π-periodic, and defined by $g_1(y) = g(y)\cos(y/2)$ on $(0, \pi)$, while g_2 is even, 2π-periodic, and defined by $g_2(y) = g(y)\sin(y/2)$ on $(0, \pi)$. (Recall Proposition 1.3.1.) Now g is bounded on $(0, \pi)$; the fact that it has a finite limit at $y = 0$ follows from a straightforward application of l'Hôpital's rule. This implies $g_1, g_2 \in \mathrm{PC}(\mathbb{T})$. So the Riemann-Lebesgue Lemma for $\mathrm{PC}(\mathbb{T})$ (Lemma 1.4.1) says $b_N(g_1), a_N(g_2) \to 0$ as $N \to \infty$; so by (6.139) and (6.141),

$$\lim_{b \to \infty} \int_0^b \operatorname{sinc} x \, dx = \frac{\pi}{2}, \qquad (6.142)$$

and we're done. □

As an application we have the following, which is entirely analogous to Corollary 1.4.1(a).

Theorem 6.6.2 (Fourier inversion on $\mathrm{PS}(\mathbb{R}) \cap L^1(\mathbb{R})$) *If $f \in \mathrm{PS}(\mathbb{R}) \cap L^1(\mathbb{R})$, then*

$$\frac{1}{2}(f(x^-) + f(x^+)) = \lim_{b \to \infty} \int_{-b}^b \widehat{f}(s) e^{2\pi i s x} \, ds \qquad (6.143)$$

for each $x \in \mathbb{R}$.

Proof. Let's fix $x \in \mathbb{R}$. We have

$$\int_{-b}^b \widehat{f}(s) e^{2\pi i s x} \, ds = \int_{-\infty}^\infty \widehat{f}(s) \chi_{[-b,b]}(s) e^{2\pi i s x} \, ds$$
$$= \int_{-\infty}^\infty f(x-y) \, 2b \operatorname{sinc} 2b\pi y \, dy \qquad (6.144)$$

by Example 6.1.1, the composition theorem (Proposition 6.2.1(d)), and the change of variable $y = x - v$. Since, as we noted in Section 6.1, $\operatorname{sinc} \notin L^1(\mathbb{R})$, we can't apply

FOURIER INVERSION OF PIECEWISE SMOOTH, INTEGRABLE FUNCTIONS

Proposition 5.7.1 here. So we take a different approach—one similar, in many ways, to the proof of Theorem 1.4.1.

Specifically, we pick an $M \in \mathbb{Z}^+$ and use (6.144) to write

$$\left| \int_{-b}^{b} \widehat{f}(s) e^{2\pi i s x} \, ds - \frac{1}{2}\left(f(x^-) + f(x^+)\right) \right|$$

$$= \left| \left[\int_{-M}^{M} f(x-y) \, 2b \operatorname{sinc} 2\pi b y \, dy - \frac{1}{2}\left(f(x^-) + f(x^+)\right) \right] \right.$$

$$\left. + \int_{|y|>M} f(x-y) \, 2b \operatorname{sinc} 2\pi b y \, dy \right|$$

$$\leq \left| \left[\int_{-M}^{M} f(x-y) \, 2b \operatorname{sinc} 2\pi b y \, dy - \frac{1}{2}\left(f(x^-) + f(x^+)\right) \right] \right|$$

$$+ \int_{|y|>M} |f(x-y)| \, dy, \qquad (6.145)$$

the last step because $|2b \operatorname{sinc} 2\pi b y| = |\sin 2\pi b y/(\pi y)| \leq |\pi y|^{-1} < |\pi M|^{-1} < 1$ for $|y| > M \geq 1$. By the squeeze law II(b) (Proposition 3.4.1(b)), it will suffice to show that the integral on the far right approaches zero as $M \to \infty$ and that, for any fixed M, the square bracketed quantity on the right approaches zero as $b \to \infty$.

The first task is straightforward: The integral in question represents the tail ends of the integral $\int_{-\infty}^{\infty} |f(x-y)| \, dy$, which converges since $f \in L^1(\mathbb{R})$. These tail ends must therefore go to zero as $M \to \infty$, which is what we want.

So now, we need only show that, for M fixed,

$$\lim_{b \to \infty} \int_{-M}^{M} f(x-y) \, 2b \operatorname{sinc} 2\pi b y \, dy = \frac{1}{2}\left(f(x^-) + f(x^+)\right). \qquad (6.146)$$

Let's break the integral up into one over $[-M, 0]$ plus one over $[0, M]$. If we can show that the first tends to $f(x^+)/2$ and the second to $f(x^-)/2$ as $b \to \infty$, then we'll be done. Let's show this. Actually, we'll only consider the second limit in question. We'll handle it using the second equality in Lemma 6.6.1; the first limit may be treated similarly using the first equality there.

We write

$$\int_0^M f(x-y) \, 2b \operatorname{sinc} 2\pi b y \, dy = \int_0^M \left[f(x-y) - f(x^-) \right] 2b \operatorname{sinc} 2\pi b y \, dy$$

$$+ f(x^-) \int_0^M 2b \operatorname{sinc} 2\pi b y \, dy. \qquad (6.147)$$

The summand on the far right equals

$$f(x^-) \int_0^M \frac{\sin 2\pi b y}{\pi y} \, dy = \frac{f(x^-)}{\pi} \int_0^{2\pi b M} \frac{\sin v}{v} \, dv \qquad (6.148)$$

(we've put $v = 2\pi by$), which, by Lemma 6.6.1, tends to $f(x^-)/2$ as $b \to \infty$. So to show, as again we need to, that the left side of (6.147) has this same limit, we need only show that

$$\lim_{b \to \infty} \int_0^M \left[f(x-y) - f(x^-) \right] 2b \operatorname{sinc} 2\pi by \, dy = 0. \tag{6.149}$$

Let's.

The integral in (6.149) equals

$$\int_0^M \left[\frac{f(x-y) - f(x^-)}{\pi y} \right] \sin 2\pi by \, dy$$

$$= \int_0^M \left[\frac{f(x-y) - f(x^-)}{2\pi i y} \right] \left(e^{2\pi i b y} - e^{-2\pi i b y} \right) dy = \widehat{h}(b) - \widehat{h}(-b), \tag{6.150}$$

where

$$h(y) = \chi_{(0,M)}(y) \left[\frac{f(x-y) - f(x^-)}{2\pi i y} \right]. \tag{6.151}$$

So by Proposition 6.6.1, the right side of (6.150) has limit zero as $b \to \infty$, and therefore (6.149) holds, and therefore we're done, *provided* h is in $L^1(\mathbb{R})$.

To show that it is, we observe that it's supported on $(0, M)$, so it suffices (by Proposition 3.5.2) to show that it's bounded there. Now the numerator is piecewise smooth on $(0, M)$, so we need only show that $h(0^+)$ exists. We do this using l'Hôpital's rule as follows:

$$h(0^+) = \chi_{(0,M)}(0^+) \lim_{y \to 0^+} \frac{f(x-y) - f(x^-)}{2\pi i y} = \lim_{y \to 0^+} \frac{-f'(x-y)}{2\pi i} = \frac{-f'(x^-)}{2\pi i}, \tag{6.152}$$

which exists by the piecewise smoothness of f. We *are* done. □

For example, let $f = \chi_{[-b,b]}$, where $b > 0$. Then f does *not* satisfy the hypotheses of Theorem 6.6.1 since f is discontinuous. Nor does it satisfy the conclusions: $\widehat{f} \notin L^1(\mathbb{R})$, as noted near the end of Section 6.1. But it *does* satisfy the hypotheses and therefore the conclusions of Theorem 6.6.2. Note that, in this case, $(f(x^-) + f(x^+))/2$ equals $f(x)$ everywhere except at $x = \pm b$, where it equals $\frac{1}{2}$.

Exercises

6.6.1 a. Show that the integrals

$$\int_{-N}^{N} \frac{\sin 2\pi s}{\pi s} e^{-2\pi i s x} \, ds$$

(as functions of x) converge in norm to $\chi_{[-1,1]}$ as $N \to \infty$. Hint: Example 6.5.1.

b. Use Theorem 6.6.2 to evaluate, for each $x \in \mathbb{R}$, the limit as $N \to \infty$ of the integrals in the previous exercise. At which points x do these integrals *not* converge pointwise to $\chi_{[-1,1]}(x)$?

6.6.2 Use the result of Example 6.3.3 along with Theorem 6.6.2 to evaluate
$$\int_{-\infty}^{\infty} \frac{v^2 - 2(1 - \cos v)}{v^4} \cos v \, dv.$$
It might help to first put $s = v/2\pi$.

6.6.3 The conclusion of the Riemann-Lebesgue lemma (Proposition 6.6.1) does *not* apply to all $f \in L^2(\mathbb{R})$. To see this, let
$$g(x) = \sum_{n=1}^{\infty} \chi_{(n, n+1/n)}(x) \quad (x \in \mathbb{R}).$$

a. Show that $g \notin L^1(\mathbb{R})$.
b. Show that $g \in L^2(\mathbb{R})$. Hint: The $\chi_{(n,n+1/n)}$'s are orthogonal (as the intervals $(n, n+1/n)$ are nonoverlapping). Use the Pythagorean theorem (3.17) on $L^2(\mathbb{R})$.
c. Let $f = \widehat{g}$. Show that $\lim_{s \to \infty} \widehat{f}(s) \neq 0$. Actually, the Fourier transform of a function in $L^2(\mathbb{R})$ is an equivalence class—see Section 6.5, especially its last paragraph. So what you really should show is that *no* element F of this equivalence class satisfies $\lim_{s \to \infty} F(s) = 0$. You can show this by finding an element F of this class such that $|F(s)| > \frac{1}{2}$, say, on a *nonnegligible* subset of (N, ∞), for any $N \in \mathbb{Z}^+$. But such a function exists by definition of f, and by Fourier inversion.

6.6.4 Prove the following partial converse to Proposition 6.6.2: If $f \in C(\mathbb{R}) \cap L^1(\mathbb{R})$ and, for some constants $C > 0$, $\lambda > 1$, and $k \in \mathbb{N}$, we have $|\widehat{f}(s)| \leq C|s|^{-(k+\lambda)}$ for all s, then f belongs to the space $C^k(\mathbb{R})$ of functions on $\mathbb{R}$ with continuous derivatives $f, f', \ldots f^{(k)}$. Hint: from the stated hypotheses we find that $s^j \widehat{f}(s)$ defines a function in $L^1(\mathbb{R})$ for $1 \leq j \leq k$. (why?) Now use Proposition 6.2.1(c)(ii) (with the roles of f and $\widehat{f}$ reversed) repeatedly.

6.7 FOURIER COSINE AND SINE TRANSFORMS

Recall that, in Section 1.12, we used even and odd extensions of functions on $(0, \ell)$ to derive Fourier cosine and sine series for such functions. We now apply analogous arguments to, and deduce analogous results concerning, functions on $\mathbb{R}^+ = (0, \infty)$.

Theorem 6.7.1 For $f \in L^1(\mathbb{R}^+)$, define the Fourier cosine and sine transforms $\widehat{f}_{\cos} = F_{\cos}[f(x)]$ and $\widehat{f}_{\sin} = F_{\sin}[f(x)]$ of f by
$$\widehat{f}_{\cos}(s) = 2 \int_0^{\infty} f(x) \cos 2\pi s x \, dx, \quad \widehat{f}_{\sin}(s) = 2 \int_0^{\infty} f(x) \sin 2\pi s x \, dx, \quad (6.153)$$

respectively, for $s \in \mathbb{R}^+$. If $f \in \mathrm{PS}(\mathbb{R}^+)$ as well then, for each $x \in \mathbb{R}^+$, we have

$$\lim_{b\to\infty} 2\int_0^b \widehat{f_{\cos}}(s)\cos 2\pi sx\, ds = \frac{1}{2}\bigl(f(x^-) + f(x^+)\bigr)$$

$$= \lim_{b\to\infty} 2\int_0^b \widehat{f_{\sin}}(s)\sin 2\pi sx\, ds. \qquad (6.154)$$

Proof. We prove only the first equality in (6.154); we leave the second to the exercises.
So let $f \in \mathrm{PS}(\mathbb{R}^+) \cap L^1(\mathbb{R}^+)$: The even extension f_{even} of f is clearly in $\mathrm{PS}(\mathbb{R}) \cap L^1(\mathbb{R})$ (Fig. 6.3).

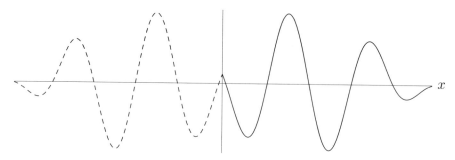

Fig. 6.3 $f \in \mathrm{PS}(\mathbb{R}^+) \cap L^1(\mathbb{R}^+)$ (solid) and its even extension f_{even} (dashed and solid combined)

So by the definition of f_{even} and by Theorem 6.6.2, we have, for $x \in \mathbb{R}^+$,

$$\frac{1}{2}\bigl(f(x^-) + f(x^+)\bigr) = \frac{1}{2}\bigl(f_{\mathrm{even}}(x^-) + f_{\mathrm{even}}(x^+)\bigr) = \lim_{b\to\infty}\int_{-b}^{b} \widehat{f_{\mathrm{even}}}(s)e^{2\pi isx}\, ds$$

$$= \lim_{b\to\infty}\int_{-b}^{0} \widehat{f_{\mathrm{even}}}(s)e^{2\pi isx}\, ds + \lim_{b\to\infty}\int_{0}^{b} \widehat{f_{\mathrm{even}}}(s)e^{2\pi isx}\, ds$$

$$= \lim_{b\to\infty}\int_{0}^{b} \widehat{f_{\mathrm{even}}}(-t)e^{-2\pi itx}\, dt + \lim_{b\to\infty}\int_{0}^{b} \widehat{f_{\mathrm{even}}}(s)e^{2\pi isx}\, ds$$

$$= \lim_{b\to\infty}\int_{0}^{b} \widehat{f_{\mathrm{even}}}(s)\bigl(e^{-2\pi isx} + e^{2\pi isx}\bigr)\, ds$$

$$= \lim_{b\to\infty} 2\int_{0}^{b} \widehat{f_{\mathrm{even}}}(s)\cos 2\pi sx\, ds; \qquad (6.155)$$

for the next to the last equality we used the fact that, since f_{even} is even, so is $\widehat{f_{\mathrm{even}}}$ (cf. Proposition 6.2.1(b)(ii)). Since, for $s \in \mathbb{R}^+$,

$$\widehat{f_{\mathrm{even}}}(s) = \int_{-\infty}^{\infty} f_{\mathrm{even}}(x)e^{-2\pi isx}\, dx = \int_{0}^{\infty} f_{\mathrm{even}}(x)\bigl(e^{2\pi isx} + e^{-2\pi isx}\bigr)\, dx$$

$$= 2\int_{0}^{\infty} f_{\mathrm{even}}(x)\cos 2\pi sx\, dx = 2\int_{0}^{\infty} f(x)\cos 2\pi sx\, dx = \widehat{f_{\cos}}(s)$$

$$\qquad (6.156)$$

(the second equality in (6.156) is by steps similar to those employed in (6.155)), we find that (6.155) is our desired inversion formula. □

Suppose f is piecewise smooth, integrable, and *continuous* on $\mathbb{R}^+$ and that $f' \in L^1(\mathbb{R}^+) \cap L^2(\mathbb{R}^+)$. It's readily seen, then, that f_{even} is piecewise smooth, integrable, and continuous on $\mathbb{R}$ and that $(f_{\text{even}})' \in L^1(\mathbb{R}) \cap L^2(\mathbb{R})$. From this it may be deduced, much as in the proof of Theorem 6.6.1, that

$$f(x) = 2\int_0^\infty \widehat{f}_{\cos}(s) \cos 2\pi s x \, ds \tag{6.157}$$

for all $x \in \mathbb{R}^+$ and that the integral converges uniformly on $\mathbb{R}^+$. We have to be careful with $\widehat{f}_{\sin}$, though. Suppose $f(0^+) \neq 0$. Then the odd extension f_{odd} of f is *discontinuous* (at zero). As it turns out, then, $\widehat{f}_{\sin} \notin L^1(\mathbb{R}^+)$, so no analog of Theorem 6.6.1 applies. The remarks of this paragraph should be compared with those of the two paragraphs following Example 1.12.1.

Finally, we remark that Fourier cosine and sine transforms of elements of $L^2(\mathbb{R}^+)$ satisfy their own inversion and Plancherel theorems. Specifically, these transforms may be defined as norm limits of "truncated" integrals over $[0, N]$. One then finds, for $f \in L^2(\mathbb{R}^+)$, that $\widehat{f}_{\cos}, \widehat{f}_{\sin} \in L^2(\mathbb{R}^+)$ as well, that

$$(\widehat{f}_{\cos})_{\cos} = f = (\widehat{f}_{\sin})_{\sin} \tag{6.158}$$

in $L^2(\mathbb{R})$, that

$$||f|| = ||\widehat{f}_{\cos}|| = ||\widehat{f}_{\sin}||, \tag{6.159}$$

and that, for g also in $L^2(\mathbb{R}^+)$,

$$\langle f, g \rangle = \langle \widehat{f}_{\cos}, \widehat{g}_{\cos} \rangle = \langle \widehat{f}_{\sin}, \widehat{g}_{\sin} \rangle. \tag{6.160}$$

All this may be deduced from results already obtained above for $L^2(\mathbb{R})$, from relations between f and f_{even} or f_{odd}, and from similar ones between $\widehat{f}$ and $\widehat{f}_{\cos}$ or $\widehat{f}_{\sin}$.

Exercises

6.7.1 **a.** Prove that, for $f \in L^1(\mathbb{R}^+)$ and $s \in \mathbb{R}^+$, $\widehat{f_{\text{odd}}}(s) = -i\widehat{f}_{\sin}(s)$.
  b. Prove the second equality in (6.154).

6.7.2 Following the method of proof of Proposition 6.2.1(d) and using trig identities (cf. Exercise A.0.16) when necessary, prove the following *composition theorems* for Fourier cosine and sine transforms:

$$\int_0^\infty \widehat{f}_{\cos}(s) g(s) \cos 2\pi s x \, ds = \frac{1}{2}\int_0^\infty f(v)\left[\widehat{g}_{\cos}(v-x) + \widehat{g}_{\cos}(v+x)\right] dv,$$

$$\int_0^\infty \widehat{f}_{\sin}(s) g(s) \sin 2\pi s x \, ds = \frac{1}{2}\int_0^\infty f(v)\left[\widehat{g}_{\cos}(v-x) - \widehat{g}_{\cos}(v+x)\right] dv$$

($f, g \in L^1(\mathbb{R}^+)$). (Remark: Here, in general, $v - x \in \mathbb{R}$. We've extended the definition (6.153) of $\widehat{f}_{\cos}(s)$ to all of $\mathbb{R}$.)

6.8 MULTIVARIABLE FOURIER TRANSFORMS AND INVERSION

The Fourier theory on $\mathbb{R}^M$ ($M \in \mathbb{Z}^+$) generalizes quite naturally that on $\mathbb{R}$. Here we discuss briefly this generalization. Throughout, an element x of $\mathbb{R}^M$ is understood to have Cartesian coordinates $(x_1, x_2, \ldots, x_M)$. Also, we'll attach the subscript "2" to norms and inner products on $L^2(\mathbb{R}^M)$, so as to distinguish these quantities from analogous ones, to be defined shortly, on $\mathbb{R}^M$ itself.

For $f \in L^1(\mathbb{R}^M)$, we define the *Fourier transform* $\widehat{f}$ of f by

$$\widehat{f}(s) = \int_{\mathbb{R}^M} f(x) e^{-2\pi i \langle s, x \rangle} \, dx, \tag{6.161}$$

where the inner product $\langle \cdot, \cdot \rangle$ on $\mathbb{R}^M$ is given by

$$\langle s, x \rangle = s_1 x_1 + s_2 x_2 + \cdots + s_M x_M \tag{6.162}$$

and dx denotes $dx_1 \, dx_2 \cdots dx_M$. Then the following, which generalizes Propositions 6.1.1 and 6.2.1 (and then some), is true.

Proposition 6.8.1 *Let $f, g \in L^1(\mathbb{R}^M)$ and $s \in \mathbb{R}^M$.*

(a) *$\widehat{f}(s)$ exists, $\widehat{f}$ belongs to the space $\mathrm{BC}(\mathbb{R}^M)$ of bounded, continuous functions on $\mathbb{R}^M$, and more specifically,*

$$\|\widehat{f}\|_\infty = \sup_{s \in \mathbb{R}^m} |\widehat{f}(s)| \leq \int_{\mathbb{R}^M} |f(x)| \, dx = \|f\|_1. \tag{6.163}$$

We also have the following. (Here, by addition, negation, and multiplication of elements of $\mathbb{R}^M$ we mean the corresponding termwise operations.)

(b) *For any $a \in \mathbb{R}^M$,*
   (i) *$F\left[e^{2\pi i \langle a, x \rangle} f(x)\right](s) = \widehat{f}(s - a)$ and*
   (ii) *$F[f(x - a)](s) = e^{-2\pi i \langle s, a \rangle} \widehat{f}(s)$.*

(c) (i) *$F[f(bx)](s) = |b_1 b_2 \cdots b_M|^{-1} \widehat{f}(b^{-1} s)$ for any $b \in \mathbb{R}^M$ none of whose coordinates is zero.*
   (ii) *$F[f(-x)](s) = \widehat{f}(-s)$.*
   (iii) *$F[\overline{f(x)}](s) = \overline{\widehat{f}(-s)}$.*

(d) *For $1 \leq m \leq M$:*
   (i) *$F[\partial f(x)/\partial x_m](s) = 2\pi i s_m \widehat{f}(s)$, provided f is, as a function of x_m, in $\mathrm{PSC}(\mathbb{R})$ for each choice of the other x_j's, and that $\partial f / \partial x_m \in L^1(\mathbb{R}^M)$;*
   (ii) *$F[x_m f(x)](s) = (-2\pi i)^{-1} \partial \widehat{f}(s)/\partial s_m$, provided the function h defined by $h(x) = x_m f(x)$ is in $L^1(\mathbb{R}^M)$.*

(e) (the composition theorem) For $x \in \mathbb{R}^M$,

$$\int_{\mathbb{R}^M} \widehat{f}(s)g(s)e^{2\pi i \langle s, x \rangle}\, ds = \int_{\mathbb{R}^M} f(v)\widehat{g}(v-x)\, dx, \qquad (6.164)$$

where convolution of functions on $\mathbb{R}^M$ is just as in Definition 5.1.1, except that the integration is over $\mathbb{R}^M$. In particular,

$$\int_{\mathbb{R}^M} \widehat{f}(s)g(s)\, ds = \int_{\mathbb{R}^M} f(v)\widehat{g}(v)\, dv. \qquad (6.165)$$

(f) (the convolution theorem) We have $f * g \in L^1(\mathbb{R}^M)$ and

$$\widehat{f * g}(s) = \widehat{f}(s)\widehat{g}(s). \qquad (6.166)$$

(g) For $x \in \mathbb{R}^M$ and $1 \le m \le M-1$, define

$$x_{(m)} = (x_1, x_2, \ldots, x_m), \quad {}^\perp x_{(m)} = (x_{m+1}, x_{m+2}, \ldots, x_M). \qquad (6.167)$$

If $f(x) = g(x_{(m)})h({}^\perp x_{(m)})$ where $g \in L^1(\mathbb{R}^m)$ and $h \in L^1(\mathbb{R}^{M-m})$, then $\widehat{f}(s) = \widehat{g}(s_{(m)})\widehat{h}({}^\perp s_{(m)})$.

(h) If R is any rotation of $\mathbb{R}^M$, then $F[f(Rx)](s) = \widehat{f}(Rs)$.

Proof. The proofs of parts (a)–(f) generalize, in straightforward fashions, those of the corresponding results on $L^1(\mathbb{R})$. For part (g) we note that, by (6.162),

$$\langle s, x \rangle = \langle s_{(m)}, x_{(m)} \rangle + \langle {}^\perp s_{(m)}, {}^\perp x_{(m)} \rangle \qquad (6.168)$$

(the inner products on the right are on $\mathbb{R}^m$ and $\mathbb{R}^{M-m}$, respectively), so by (6.161),

$$\widehat{f}(s) = \int_{\mathbb{R}^M} g(x_{(m)})h({}^\perp x_{(m)})e^{-2\pi i \langle s_{(m)}, x_{(m)} \rangle} e^{-2\pi i \langle {}^\perp s_{(m)}, {}^\perp x_{(m)} \rangle}\, dx$$

$$= \left(\int_{\mathbb{R}^m} g(x_{(m)}) e^{-2\pi i \langle s_{(m)}, x_{(m)} \rangle}\, dx_{(m)} \right)$$

$$\cdot \left(\int_{\mathbb{R}^{M-m}} h({}^\perp x_{(m)}) e^{-2\pi i \langle {}^\perp s_{(m)}, {}^\perp x_{(m)} \rangle}\, d^\perp x_{(m)} \right)$$

$$= \widehat{g}(s_{(m)})\widehat{h}({}^\perp s_{(m)}). \qquad (6.169)$$

Finally, for part (g), we use the fact that inner products and the element dx are invariant under rotations. That is, if R is a rotation of $\mathbb{R}^M$, then $\langle Rs, Rx \rangle = \langle s, x \rangle$ for any $s, x \in \mathbb{R}^M$ and

$$\int_{\mathbb{R}^M} g(Rx)\, dx = \int_{\mathbb{R}^M} g(x)\, dx \qquad (6.170)$$

for any $g \in L^1(\mathbb{R}^M)$. Geometrically, this means rotations preserve angles between vectors as well as infinitesimal volumes; this is intuitively plausible, though we won't prove it. At any rate, the consequence is that

$$F[f(Rx)](s) = \int_{\mathbb{R}^M} f(Rx) e^{-2\pi i \langle s, x \rangle} \, dx = \int_{\mathbb{R}^M} f(Rx) e^{-2\pi i \langle Rs, Rx \rangle} \, dx$$
$$= \int_{\mathbb{R}^M} f(x) e^{-2\pi i \langle Rs, x \rangle} \, dx = F[f(x)](Rs), \qquad (6.171)$$

as claimed. □

Example 6.8.1 Let $f(x_1, x_2) = \chi_{[-c,c]}(x_1) \chi_{[-d,d]}(x_2)$ $(c, d > 0)$. Find:

(a) $\widehat{f}$;

(b) $F[f(x_1 - x_2 + 2, x_1 + x_2 - 1)]$.

Solution. Part (a): By Proposition 6.8.1(g) and Example 6.1.1,

$$\widehat{f}(s) = F[\chi_{[-c,c]}(x_1)](s_1) \, F[\chi_{[-d,d]}(x_2)](s_2) = 4cd \operatorname{sinc} 2\pi c s_1 \operatorname{sinc} 2\pi d s_2$$
$$= \frac{\sin 2\pi c s_1 \sin 2\pi d s_2}{\pi^2 s_1 s_2}. \qquad (6.172)$$

For part (b), we could compute the relevant double integral directly, but instead we take an approach that makes use of, and helps illuminate, some results above. Namely, we note that

$$\begin{pmatrix} x_1 - x_2 + 2 \\ x_1 + x_2 - 1 \end{pmatrix} = \begin{pmatrix} \sqrt{2} & 0 \\ 0 & \sqrt{2} \end{pmatrix} \begin{pmatrix} \cos(-\pi/4) & \sin(-\pi/4) \\ -\sin(-\pi/4) & \cos(-\pi/4) \end{pmatrix} \begin{pmatrix} x_1 + 1/2 \\ x_2 - 3/2 \end{pmatrix} \qquad (6.173)$$

(DIY: check this). The second 2×2 matrix on the right represents a rotation $R_{-\pi/4}$ through an angle of $-\pi/4$ radians; the first represents componentwise multiplication by $\sqrt{2}$. So by parts (b)(ii), (c)(i), and (h) of Proposition 6.8.1, with $a = (-\frac{1}{2}, \frac{3}{2})$, $b = (\sqrt{2}, \sqrt{2})$, and $R = R_{-\pi/4}$,

$$F[f(x_1 - x_2 + 2, x_1 + x_2 - 1)](s)$$
$$= F[f(b \, R_{-\pi/4}(x - a))](s) = e^{-2\pi i \langle a, s \rangle} F[f(b \, R_{-\pi/4}(x))](s)$$
$$= \frac{1}{2} e^{\pi i (s_1 - 3 s_2)} F[f(R_{-\pi/4}(x))](2^{-1/2} s) = \frac{1}{2} e^{\pi i (s_1 - 3 s_2)} \widehat{f}(R_{-\pi/4}(2^{-1/2} s))$$
$$= \frac{1}{2} e^{\pi i (s_1 - 3 s_2)} \widehat{f}\left(\frac{s_1 - s_2}{2}, \frac{s_1 + s_2}{2}\right)$$
$$= \frac{2 e^{\pi i (s_1 - 3 s_2)} \sin \pi c (s_1 - s_2) \sin \pi d (s_1 + s_2)}{\pi^2 (s_1^2 - s_2^2)}. \qquad (6.174)$$

One would also expect Fourier inversion results in M dimensions, and one's expectations would be met quite nicely. Let's consider, in particular, some of our more elegant inversion theorems in the case $M = 1$—namely, Theorems 6.4.1 and 6.5.1. It turns out that these also generalize most elegantly. Here are those generalizations and some other important, related ones.

Theorem 6.8.1 (a) **(Fourier inversion on $FL^1(\mathbb{R}^M)$)** If f belongs to the space

$$FL^1(\mathbb{R}^M) = \{f \in L^1(\mathbb{R}^M): \widehat{f} \in L^1(\mathbb{R}^M)\}, \tag{6.175}$$

then

$$f(x) = \int_{\mathbb{R}^M} \widehat{f}(s) e^{2\pi i \langle s, x \rangle} \, ds \tag{6.176}$$

for almost all $x \in \mathbb{R}^M$. The integral on the right side is uniformly convergent; that is,

$$I_b(x) = \int_{\|x\| \leq b} \widehat{f}(s) e^{2\pi i s x} \, ds, \tag{6.177}$$

where $\|b\|$ denotes $\sqrt{\langle b, b \rangle}$, converges uniformly on $\mathbb{R}^M$ as $b \to \infty$.

If, in addition, f is continuous on $\mathbb{R}^M$, then (6.176) is true for all $x \in \mathbb{R}^M$ (and the integral there converges uniformly to f on $\mathbb{R}^M$).

(b) **(the Riemann-Lebesgue lemma for $L^1(\mathbb{R}^M)$)** If $f \in L^1(\mathbb{R}^M)$, then

$$\lim_{\|s\| \to \infty} \widehat{f}(s) = 0. \tag{6.178}$$

(c) **(Fourier inversion on $L^2(\mathbb{R}^M)$)** If $f \in L^2(\mathbb{R}^M)$, then so is the Fourier transform $\widehat{f}$ defined by

$$\widehat{f}(s) = \int_{\mathbb{R}^M} f(x) e^{-2\pi i \langle s, x \rangle} \, dx, \tag{6.179}$$

where the integral on the right denotes the norm limit of the corresponding integrals over $\{x \in \mathbb{R}^M : \|x\| \leq N\}$. Moreover, for such f we have

$$f(x) = \int_{\mathbb{R}^M} \widehat{f}(s) e^{2\pi i \langle s, x \rangle} \, ds, \tag{6.180}$$

where again the integral on the right denotes a norm limit of truncated integrals.

(d) For $f, g \in L^2(\mathbb{R}^M)$, we have

(i) **(the convolution-inversion theorem)** $f * g(x) = (\widehat{\widehat{f}\widehat{g}})(-x)$ for all $x \in \mathbb{R}^M$;

(ii) **(the product theorem)** $\widehat{fg}(s) = \widehat{f} * \widehat{g}(s)$ *for all* $s \in \mathbb{R}^M$;

(iii) **(the Plancherel theorem)** $\langle f, g \rangle = \langle \widehat{f}, \widehat{g} \rangle$; *in particular,* $||f||_2 = ||\widehat{f}||_2$.

We conclude with a discussion of *radial* functions—those dependent only on the norm $||x||$ of the independent variable—and of their Fourier transforms. We remark first of all that, in one variable, a radial function is just an *even* function. Note that if f is radial and in $L^1(\mathbb{R})$, then $\widehat{f}$ is radial too, by Proposition 6.2.1(b)(ii). The analogous result in fact obtains for any $M \in \mathbb{Z}^+$. Indeed, suppose $M \geq 1$ and $f \in L^1(\mathbb{R}^M)$ is radial; let $s, s' \in \mathbb{R}^M$ satisfy $||s|| = ||s'||$. There exists a rotation R such that $s' = Rs$; then by Proposition 6.8.1(h),

$$\widehat{f}(s') = \widehat{f}(Rs) = F[f(Rx)](s) = F[f(x)](s) = \widehat{f}(s), \quad (6.181)$$

so $\widehat{f}$ is radial.

We now consider some specifics of the two- and three-dimensional situations. (Our discussions may be abstracted readily to the case of general $M \in \mathbb{Z}^+$; see Section 2.10 in [15].)

Proposition 6.8.2 (a) *If* $f \in L^1(\mathbb{R}^2)$ *is radial—say* $f(x) = g(||x||)$—*and* $s \in \mathbb{R}^2$, *then*

$$\widehat{f}(s) = 2\pi \int_0^\infty \rho g(\rho) J_0(2\pi ||s|| \rho) \, d\rho, \quad (6.182)$$

with J_0, *the Bessel function of order zero, given by* (4.80).

(b) *If* $f \in L^1(\mathbb{R}^3)$ *is radial—say* $f(x) = g(||x||)$—*and* $s \in \mathbb{R}^3$, *then*

$$\widehat{f}(s) = \frac{2}{||s||} \int_0^\infty r g(r) \sin 2\pi ||s|| r \, dr. \quad (6.183)$$

Proof. Part (a): For f as stipulated, $\widehat{f}$ is again radial, so

$$\widehat{f}(s) = \widehat{f}(0, ||s||) = \int_{\mathbb{R}^2} f(x) e^{-2\pi i (0 \cdot x_1 + ||s|| x_2)} \, dx = \int_{\mathbb{R}^2} g(||x||) e^{-2\pi i ||s|| x_2} \, dx$$

$$= \int_{-\pi}^{\pi} \int_0^\infty g(\rho) e^{-2\pi i ||s|| \rho \sin \theta} \rho \, d\rho \, d\theta$$

$$= \int_0^\infty \rho g(\rho) \left(\int_{-\pi}^{\pi} e^{-2\pi i ||s|| \rho \sin \theta} \, d\theta \right) d\rho, \quad (6.184)$$

the next to the last step by the polar coordinate substitution $x_1 = \rho \cos \theta, x_2 = \rho \sin \theta$. But

$$\int_{-\pi}^{\pi} e^{-2\pi i ||s|| \rho \sin \theta} \, d\theta = 2\pi J_0(2\pi ||s|| \rho) \quad (6.185)$$

by Exercise 6.8.3, so we're done with part (a).

Part (b) is similar: Here, we utilize the spherical coordinate change of variable

$$x_1 = r\cos\theta\sin\phi, \quad x_2 = r\sin\theta\sin\phi, \quad x_3 = r\cos\phi. \tag{6.186}$$

This takes $dx = dx_1\, dx_2\, dx_3$ to $r^2 \sin\phi\, dr\, d\theta\, d\phi$ so, for $f \in L^1(\mathbb{R}^3)$ radial,

$$\widehat{f}(s) = \widehat{f}(0,0,\|s\|) = \int_{\mathbb{R}^3} f(x) e^{-2\pi i(0\cdot x_1 + 0\cdot x_2 + \|s\|x_3)}\, dx$$

$$= \int_{\mathbb{R}^3} g(\|x\|)\, e^{-2\pi i\|s\|x_3}\, dx$$

$$= \int_0^\pi \int_{-\pi}^\pi \int_0^\infty g(r) e^{-2\pi i\|s\|r\cos\phi} r^2 \sin\phi\, dr\, d\theta\, d\phi$$

$$= \int_0^\infty r^2 g(r) \left(\int_{-\pi}^\pi d\theta \right) \left(\int_0^\pi e^{-2\pi i\|s\|r\cos\phi} \sin\phi\, d\phi \right) dr$$

$$= 2\pi \int_0^\infty r^2 g(r) \left(\int_0^\pi e^{-2\pi i\|s\|r\cos\phi} \sin\phi\, d\phi \right) dr. \tag{6.187}$$

Now by the substitution $u = \cos\phi$,

$$\int_0^\pi e^{-2\pi i\|s\|r\cos\phi} \sin\phi\, d\phi = \int_{-1}^1 e^{-2\pi i\|s\|ru}\, du$$

$$= \left. \frac{e^{-2\pi i\|s\|ru}}{-2\pi i\|s\|r} \right|_{-1}^1 = \frac{\sin 2\pi\|s\|r}{\pi\|s\|r}. \tag{6.188}$$

Plugging this into (6.187) gives us part (b) of our proposition, and our proof is complete. □

The right side of (6.182) is called a *Hankel transform* of g, or of f; the transform in (6.183) has no special name, but note that it's essentially a Fourier sine transform of $\|x\| g(\|x\|)$. It should also be observed that $\widehat{f}_{\cos}$ is the one-dimensional analog of both transforms defined in the above proposition. More precisely, if $f \in L^1(\mathbb{R})$ is radial—meaning, again, even—then $\widehat{f}(s) = \widehat{f}_{\cos}(\|s\|)$, by (6.156).

Example 6.8.2 Find the Fourier transform of $f(x) = e^{-2\pi\|x\|}$ in

(a) two dimensions;

(b) three dimensions.

Solution. For part (a), we use Proposition 6.8.2 in conjunction with (6.185) to get

$$\widehat{f}(s) = \int_0^\infty \rho e^{-2\pi\rho} \int_{-\pi}^\pi e^{-2\pi i\|s\|\rho\sin\theta}\, d\theta\, d\rho$$

$$= \int_{-\pi}^\pi \left(\int_0^\infty \rho e^{-2\pi\rho(1+i\|s\|\sin\theta)}\, d\rho \right) d\theta = \frac{1}{4\pi^2} \int_{-\pi}^\pi \frac{d\theta}{(1+i\|s\|\sin\theta)^2}, \tag{6.189}$$

the last step by the integration-by-parts formula (1.41). The rightmost integral may be evaluated by some changes of variable (see Exercise 6.8.4). The result is

$$\widehat{f}(s) = \frac{1}{2\pi(1+||s||^2)^{3/2}}. \tag{6.190}$$

Part (b): By Proposition 6.8.2(b),

$$\widehat{f}(s) = \frac{2}{||s||} \int_0^\infty r e^{-2\pi r} \sin 2\pi ||s|| r \, dr = \frac{1}{||s||} \widehat{h}_{\sin}(||s||), \tag{6.191}$$

where $h(r) = r e^{-2\pi r}$ and $\widehat{h}_{\sin}$ is as in (6.153). Now by the Fourier sine transform analog of (6.156) (see Exercise 6.7.1), by Proposition 6.2.1(c)(ii), and by Example 6.1.2,

$$\widehat{h}_{\sin}(||s||) = F_{\sin}\left[r e^{-2\pi r}\right](||s||) = i F\left[r e^{-2\pi |r|}\right](||s||)$$

$$= -\frac{1}{2\pi} F\left[e^{-2\pi |r|}\right]'(||s||) = \frac{||s||}{\pi^2(1+||s||^2)^2}, \tag{6.192}$$

so by (6.191),

$$\widehat{f}(s) = \frac{1}{\pi^2(1+||s||^2)^2}. \tag{6.193}$$

Compare the above with Example 6.1.2.

Exercises

6.8.1 Find $\widehat{f}$, in the cases $M = 2$ and $M = 3$, if $f = \chi_{\{x:||x||\leq 1\}}$ (that is, $f(x) = 1$ if $||x|| \leq 1$ and $f(x) = 0$ otherwise). You may want to use (4.107).

6.8.2 Let $f \in L^1(\mathbb{R}^3)$ be defined by $f(x) = e^{-\pi ||x||^2}$.
   a. Compute $\widehat{f}$ using the fact that f is a product of single-variable Gaussians.
   b. Compute $\widehat{f}$ using Proposition 6.8.2(b). Hint: Gaussians are even, so the integral in Proposition 6.8.2(b), when g is a Gaussian, can be written as a certain Fourier transform of r times a Gaussian in r. Use Proposition 6.2.1(c)(ii).

6.8.3 Deduce, from part a of Exercise 5.6.9 and the substitution $t = \sin \theta$, the formula (6.185) for $J_0(2\pi ||s|| \rho)$. (The result of this substitution will give you an integral over $[0, \pi/2]$; use properties of cosine and sine to rewrite this an an integral over $[-\pi, \pi]$.)

6.8.4 Fill in the ingredient missing from our calculations in Example 6.8.2(a) by showing that

$$\int_{-\pi}^{\pi} \frac{d\theta}{(1+i||s||\sin\theta)^2} = \frac{2\pi}{(1+||s||^2)^{3/2}},$$

as follows.

a. Divide the integral into one over $[0, \pi]$ plus one over $[-\pi, 0]$; make a change of variable in the second integral so that it has the same domain of integration as the first; and combine the two integrals, using a common denominator, to get

$$\int_{-\pi}^{\pi} \frac{d\theta}{(1 + i||s|| \sin \theta)^2} = 2 \int_0^{\pi} \frac{(1 - ||s||^2 \sin^2 \theta)\, d\theta}{(1 + ||s||^2 \sin^2 \theta)^2}$$

$$= 2 \int_0^{\pi} \frac{(\csc^2 \theta - ||s||^2) \csc^2 \theta \, d\theta}{(\csc^2 \theta + ||s||^2)^2}.$$

b. Put $u = \cot \theta$, so that $du = -\csc^2 \theta \, d\theta$ and $1 + u^2 = \csc^2 \theta$, into the above to get

$$\int_{-\pi}^{\pi} \frac{d\theta}{(1 + i||s|| \sin \theta)^2} = 2 \int_{-\infty}^{\infty} \frac{(1 + u^2 - ||s||^2)\, du}{(1 + u^2 + ||s||^2)^2}.$$

c. The integral on the right in part b can be evaluated by standard trig substitution techniques. (Namely, put $u = \sqrt{1 + ||s||^2} \tan v$.) Use these techniques to do the computations and complete the problem.

6.8.5 Find the Fourier transform of the function $f \in L^1(\mathbb{R}^M)$ defined by $f(x) = e^{-2\pi ||x||}/||x||$ for $M = 2$ and $M = 3$. For the former case, you might want to proceed as in Example 6.8.2 and use arguments like those in Exercise 6.8.4 to do the resulting integral in θ.

6.9 TEMPERED DISTRIBUTIONS: A HOME FOR THE DELTA SPIKE

Let's return to the single variable setting (though everything in this section generalizes readily to the multivariable context). We made use earlier, and will make further use later, of a mythical entity δ satisfying:

Delta property 6.9.1 $f * \delta = f$ *for all "nice enough"* f.

Delta property 6.9.2 *For any approximate identity* $\{g_{[\varepsilon]} : \varepsilon > 0\}$,

$$\delta = \lim_{\varepsilon \to 0^+} g_{[\varepsilon]}. \tag{6.194}$$

Delta property 6.9.3 $\widehat{\delta} = 1$ *and* $\widehat{1} = \delta$.

We've argued that no *function* δ can satisfy these properties. Here, we wish to show there's a *generalized* function δ that can!

To initiate our discussion of generalized functions, we first introduce a fresh perspective on ordinary, familiar functions. We're used to thinking of these in terms of what they do to the input variable; that is, we understand a function f through its values $f(x)$. But here's another way: Let's comprehend f by how it *integrates* against other functions!

In other words, let's think of a function f as a rule that associates, to a given input function ϕ, a complex number output, let's call it $T_f[\phi]$, defined by

$$T_f[\phi] = \int_{-\infty}^{\infty} f(y)\phi(y)\,dy \qquad (6.195)$$

provided the integral exists. We are thus thinking of f as a *functional*, meaning a function of functions.

We note that this new interpretation of f *completely determines* the old one, at least for sufficiently reasonable f. Specifically, let's say, to fix ideas, that $f \in \mathrm{BC}(\mathbb{R})$. If we define

$$\phi_{\varepsilon,x}(y) = G_{[\varepsilon]}(x-y) \qquad (6.196)$$

for $\varepsilon > 0$, $x \in \mathbb{R}$, G the Gaussian $G(v) = e^{-\pi v^2}$, and $G_{[\varepsilon]}(v) = \varepsilon^{-1} G(\varepsilon^{-1} v)$ as usual, then

$$\lim_{\varepsilon \to 0^+} T_f[\phi_{\varepsilon,x}] = \lim_{\varepsilon \to 0^+} \int_{-\infty}^{\infty} f(y)\phi_{\varepsilon,x}(y)\,dy$$

$$= \lim_{\varepsilon \to 0^+} \int_{-\infty}^{\infty} f(y) G_\varepsilon(x-y)\,dy = \lim_{\varepsilon \to 0^+} f * G_\varepsilon(x) = f(x)$$
$$(6.197)$$

by Proposition 5.7.1(a). So indeed, knowing how f integrates against a large enough class of functions tells us exactly what f is.

The above perspective leads us to our desired generalization of the notion of function, in the following way. The functional defined by (6.195) has a certain particularly pleasant nature, corresponding to the fact that the definite integral used to define it has certain particularly pleasant properties. We wish to *abstract* this nature, that is, to consider the collection, we'll call it $\mathrm{TD}(\mathbb{R})$, of *all* functionals of such a nature. Then every nice enough function f *is*, essentially, an element of $\mathrm{TD}(\mathbb{R})$, again by (6.195), so it's fair enough to say that elements T of $\mathrm{TD}(\mathbb{R})$ *are* generalizations of (nice enough) functions.

Our ultimate goal is to show that $\mathrm{TD}(\mathbb{R})$ contains an element δ satisfying, in a sense to be made clear along the way, Delta properties 6.9.1, 6.9.2, 6.9.3. But first things first: We need to say which kind of functions ϕ will serve as "inputs" to functionals $T \in \mathrm{TD}(\mathbb{R})$ and to specify the "particularly pleasant nature" we want these functionals to have. Let's.

Definition 6.9.1 (a) A *Schwartz function* is a function ϕ on $\mathbb{R}$ that is:

(i) *infinitely differentiable*, meaning $\phi^{(m)}(y)$ exists for all $m \in \mathbb{N}$ and $y \in \mathbb{R}$;

(ii) *of rapid decay*, meaning for any polynomial P on $\mathbb{R}$ and any $m \in \mathbb{N}$, the function $P\phi^{(m)}$ is bounded on $\mathbb{R}$.

The vector space of Schwartz functions is denoted $S(\mathbb{R})$.

(b) A *tempered distribution* is a functional $T: S(\mathbb{R}) \to \mathbb{C}$ that is:

(i) *linear*, meaning $T[a\phi + b\psi] = aT[\phi] + bT[\psi]$ for all $\phi, \psi \in S(\mathbb{R})$ and $a, b \in \mathbb{C}$;

(ii) *continuous*, meaning the following. If ϕ_N is a sequence of functions in $S(\mathbb{R})$ such that

$$\lim_{N \to \infty} ||P\phi_N^{(m)}||_\infty = 0 \qquad (6.198)$$

for every polynomial P on $\mathbb{R}$ and every $m \in \mathbb{N}$—such a sequence is said to *Schwartz-converge to zero*—then

$$\lim_{N \to \infty} T[\phi_N] = 0. \qquad (6.199)$$

The vector space of tempered distributions is denoted $\mathrm{TD}(\mathbb{R})$.

Perhaps some elements of the above definition seem less than completely natural; the investigations and results that follow should shed some light on the reasons behind them.

We'll need:

Lemma 6.9.1 (a) *If $\phi \in S(\mathbb{R})$, then so is $Q\phi^{(k)}$ for each polynomial Q on $\mathbb{R}$ and each $k \in \mathbb{N}$.*

(b) *If $\phi \in S(\mathbb{R})$ then, for each polynomial Q on $\mathbb{R}$ and each $k \in \mathbb{N}$, $Q\phi^{(k)}$ is in $L^1(\mathbb{R}) \cap L^2(\mathbb{R}) \cap \mathrm{BC}(\mathbb{R})$.*

(c) *If ϕ_N Schwartz-converges to zero and $p = 1$ or 2, then*

$$\lim_{N \to \infty} ||Q\phi_N^{(k)}||_p = 0 \qquad (6.200)$$

for each polynomial Q on $\mathbb{R}$ and each $k \in \mathbb{N}$.

Proof. Part (a) follows directly from the definition of $S(\mathbb{R})$. For part (b) we note that, by part (a), we need only show that $S(\mathbb{R}) \subset L^1(\mathbb{R}) \cap L^2(\mathbb{R}) \cap \mathrm{BC}(\mathbb{R})$. We do so as follows: Let $\phi \in S(\mathbb{R})$. Then certainly ϕ is continuous. Moreover, by Definition 6.9.1(a)(ii) with $m = 0$ and $P(y) = 1 + y^2$, there's a constant M such that $|\phi(y)| \leq M(1 + y^2)^{-1}$ for all y. Since $(1 + y^2)^{-1}$ is bounded, is in $L^1(\mathbb{R})$, and is in $L^2(\mathbb{R})$, Proposition 3.5.2 tell us that ϕ is too.

For part (c), suppose Q and k and a sequence of ϕ_N's Schwartz-converging to zero are given. Then (6.198) and the fact that $(1 + y^2)Q(y)$ is a polynomial tell us that, for N large enough,

$$\left|(1 + y^2)Q(y)\phi_N^{(k)}(y)\right| \leq 1 \qquad (6.201)$$

for all $y \in \mathbb{R}$. Thus

$$\left|Q(y)\phi_N^{(k)}(y)\right| \leq \frac{1}{1+y^2} \qquad (6.202)$$

for such N and y. The right side and its square are, again, both integrable on $\mathbb{R}$; the left side converges uniformly and therefore pointwise to zero as $N \to \infty$, by assumption on the ϕ_N's. So by the Lebesgue dominated convergence theorem (Proposition 3.5.1), $\|Q\,\phi_N^{(k)}\|_p \to 0$ as $N \to \infty$ for $p = 1$ or 2, as required. □

We've already encountered some Schwartz functions, namely, the Gaussian G and the compactly supported, infinitely differentiable function g of (5.110)/(5.111). See Exercises 6.9.1 and 5.7.5. In fact, Schwartz functions are abundant: It may be shown that $S(\mathbb{R})$ is *dense* in $L^1(\mathbb{R})$ with respect to the mean norm and in $L^2(\mathbb{R})$ with respect to the mean square norm.

The following proposition tells us that we've already seen some tempered distributions, too.

Proposition 6.9.1 *If there is a polynomial Q on $\mathbb{R}$ such that $f/Q \in L^1(\mathbb{R})$, or $f/Q \in L^2(\mathbb{R})$, or f/Q is bounded and measurable, then (6.195) defines a tempered distribution.*

Proof. We note from (6.195) that

$$T_f[\phi] = \int_{-\infty}^{\infty} (f/Q)(y)(Q\,\phi)(y)\,dy = \int_{-\infty}^{\infty} (Q\,\phi)^{-}(0-y)(f/Q)(y)\,dy$$
$$= (Q\,\phi)^{-} * (f/Q)(0) \qquad (6.203)$$

provided the convolution on the right exists.

Now suppose f and Q are as stipulated. Let ϕ belong to $S(\mathbb{R})$; then clearly so does ϕ^-, so by part (b) of the above lemma, $(Q\,\phi)^{-} \in L^1(\mathbb{R}) \cap L^2(\mathbb{R}) \cap \mathrm{BC}(\mathbb{R})$. Therefore, by Proposition 5.3.1, $(Q\,\phi)^{-} * (f/Q)(0)$ is defined for $f \in L^1(\mathbb{R})$ or $f \in L^2(\mathbb{R})$ or f bounded. So by (6.203) $T_f[\phi]$ is defined for such f and ϕ.

To show that T_f is continuous, in this situation, we assume ϕ_N Schwartz-converges to zero; then certainly (6.198) holds for $P = Q$ and $m = 0$. So $\|(Q\,\phi_N)^{-}\|_\infty = \|Q\,\phi_N\|_\infty$ approaches zero as $N \to \infty$. By Lemma 6.9.1(c), so does $\|(Q\,\phi_N)^{-}\|_p$ for $p = 1$ or 2. So by the latter part of Proposition 5.3.1, $T_f[\phi_N] = (Q\,\phi_N)^{-} * (f/Q)(0)$ itself approaches zero as $N \to \infty$. So T_f is continuous.

That T_f is linear follows directly from its definition. □

If f satisfies the hypotheses of the above proposition, then T_f is called *the tempered distribution associated with f*. Here is a tempered distribution that arises in a different way.

Example 6.9.1 Show that the functional δ from $S(\mathbb{R})$ to $\mathbb{C}$ given by

$$\delta[\phi] = \phi(0) \qquad (6.204)$$

is a tempered distribution.

Solution. Linearity is straightforward:

$$\delta[a\phi + b\psi] = (a\phi + b\psi)(0) = a\phi(0) + b\psi(0) = a\delta[\phi] + b\delta[\psi]. \quad (6.205)$$

So is continuity: If the sequence ϕ_N Schwartz-converges to zero, then by Proposition 1.7.1 it converges pointwise to zero, so

$$\lim_{N\to\infty} \delta[\phi_N] = \lim_{N\to\infty} \phi_N(0) = 0. \quad (6.206)$$

It's no accident that we've called the above tempered distribution δ: This δ *does* satisfy Delta properties 6.9.1, 6.9.2, 6.9.3. We aim to prove this, but first we must explain it: What do the relevant operations—the convolution, the limit, and the Fourier transforms—*mean* when the entities being operated on are not functions but tempered distributions?

Let's answer: We begin by making sense of Delta property 6.9.1, in the following way. We describe a natural means of convolving a tempered distribution with a Schwartz function and show that δ is an identity for this convolution.

To motivate our new type of convolution, let's take another look at the old one. For appropriate functions f and ψ, we have

$$f * \psi(x) = \psi * f(x) = \int_{-\infty}^{\infty} \psi(x-y) f(y)\, dy = T_f[\psi_x], \quad (6.207)$$

where for any $x \in \mathbb{R}$, ψ_x is the function on $\mathbb{R}$ defined by

$$\psi_x(y) = \psi(x - y) \quad (6.208)$$

and $T_f[\psi_x]$ is as described by (6.195). Note that each ψ_x is in $S(\mathbb{R})$ whenever ψ is. So a sensible way to define $T * \psi$, for T *any* element of TD($\mathbb{R}$) and $\psi \in S(\mathbb{R})$, would be to simply replace T_f by T on the far right side of (6.207)! Let's be sensible.

Definition 6.9.2 If $T \in \text{TD}(\mathbb{R})$ and $\psi \in S(\mathbb{R})$, then the *convolution* $T * \psi$ is the function on $\mathbb{R}$ defined by

$$T * \psi(x) = T[\psi_x], \quad (6.209)$$

where ψ_x is given by (6.208).

We've set things up so that, if $T = T_f$ for some suitable function f, then Definition 6.9.2 yields the same thing as Definition 5.1.1. But now, let's consider a situation involving a $T \in \text{TD}(\mathbb{R})$ not of that form.

Example 6.9.2 Show that the tempered distribution δ is an identity for the convolution of Definition 6.9.2.

Solution. For $\psi \in S(\mathbb{R})$ and $x \in \mathbb{R}$,

$$\delta * \psi(x) = \delta[\psi_x] = \psi_x(0) = \psi(x-0) = \psi(x), \tag{6.210}$$

the second equality by definition of δ and the third by that of ψ_x.

This takes care of Delta property 6.9.1. To consider the next one, we need the following, which gives a simple, natural, and quite useful notion of convergence in $TD(\mathbb{R})$.

Definition 6.9.3 We say a collection $\{T_\varepsilon : \varepsilon > 0\}$ (respectively, $\{T_N : N \in \mathbb{Z}^+\}$) of elements of $TD(\mathbb{R})$ *converges temperately* to $T \in TD(\mathbb{R})$ if

$$\lim_{\varepsilon \to 0^+} T_\varepsilon[\phi] = T[\phi] \tag{6.211}$$

(respectively, $\lim_{N \to \infty} T_N[\phi] = T[\phi]$) for all Schwartz functions ϕ.

Example 6.9.3 Show that Delta property 6.9.2 holds in the following sense. Let $\{g_{[\varepsilon]} : \varepsilon > 0\}$ be an approximate identity for convolution, as described in Section 5.7. Show that the tempered distributions $T_{g_{[\varepsilon]}}$ converge temperately to the tempered distribution δ.

Solution. Let $\phi \in S(\mathbb{R})$; then certainly $\phi^- \in BC(\mathbb{R})$. So if $\{g_{[\varepsilon]} : \varepsilon > 0\}$ is an approximate identity, then by (6.195) and Proposition 5.7.1(a),

$$\lim_{\varepsilon \to 0^+} T_{g_{[\varepsilon]}}[\phi] = \lim_{\varepsilon \to 0^+} \phi^- * g_{[\varepsilon]}(0) = \phi^-(0) = \phi(0) = \delta[\phi], \tag{6.212}$$

as required.

So we are down to consideration of Delta property 6.9.3. Here is the key.

Lemma 6.9.2 *If* $\phi \in S(\mathbb{R})$, *then* $\widehat{\phi} \in S(\mathbb{R})$.

Proof. Let $\phi \in S(\mathbb{R})$: By Lemma 6.9.1(b), $x^m \phi^{(k)}(x) \in L^1(\mathbb{R})$ for all $k, m \in \mathbb{N}$. Now by repeated application of Proposition 6.2.1(c)(i,ii),

$$F[x^m \phi^{(k)}(x)](s) = (2\pi i)^{k-m}(-1)^m s^k (\widehat{\phi})^{(m)}(s). \tag{6.213}$$

In particular, $(\widehat{\phi})^{(m)}(s)$ exists for all $m \in \mathbb{N}$ and $s \in \mathbb{R}$, so $\widehat{\phi}$ is infinitely differentiable. Moreover, by Proposition 6.1.1, the left and therefore the right side of (6.213) is bounded in s; so clearly $P(s)(\widehat{\phi})^{(m)}(s)$ is too, for any polynomial P and any $m \in \mathbb{N}$, so $\widehat{\phi}$ has rapid decay. So $\widehat{\phi} \in S(\mathbb{R})$. $\square$

The above lemma permits the following.

Definition 6.9.4 The *Fourier transform* of a tempered distribution T is the functional $\widehat{T} : S(\mathbb{R}) \to \mathbb{C}$ defined by

$$\widehat{T}[\phi] = T[\widehat{\phi}] \tag{6.214}$$

for $\phi \in S(\mathbb{R})$.

We check that $\widehat{T}$ behaves as it should:

Proposition 6.9.2 (a) $\widehat{T} \in \text{TD}(\mathbb{R})$ *for each* $T \in \text{TD}(\mathbb{R})$.

(b) *Let* $f \in L^1(\mathbb{R})$ *or* $f \in L^2(\mathbb{R})$. *Then* $T_{\widehat{f}} = \widehat{T_f}$; *that is, the tempered distribution associated with the Fourier transform of* f *equals the Fourier transform of the tempered distribution associated with* f.

Proof. Part (a): Let $T \in \text{TD}(\mathbb{R})$. Then for $\phi, \psi \in S(\mathbb{R})$ and $a, b \in \mathbb{C}$,

$$\widehat{T}[a\phi + b\psi] = T[F[a\phi(x) + b\psi(x)]] = T\left[a\widehat{\phi} + b\widehat{\psi}\right] = aT\left[\widehat{\phi}\right] + bT\left[\widehat{\psi}\right]$$
$$= a\widehat{T}[\phi] + b\widehat{T}[\psi] \tag{6.215}$$

by linearity of T and of the usual Fourier transform on $L^1(\mathbb{R})$. So $\widehat{T}$ is linear. To show that it's continuous, let ϕ_N Schwartz-converge to zero; let $k, m \in \mathbb{N}$, and let $P_m(x) = x^m$. Then by Lemma 6.9.1(c), $||P_m \phi_N^{(k)}||_1 \to 0$ as $N \to \infty$, so by (6.213) and Proposition 6.1.1, $||P_k(\widehat{\phi_N})^{(m)}||_\infty$ goes to zero as $N \to \infty$, whence, clearly, so does $||P(\widehat{\phi_N})^{(m)}||_\infty$ for any polynomial P. But then by Definition 6.9.4 and the continuity of T,

$$\lim_{N \to \infty} \widehat{T}[\phi_N] = \lim_{N \to \infty} T[\widehat{\phi_N}] = 0. \tag{6.216}$$

So $\widehat{T}$ is continuous, as claimed.

Part (b): Let $\phi \in S(\mathbb{R})$. Then $\phi \in L^1(\mathbb{R}) \cap L^2(\mathbb{R})$ so, if also $f \in L^1(\mathbb{R}) \cup L^2(\mathbb{R})$, then by the composition theorem (Proposition 6.2.1(d)(ii), which generalizes to the situation at hand),

$$\int_{-\infty}^{\infty} \widehat{f}(y)\phi(y)\, dy = \int_{-\infty}^{\infty} f(y)\widehat{\phi}(y)\, dy. \tag{6.217}$$

The left side, according to (6.195), equals $T_{\widehat{f}}[\phi]$; the right side, according to (6.195) and Definition 6.9.4, equals $\widehat{T_f}[\phi]$. We're done. □

Now, we consider the Fourier transform of two tempered distributions, one associated with a function neither in $L^1(\mathbb{R})$ nor in $L^2(\mathbb{R})$ and one not associated with any function at all. In so doing, we take care of Delta property 6.9.3.

Example 6.9.4 Show that $\widehat{\delta} = T_1$ and $\widehat{T_1} = \delta$.
Solution. Let $\phi \in S(\mathbb{R})$. Then by Definition 6.9.4,

$$\widehat{\delta}[\phi] = \delta[\widehat{\phi}] = \widehat{\phi}(0) = \int_{-\infty}^{\infty} \phi(y) e^{-2\pi i (0) y}\, dy = \int_{-\infty}^{\infty} 1 \cdot \phi(y)\, dy = T_1[\phi],$$
$$\tag{6.218}$$

so $\widehat{\delta} = T_1$. Moreover,

$$\widehat{T_1}[\phi] = T_1[\widehat{\phi}] = \int_{-\infty}^{\infty} 1 \cdot \widehat{\phi}(s)\, ds = \int_{-\infty}^{\infty} \widehat{\phi}(s) e^{2\pi i(0)s}\, ds = \phi(0) = \delta[\phi] \tag{6.219}$$

(the next to the last step by Theorem 6.4.1(c)), so $\widehat{T_1} = \delta$.

What would the Fourier transform on $\mathrm{TD}(\mathbb{R})$ be without a Fourier inversion result there? It's a rhetorical question, in light of:

Theorem 6.9.1 (Fourier inversion on $\mathrm{TD}(\mathbb{R})$) *If $T \in \mathrm{TD}(\mathbb{R})$, then $\widehat{\widehat{T}} = T^-$, where $T^- \in \mathrm{TD}(\mathbb{R})$ is defined by*

$$T^-[\phi] = T[\phi^-] \quad (\phi \in S(\mathbb{R})). \tag{6.220}$$

Proof. For T and ϕ as stipulated,

$$\widehat{\widehat{T}}[\phi] = \widehat{T}[\widehat{\phi}] = T[\widehat{\widehat{\phi}}] = T[\phi^-] = T^-[\phi] \tag{6.221}$$

by Theorem 6.4.2 (or Theorem 6.5.1). □

Exercises

6.9.1 Show that $G \in S(\mathbb{R})$ (as usual, $G(x) = e^{-\pi x^2}$). Hint: First show, by induction on k, that $G^{(k)}(x) = p_k(x)G(x)$, where p_k is a polynomial. Then use l'Hôpital's rule and induction to show that $x^m G(x)$ is bounded for any $m \in \mathbb{Z}^+$.

6.9.2 Let $T \in \mathrm{TD}(\mathbb{R})$; define the *derivative T'* by

$$T'[\phi] = -T[\phi'] \quad (\phi \in S(\mathbb{R})). \tag{6.222}$$

a. Show that $T' \in \mathrm{TD}(\mathbb{R})$. Hint: If a sequence ϕ_N Schwartz-converges to zero (see Definition 6.9.1), then so clearly does ϕ'_N.

b. Show that the definition of T' is logical, in the sense that, if $f \in \mathrm{PSC}(\mathbb{R}) \cap L^1(\mathbb{R})$ and $f' \in L^1(\mathbb{R})$, say, then $(T_f)'[\phi] = T_{f'}[\phi]$ for all $\phi \in \mathbb{R}$. (T_f is, again, the tempered distribution associated with f, cf. (6.195); similarly for $T_{f'}$.) Hint: Integrate by parts in the integral defining $T_{f'}[\phi]$. (In cleaning up after the integration by parts, some technical details from the proof of Proposition 6.2.1(c)(i) may be of use.)

c. Show that "the derivative of $\chi_{[0,\infty)}$ is δ," in the sense that

$$(T_{\chi_{[0,\infty)}})'[\phi] = \delta[\phi] \quad (\phi \in S(\mathbb{R})).$$

(The tempered distribution δ is as in Example 6.9.1.)

d. Show that, if $T_1, T_2, \ldots$ and T are elements of $\mathrm{TD}(\mathbb{R})$ such that the T_N's converge temperately to T, then the T_N''s converge temperately to T'.

6.9.3 Suppose that $\phi \in S(\mathbb{R})$ and that g is infinitely differentiable on $\mathbb{R}$, with $g^{(k)}$ of at most polynomial growth (that is, $g^{(k)}$ divided by some polynomial is bounded) for all $k \in \mathbb{N}$.

    a. Show that the product $g\phi$ is in $S(\mathbb{R})$.

    b. Show that, if T is in $\mathrm{TD}(\mathbb{R})$, then so is the functional gT defined by $gT[\phi] = T[g\phi]$.

    c. Show that this definition of gT is logical in the sense that, if T_f is the tempered distribution (6.195) associated with a suitable function f, then $gT_f = T_{gf}$.

    d. Verify the product rule $(gT)' = g'T + gT'$, where the derivative of a tempered distribution is as defined in Exercise 6.9.2.

    e. Let $X(x) = x$. Verify the analogs

$$\widehat{T'} = 2\pi i X\widehat{T}, \quad \widehat{XT} = -(2\pi i)^{-1}\left(\widehat{T}\right)' \quad (T \in \mathrm{TD}(\mathbb{R}))$$

of Proposition 6.2.1(c), where the Fourier transform of a tempered distribution is given by Definition 6.9.4.

6.9.4 Suppose that functions f_N and f satisfy the hypotheses of Proposition 6.9.1, that the f_N's converge pointwise almost everywhere to f, and that the f_N's are all bounded by the same constant K. Show that the T_{f_N}'s converge temperately to T_f. Hint: Proposition 3.5.1.

6.9.5 Prove that, if $T_N \to T$ temperately as $N \to \infty$ for $T_N, T \in \mathrm{TD}(\mathbb{R})$, then $\widehat{T_N} \to \widehat{T}$ temperately as well.

6.9.6 Write $R(y) = 1/y$. We'd like to define the tempered distribution T_R in the usual way: $T_R[\phi] = \int_{-\infty}^{\infty} y^{-1}\phi(y)\, dy$. But this integral need not converge, since y^{-1} blows up at $y = 0$. So we modify our approach somewhat: We define

$$T_R[\phi] = \lim_{\varepsilon \to 0^+} \int_{\varepsilon < |y|} \frac{\phi(y)}{y}\, dy.$$

    a. Show that this definition makes sense for any $\phi \in S(\mathbb{R})$. Hint: Integrability of $y^{-1}\phi(y)$ over $|y| > b$, for any positive b, is clear from the rapid decay of ϕ. So we need only check existence of $\lim_{\varepsilon \to 0^+} \int_{\varepsilon < |y| < b} y^{-1}\phi(y)\, dy$. Write

$$\int_{\varepsilon < |y| < b} \frac{\phi(y)}{y}\, dy = \int_{\varepsilon < |y| < b} \frac{\phi(y) - \phi(0)}{y}\, dy + \int_{\varepsilon < |y| < b} \frac{\phi(0)}{y}\, dy.$$

Use existence of $\phi'(0)$ to show integrability of $(\phi(y) - \phi(0))/y$ on $(-b, b)$ for any $b > 0$. Also, actually evaluate the rightmost integral for any ε, b; now take limits.

    b. Show $T_R \in \mathrm{TD}(\mathbb{R})$.

6.9.7 **a.** Show that "the Fourier transform of the reciprocal function is πi times the sign function." In other words, show $\widehat{T_R} = \pi i\, T_{\text{sgn}}$, where the sign function sgn y is given by (A.12). Hint:

$$\widehat{T_R}[\phi]$$

$$= \lim_{\varepsilon \to 0^+} \int_{\varepsilon < |y|} \frac{\widehat{\phi}(y)}{y}\, dy = \lim_{\varepsilon \to 0^+} \lim_{b \to \infty} \int_{\varepsilon < |y| < b} \left(\int_{-\infty}^{\infty} \phi(x) e^{-2\pi i x y}\, dx \right) \frac{1}{y}\, dy.$$

Switch the order of integration (justified because ϕ is of rapid decay and $|y|^{-1}$ is bounded on $(-b, -\varepsilon) \cup (\varepsilon, b)$). Write the complex exponential out in terms of cosine and sine; show that the integral in y of the terms involving cosine is zero. Show that you can bring the limits inside the resulting integral in x; evaluate the resulting limit of integrals in y. Here Lemma 6.6.1 may help.

b. What is the Fourier transform of the sign function?

7

Special Topics and Applications

Here we consider a variety of implementations and extensions of the ideas and results of the previous two chapters. In so doing we provide a small, but we hope suggestive, glimpse of the incredibly vast array of questions, problems, and disciplines to which Fourier transform concepts and methods are of fundamental relevance.

7.1 HERMITE FUNCTIONS

Fourier transform notions are useful in many situations where eigenvalue problems on unbounded intervals are at issue. In this section we apply some of these notions to analysis of the *Hermite equation*

$$h''(x) - x^2 h(x) + \lambda h(x) = 0 \quad (-\infty < x < \infty), \tag{7.1}$$

which arises in the study of the Schrödinger equation in quantum physics, as we'll discuss a bit later in this section.

We have:

Proposition 7.1.1 (a) *For each $n \in \mathbb{N}$, $\lambda_n = 2n + 1$ is an eigenvalue of (7.1), with corresponding eigenfunction given by the nth Hermite function*

$$h_n(x) = (-1)^n e^{x^2/2} \frac{d^n}{dx^n} e^{-x^2}. \tag{7.2}$$

(b) $\{h_n \colon n \in \mathbb{N}\}$ *is an orthogonal basis for $L^2(\mathbb{R})$, and for each n,*

$$||h_n||^2 = 2^n n! \sqrt{\pi}. \tag{7.3}$$

Proof. Part (a): We deduce from (7.2) and the product rule that

$$h'_n(x) = (-1)^n \left[x\, e^{x^2/2} \frac{d^n}{dx^n} e^{-x^2} + e^{x^2/2} \frac{d^n}{dx^n} \left(\frac{d}{dx} e^{-x^2} \right) \right]$$

$$= x h_n(x) - 2(-1)^n e^{x^2/2} \frac{d^n}{dx^n} \left(x e^{-x^2} \right). \qquad (7.4)$$

Now by Leibniz' rule (4.155) and the fact that the mth derivative of x is zero for $m > 1$,

$$\frac{d^n}{dx^n} \left(x e^{-x^2} \right) = x \frac{d^n}{dx^n} e^{-x^2} + n \frac{d^{n-1}}{dx^{n-1}} e^{-x^2}, \qquad (7.5)$$

so (7.4) reads

$$h'_n(x) = x h_n(x) - 2(-1)^n e^{x^2/2} \left[x \frac{d^n}{dx^n} e^{-x^2} + n \frac{d^{n-1}}{dx^{n-1}} e^{-x^2} \right]$$

$$= -x h_n(x) - 2n(-1)^n e^{x^2/2} \frac{d^{n-1}}{dx^{n-1}} e^{-x^2}. \qquad (7.6)$$

(At the last step we've applied the definition (7.2) of h_n again.) Differentiating both sides of (7.6) gives

$$h''_n(x) = -h_n(x) - x h'_n(x) - 2n(-1)^n \left[x\, e^{x^2/2} \frac{d^{n-1}}{dx^{n-1}} e^{-x^2} + e^{x^2/2} \frac{d^n}{dx^n} e^{-x^2} \right]$$

$$= -(2n+1) h_n(x) - x h'_n(x) - 2n(-1)^n x\, e^{x^2/2} \frac{d^{n-1}}{dx^{n-1}} e^{-x^2}$$

$$= -(2n+1) h_n(x) + x^2 h_n(x); \qquad (7.7)$$

the second step is by (7.2) and the last by (7.6). This proves part (a) of our proposition.

For part (b), we observe that the nth derivative of $(-1)^n e^{-x^2}$ is a *polynomial* $H_n(x)$ times e^{-x^2}. In fact, H_n has degree n and leading term $(2x)^n$; this is essentially because of the factor of $-2x$ coming from each differentiation of e^{-x^2}. By definition (7.2) of h_n, we have

$$h_n(x) = e^{-x^2/2} H_n(x). \qquad (7.8)$$

The rapid decay of the exponential factor $e^{-x^2/2}$ and the comparatively slow growth of the polynomial $H_n(x)$ then assure that h_n is bounded, and is in fact in $L^2(\mathbb{R})$.

Now let's evaluate the inner product $\langle h_k, h_n \rangle$. This equals $\langle h_n, h_k \rangle$, so it suffices to consider the case $k \le n$. By (7.2) and (7.8), we have

$$\langle h_k, h_n \rangle = \int_{-\infty}^{\infty} h_k(x) h_n(x)\, dx = (-1)^n \int_{-\infty}^{\infty} h_k(x) e^{x^2/2} \frac{d^n}{dx^n} e^{-x^2}\, dx$$

$$= (-1)^n \int_{-\infty}^{\infty} H_k(x) \frac{d^n}{dx^n} e^{-x^2}\, dx. \qquad (7.9)$$

We integrate by parts n times, the first with $u = H_k(x)$ and $dv = (d^n/dx^n)e^{-x^2}\,dx$, the second with $u = H_k'(x)$ and $dv = (d^{n-1}/dx^{n-1})e^{-x^2}\,dx$, and so on. We note that, each time, the term $uv\big|_{-\infty}^{\infty}$ evaluates to zero because, again, e^{-x^2} and its derivatives decay rapidly as $|x| \to \pm\infty$. So we get

$$\langle h_k, h_n \rangle = \int_{-\infty}^{\infty} H_k^{(n)}(x) e^{-x^2}\,dx = 2^n n!\, \delta_{k,n} \int_{-\infty}^{\infty} e^{-x^2}\,dx; \tag{7.10}$$

the last step is because, since H_k has leading term $(2x)^k$, its nth derivative is zero if $k < n$ and $(-2)^n n!$ if $k = n$. By Lemma 6.2.1 and a change of variable, we find that the integral on the right side of (7.10) equals $\sqrt{\pi}$; so we've demonstrated not only the orthogonality of the h_n's but also that their norms have the stated values.

It remains to show that these Hermite functions form a basis for $L^2(\mathbb{R})$. We argue as follows: We pick $f \in L^2(\mathbb{R})$ and let

$$g = f - \sum_{n=0}^{\infty} \frac{\langle f, h_n \rangle}{||h_n||^2} h_n. \tag{7.11}$$

(By Proposition 3.7.2(b), the sum converges in norm.) If we can show that $g = 0$, then we'll be done by Corollary 3.6.2.

By Lemma 3.6.1, we have

$$\langle g, h_k \rangle = \left\langle f - \sum_{n=0}^{\infty} \frac{\langle f, h_n \rangle}{||h_n||^2} h_n, h_k \right\rangle$$

$$= \langle f, h_k \rangle - \sum_{n=0}^{\infty} \frac{\langle f, h_n \rangle}{||h_n||^2} \langle h_n, h_k \rangle = \langle f, h_k \rangle - \sum_{n=0}^{\infty} \frac{\langle f, h_n \rangle}{||h_n||^2} ||h_n||^2 \delta_{n,k}$$

$$= \langle f, h_k \rangle - \frac{\langle f, h_k \rangle}{||h_k||^2} ||h_k||^2 = 0 \tag{7.12}$$

for $k \in \mathbb{N}$. So g is orthogonal to all h_k's. Consequently, to prove g is zero, we need only show that *any element g of $L^2(\mathbb{R})$ that's orthogonal to all h_k's is zero.* Further, to demonstrate the latter, it suffices to show that *any element g of $L^2(\mathbb{R})$ that's orthogonal to $L_m(x) = x^m e^{-x^2/2}$ for each $m \in \mathbb{N}$ is zero*. Why does this suffice? Because each h_k and each L_m is a polynomial times $e^{-x^2/2}$, so just as in the proof of Theorem 4.6.1(a), we may write

$$L_m(x) = \sum_{k=0}^{\infty} d_k h_k(x) \tag{7.13}$$

for contants d_k. Thus any $g \in L^2(\mathbb{R})$ orthogonal to all h_k's is orthogonal to all L_m's. So indeed the second italicized statement above implies the first.

So let's prove the second: Here's where the Fourier transform and Fourier inversion play a role. Suppose $g \in L^2(\mathbb{R})$ satisfies $\langle g, L_m \rangle = 0$ for each m. Let $s \in \mathbb{R}$ and

356   SPECIAL TOPICS AND APPLICATIONS

$N \in \mathbb{N}$; then

$$0 = \sum_{m=0}^{N} \frac{(-2\pi i s)^m}{m!} \langle g, L_m \rangle = \sum_{m=0}^{N} \frac{(-2\pi i s)^m}{m!} \int_{-\infty}^{\infty} g(x) x^m e^{-x^2/2} \, dx$$

$$= \int_{-\infty}^{\infty} g(x) \left(\sum_{m=0}^{N} \frac{(-2\pi i s x)^m}{m!} \right) e^{-x^2/2} \, dx. \qquad (7.14)$$

Now

$$\left| \sum_{m=0}^{N} \frac{(-2\pi i s x)^m}{m!} \right| < \sum_{m=0}^{N} \left| \frac{(-2\pi i s x)^m}{m!} \right| < \sum_{m=0}^{\infty} \frac{|2\pi s x|^m}{m!} = e^{2\pi|sx|}; \qquad (7.15)$$

also, from the rapid decay of $e^{-x^2/2} e^{2\pi|sx|}$ and the fact that $g \in L^2(\mathbb{R})$, we find readily that $|g(x)| e^{-x^2/2} e^{|2\pi s x|}$ is integrable on $\mathbb{R}$. So the Lebesgue dominated convergence theorem (Proposition 3.5.1) applies to (7.14); the result is

$$0 = \lim_{N \to \infty} \int_{-\infty}^{\infty} g(x) \left(\sum_{m=0}^{N} \frac{(-2\pi i s x)^m}{m!} \right) e^{-x^2/2} \, dx$$

$$= \int_{-\infty}^{\infty} g(x) \left(\sum_{m=0}^{\infty} \frac{(-2\pi i s x)^m}{m!} \right) e^{-x^2/2} \, dx = \int_{-\infty}^{\infty} g(x) e^{-2\pi i s x} e^{-x^2/2} \, dx$$

$$= F[g(x) e^{-x^2/2}](s), \qquad (7.16)$$

the last step by Definition 6.1.1 of the Fourier transform. But if the Fourier transform of a function in $L^1(\mathbb{R})$ is zero, then so is that function, by Fourier inversion. So $g(x) e^{-x^2/2}$ is the zero function; hence g is too, and we're done. $\square$

In quantum mechanics, a *simple, isotropic harmonic oscillator* of angular frequency $\omega > 0$ is a single-particle system of potential energy

$$U(x, y, z) = \frac{m\omega^2}{2}(x^2 + y^2 + z^2), \qquad (7.17)$$

where m is the mass of the particle. The so-called *observable energy levels* of this system are the eigenvalues of the Schrödinger equation

$$\frac{\hbar^2}{2m} \nabla_3^2 \psi - U\psi + E\psi = 0, \qquad (7.18)$$

where $\hbar$ is Planck's constant. Let's find these energy levels.

We do so by separating variables. Specifically, if X, Y, and Z solve, respectively, the Sturm-Liouville problems

$$\frac{\hbar^2}{2m} X''(x) - \frac{m\omega^2}{2} x^2 X(x) + E_x X(x) = 0, \qquad (7.19)$$

$$\frac{\hbar^2}{2m} Y''(y) - \frac{m\omega^2}{2} y^2 Y(y) + E_y Y(y) = 0, \qquad (7.20)$$

$$\frac{\hbar^2}{2m} Z''(z) - \frac{m\omega^2}{2} z^2 Z(z) + E_z Z(z) = 0, \qquad (7.21)$$

then as is readily shown, $\psi = XYZ$ solves (7.18) for U as in (7.17) and $E = E_x + E_y + E_z$.

Now it's straightforward to check (DIY) that the function h satisfies the Hermite equation (7.1) if and only if the function $X(x) = h(\sqrt{m\omega}\,x/\sqrt{\hbar})$ satisfies (7.19), with $E_x = \lambda\hbar\omega/2$. Similarly for Y and Z. So, since the Hermite equation has an eigenvalue $\lambda = (2n+1)$ for each $n \in \mathbb{N}$, (7.18)—again, with U as in (7.17)—has an eigenvalue

$$E_{n,k,j} = \frac{(2n+1)\hbar\omega}{2} + \frac{(2k+1)\hbar\omega}{2} + \frac{(2j+1)\hbar\omega}{2} = \left(n+k+j+\frac{3}{2}\right)\hbar\omega \tag{7.22}$$

for each triple $(n,k,j) \in \mathbb{N}^3$.

We claim that these are the *only* eigenvalues. To prove this, suppose F were another one. By an argument much like that of the proof of Theorem 4.1.1(b), concerning Sturm-Liouville problems, it would follow that, for each $(n,k,j) \in \mathbb{N}^3$,

$$\int_{-\infty}^{\infty}\int_{-\infty}^{\infty}\int_{-\infty}^{\infty} \psi_{n,k,j}(x,y,z)\overline{\psi(x,y,z)}\,dx\,dy\,dz = 0, \tag{7.23}$$

where ψ is the eigenfunction corresponding to F and

$$\psi_{n,k,j}(x,y,z) = h_n\left(\sqrt{\frac{m\omega}{\hbar}}x\right) h_k\left(\sqrt{\frac{m\omega}{\hbar}}y\right) h_j\left(\sqrt{\frac{m\omega}{\hbar}}z\right) \tag{7.24}$$

the eigenfunction corresponding to $E_{n,k,j}$. From this it would follow in turn that ψ is zero, much as the orthogonality of a function g to all h_k's implied g was zero in the proof of Proposition 7.1.1(b) directly above. But eigenfunctions are nonzero, so we have a contradiction to the assumption that there be eigenvalues other than the $E_{n,k,j}$'s, and our claim is proved.

To finish we note that, as n, k, and j range over $\mathbb{N}$, so does their sum. So we have the following.

Proposition 7.1.2 *The observable energy levels of the simple, isotropic harmonic oscillator given by* (7.17), (7.18) *are precisely the numbers*

$$\left(M+\frac{3}{2}\right)\hbar\omega \quad (M \in \mathbb{N}). \tag{7.25}$$

The methods of this section may also be used to study *Laguerre polynomials*, which arise in the study of quantum mechanical systems consisting of a single proton and electron. See, for example, Section 6.5 in [19].

Exercises

7.1.1 Compute explicitly the first six Hermite functions $h_0, h_1, \ldots, h_5$.

7.1.2 Use the definition of h_n to show that $h_{n+1}(x) = xh_n(x) - h'_n(x)$ for all n.

7.1.3 The *Hermite polynomials* H_n ($n \in \mathbb{N}$) are the poynomials defined by (7.8). Use the definition of h_n and Proposition 7.1.1 to show that, for any $n \in \mathbb{N}$:

a. $H_n(x) = (-1)^n e^{x^2} \frac{d^n}{dx^n} e^{-x^2}$.

b. $d(e^{-x^2} H_n(x))/dx = -e^{-x^2} H_{n+1}(x)$.

c. H_n is an eigenfunction of the problem

$$\left[e^{-x^2} H'(x)\right]' + \lambda e^{-x^2} H(x) = 0 \quad (-\infty < x < \infty),$$

with eigenvalue $\lambda = 2n$.

d. The H_n's form an orthogonal basis for $L^2_w(a,b)$, where $w(x) = e^{-x^2}$. (See Definition 4.1.2.) What is $\|H_n\|_w$?

7.1.4 Expand $f(x) = x^3 + x^4$ into a series of Hermite polynomials. Hint: See the beginning of the proof of Theorem 4.6.1.

7.1.5 Show that, for any $z \in \mathbb{R}$,

$$e^{2xz-z^2} = \sum_{n=0}^{\infty} H_n(x) \frac{z^n}{n!}.$$

Hint: Start with the Maclaurin series

$$e^{-(x-z)^2} = \sum_{n=0}^{\infty} \frac{z^n}{n!} \left[\frac{d^n}{dz^n} e^{-(x-z)^2}\right]_{z=0}.$$

Compute the derivatives on the right using Exercise 7.1.3a and the chain rule; multiply both sides by e^{x^2}.

7.1.6 Use the above exercise to expand $f(x) = e^x$ into a series of Hermite polynomials in x.

7.1.7 Show that $H_n(0) = 0$ for all odd n and $H_{2m}(0) = (-1)^m (2m)!/m!$ for $m \in \mathbb{N}$. Hint: Plug $x = 0$ into the result of Exercise 7.1.5. Compare with the Maclaurin series for e^{-z^2} (which you can get by plugging $x = -z^2$ into (A.20)).

7.1.8 Expand $f(x) = \chi_{[0,\infty)}$ into a series of Hermite polynomials. Hint: Exercise 7.1.3b, Lemma 6.2.1, and Exercise 7.1.7. ANSWER:

$$f = \frac{1}{2} H_0 + \frac{1}{2\sqrt{\pi}} \sum_{m=0}^{\infty} \frac{(-1)^m}{2^{2m}(2m+1)m!} H_{2m+1}.$$

7.2 BOUNDARY VALUE PROBLEMS

Here we study some boundary value problems that lead to eigenvalue problems on unbounded intervals. These eigenvalue problems share, as we'll see, an interesting feature: The *spectrum*, or set of eigenvalues, of each is uncountable. This is to be contrasted with the Hermite equation (7.1), for example, which had a countable spectrum.

Example 7.2.1 Solve the boundary value problem

$$k\, u_{xx} = u_t \quad (-\infty < x < \infty,\ t > 0), \tag{7.26}$$
$$u(x,0) = f(x) \quad (-\infty < x < \infty), \tag{7.27}$$

representing heat flow in a uniform bar of infinite extent.

Solution. We separate variables as in Example 2.3.1. We find that $u = XT$ solves the differential equation (7.26) provided

$$X''(x) + \lambda X(x) = 0 \quad (-\infty < x < \infty), \quad T'(t) + \lambda k\, T(t) = 0 \quad (t > 0). \tag{7.28}$$

These ordinary differential equations have the following familiar, simultaneous solutions:

$$X(x) = \gamma + \delta x \text{ and } T(t) = \sigma \text{ if } \lambda = 0, \tag{7.29}$$
$$X(x) = \gamma e^{i\sqrt{\lambda}\,x} + \delta e^{-i\sqrt{\lambda}\,x} \text{ and } T(t) = \sigma e^{-\lambda k t} \text{ if } \lambda \neq 0. \tag{7.30}$$

In Example 2.3.1 we wrote our X's in terms of cosines and sines, but here, for reasons to become clear shortly, complex exponentials are preferable.

Now if λ is real and positive, say $\lambda = (2\pi s)^2$ with $s \in \mathbb{R}^+$ (the factor of 2π is to prepare us for use of the Fourier transform), then the function X of (7.30) is a linear combination of complex exponentials in x and is therefore bounded on $\mathbb{R}$. Moreover, for such λ, the function T of (7.30) is itself bounded for $t > 0$. That is, we have a bounded separated solution

$$u_s(x,t) = \left(D_s e^{2\pi i s x} + E_s e^{-2\pi i s x}\right) e^{-(2\pi s)^2 k t} \tag{7.31}$$

to the differential equation (7.26) for each $s \in \mathbb{R}^+$. Of course we might also consider (7.30) in the case $\lambda \notin \mathbb{R}^+$, but such λ yield *exponentially growing* solutions X (See Exercise 7.2.1), which we reject because they are physically implausible. (And because we won't need them!)

We are thereby led to consideration of the *continuous superposition*

$$u(x,t) = \int_0^\infty u_s(x,t)\, ds = \int_0^\infty \left(D_s e^{2\pi i s x} + E_s e^{-2\pi i s x}\right) e^{-(2\pi s)^2 k t}\, ds$$
$$= \int_{-\infty}^\infty C_s e^{-(2\pi s)^2 k t} e^{2\pi i s x}\, ds. \tag{7.32}$$

Here, for the last equality, we wrote the integral as a sum of two such; replaced s by $-s$ in the second of these; and wrote C_s for the quantity that equals D_s for $s > 0$ and equals E_{-s} for $s < 0$. Also, because integrating over $\mathbb{R} - \{0\}$ is the same doing so over $\mathbb{R}$, we ignored (7.29).

Since the differential equation in question is linear and homogeneous, the above superposition should, at least formally, solve it as well. Now of course, we want this superposition also to satisfy the inhomogeneous condition (7.27); that is, we want

$$f(x) = u(x,0) = \int_{-\infty}^{\infty} C_s e^{-(2\pi s)^2 k(0)} e^{2\pi i s x} \, ds = \int_{-\infty}^{\infty} C_s e^{2\pi i s x} \, ds. \quad (7.33)$$

But the Fourier inversion formula (6.3) tells us that, at least for nice enough f, the far left and right sides will be equal if we put $C_s = \widehat{f}(s)$!

Then (7.32) gives

$$u(x,t) = \int_{-\infty}^{\infty} \widehat{f}(s) e^{-(2\pi s)^2 kt} e^{2\pi i s x} \, ds = \int_{-\infty}^{\infty} f(v) F[e^{-(2\pi s)^2 kt}](v-x) \, dv, \quad (7.34)$$

the last step by the composition theorem (Proposition 6.2.1(d)). Now one computes, using Proposition 6.2.1(b)(i) and Example 6.2.3, that

$$F[e^{-(2\pi s)^2 kt}](v-x) = (2\sqrt{\pi kt})^{-1} e^{-(v-x)^2/(4kt)}, \quad (7.35)$$

so we arrive finally at:

Proposition 7.2.1 *The boundary value problem (7.26), (7.27) has formal solution*

$$u(x,t) = \frac{1}{2\sqrt{\pi kt}} \int_{-\infty}^{\infty} f(v) e^{-(v-x)^2/(4kt)} \, dv = f * G_{[2\sqrt{\pi kt}]}(x), \quad (7.36)$$

where, as usual, $G(x) = e^{-\pi x^2}$ and $G_{[\varepsilon]}(x) = \varepsilon^{-1} G(\varepsilon^{-1} x)$.

All sorts of familiar results regarding convolutions may now be invoked to determine circumstances under which this formal solution is bona fide. For example, suppose $f \in L^1(\mathbb{R}) \cap BC(\mathbb{R})$. Then by Proposition 5.7.1(b), we do indeed have $u(x, 0^+) = f(x)$ for each $x \in \mathbb{R}$. So the boundary condition (7.27) holds. (More exactly, the implicit requirement underlying (7.27) of continuity at the boundary is satisfied. The right side of (7.36) does not, strictly speaking, make sense *at $t = 0$*, so we *define* $u(x, 0)$ to equal $u(x, 0+)$. Then (7.27) also holds *per se*.) We may also, for such f, differentiate under the integral sign in (7.36). Specifically, we may do so (any number of times) with respect to x using Proposition 5.4.1, and with respect to t using similar results. When we do, we find that the right side of (7.36) satisfies the differential equation (7.26), as desired.

Example 7.2.2 Solve the Dirichlet problem

$$u_{xx} + u_{yy} = 0 \quad (-\infty < x < \infty, \ y > 0), \quad (7.37)$$
$$u(x,0) = f(x) \quad (-\infty < x < \infty) \quad (7.38)$$

for the half-plane $H^2 = \{(x,y) \in \mathbb{R}^2 : y > 0\}$.

Solution. We begin much as in Example 2.1.1. We obtain separated solutions

$$u(x, y) = X(x)Y(y) = (\gamma e^{i\sqrt{\lambda}x} + \delta e^{-\sqrt{\lambda}x})(\sigma e^{\sqrt{\lambda}y} + \tau e^{-\sqrt{\lambda}y}) \quad (7.39)$$

to the Laplace equation (7.37) for each nonzero $\lambda \in \mathbb{C}$. (Again we use complex exponentials, in anticipation of our application of the Fourier transform, rather than cosines and sines.)

As in our previous example, we avoid solutions of rapid growth by considering only $\lambda = (2\pi s)^2$, with $s \in \mathbb{R}^+$. Then for the same reason, we must choose the coefficient σ, in the above formula for $u(x, y)$, to be zero. Arguing much as in that example, we arrive at the following formal solutions to (7.37):

$$u(x, y) = \int_{-\infty}^{\infty} C_s e^{-2\pi|s|y} e^{2\pi i s x} \, ds. \quad (7.40)$$

Imposing our inhomogeneous condition (7.38) gives

$$f(x) = \int_{-\infty}^{\infty} C_s e^{2\pi i s x} \, ds, \quad (7.41)$$

so as before, we choose $C_s = \widehat{f}(s)$. Then

$$u(x, y) = \int_{-\infty}^{\infty} \widehat{f}(s) e^{-2\pi|s|y} e^{2\pi i s x} \, ds = \int_{-\infty}^{\infty} f(v) F[e^{-2\pi|s|y}](v - x) \, dv \quad (7.42)$$

by the composition theorem (Proposition 6.2.1(d)). The Fourier transform in the integrand may be computed (DIY) using the result of Example 6.1.2 together with Proposition 6.2.1(b)(i) in the case $b = y$. The result is:

Proposition 7.2.2 *The Dirichlet problem* (7.37), (7.38) *on the half-plane has formal solution*

$$u(x, y) = \frac{y}{\pi} \int_{-\infty}^{\infty} \frac{f(v)}{(v - x)^2 + y^2} \, dv. \quad (7.43)$$

Our next example makes use of the Fourier sine transform $\widehat{f}_{\sin}$ of Section 6.7 to complete a study begun earlier.

Example 7.2.3 Solve the Dirichlet problem indicated by Figure 7.1.

Solution. The problem in question is just the boundary value problem (4.60)–(4.62) of Section 4.2. Let's recall our discussions surrounding that problem. We noted near the end of that section (see (4.54) and (4.64)) that

$$u_\lambda(\rho, \theta) = (\alpha e^{\sqrt{\lambda}\theta} + \beta e^{-\sqrt{\lambda}\theta}) \sin(\sqrt{\lambda} \ln \rho) \quad (7.44)$$

solves the homogeneous parts of this problem for each $\lambda \in \mathbb{C}$. We consider only $\lambda \in \mathbb{R}^+$, because otherwise u_λ is exponentially unbounded. Then, writing $\lambda = (2\pi s)^2$

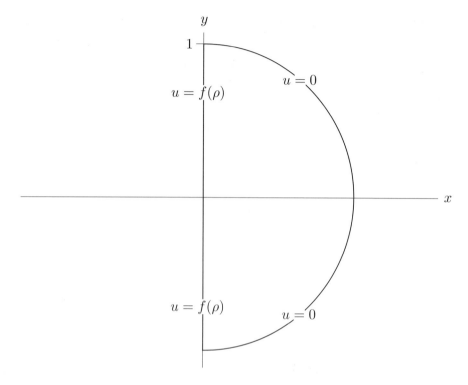

Fig. 7.1 A Dirichlet problem on a semidisk

with $s \in \mathbb{R}^+$, we're led to consideration of integral solutions

$$u(\rho, \theta) = \int_0^\infty u_{(2\pi s)^2}(\rho, \theta)\, ds = \int_0^\infty (D_s e^{2\pi s\theta} + E_s e^{-2\pi s\theta}) \sin(2\pi s \ln \rho)\, ds. \tag{7.45}$$

We could instead express these solutions as integrals over all of $\mathbb{R}$ in the manner of our previous two examples, but we don't for the following reason. The separated solutions in those examples contained complex exponential factors, which suggested the use of the Fourier transform $\widehat{f}$ and the corresponding inversion formula; these entail integration over $\mathbb{R}$. On the other hand, here our separated solutions involve sines (and no cosines); thus we anticipate that $\widehat{f}_{\sin}$ and its inversion formula, both of which entail integration over $\mathbb{R}^+$, will be relevant. And they will, as we'll soon see.

Let's determine D_s and E_s. Into the above expression for $u(\rho, \theta)$ we put the inhomogeneous conditions $u(\rho, -\pi/2) = u(\rho, \pi/2) = f(\rho)$ of (4.62); we get

$$f(\rho) = \int_0^\infty (D_s e^{-\pi^2 s} + E_s e^{\pi^2 s}) \sin(2\pi s \ln \rho)\, ds$$

$$= \int_0^\infty (D_s e^{\pi^2 s} + E_s e^{-\pi^2 s}) \sin(2\pi s \ln \rho)\, ds. \tag{7.46}$$

BOUNDARY VALUE PROBLEMS   363

Let's rewrite these by putting $x = -\ln \rho$, so that $0 < \rho < 1$ means $0 < x < \infty$. Then for such x,

$$f(e^{-x}) = -\int_0^\infty (D_s e^{-\pi^2 s} + E_s e^{\pi^2 s}) \sin(2\pi sx)\, ds,$$

$$= -\int_0^\infty (D_s e^{\pi^2 s} + E_s e^{-\pi^2 s}) \sin(2\pi sx)\, ds. \tag{7.47}$$

The inversion formula in (6.154) for $\widehat{f}_{\sin}$ tells us that we should choose

$$D_s e^{\pi^2 s} + E_s e^{-\pi^2 s} = D_s e^{-\pi^2 s} + E_s e^{\pi^2 s} = -2F_{\sin}[f(e^{-x})](s), \tag{7.48}$$

or

$$D_s = E_s = -\frac{F_{\sin}[f(e^{-x})](s)}{\cosh \pi^2 s}, \tag{7.49}$$

where again $\cosh y = (e^y + e^{-y})/2$. (For the moment we're seeking formal solutions, so let's not worry about the limiting, or averaging, processes in this inversion formula.) Then by the definition of $\widehat{f}_{\sin}$ in (6.153) and by the substitution $v = e^{-x}$,

$$D_s = E_s = -\frac{2}{\cosh \pi^2 s} \int_0^\infty f(e^{-x}) \sin 2\pi sx\, dx$$

$$= \frac{2}{\cosh \pi^2 s} \int_0^1 f(v) \sin(2\pi s \ln v) \frac{dv}{v}. \tag{7.50}$$

So (7.45) gives

$$u(\rho, \theta) = 4 \int_0^\infty \int_0^1 f(v) \frac{\cosh 2\pi s\theta}{\cosh \pi^2 s} \sin(2\pi s \ln \rho) \sin(2\pi s \ln v) \frac{dv}{v}\, ds. \tag{7.51}$$

To simplify, we switch the order of integration and apply to the result the trig identity $\sin A \sin B = (\cos(A - B) - \cos(A + B))/2$, as well as the fact that $\ln C \pm \ln D = \ln CD^{\pm 1}$. We get

$$u(\rho, \theta) = 2 \int_0^1 f(v) \int_0^\infty \frac{\cosh 2\pi s\theta}{\cosh \pi^2 s} \left(\cos\left(2\pi s \ln \frac{\rho}{v}\right) - \cos(2\pi s \ln \rho v) \right) ds\, \frac{dv}{v}. \tag{7.52}$$

To the integral in s, we apply the integration formula

$$\int_0^\infty \frac{\cosh \beta s}{\cosh \gamma s} \cos as\, ds = \frac{\pi}{\gamma} \frac{\cos(\pi\beta/(2\gamma)) \cosh(\pi a/(2\gamma))}{\cos(\pi\beta/\gamma) + \cosh(\pi a/\gamma)}, \tag{7.53}$$

valid as long as $|\beta| < \gamma$. (See formula 3.981(10) in [25].) It gives

$$\int_0^\infty \frac{\cosh 2\pi s\theta}{\cosh \pi^2 s} \left(\cos\left(2\pi s \ln \frac{\rho}{v}\right) - \cos(2\pi s \ln \rho v) \right) ds$$

$$= \frac{\cos \theta}{\pi} \left(\frac{\cosh(\ln(\rho/v))}{\cos 2\theta + \cosh(2\ln(\rho/v))} - \frac{\cosh(\ln \rho v)}{\cos 2\theta + \cosh(2\ln \rho v)} \right)$$

$$= \frac{\cos \theta}{\pi} \left(\frac{\rho v(\rho^2 + v^2)}{\rho^4 + v^4 + 2\rho^2 v^2 \cos 2\theta} - \frac{\rho v(\rho^2 v^2 + 1)}{\rho^4 v^4 + 1 + 2\rho^2 v^2 \cos 2\theta} \right). \tag{7.54}$$

The last step is by some algebra: We wrote out all "cosh" terms using the fact that

$$\cosh(n \ln y) = \frac{\left(e^{\ln y}\right)^n + \left(e^{-\ln y}\right)^n}{2} = \frac{y^n + y^{-n}}{2} \tag{7.55}$$

and multiplied numerators and denominators of the result by $\rho^2 v^2$ to clean up.
Putting (7.54) into (7.52) gives us our final answer.

Proposition 7.2.3 *The Dirichlet problem for the semidisk of Figure 7.1 has formal solution*

$$u(\rho,\theta)$$
$$= \frac{2\rho\cos\theta}{\pi}\int_0^1 f(v)\left(\frac{\rho^2+v^2}{\rho^4+v^4+2\rho^2v^2\cos 2\theta} - \frac{\rho^2v^2+1}{\rho^4v^4+1+2\rho^2v^2\cos 2\theta}\right)dv. \tag{7.56}$$

It's interesting to compare the above result to Proposition 2.5.3.

Exercises

7.2.1 Show that the function $X(x)$ of (7.30) is unbounded, and in fact grows exponentially as $x \to \pm\infty$, whenever $\lambda \notin [0,\infty)$. Exercise A.0.4 may be of use.

7.2.2 Using Proposition 7.2.1, solve the heat problem of Example 7.2.1 for each of the following initial temperature functions.
  a. $f(x) = \cos x$.
  b. $f(x) = G(x)(= e^{-\pi x^2})$.
Lemma 6.2.1 and Example 6.2.3 should help here.

7.2.3 The initial temperature function f of part a of the above exercise is not in $L^1(\mathbb{R})$ or $L^2(\mathbb{R})$, so one might be skeptical of the result given by Proposition 7.2.1. (That proposition was proved by way of the Fourier transform $\hat{f}$.) Show directly that, nevertheless, the solution you obtained in that exercise really *does* solve the problem (7.26)/(7.27).

7.2.4 Show, using the general method of Example 7.2.1 and variation of parameters (cf. Section 2.11, in particular Exercise 2.11.8), that the inhomogeneous heat problem

$$k\,u_{xx} = u_t - F(x,t) \quad (-\infty < x < \infty,\ t > 0),$$
$$u(x,0) = f(x) \quad (-\infty < x < \infty)$$

has formal solution

$$u(x,t) = f * G_{\left[2\sqrt{\pi k t}\right]}(x) + \int_{-\infty}^{\infty}\int_0^t F(v,s) G_{\left[2\sqrt{\pi k (t-s)}\right]}(x-v)\,ds\,dv.$$

7.2.5 Using the Fourier sine transform (Section 6.7), show that the problem

$$k u_{xx} = u_t \quad (0 < x < \infty, \ t > 0),$$
$$u(0, t) = 0 \quad (t > 0),$$
$$u(x, 0) = f(x) \quad (0 < x < \infty),$$

representing heat flow in a semi-infinite, uniform bar with constant temperature zero at one end, has formal solution

$$u(x, t) = \frac{1}{2\sqrt{\pi k t}} \int_0^\infty f(v) \left[e^{-(v-x)^2/(4kt)} - e^{-(v+x)^2/(4kt)} \right] dv.$$

Hint: First find the bounded, separated solutions $u(x,t) = e^{-2\pi sy} \sin 2\pi s x$ ($s \in \mathbb{R}^+$) to the homogeneous parts of the problem. Then proceed as in Example 7.2.2, but with Fourier sine transforms instead of Fourier transforms. Also, Exercise 6.7.2 should help. (Remark: Gaussians are even, so the Fourier cosine transform of any one equals its Fourier transform.)

7.2.6 Formally solve the radial heat equation $k \nabla_3^2 u = u_t$ in $\mathbb{R}^3$, assuming radial initial conditions $u(r, 0) = f(r)$. ANSWER:

$$u(r, t) = \frac{e^{-r^2/(4kt)}}{r\sqrt{\pi k t}} \int_0^\infty v f(v) e^{-v^2/(4kt)} \sinh \frac{rv}{2kt} \, dv.$$

Some of the hints from the previous exercise may help.

7.2.7 What is the analog of the result of Exercise 7.2.5 in the case where the one end of the bar is insulated?

7.2.8 Using Proposition 7.2.2, solve the Dirichlet problem of Example 7.2.2 for each of the following initial temperature functions.
    a. $f(x) = \sin x$.
    b. $f(x) = 1/(1 + x^2)$.
Example 6.4.2 and Exercise 6.4.7 should help here.

7.2.9 The boundary condition function f of part a of the above exercise is not in $L^1(\mathbb{R})$ or $L^2(\mathbb{R})$, so one should be skeptical of the result given by Proposition 7.2.2. (That proposition was proved by way of the Fourier transform $\widehat{f}$.) Show directly that, nevertheless, the solution you obtained in that exercise really *does* solve the problem (7.37)/(7.38).

7.3 MULTIDIMENSIONAL FOURIER TRANSFORMS AND WAVE EQUATIONS

Our next two examples are of considerable interest in the acoustical and geological sciences and elsewhere. (These examples should be compared with Exercise 7.3.1.)

Example 7.3.1 Solve the three-dimensional wave problem
$$c^2(u_{x_1x_1} + u_{x_2x_2} + u_{x_3x_3}) = u_{tt} \quad (x = (x_1, x_2, x_3) \in \mathbb{R}^3, \ t > 0), \quad (7.57)$$
$$u(x,0) = f(x), \quad u_t(x,0) = g(x) \quad (x \in \mathbb{R}^3). \quad (7.58)$$

Solution. Writing $u = XYZT$ we find that, to solve the above differential equation, it suffices to solve $X'' = -\lambda X, Y'' = -\mu Y, Z'' = -\nu Z, T'' = -(\lambda + \mu + \nu)c^2 T$. We obtain, by familiar arguments, the separated solutions

$$u(x,t) = (\sigma e^{i\sqrt{\lambda_1+\lambda_2+\lambda_3}\,ct} + \tau e^{-i\sqrt{\lambda_1+\lambda_2+\lambda_3}\,ct}) \prod_{j=1}^{3}(\gamma_j e^{i\sqrt{\lambda_j}\,x_j} + \delta_j e^{-i\sqrt{\lambda_j}\,x_j})$$

(7.59)

to that equation. From these we reject the unbounded solutions; for the remaining ones, we write $\lambda_i = (2\pi s_i)^2$ for $1 \leq i \leq 3$ and form the integral superposition

$$u(x,t) = \int_{\mathbb{R}^3} \left(D_s e^{2\pi i ||s|| ct} + E_s e^{-2\pi i ||s|| ct} \right) e^{2\pi i \langle s, x \rangle} ds \quad (7.60)$$

(where $s = (s_1, s_2, s_3)$ and $||s|| = \sqrt{s_1^2 + s_2^2 + s_3^2}$, as in Section 6.8).

Putting into this the inhomogeneous boundary conditions (7.58), we get

$$f(x) = \int_{\mathbb{R}^3} (D_s + E_s) e^{2\pi i \langle s,x\rangle} ds, \quad g(x) = \int_{\mathbb{R}^3} 2\pi i ||s|| c (D_s - E_s) e^{2\pi i \langle s,x\rangle} ds.$$

(7.61)

So we should choose $D_s + E_s = \widehat{f}(s)$ and $2\pi i ||s|| c (D_s - E_s) = \widehat{g}(s)$. Let's. Solving for D_s and E_s and putting the result into (7.60) gives (DIY)

$$u(x,t) = \int_{\mathbb{R}^3} \left(\widehat{f}(s) \cos 2\pi ||s|| ct + \widehat{g}(s) \frac{\sin 2\pi ||s|| ct}{2\pi ||s|| c} \right) e^{2\pi i \langle s,x\rangle} ds. \quad (7.62)$$

At this point, our standard operating procedure would involve the application of the composition theorem (this time in its multivariable incarnation, cf. Proposition 6.8.1(e)), but that procedure won't work here! The problem is that, as is straightforward to show, neither $h_1(s) = \cos 2\pi ||s|| ct$ nor $h_2(s) = \sin 2\pi ||s|| ct / (2\pi ||s|| c)$ is in $L^1(\mathbb{R}^3)$ or in $L^2(\mathbb{R}^3)$. So taking the Fourier transform of either function is problematic.

So we take a different approach. We note that, by (6.188) with $r = ct$, and the fact that $\int_0^{2\pi} d\theta = 2\pi$,

$$\frac{\sin 2\pi ||s|| ct}{2\pi ||s|| c} = \frac{t}{2} \int_0^{\pi} e^{-2\pi i ||s|| ct \cos \phi} \sin \phi \, d\phi$$

$$= \frac{t}{4\pi} \int_0^{\pi} \int_{-\pi}^{\pi} e^{-2\pi i ||s|| ct \cos \phi} \sin \phi \, d\theta \, d\phi$$

$$= \frac{t}{4\pi} \int_0^{\pi} \int_{-\pi}^{\pi} e^{-2\pi i ct \langle (0,0,||s||),(\cos\theta\sin\phi,\sin\theta\sin\phi,\cos\phi)\rangle} \sin \phi \, d\theta \, d\phi.$$

(7.63)

(The inner product in the rightmost integral equals $||s||\cos\phi$ by direct computation.) The integral on the right side is an integral over the unit sphere $O = \{o \in \mathbb{R}^3 : ||o|| = 1\}$; this is readily seen by letting r equal 1 in our discussions, in Section 2.6 and in the proof of Proposition 6.8.2(b), of spherical coordinates. By those same discussions, $\sin\phi \, d\theta \, d\phi$ represents the infinitesimal surface area element do on that sphere, and the element $(\cos\theta\sin\phi, \sin\theta\sin\phi, \cos\phi)$ of the inner product in the integral is just the generic element o of O, in spherical coordinates. So (7.63) takes the following qualitatively more suggestive (though perhaps quantitatively less explicit) form:

$$\frac{\sin 2\pi ||s||ct}{2\pi ||s||c} = \frac{t}{4\pi} \int_{||o||=1} e^{-2\pi i ct \langle (0,0,||s||), o\rangle} \, do. \tag{7.64}$$

Actually, we can do even better: Rotations preserve inner products, as noted in Section 6.8; they also preserve areas of regions on spheres centered at the origin, and consequently they preserve the surface area element do. So if R is the rotation taking the vector $(0, 0, ||s||)$ to $(s_1, s_2, s_3) = s$, then

$$\int_{||o||=1} e^{-2\pi i ct \langle (0,0,||s||), o\rangle} \, do = \int_{||o||=1} e^{-2\pi i ct \langle R(0,0,||s||), Ro\rangle} \, do$$
$$= \int_{||o||=1} e^{-2\pi i ct \langle s, Ro\rangle} \, do = \int_{||o||=1} e^{-2\pi i ct \langle s, o\rangle} \, do. \tag{7.65}$$

So (7.64) yields the formula

$$\frac{\sin 2\pi ||s||ct}{2\pi ||s||c} = \frac{t}{4\pi} \int_{||o||=1} e^{-2\pi i ct \langle s, o\rangle} \, do. \tag{7.66}$$

One more change of variable will help: We write $y = ct \, o$. Then y is a generic element of the sphere $\{y \in \mathbb{R}^3 : ||y|| = ct\}$, and since a linear rescaling by a factor of ct changes area by a factor of $c^2 t^2$, we have $dy = c^2 t^2 \, do$. So (7.66) gives, FINALLY,

$$\frac{\sin 2\pi ||s||ct}{2\pi ||s||c} = \frac{1}{4\pi c^2 t} \int_{||y||=ct} e^{-2\pi i \langle s, y\rangle} \, dy. \tag{7.67}$$

This is the formula that we will find most useful.

Let's use it. It and Fourier inversion yield

$$\int_{\mathbb{R}^3} \widehat{g}(s) \frac{\sin 2\pi ||s||ct}{2\pi ||s||c} e^{2\pi i \langle s, x\rangle} \, ds = \frac{1}{4\pi c^2 t} \int_{||y||=ct} \int_{\mathbb{R}^3} \widehat{g}(s) e^{2\pi i \langle s, x-y\rangle} \, ds \, dy$$
$$= \frac{1}{4\pi c^2 t} \int_{||y||=ct} g(x-y) \, dy. \tag{7.68}$$

Similarly, since

$$\cos 2\pi ||s||ct = \frac{d}{dt} \frac{\sin 2\pi ||s||ct}{2\pi ||s||c}, \tag{7.69}$$

we have

$$\int_{\mathbb{R}^3} \hat{f}(s) \cos 2\pi \|s\| ct \, e^{2\pi i \langle s, x \rangle} \, ds = \frac{\partial}{\partial t} \frac{1}{4\pi c^2 t} \int_{\|y\|=ct} f(x-y) \, dy. \quad (7.70)$$

Combining (7.62), (7.68), and (7.70) gives us our desired result.

Proposition 7.3.1 *The three-dimensional wave problem (7.57), (7.58) has formal solution*

$$u(x,t) = \frac{\partial}{\partial t} \frac{1}{4\pi c^2 t} \int_{\|y\|=ct} f(x-y) \, dy + \frac{1}{4\pi c^2 t} \int_{\|y\|=ct} g(x-y) \, dy. \quad (7.71)$$

The proposition admits a very interesting physical interpretation. To discuss this, let's assume the initial position f to be zero and imagine g to be the three-dimensional delta function δ; then the above result reads

$$u(x,t) = \frac{1}{4\pi c^2 t} \int_{\|y\|=ct} \delta(x-y) \, dy = \frac{1}{4\pi c^2 t} \int_{\mathbb{R}^3} \delta(x-y) \chi_{\{y:\|y\|=ct\}}(y) \, dy$$

$$= \frac{1}{4\pi c^2 t} \delta * \chi_{\{y:\|y\|=ct\}}(x) = \frac{1}{4\pi c^2 t} \chi_{\{y:\|y\|=ct\}}(x). \quad (7.72)$$

Here the convolution is in $\mathbb{R}^3$ (cf. Section 6.8), $\chi_{\{y:\|y\|=ct\}}$ denotes the characteristic function of the sphere $\{y : \|y\| = ct\}$ in $\mathbb{R}^3$, and we've used the "fact" that $\delta * h = h$. (Recall (5.115) and the surrounding discussions. That "fact" and those discussions carry over readily to multidimensional situations.) In sum, for $f = 0$ and $g = \delta$,

$$u(x,t) = \frac{1}{4\pi c^2 t} \chi_{\{y:\|y\|=ct\}}(x) = \begin{cases} \frac{1}{4\pi c^2 t} & \text{if } \|x\| = ct, \\ 0 & \text{if not.} \end{cases} \quad (7.73)$$

Of course there are no delta functions, but there are approximate ones, meaning disturbances sharply concentrated near the origin. So our discussion above really says this: If g is such a disturbance, then an observer $\|x\|$ units away from that origin will experience the wave generated by g only for a brief (that is, an "approximately instantaneous") time interval around the time $t = \|x\|/c$.

It's a good thing that we live in a three-dimensional world: Sound and other waves affect us relatively briefly, so that we may process and distinguish large and varied quantities of them. The next example—for which we'll use results from the one just completed—shows that life in two dimensions would be far more cacaphonous.

Example 7.3.2 Solve the two-dimensional wave problem

$$c^2(z_{x_1 x_1} + z_{x_2 x_2}) = z_{tt} \quad (x = (x_1, x_2) \in \mathbb{R}^2, \, t > 0), \quad (7.74)$$

$$z(x,0) = f(x), \quad z_t(x,0) = g(x) \quad (x \in \mathbb{R}^2). \quad (7.75)$$

Solution. The analyses are, at the outset, very similar to those in the previous example. Here they lead us all the way to a two-dimensional version of (7.62):

$$z(x,t) = \int_{\mathbb{R}^2} \left(\widehat{f}(s) \cos 2\pi \sqrt{s_1^2 + s_2^2}\, ct + \widehat{g}(s) \frac{\sin 2\pi \sqrt{s_1^2 + s_2^2}\, ct}{2\pi \sqrt{s_1^2 + s_2^2}\, c} \right) e^{2\pi i \langle s, x \rangle}\, ds.$$
(7.76)

We've written out the norm $\sqrt{s_1^2 + s_2^2}$ explicitly to distinguish it from the corresponding quantity in $\mathbb{R}^3$, because we now wish to apply some observations from the previous example to the present one.

Specifically, putting $s = (s_1, s_2, 0)$ into the formula (7.63) gives

$$\frac{\sin 2\pi \sqrt{s_1^2 + s_2^2}\, ct}{2\pi \sqrt{s_1^2 + s_2^2}\, c}$$

$$= \frac{t}{4\pi} \int_0^\pi \int_{-\pi}^\pi e^{-2\pi i ct \langle (0,0,\sqrt{s_1^2+s_2^2}),(\cos\theta \sin\phi, \sin\theta \sin\phi, \cos\phi) \rangle} \sin\phi\, d\theta\, d\phi$$

$$= \frac{t}{4\pi} \int_0^\pi \int_{-\pi}^\pi e^{-2\pi i ct \langle (s_1, s_2, 0),(\cos\theta \sin\phi, \sin\theta \sin\phi, \cos\phi) \rangle} \sin\phi\, d\theta\, d\phi$$

$$= \frac{t}{4\pi} \int_0^\pi \int_{-\pi}^\pi e^{-2\pi i ct \sin\phi (s_1 \cos\theta + s_2 \sin\theta)} \sin\phi\, d\theta\, d\phi$$

$$= \frac{t}{2\pi} \int_0^{\pi/2} \int_{-\pi}^\pi e^{-2\pi i ct \sin\phi (s_1 \cos\theta + s_2 \sin\theta)} \sin\phi\, d\theta\, d\phi.$$
(7.77)

The second equality is because the inner product and the area element are invariant under the rotation taking $(0, 0, \sqrt{s_1^2 + s_2^2})$ to $(s_1, s_2, 0)$; the third is by simple evaluation of the inner product; the last is by breaking the integral in ϕ into one over $[0, \pi/2]$ plus one over $[\pi/2, \pi]$, putting $u = \pi - \phi$ into the latter, and combining the results.

Into (7.77) we substitute $r = ct \sin\phi$. This means $dr = ct \cos\phi\, d\phi$, so

$$\sin\phi\, d\phi = \left(\frac{r}{ct}\right) \cdot \left(\frac{dr}{ct \cos\phi}\right) = \frac{r\, dr}{c^2 t^2 \sqrt{1 - \sin^2\phi}} = \frac{r\, dr}{c^2 t^2 \sqrt{1 - r^2/(c^2 t^2)}}$$

$$= \frac{r\, dr}{ct\sqrt{c^2 t^2 - r^2}},$$
(7.78)

so

$$\frac{\sin 2\pi \sqrt{s_1^2 + s_2^2}\, ct}{2\pi \sqrt{s_1^2 + s_2^2}\, c} = \frac{1}{2\pi c} \int_{-\pi}^\pi \int_0^{ct} e^{-2\pi i (s_1 r \cos\theta + s_2 r \sin\theta)} \frac{r\, dr\, d\theta}{\sqrt{c^2 t^2 - r^2}}$$

$$= \frac{1}{2\pi c} \int_{\|y\| \leq ct} e^{-2\pi i \langle s, y \rangle} \frac{dy}{\sqrt{c^2 t^2 - \|y\|^2}},$$
(7.79)

where dy is the usual infinitesimal area element of the plane and the norm is the one on $\mathbb{R}^2$. (That is, the last equality is by the polar coordinate substitution $y = (y_1, y_2) =$

$(r\cos\theta, r\sin\theta)$.) We apply (7.79) and the formula

$$\frac{\sin 2\pi\sqrt{s_1^2+s_2^2}\,ct}{2\pi\sqrt{s_1^2+s_2^2}\,c} = \frac{d}{dt}\cos 2\pi\sqrt{s_1^2+s_2^2}\,ct \qquad (7.80)$$

to the prescription (7.76) for $z(x,t)$. The procedure here is almost exactly as in the previous example; here is the result.

Proposition 7.3.2 *The two-dimensional wave problem (7.74), (7.75) has formal solution*

$$z(x,t) = \frac{1}{2\pi c}\frac{\partial}{\partial t}\int_{\|y\|\le ct}\frac{f(x-y)\,dy}{\sqrt{c^2t^2-\|y\|^2}} + \frac{1}{2\pi c}\int_{\|y\|\le ct}\frac{g(x-y)\,dy}{\sqrt{c^2t^2-\|y\|^2}}. \qquad (7.81)$$

The salient difference between the three-dimensional wave $u(x,t)$ of (7.71) and the above two-dimensional wave $z(x,t)$ is this: The former is obtained by integrating over the thin spherical shell $\|y\| = ct$ in three space, and the latter by integrating over the entire disk $\|y\| \le ct$ in two space. This has dramatic consequences.

In particular, let's consider what happens in two dimensions when the initial position f is zero and the initial velocity g is the two-dimensional delta function δ. We write $W(y) = (c^2t^2 - \|y\|^2)^{-1/2}$; we get, by the above proposition and by arguments much like those following Proposition 7.3.1,

$$z(x,t) = \frac{1}{2\pi c}\int_{\|y\|\le ct}\frac{\delta(x-y)\,dy}{\sqrt{c^2t^2-\|y\|^2}} = \frac{1}{2\pi c}\delta * (W\chi_{\{y:\|y\|\le ct\}})(x)$$

$$= \frac{1}{2\pi c}(W\chi_{\{y:\|y\|\le ct\}})(x) = \begin{cases} \dfrac{1}{2\pi c\sqrt{c^2t^2-\|x\|^2}} & \text{if } \|x\| \le ct, \\ 0 & \text{if not.} \end{cases} \qquad (7.82)$$

In other words, effects from a delta disturbance are felt by an observer $\|x\|$ units from the origin at time $t = \|x\|/c$ *and forever thereafter!* These effects decay with time, because of the factor of $1/\sqrt{c^2t^2-\|x\|^2}$, but still, they do not depart with anything like the immediacy of the analogous three-dimensional ones. In two dimensions, we'd still be hearing the Bay City Rollers!

Exercises

7.3.1 a. Proceed as we did at the outset of Example 7.3.1 or Example 7.3.2 to show that the one-dimensional (infinite) wave problem

$$c^2 y_{xx} = y_{tt} \quad (x \in \mathbb{R},\ t > 0),$$
$$y(x,0) = f(x), \quad y_t(x,0) = g(x) \quad (x \in \mathbb{R})$$

has formal solution

$$y(x,t) = \int_{\mathbb{R}}\left(\widehat{f}(s)\cos 2\pi sct + \widehat{g}(s)\frac{\sin 2\pi sct}{2\pi sc}\right)e^{2\pi isx}\,ds.$$

(The above might be called *Fourier's form* of the solution.)

b. From part a above, from Fourier inversion, from properties of the Fourier transform (see (6.35)), and from the composition theorem (Proposition 6.2.1(d)), formally deduce *d'Alembert's form*

$$y(x,t) = \frac{1}{2}[f(x+ct) + f(x-ct)] + \frac{1}{2c}\int_{x-ct}^{x+ct} g(v)\,dv$$

of the solution to the same wave problem. (Compare this last result with Proposition 2.9.1.)

7.3.2 Show that Dirichlet's problem

$$u_{x_1 x_1} + u_{x_2 x_2} + u_{x_3 x_3} = 0 \quad (-\infty < x_j < \infty \text{ for } 1 \le j \le 3),$$
$$u(x_1, x_2, 0) = f(x_1, x_2) \quad (-\infty < x_j < \infty \text{ for } 1 \le j \le 2)$$

for the half-space $H^3 = \{(x_1, x_2, x_3) \in \mathbb{R}^2 : x_3 > 0\}$ has formal solution

$$u(x_1, x_2, x_3) = \frac{x_3}{2\pi} \int_{\mathbb{R}^2} \frac{f(u_1, u_2)}{((x_1 - u_1)^2 + (x_2 - u_2)^2 + x_3^2)^{3/2}} \, du_1 \, du_2.$$

Hints: Take the general approach of Example 7.2.3. At some point, you'll also want to use the multidimensional composition theorem (Proposition 6.8.1(e)) and the result of Example 6.8.2(a).

7.3.3 Solve the infinite wave problem in two and in three dimensions, assuming constant initial position z_0 and velocity v_0. Note: In three dimensions, substitute $y_1 = ct \cos\theta \sin\phi$, $y_2 = ct \sin\theta \sin\phi$, $y_3 = ct \cos\phi$; integration over the sphere of radius ct, with respect to the surface area element dy, then becomes integration over $(\theta, \phi) \in (-\pi, \pi) \times (0, \pi)$, with area element $c^2 t^2 \sin\phi\, d\theta\, d\phi$. Similarly, in two dimensions, substitute $y_1 = \rho\cos\theta$, $y_2 = \rho\sin\theta$; integration over the disk of radius ct then becomes integration over $(\rho, \theta) \in (0, ct) \times (-\pi, \pi)$ and dy becomes $\rho\, d\rho\, d\theta$.

7.3.4 Solve the infinite wave problem in two dimensions, assuming initial position $f(x) = 0$ and initial velocity $g(x) = 1/(1 + ||x||^2)$. Hint: Make the substitutions delineated in the previous exercise. Now since f and g are radial, u will be too, so $u(x_1, x_2, t) = u(||x||, 0, t)$. Note that

$$g((||x||, 0) - (\rho\cos\theta, \rho\sin\theta)) = \frac{1}{1 + (||x|| - \rho\cos\theta)^2 + (0 - \rho\sin\theta)^2}$$
$$= \frac{1}{1 + ||x||^2 - 2||x||\rho\cos\theta + \rho^2}.$$

If you now put $u = 1/\sqrt{c^2 t^2 - \rho^2}$, you should end up with an integral of the form

$$\int_{-\pi}^{\pi} \int_0^{ct} \frac{du\, d\theta}{A\cos\theta + B},$$

where A and B depend on u. Integrate in θ using techniques similar to those of Exercise 6.8.4a. You should be left with an integral in u that may be evaluated in standard ways.

7.4 BANDLIMITED FUNCTIONS AND THE SHANNON SAMPLING THEOREM

The human ear can detect frequencies as high as about 22 kHz (kHz is for "kilohertz," meaning one thousand cycles per second). Music recorded for CD is sampled at roughly 44 kHz. The following theorem, which is of tremendous import in signal processing applications, explains the connection. The theorem—whose proof exploits a very nifty interplay between Fourier integrals and series—says: A signal containing frequencies only up to a certain rate can, in theory, be *perfectly* reproduced from samples taken at twice that rate.

Theorem 7.4.1 (the Shannon sampling theorem, or the bandlimited Fourier inversion theorem) *Let $f \in L^2(\mathbb{R})$ be continuous and Ω-bandlimited; the latter means $\widehat{f}(s) = 0$ for almost all s outside of $[-\Omega, \Omega]$. Then for any $t \in \mathbb{R}$,*

$$f(t) = \sum_{m=-\infty}^{\infty} f\left(\frac{m}{2\Omega}\right) \operatorname{sinc} \pi(m - 2\Omega t). \tag{7.83}$$

In particular, f may be reconstructed from its values at the points $t_m = m/2\Omega$ ($m \in \mathbb{Z}$).

Proof. By Theorem 6.5.2, f is the norm limit of the integrals

$$\int_{-N}^{N} \widehat{f}(s) e^{2\pi i s t}\, ds. \tag{7.84}$$

Since $\widehat{f}$ is Ω-bandlimited, the integral over $[-N, N]$ is, for all N large enough, equal to the integral over $[\Omega, \Omega]$; we conclude that

$$f(t) = \int_{-\Omega}^{\Omega} \widehat{f}(s) e^{2\pi i s t}\, ds. \tag{7.85}$$

The equality is in $L^2(\mathbb{R})$, which is to say almost everywhere. But the right side is continuous, by essentially the proof of Proposition 6.1.1, and the left side is too, by assumption, so in fact the two sides are equal everywhere.

Note that the integral on the right is just the inner product, in $L^2(-\Omega, \Omega)$, of $\widehat{f}$, considered as a function in that space, with $e^{-2\pi i s t}$, also considered as a function (of s, for any given t) in that space. Now $\widehat{f}$ is the norm limit of its Fourier series, by Theorem 3.6.3, and norm limits may be brought outside inner products, by Lemma 3.6.1. So the expression (7.85) for $f(t)$ gives

$$\begin{aligned}f(t) &= \sum_{n=-\infty}^{\infty} c_n(\widehat{f}) \int_{-\Omega}^{\Omega} e^{2\pi i n s/(2\Omega)} e^{2\pi i s t}\, ds \\ &= \sum_{m=-\infty}^{\infty} c_{-m}(\widehat{f}) \int_{-\Omega}^{\Omega} e^{-2\pi i s(m/(2\Omega) - t)}\, ds \end{aligned} \tag{7.86}$$

BANDLIMITED FUNCTIONS AND THE SHANNON SAMPLING THEOREM 373

(we substituted $m = -n$). The integral on the right equals $2\Omega \operatorname{sinc} \pi(m - 2\Omega t)$, by the result of Example 6.1.1 and some algebra; moreover,

$$c_{-m}(\widehat{f}) = \frac{1}{2\Omega} \int_{-\Omega}^{\Omega} \widehat{f}(s) e^{2\pi i m s/(2\Omega)} \, ds = \frac{1}{2\Omega} f\left(\frac{m}{2\Omega}\right). \tag{7.87}$$

The first equality is by formula (1.18) for Fourier coefficients, and the second by (7.85) again. Combining (7.86) and (7.87) gives the desired formula (7.83), so we're done. □

Thus, for complete knowledge of a (sufficiently reasonable) Ω-bandlimited function, it's enough to know what it does at the points $t_m = m/2\Omega$ for $m \in \mathbb{Z}$. And these points are $1/(2\Omega)$ units apart; so, *to completely determine a (sufficiently reasonable) Ω-bandlimited function, it's sufficient to sample it at a rate of 2Ω* (*times per unit of the independent variable*). The latter rate is called the *Nyquist rate* or the *Nyquist frequency*.

We now wish to discuss some crucial issues associated with the application of the Shannon theorem. To do so, we need to examine what happens when one attempts to reconstruct a function that's *not* Ω-bandlimited from samples of that function taken at the Nyquist frequency. Let's. We begin with:

Proposition 7.4.1 *Suppose $f \in C(\mathbb{R}) \cap L^2(\mathbb{R})$. Let $f_{2\Omega}$ be the "2Ω-sampling of f"; that is, $f_{2\Omega}(t)$ is defined to be the sum on the right side of (7.83). Then $f_{2\Omega}$ is Ω-bandlimited (even if f is not).*

Proof. We have

$$\widehat{f_{2\Omega}} = F\left[\sum_{m=-\infty}^{\infty} f\left(\frac{m}{2\Omega}\right) \operatorname{sinc} \pi(m - 2\Omega t)\right]$$

$$= \sum_{m=-\infty}^{\infty} f\left(\frac{m}{2\Omega}\right) F\left[\operatorname{sinc} \pi(m - 2\Omega t)\right] = \frac{\chi_{[-\Omega,\Omega]}}{2\Omega} \sum_{m=-\infty}^{\infty} f\left(\frac{m}{2\Omega}\right) e_{-m/(2\Omega)}, \tag{7.88}$$

the last step by the formula (see (6.35)) for the Fourier transform of a shift and the formula (see Example 6.5.1) for the Fourier transform of a sinc. The right side of (7.88) is supported on $[-\Omega, \Omega]$ since $\chi_{[-\Omega,\Omega]}$ is, and we're done. □

The moral is that the 2Ω-sampling of a function with components of frequency larger than Ω itself has no such components. What happens to them? Do they just completely vanish? In many situations, such vanishing would not be a serious problem *per se*. For example, since, as noted above, we hear (roughly speaking) only frequencies in the range 0–22 kHz, there's no need for sound recording technology to retain frequency content beyond that range.

Unfortunately, they don't just disappear. Instead, these higher frequency components get "folded back" into the lower frequencies—those less than or equal to

374 SPECIAL TOPICS AND APPLICATIONS

Ω—producing spurious frequency content in that range. This is the phenomenon of *aliasing*. Let's now be more precise about what aliasing is and how it works. We'll do so by focusing our attention, at first, on functions f whose frequencies are between Ω and 2Ω only. We'll discuss the more general situation a bit later.

We have:

Proposition 7.4.2 *Suppose* $f \in C(\mathbb{R}) \cap L^2(\mathbb{R})$. *Also suppose that* $\widehat{f}(s) = 0$ *for* $s \notin [-2\Omega, -\Omega] \cup [\Omega, 2\Omega]$. *(See Fig. 7.2, top.) If we let g be the function defined by*

$$\widehat{g}(s) = \begin{cases} \widehat{f}(s + 2\Omega) & \text{if } -\Omega \leq s \leq 0, \\ \widehat{f}(s - 2\Omega) & \text{if } 0 \leq s \leq \Omega, \\ 0 & \text{otherwise} \end{cases} \quad (7.89)$$

(that is, we define $g = \widehat{g}^{-}$, with $\widehat{g}$ as above), then $f_{2\Omega} = g$. (See Fig. 7.2, bottom.)

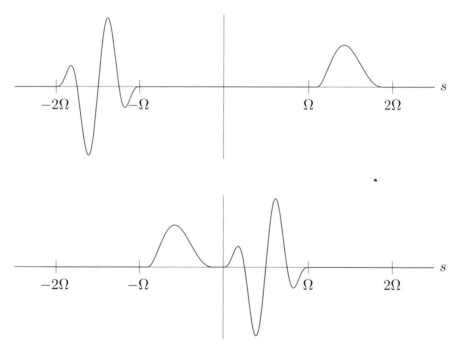

Fig. 7.2 Top: $\widehat{f}$ is supported on $[-2\Omega, -\Omega] \cup [\Omega, 2\Omega]$. Bottom: What happens to the frequency content of f under 2Ω-sampling

Proof. By definition of $\widehat{g}$ and by our assumptions on $\widehat{f}$, we check readily that $\widehat{g} \in L^2(\mathbb{R})$ and moreover that $\widehat{g}(s) = 0$ for $|s| > \Omega$. Then g, being the inverse Fourier transform of $\widehat{g}$, is also in $L^2(\mathbb{R})$. Further, $\widehat{g}$ is compactly supported and therefore in $L^1(\mathbb{R})$ by Corollary 3.5.1(c); so g is continuous by Proposition 6.1.1. So by the

Shannon theorem,

$$g(t) = \sum_{m=-\infty}^{\infty} g\left(\frac{m}{2\Omega}\right) \operatorname{sinc} \pi(m - 2\Omega t). \qquad (7.90)$$

On the other hand,

$$\begin{aligned}
g\left(\frac{m}{2\Omega}\right) &= \int_{-\infty}^{\infty} \widehat{g}(s) e^{2\pi i s m/(2\Omega)} \, ds = \int_{-\Omega}^{\Omega} \widehat{g}(s) e^{\pi i s m/\Omega} \, ds \\
&= \int_{-\Omega}^{0} \widehat{f}(s+2\Omega) e^{\pi i s m/\Omega} \, ds + \int_{0}^{\Omega} \widehat{f}(s-2\Omega) e^{\pi i s m/\Omega} \, ds \\
&= \int_{\Omega}^{2\Omega} \widehat{f}(u) e^{\pi i(u-2\Omega) m/\Omega} \, du + \int_{-2\Omega}^{-\Omega} \widehat{f}(u) e^{\pi i(u+2\Omega) m/\Omega} \, du \\
&= e^{-2\pi i m} \int_{\Omega}^{2\Omega} \widehat{f}(u) e^{\pi i u m/\Omega} \, du + e^{-2\pi i m} \int_{-2\Omega}^{-\Omega} \widehat{f}(u) e^{\pi i u m/\Omega} \, du \\
&= \int_{-\infty}^{\infty} \widehat{f}(u) e^{2\pi i u(m/2\Omega)} \, du = f\left(\frac{m}{2\Omega}\right). \qquad (7.91)
\end{aligned}$$

The second equality is because of the support of g; the third by the definition (7.89) of $\widehat{g}$; the fourth by changes of variable $u = s+2\Omega$ and $u = s-2\Omega$ in respective integrals; the fifth by pulling u-independent complex exponentials out of those integrals; the sixth by the fact that those complex exponentials equal 1, and by assumption on the support of $\widehat{f}$; the last by Fourier inversion.

So (7.90) and the definition of $f_{2\Omega}$ give

$$g(t) = \sum_{m=-\infty}^{\infty} f\left(\frac{m}{2\Omega}\right) \operatorname{sinc} \pi(m - 2\Omega t) = f_{2\Omega}(t), \qquad (7.92)$$

and we're done. □

The upshot of the above proposition is this: If we 2Ω-sample a function f with frequencies in $[\Omega, 2\Omega]$ only, we end up squeezing these frequencies into the range $[0, \Omega]$. Or actually not squeezing, but *sliding*: That part of $\widehat{f}$ that lives on $[-2\Omega, -\Omega]$ is slid to the right 2Ω units and ends up in $[0, \Omega]$, while that part on $[\Omega, 2\Omega]$ is slid to the left 2Ω units and ends up in $[-\Omega, 0]$. (See again Fig. 7.2.) This is how aliasing works, at least for such functions f.

For more general functions, it works similarly. Specifically, if f is a general 2Ω-bandlimited function, then the part of f that's Ω-bandlimited will be preserved by 2Ω-sampling, while the part with frequencies in $[\Omega, 2\Omega]$ will be aliased as above. If f also has frequencies in $[2\Omega, 3\Omega]$, then 2Ω-sampling will slide these over 4Ω units (more precisely, the part of $\widehat{f}$ on $[2\Omega, 3\Omega]$ will be slid left and the part on $[-3\Omega, -2\Omega]$ will be slid right), and so on. We omit proofs of these more general results; those proofs are similar to ones just seen.

So if, for example, we attempt to reproduce sound by 2Ω-sampling, and if the signal being sampled in fact contains frequency content *outside* that range, then the recorded result will contain aliased frequency content that we *can* hear but *shouldn't*! How do we avoid this pitfall?

The solution is simple: We *filter* the sound, *before* we sample it. That is, we first remove from the signal all frequency components outside the desired, audible range. Hey, they're inaudible. The result is a signal with *all*, and *only*, the desired frequency content and which can be faithfully, digitally reproduced by sampling at 44 kHz. We'll discuss the filtering process in Section 7.7.

A few more remarks are in order. First, the Shannon theorem says essentially that, if we define

$$L^2(\Omega; \mathbb{R}) = \{f \in L^2(\mathbb{R}) : f \text{ is } \Omega\text{-bandlimited}\} \tag{7.93}$$

and

$$\sigma_m(t) = \operatorname{sinc} \pi(m - 2\Omega t), \tag{7.94}$$

then $\mathcal{S} = \{\sigma_m : m \in \mathbb{Z}\}$ is a basis for $L^2(\Omega, \mathbb{R})$. Indeed, it's an orthogonal one, and moreover the norms of the σ_m's may be computed explicitly. See Exercise 7.4.3.

Next, the right side of (7.83) is an infinite series, which in real life is impractical and generally approximated by a partial sum thereof. Under mild conditions on f, reasonable bounds may be obtained on the error in such an approximation; see Section 3.8.A in [50]. On the other hand, the σ_m's limit, in a sense, how good such bounds can be: For a fixed $t \in \mathbb{R}$, $\sigma_m(t)$ decays only about like a constant times $1/m$, which as we've noted in not very quickly, relatively speaking. This means that a rather large number of summands may be required to get a decent approximation to $f(t)$. (How many are required will depend on what's considered "decent" and on the rate of decay of f itself.) Things would be nice were there sampling theorems involving more rapidly decaying functions τ_m in place of the σ_m's. Fortunately, things *are* nice—there *are* such theorems, which are useful in a variety of practical applications. These theorems entail *oversampling*—sampling at a rate higher than the Nyquist rate. See Exercise 7.4.4.

Finally we remark that, were things *really* nice, there would be lots of worthwhile, *compactly supported*, bandlimited functions out there. What would be so nice about that? Well if f is (reasonable and) bandlimited, then it has a representation of the form (7.83), but if it's also compactly supported, then all but finitely many of the summands there are *zero*. So such a function is *completely* determined by finitely many of its values!

Unfortunately, there *aren't* many functions that behave so nicely; in fact, it turns out that any such function is zero almost everywhere. This perhaps seems to contradict real life; in real life, nothing lasts forever, meaning everything is really of FINITE extent. But unless f is nothing, then both f and $\hat{f}$ are something, so both should be of finite extent. However, as the following proposition attests, if the latter happens, then f really *is* nothing (almost everywhere).

Proposition 7.4.3 *If $f \in L^1(\mathbb{R}) \cup L^2(\mathbb{R})$, then f and $\widehat{f}$ cannot both be compactly supported unless $f(x) = 0$ almost everywhere.*

Proof. Let $f \in L^1(\mathbb{R}) \cup L^2(\mathbb{R})$ be supported on $(-R, R)$. Then in fact $f \in L^1(\mathbb{R})$ by Corollary 3.5.1(c). Consequently, for any $n \in \mathbb{N}$,

$$\int_{-\infty}^{\infty} |t^n f(t)|\, dt = \int_{-R}^{R} |t^n f(t)|\, dt \leq R^n \int_{-R}^{R} |f(t)|\, dt < \infty, \qquad (7.95)$$

so $t^n f(t)$ defines a function in $L^1(\mathbb{R})$ as well. By Proposition 6.2.1(c)(ii), then,

$$F[t^n f(t)](s) = (-2\pi i)^{-n}(\widehat{f})^{(n)}(s). \qquad (7.96)$$

Now assume $\widehat{f}$ is also compactly supported, say $\widehat{f}(s) = 0$ for $|t| \geq R^*$; then the same is true of any derivative of $\widehat{f}$. So for $n \in \mathbb{N}$,

$$0 = (\widehat{f})^{(n)}(R^*) = F[(-2\pi i t)^n f(t)](R^*) = \int_{-\infty}^{\infty} (-2\pi i t)^n f(t) e^{-2\pi i R^* t}\, dt$$

$$= \int_{-R}^{R} (-2\pi i t)^n f(t) e^{-2\pi i R^* t}\, dt, \qquad (7.97)$$

the last step because, again, f is supported on $(-R, R)$.

Let's multiply both sides of (7.97) by $(s - R^*)^n / n!$, where s is an arbitrary real number, and then sum over all $n \in \mathbb{N}$. We get

$$\sum_{n=0}^{\infty} \frac{0 \cdot (s - R^*)^n}{n!} = \sum_{n=0}^{\infty} \frac{(s - R^*)^n}{n!} \int_{-R}^{R} (-2\pi i t)^n f(t) e^{-2\pi i R^* t}\, dt$$

$$= \int_{-R}^{R} \left(\sum_{n=0}^{\infty} \frac{(2\pi i (R^* - s) t)^n}{n!} \right) f(t) e^{-2\pi i R^* t}\, dt$$

$$= \int_{-R}^{R} e^{2\pi i (R^* - s) t} f(t)\, e^{-2\pi i R^* t}\, dt$$

$$= \int_{-R}^{R} f(t) e^{-2\pi i s t}\, dt = \widehat{f}(s). \qquad (7.98)$$

The second equality is readily justified by the Lebesgue dominated convergence theorem (Proposition 3.5.1); the third is by the Maclaurin series for e^z; the fourth is by combining the complex exponentials; the fifth is by Fourier inversion and because, once more, f is supported on $(-R, R)$. The left side of (7.98) is identically zero, so the right side is too. But $\widehat{f}$ being identically zero is equivalent, by Fourier inversion, to f being zero almost everywhere, so we're done. □

The above proposition is another manifestation of the local/global principle, relating concentration in the original domain to spread in the frequency domain. Indeed, the proposition says $f \in L^1(\mathbb{R}) \cup L^2(\mathbb{R})$ can't be completely concentrated on a finite portion of either domain without being spread all the way across the other.

Again this seems counterintuitive, perhaps. But again math, while modeling real life, isn't real life. And the bandlimited model suits many practical situations.

In the next section, we'll discuss another model of absolutely huge practical importance.

Exercises

7.4.1 Use Theorem 7.4.1 to express $\operatorname{sinc} 2\pi\beta\Omega t$, for any $\beta \in (0, 1)$, as an infinite linear combination of shifts of the function $\operatorname{sinc} 2\pi\Omega t$. Explain why application of the theorem is valid here.

7.4.2 Use the previous exercise to evaluate

$$\sum_{m=-\infty}^{\infty} \frac{(-1)^m \sin \pi m \beta}{m(m-\beta)}$$

for $\beta \in (0, 1)$.

7.4.3 **a.** Show that the σ_m's of (7.94) form an orthogonal set in $L^2(\mathbb{R})$ and that $\|\sigma_m\| = (2\Omega)^{-1}$ for each m. Hint: Use the Plancherel theorem.
b. Show that these σ_m's form an orthogonal basis for the subspace $L^2(\Omega; \mathbb{R})$ of $L^2(\mathbb{R})$ defined by (7.93). Hints: The partial sums of the series on the right side of (7.83) converge in norm to f, for $f \in C(\mathbb{R}) \cap L^2(\Omega; \mathbb{R})$, by a very slight modification of our proof of Theorem 7.4.1. Also, $C(\mathbb{R}) \cap L^2(\Omega; \mathbb{R})$ is dense in $L^2(\Omega; \mathbb{R})$. Assuming these facts, use Corollary 3.7.2 and the transitivity of norm density (Exercise 3.2.6, generalized to $L^2(\mathbb{R})$) to finish.

7.4.4 Let $\Omega, \gamma > 0$; let

$$h(s) = (\Omega\gamma)^{-1} \chi_{[-\Omega(1+\gamma/2), \Omega(1+\gamma/2)]} * \chi_{[-\Omega\gamma/2, \Omega\gamma/2]}(s).$$

a. Sketch the graph of h. Proposition 5.2.1 should help.
b. Show that

$$\widehat{h}(t) = \frac{\cos 2\pi\Omega t - \cos 2\pi\Omega(1+\gamma)t}{2\Omega\gamma \pi^2 t^2}.$$

The convolution theorem (Proposition 6.2.1(e)) and a trig identity (cf. Exercise A.0.16) should help.

c. Prove the following sampling theorem: If $f \in C(\mathbb{R}) \cap L^2(\mathbb{R})$ is Ω-bandlimited, then for any $t \in \mathbb{R}$,

$$f(t) = \frac{1}{2\Omega(1+\gamma)} \sum_{m=-\infty}^{\infty} f\left(\frac{m}{2\Omega(1+\gamma)}\right) \widehat{h}\left(t - \frac{m}{2\Omega(1+\gamma)}\right).$$

Hints: Since $\widehat{f}$ is supported on $[-\Omega, \Omega]$ and $h(s) = 1$ there, we have $\widehat{f}(s) = h(s)\widehat{f}(s)$ always. Arguing as in our proof of Theorem 7.4.1, then, and using the fact that h is supported on $[-\Omega(1+\gamma), \Omega(1+\gamma)]$ (see part a of this exercise), we get

$$f(t) = \int_{-\Omega(1+\gamma)}^{\Omega(1+\gamma)} h(s)\widehat{f}(s)e^{2\pi i s t}\, ds$$

instead of (7.85). Write $\widehat{f}(s)$, *considered as a function on* $[-\Omega(1+\gamma), \Omega(1+\gamma)]$, out as a Fourier series, bring the sum outside the integral, and integrate; then evaluate the $c_{-m}(\widehat{f}\,)$'s, as we did in our proof above.

d. What advantage does the above theorem offer over Shannon's theorem? (Recall the discussion following (7.94).)

7.5 THE DISCRETE FOURIER TRANSFORM

Real-world phenomena are not always amenable to modeling by functions of continuous real variables. Even when they are, definite integrals of such functions or of these functions times complex exponentials can be exceptionally difficult or impossible (for humans or machines) to evaluate algebraically. For these reasons we'd like a *numerical* Fourier theory that applies to finite sets of *observations* of phenomena.

To introduce this theory, let's suppose we have such a set: We understand its elements to be successive values $f(t_0), f(t_1), \ldots, f(t_{N-1})$, say, of an underlying, sufficiently reasonable function f, which again we don't necessarily know or want to know explicitly. (We're imagining, then, that f is a function of a *single* real variable. Multivariable Fourier transforms are just successions of single-variable ones; multivariable discrete Fourier transforms may be defined iteratively once the one-dimensional constructs are in place.) The t_k's will lie within some interval $[a, a + \ell]$ on the t axis; by appropriate choice of coordinates (or by replacing f with a shift thereof), we may assume that $a = 0$.

We recall from Section 1.11 that, as a function on $[0, \ell]$, f comprises the frequencies n/ℓ ($n \in \mathbb{N}$), in amounts given by the Fourier coefficients

$$c_n(f) = \frac{1}{\ell}\int_0^\ell f(t)e^{-2\pi i n t/\ell}\, dt. \tag{7.99}$$

Actually both $c_{-n}(f)$ and $c_n(f)$ are, as we've seen, required to build the component of frequency n/ℓ. On the other hand, if f is real-valued, as is often the case in practice, then $c_{-n}(f) = \overline{c_n(f)}$. See (1.197).

Let's now assume that, in fact, the t_k's are the evenly spaced points

$$t_k = \frac{k\ell}{N} \quad (0 \leq k \leq N-1) \tag{7.100}$$

of $[0, \ell]$ and that f is continuous there. Then we may form the *left endpoint Riemann sum approximation*, which we denote by $\widetilde{f}_n$, to the right side of (7.99)—and therefore to the left side $c_n(f)$, like this:

$$\tilde{f}_n = \frac{1}{\ell}\sum_{k=0}^{N-1} f(t_k)e^{-2\pi i n t_k/\ell}\Delta t_k = \frac{1}{N}\sum_{k=0}^{N-1} f\left(\frac{k\ell}{N}\right)e^{-2\pi ikn/N} \quad (7.101)$$

($\Delta t_k = \ell/N$ is just the spacing between the successive partition points t_k in the Riemann sum).

We make one more note, namely,

$$\tilde{f}_{N+n} = \frac{1}{N}\sum_{k=0}^{N-1} f\left(\frac{k\ell}{N}\right)e^{-2\pi ik(n+N)/N} = \frac{1}{N}\sum_{k=0}^{N-1} f\left(\frac{k\ell}{N}\right)e^{-2\pi ik}e^{-2\pi ikn/N}$$

$$= \frac{1}{N}\sum_{k=0}^{N-1} f\left(\frac{k\ell}{N}\right)e^{-2\pi ikn/N} = \tilde{f}_n. \quad (7.102)$$

That is, our approximation $\tilde{f}_n$ to $c_n(f)$ is N-periodic in n. Certainly, then, the cases $0 \leq n \leq N - 1$ of (7.101) tell us everything. We are thus led inexorably to the following.

Definition 7.5.1 (a) Let a positive integer N and an interval $[0, \ell]$ be given. If f is continuous on $[0, \ell]$, then the *discrete Fourier transform* $\tilde{f}$ of f is the *vector*

$$\tilde{f} = (\tilde{f}_0, \tilde{f}_1, \ldots, \tilde{f}_{N-1}) \in \mathbb{C}^N, \quad (7.103)$$

with nth coordinate $\tilde{f}_n$ given by (7.101) for $0 \leq n \leq N - 1$.

(b) More generally, we define the discrete Fourier transform of *any* vector $a = (a_0, a_1, \ldots, a_{N-1}) \in \mathbb{C}^N$ to be the vector $\tilde{a} \in \mathbb{C}^N$ with nth coordinate

$$\tilde{a}_n = \frac{1}{N}\sum_{k=0}^{N-1} a_k\, e^{-2\pi ikn/N}. \quad (7.104)$$

The idea is that, under the circumstances discussed above, a function f is understood through its "discretization"

$$\left(f(0), f\left(\frac{\ell}{N}\right), \ldots, f\left(\frac{(N-1)\ell}{N}\right)\right) \in \mathbb{C}^N. \quad (7.105)$$

In this sense, part (a) of the above definition amounts to a special case of part (b). It should be noted that the latter part has a variety of other applications, in number theory and elsewhere. See, for example, [45]. But our primary interest is in the former part.

Example 7.5.1 Compute $\tilde{g}$ if $N = 4$, $\ell = 1$, and $g(x) = x$.

THE DISCRETE FOURIER TRANSFORM

Solution. By (7.101),

$$\tilde{g}_n = \frac{1}{4}\sum_{k=0}^{3} g\left(\frac{k}{4}\right) e^{-2\pi i k n/4} = \frac{1}{4}\sum_{k=0}^{3} \frac{k}{4} e^{-\pi i k n/2} = \frac{1}{16}\sum_{k=0}^{3} k(-i)^{kn}, \quad (7.106)$$

since $e^{-\pi i/2} = -i$. From this we compute

$$\tilde{g}_0 = \frac{1}{16}(0+1+2+3) = \frac{3}{8},$$

$$\tilde{g}_1 = \frac{1}{16}(0+(-i)+2(-i)^2+3(-i)^3) = \frac{1}{16}(0-i-2+3i) = \frac{-1+i}{8},$$

$$\tilde{g}_2 = \frac{1}{16}(0+(-i)^2+2(-i)^4+3(-i)^6) = \frac{1}{16}(0-1+2-3) = \frac{-1}{8},$$

$$\tilde{g}_3 = \frac{1}{16}(0+(-i)^3+2(-i)^6+3(-i)^9) = \frac{1}{16}(0+i-2-3i) = \frac{-1-i}{8}.$$

(7.107)

(The fact that $(-i)^4 = 1$ aids the calculations.)

Let's expand on the above example. With the aid of a computer we calculate $\tilde{g}$, again for $\ell = 1$, but this time with $N = 100$. In Figure 7.3 we compare $|\tilde{g}_n|$ with $|c_n(g)|$—the latter equals $(2\pi n)^{-1}$ for $n > 0$, as we saw in Example 1.11.2. Recall that $\tilde{g}_n$ is supposed to *approximate* $c_n(g)$; our plot should give us some idea of the quality of the approximation.

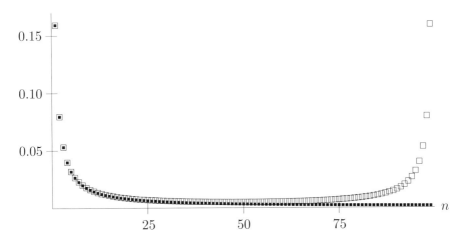

Fig. 7.3 $|\tilde{g}_n|$, with $N = 100$ (empty boxes), and $|c_n(g)|$ (smaller, filled boxes)

(To allow for a convenient viewing scale, we've omitted the case $n = 0$ from our figure. One computes that $c_0(g) = \frac{1}{2}$ and $\tilde{g}_0 = (\frac{1}{100})(0 + \frac{1}{100} + \frac{2}{100} + \cdots + \frac{99}{100}) = \frac{99}{200}$.)

Observe that, while $|\tilde{g}_n|$ does seem to approximate $|c_n(g)|$ quite well for n small, it does so rather poorly for n large. What's going on here?

What's going on may be explained as follows. Let f be any function continuous on $[0, \ell]$. Were $\tilde{f}_n$ to well approximate $c_n(f)$ for all n simultaneously, then (7.102) would imply that we could make $|c_{N+n}(f)|$ close to $|c_n(f)|$ by choosing N large enough. But the latter contradicts the fact that, by the Riemann-Lebesgue lemma (Lemma 1.4.1) (or, more exactly, by its analog for functions on $[0, \ell]$), $c_{N+n}(f)$ will generally be much *smaller* than $c_n(f)$.

The problem may be appreciated further by reflecting on the nature of Riemann sum approximations to definite integrals. For such an approximation to be accurate, the integrand in question should generally fluctuate only slightly between successive points in the partition of the sum. But for n large relative to N, the complex exponential $e^{-2\pi i n t/\ell}$ in the integral in (7.99) actually varies quite a bit between the t_k's of (7.100). So for such n, the Riemann sum defining $\tilde{f}_n$ does not necessarily do so well in approximating the integral defining $c_n(f)$.

So we use $\tilde{f}_n$ to estimate $c_n(f)$ only for n relatively small compared to N. How small? Well, as Figure 7.3 suggests, certainly no larger than $N/2$. There's some argument for using the requirement $n < N/8$ as a rule of thumb; see Section 2, Chapter 2, in [48]. At any rate N should, if possible, be chosen so that the $\tilde{f}_n$'s fade out—that is, they shrink to magnitudes insignificant to the application at hand—by some integer n_0 to the left of $N/2$.

Then practically speaking, only those $\tilde{f}_n$'s with $0 \le n < n_0$ are relevant to the frequency analysis of f, at least if f is real-valued. For such f we need not consider the case $n < 0$ because, again, $c_n(f)$ determines $c_{-n}(f)$, and therefore the entire component of frequency n/ℓ, for such f. On the other hand, if f is complex-valued, this is not so; we need a direct estimate of $c_{-n}(f)$. To this end we note that, if n is small, then $c_{-n}(f) \approx \tilde{f}_{-n} = \tilde{f}_{N-n}$. So for complex-valued functions f, both the nth and the $(N-n)$th coordinates of $\tilde{f}$ are relevant to estimation of content at frequency n/ℓ. Or: If the content of such an f becomes neglible at the frequency n_0/ℓ, then to study that content, we should consider the first *and* the last n_0 coordinates of $\tilde{f}$. (For a concrete illustration of these ideas, see Exercise 7.5.6.)

The *symmetry* of Figure 7.3, about the vertical line $n = N/2$, also bears reflection (pun intended). Such symmetry will, in fact, be present in $|\tilde{f}|$ whenever f is real-valued, since in this case $\tilde{f}_n = \overline{\tilde{f}_{-n}}$ by (7.101), so that

$$|\tilde{f}_n| = |\overline{\tilde{f}_n}| = |\tilde{f}_{-n}| = |\tilde{f}_{N-n}| \tag{7.108}$$

by N-periodicity of $\tilde{f}_n$.

In general, the periodic behavior of the discrete Fourier transform echoes quite eerily the aliasing encountered in the previous section. (See especially Proposition 7.4.2.) Indeed, this periodicity is itself referred to as the *aliasing* effect. The parallels between bandlimited and discrete situations are well worth contemplating.

THE DISCRETE FOURIER TRANSFORM

Recall that, in the former situation, there is an inversion theorem (Theorem 7.4.1). And so it is with any worthwhile Fourier theory. The discrete Fourier theory is worthwhile. Therefore, we have:

Theorem 7.5.1 (the discrete Fourier inversion theorem) *We have*

$$a_m = \sum_{n=0}^{N-1} \tilde{a}_n e^{2\pi i m n/N} \quad (0 \le m \le N-1) \tag{7.109}$$

for $a = (a_0, a_1, \ldots, a_{N-1}) \in \mathbb{C}^N$ *and* $\tilde{a} = (\tilde{a}_0, \tilde{a}_1, \ldots, \tilde{a}_{N-1})$ *as in Definition 7.5.1(b).*

In particular, if f *is as in Definition 7.5.1(a) then*

$$f\left(\frac{m\ell}{N}\right) = \sum_{n=0}^{N-1} \tilde{f}_n e^{2\pi i m n/N} \quad (0 \le m \le N-1); \tag{7.110}$$

that is, the discretization (7.105) of f *may be recovered completely from* $\tilde{f}$.

Proof. It suffices to prove (7.109); the second part of the theorem is a variation on that theme. We have, by Definition 7.5.1(b),

$$\sum_{n=0}^{N-1} \tilde{a}_n e^{2\pi i m n/N} = \sum_{n=0}^{N-1} \left[\frac{1}{N} \sum_{k=0}^{N-1} a_k e^{-2\pi i k n/N}\right] e^{2\pi i m n/N}$$

$$= \sum_{k=0}^{N-1} a_k \left[\frac{1}{N} \sum_{n=0}^{N-1} e^{2\pi i m n/N} e^{-2\pi i k n/N}\right]$$

$$= \sum_{k=0}^{N-1} a_k \left[\frac{1}{N} \sum_{n=0}^{N-1} \left(e^{2\pi i (m-k)/N}\right)^n\right]. \tag{7.111}$$

But

$$\frac{1}{N} \sum_{n=0}^{N-1} \left(e^{2\pi i (m-k)/N}\right)^n = \delta_{m,k} \tag{7.112}$$

for the following reason. If $m = k$ then $e^{2\pi i(m-k)/N} = 1$, so the left side is just $1/N$ times the sum of N 1's, and is therefore indeed equal to 1. If $m \ne k$ then, since both of these integers are between 0 and $N-1$, $|m-k|$ must be strictly smaller than N and larger than zero, so $(m-k)/N$ is not an integer, so $e^{2\pi i(m-k)/N} \ne 1$. But then the sum in (7.112) is zero, by the geometric sum formula (1.88) and the fact that $\left(e^{2\pi i(m-k)/N}\right)^N = e^{2\pi i(m-k)} = 1$.

Putting (7.112) into (7.111) gives

$$\sum_{n=0}^{N-1} \tilde{a}_n e^{2\pi i m n/N} = \sum_{k=0}^{N-1} a_k \delta_{m,k} = a_m, \tag{7.113}$$

and we're done. □

The essence of the above theorem is this: If we define
$$E_n = \left(0, e^{2\pi i n/N}, e^{2\pi i(2n)/N}, \ldots, e^{2\pi i((N-2)n)/N}, e^{2\pi i((N-1)n)/N}\right) \in \mathbb{C}^N, \tag{7.114}$$

then $\{E_n : 0 \leq n \leq N-1\}$ is an orthogonal basis for $\mathbb{C}^N$. See Exercise 7.5.5.

Example 7.5.2 Verify discrete Fourier inversion for the function g of Example 7.5.1.
Solution. We compute

$$\sum_{n=0}^{3} \widetilde{g}_n e^{2\pi i n \cdot 0/4} = \sum_{n=0}^{3} \widetilde{g}_n = \frac{3 + (-1+i) - 1 + (-1-i)}{8} = 0 = g(0),$$

$$\sum_{n=0}^{3} \widetilde{g}_n e^{2\pi i n \cdot 1/4} = \sum_{n=0}^{3} \widetilde{g}_n (i)^n = \frac{3 + (-1+i)(i) - 1(-1) + (-1-i)(-i)}{8}$$
$$= \frac{3 + (-i-1) + 1 + (i-1)}{8} = \frac{1}{4} = g\left(\frac{1}{4}\right),$$

$$\sum_{n=0}^{3} \widetilde{g}_n e^{2\pi i n \cdot 2/4} = \sum_{n=0}^{3} \widetilde{g}_n (-1)^n = \frac{3 + (-1+i)(-1) - 1(1) + (-1-i)(-1)}{8}$$
$$= \frac{3 + (1-i) - 1 + (1+i)}{8} = \frac{1}{2} = g\left(\frac{1}{2}\right),$$

$$\sum_{n=0}^{3} \widetilde{g}_n e^{2\pi i n \cdot 3/4} = \sum_{n=0}^{3} \widetilde{g}_n (-i)^n = \frac{3 + (-1+i)(-i) - 1(-1) + (-1-i)(i)}{8}$$
$$= \frac{3 + (i+1) + 1 + (-i+1)}{8} = \frac{3}{4} = g\left(\frac{3}{4}\right). \tag{7.115}$$

Cool.

Exercises

7.5.1 Repeat Examples 7.5.1 and 7.5.2 with the same N and ℓ but this time with $g(x) = x^2$.

7.5.2 Repeat Examples 7.5.1 and 7.5.2 with the same N and ℓ but this time with $g(x) = \cos \pi x$.

7.5.3 Repeat Examples 7.5.1 and 7.5.2 with the same ℓ and g but this time with $N = 6$. Hint: $e^{\pi i/3} = \frac{1}{2}(1 + i\sqrt{3})$.

7.5.4 Let a be the vector $a = (2, i, 1-i, 1) \in \mathbb{C}^4$.
   a. Compute *by hand* the discrete Fourier transform $\widetilde{a}$.

b. Show by direct computation that the discrete Fourier inversion theorem holds for this a.

7.5.5 a. Show that $\{E_n : 0 \leq n \leq N-1\}$, with E_n as in (7.114), is an orthogonal set in $\mathbb{C}^N$. That is, show $\langle E_n, E_k \rangle = \delta_{n,k}$, the inner product being defined by $\langle a, b \rangle = \sum_{n=0}^{N-1} a_n \overline{b_n}$. Hint: Use the geometric sum formula (1.88).

b. Show that

$$a = \sum_{n=0}^{N-1} \langle a, E_n \rangle E_n \qquad (*)$$

for any $a \in \mathbb{C}^N$. (Actually, this follows from part a, since any N-element orthogonal set in $\mathbb{C}^N$ is an orthogonal basis, but it's instructive to prove it directly.) Hint: Show both sides of $(*)$ have the same mth coordinate for $0 \leq m \leq N-1$ using the discrete Fourier inversion theorem.

7.5.6 Let $f \in C(\mathbb{T}_1)$ (recall that $\mathbb{T}_1$ is the torus of diameter 1) be defined by

$$f(t) = \sum_{k=-5}^{-1} e^{2\pi i k t}, \qquad (7.116)$$

so that, clearly, $c_k(f) = 1$ for $-5 \leq n \leq -1$ and $c_k(f) = 0$ otherwise. With a calculator or computer (or by hand, if you're not busy for a while), numerically evaluate the discrete Fourier transform $\tilde{f}$ for $N = 20$. Explain how your results reflect the discussion just above (7.108) of discrete Fourier transforms of complex-valued functions.

7.6 THE FAST FOURIER TRANSFORM, OR FFT, ALGORITHM

How long does it take to compute the coordinates $\tilde{a}_n$ of a discrete Fourier transform $\tilde{a}$, successively for $0 \leq n \leq N-1$? Quite long, actually. How long does it take to compute a discrete Fourier transform $\tilde{a}$ of length N? Not that long, really.

The point in that there is both a slow and a fast way to compute a discrete Fourier transform. We wish to compare these ways; here's our framework for comparison. We imagine that the complex coordinates $a_0, a_1, \ldots, a_{N-1}$ of the vector $a \in \mathbb{C}^N$ are already stored somewhere, as are the values $e^{2\pi i k n/N}$ ($0 \leq k, n \leq N-1$) of the required complex exponentials. (Again the coordinates of a will, generally, represent distinct observations of some signal, or phenomenon, or function, f.) We define an *elementary operation* to be either an addition or a multiplication of two complex numbers. We then agree to measure "speed" of an algorithm for computation of $\tilde{a}$ by the number of elementary operations that algorithm requires. This measure is simplistic to be sure and skirts a good number of technical issues. (In particular we ignore the fact that, in "real-life," real-valued situations, one actually needs only compute *half* of a discrete Fourier transform, by the symmetry documented in (7.108).) But it will be good enough for gleaning the big picture.

Let's consider first the slow algorithm, which was the commonplace one until about 1965. It entails a literal, "naive" implementation of the recipe (7.104). That is, first $\widetilde{a}_0$ is computed directly from that recipe, then $\widetilde{a}_1$, and so on. Let's "time" this algorithm. Since each multiplication of an a_k by an $e^{2\pi i k n/N}$ constitutes an elementary operation, as does each addition of a resulting summand $a_k e^{2\pi i k n/N}$ to the next one, as does the final division by N, we find that

$$N + (N-1) + 1 = 2N \tag{7.117}$$

elementary operations are needed *per coordinate* of $\widetilde{a}$. So computation of that entire vector requires $N \cdot 2N = 2N^2$ such operations.

YIKES!!! $2N^2$ is huge when N is large! To illustrate this, let's suppose we have a computer capable of a billion elementary operations per second, and a CD recording containing about 44000 samples per second. To perform a discrete Fourier transform on a selection of length $N = 65536 = 2^{16}$ samples, or just under a second and a half, that computer would require $2 \cdot 65536^2 = 8589934592$ elementary operations and would therefore take about eight and a half seconds. And if that doesn't seem so bad, consider now what discrete Fourier transformation of a mere $65536 \cdot 32 = 2^{21} = 2097152$ samples, or a little less than 48 seconds' worth, of the recording would entail. Multiplying N by 32 means multiplying N^2 by 1024; the roughly eight and a half seconds seconds would become about 8800, which is to say more than 2.4 *hours*. And a selection of length $2097152 \cdot 32 = 2^{26} = 67108864$ samples, or about 25 and a half minutes, would take roughly 104 *days* to discrete Fourier analyze! And so on. The bottom line is that the slow way is *much* too slow for practical applications.

It's quite amazing that the *fast Fourier transform*, or *FFT*, algorithm solves the problem so efficiently and so *simply*. Amazing, but true. Let's now investigate that algorithm. It requires that N factor nontrivially, in other words, that $N = N_1 N_2$, where $N_1, N_2 \in \mathbb{Z}^+$ and $1 < N_1, N_2 < N$. So let's suppose that this is so. We invoke some rather elementary number theory. Namely, given k and n in $\{0, 1, \ldots, N-1\}$, we divide N_1 into k and N_2 into n, yielding quotients and remainders. More specifically, we write

$$\begin{aligned} k &= N_1 q_1 + r_1 \quad (0 \leq q_1 \leq N_2 - 1, \ 0 \leq r_1 \leq N_1 - 1), \\ n &= N_2 q_2 + r_2 \quad (0 \leq q_2 \leq N_1 - 1, \ 0 \leq r_2 \leq N_2 - 1). \end{aligned} \tag{7.118}$$

(The fact that the quotient q_1 must be less than N_2 is because $k < N = N_1 N_2$. We have $q_2 < N_1$ for an analogous reason.) Note that (7.118) associates, with each pair $(q_1, r_1) \in \{0, 1, \ldots, N_1 - 1\} \times \{0, 1, \ldots, N_2 - 1\}$, precisely one $k \in \{0, 1, \ldots, N-1\}$, and conversely. Similarly for n, q_2, and r_2.

We then have

$$\frac{kn}{N} = \frac{(N_1 q_1 + r_1)(N_2 q_2 + r_2)}{N_1 N_2} = q_1 q_2 + \frac{q_1 r_2}{N_2} + \frac{q_2 r_1}{N_1} + \frac{r_1 r_2}{N_1 N_2}, \tag{7.119}$$

and consequently, since $e^{2\pi i q_1 q_2} = 1$ for $q_1, q_2 \in \mathbb{Z}$,

$$e^{-2\pi i k n/N} = e^{-2\pi i (q_1 r_2/N_2 + q_2 r_1/N_1 + r_1 r_2/(N_1 N_2))}. \tag{7.120}$$

Using this, we rewrite the formula (7.104) for $\tilde{a}_n$ as follows:

$$\tilde{a}_n = \frac{1}{N} \sum_{k=0}^{N-1} a_k e^{-2\pi i (q_1 r_2/N_2 + q_2 r_1/N_1 + r_1 r_2/(N_1 N_2))}$$

$$= \frac{1}{N} \sum_{r_1=0}^{N_1-1} \left[\sum_{q_1=0}^{N_2-1} a_{N_1 q_1 + r_1} e^{-2\pi i q_1 r_2/N_2} \right] e^{2\pi i (q_2 r_1/N_1 + r_1 r_2/(N_1 N_2))}. \quad (7.121)$$

Let's denote by $S(a; r_1, r_2)$ the quantity in square brackets on the right side of the above. The *key* to the fast Fourier transform algorithm is the assignment of a starring role to this quantity.

Indeed, we may find $\tilde{a}$ by way of the following two steps.

- FIRST, we compute all $S(a; r_1, r_2)$'s. There are $N_1 N_2 = N$ of them, since there are N_1 r_1's and N_2 r_2's. Moreover, each $S(a; r_1, r_2)$ entails N_2 multiplications (of an $a_{N_1 q_1 + r_1}$ by a complex exponential) and $N_2 - 1$ additions (of resulting summands to each other), which is to say $2N_2 - 1$ elementary operations. Thus we need a total of $(2N_2 - 1)N$ elementary operations for this first step.

- NEXT, we *use* the $S(a; r_1, r_2)$'s to compute the $\tilde{a}_n$'s. There are N of the latter, and each requires $2N_1$ elementary operations. (By (7.121) we need, for each n, N_1 multiplications of an $S(a; r_1, r_2)$ by a complex exponential, followed by $N_1 - 1$ additions of resulting summands, followed by a multiplication by $1/N$.) So this step requires $2N_1 N$ elementary operations.

The above algorithm therefore necessitates a GRAND TOTAL of

$$(2N_2 - 1)N + 2N_1 N = (2N_1 + 2N_2 - 1)N \quad (7.122)$$

elementary operations. For N large, this is far fewer than the $2N^2$ required by the slow algorithm. In particular, suppose N is a perfect square and $N_1 = N_2 = \sqrt{N}$; this factorization is optimal, in that it minimizes (7.122). (See Exercise 7.6.2.) Then $(2N_1 + 2N_2 - 1)N = 4N^{3/2} - N$.

Thus, instead of taking eight and a half seconds to analyze about a second and a half of sound, our computer from before would require only $(4 \cdot 65536^{3/2} - 65536)/10^9$, or about seven-hundredths, of a second. Not bad. Even better, the 48 second selection would now take $(4 \cdot 2097152^{3/2} - 2097152)/10^9 \approx 12$ seconds, rather than two and a half hours. And for the 25 and a half minute segment we'd need only $(4 \cdot 67108864^{3/2} - 67108864)/10^9 \approx 2200$ seconds, or slightly less than 37 minutes, as opposed to 104 days!

But wait, we can do still *better*. Much better. We can *iterate*. Specifically, if $N = N_1 N_2 \cdots N_J$, then by applying the above argument repeatedly, we find that $\tilde{a}$ may be evaluated in $(2N_1 + 2N_2 + \cdots 2N_J - 1)N$ elementary operations. In particular, if $N_j = 2$ for $1 \leq j \leq n$, so that $N = 2^n$, then we need only

$$(2 \cdot 2 + 2 \cdot 2 + \cdots + 2 \cdot 2 - 1)N = (4n - 1)N = 4N \log_2 N - N \quad (7.123)$$

388 SPECIAL TOPICS AND APPLICATIONS

such operations. Compared to $2N^2$, or even to $4N^{3/2} - N$, (7.123) is, generally speaking, *tiny*.

For example, discrete Fourier analysis of our one and one half second recording, again at a billion elementary operations per second, now takes not eight and a half seconds or seven-hundredths of a second, but only $(4 \cdot 65536 \cdot \log_2 65536 - 65536)/10^9 \approx$ four *thousandths* of a second. The 48 second selection requires not two and a half hours hours, or 12 seconds, but only $(4 \cdot 2097152 \cdot \log_2 2097152 - 2097152)/10^9 \approx$ two-tenths of a second. Finally, instead of 104 days or 37 minutes, we may now analyze our 25 and a half minute segment in $(4 \cdot 67108864 \cdot \log_2 67108864 - 67108864)/10^9 \approx$ *seven seconds*. Astounding!

Equally astounding is that the FFT algorithm allows us to study *local* frequency content in essentially real time. That is, instead of performing a discrete Fourier transformation on the entirety of a long audio signal, say, we may analyze a piece of it, display the results of that analysis, then analyze the next piece, and so on, rapidly enough that frequency content is seen to evolve with time in an animation that appears concurrent with the signal itself. See Figure 7.4.

Actually, the study of local frequency content constitutes a vast and complex industry all by itself. We'll examine this industry, and the intimately related one of *wavelets*, in Chapter 8.

The first known implementation of the idea behind the fast Fourier transform was by Gauss, in 1805, in his study of the asteroid Pallas. See [27] for an interesting discussion. It was then largely forgotten until 1965, when J. W. Cooley and J. W. Tukey published the paper An algorithm for the machine calculation of complex Fourier series [12], highlighting its utility. The timing was right, technologically speaking, and the article ignited a revolution. Its appearance may fairly be considered one of the most significant events in the history of Fourier analysis since Joseph Fourier, not to mention one of the greatest milestones in the annals of computing since the computer. As testament to this, we note that it is still among the most frequently cited papers in mathematics.

Exercises

7.6.1 Repeat Example 7.5.1 using the fast Fourier transform algorithm. (Here $4 = N = N_1 N_2$, where $N_1 = N_2 = 2$.) Specifically, first compute and write down all relevant $S(a; r_1, r_2)$'s, as described following (7.121), where a is the sampled version of g from the exercise in question. Then use these $S(a; r_1, r_2)$'s and (7.121) itself to compute the coordinates of the discrete Fourier transform of a.

7.6.2 Show that, if $N \in \mathbb{Z}^+$ is a perfect square, then the number (7.122) of elementary operations required for (a single iteration of) the FFT algorithm is smallest when $N_1 = N_2 = \sqrt{N}$. Hint: Minimize $f(x) = (2x + 2(N/x) - 1)N$, as a function of x, on $[1, N]$.

7.6.3 **a.** Repeat the argument preceding (7.123) to show that, if N is a power of $r \in \mathbb{Z}^+$ and $r, N > 1$, then the fast Fourier transform algorithm requires $T(r, N) = 2rN \log_r N - N$ elementary operations.

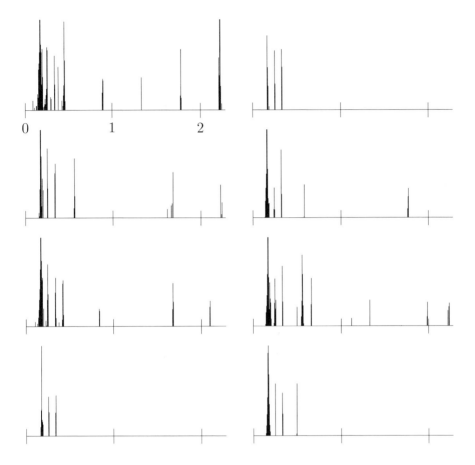

Fig. 7.4 Local frequency content of an excerpt from the author's composition "Col de Torrent." (Time evolves from top to bottom, left to right.) The horizontal axis is in kilohertz (kHz); the vertical scale is logarithmic

b. Let's for the moment let r be any real number larger than 1. Show that, for any *fixed* N, the above quantity $T(r, N)$ attains its minimum when $r = e$. Hint: In general, $\log_a x / \log_b x = \ln b / \ln a$.

c. Show further that, for fixed $N > 1$, $T(3, N) < T(2, N)$.

d. A positive integer power of 2 can never equal a positive integer power of 3, but sometimes two such powers can be "close"; for example, $N = 3^{12}$ and $M = 2^{19}$ are within 2% of each other. Show that, even though $N > M$, we have $T(3, N) < T(2, M)$. (So in some theoretical sense powers of 3 may seem to work better, but because computers are binary, powers of 2 are, in fact, generally preferable. See [12].)

7.7 FILTERING

We focus again on functions of continuous real variables—functions in $L^1(\mathbb{R}^M)$, $L^2(\mathbb{R}^M)$, and so on. Here general ideas and basic theories are more readily formulated and clearly conveyed. Of course, like the Fourier transform itself, these theories and ideas all admit reformulation in the discrete language of real-life applications. We'll address questions of numerical computation sporadically and then, for the most part, briefly. In this section we consider the art and science of filtering, which we introduce by way of the following problem. Suppose we receive a "corrupted" signal $g + r$, where g denotes some original, "true" signal in which we're interested and r denotes *noise* that is acquired between transmission and reception. How can we recover g, or an acceptable approximation thereof, with explicit knowledge of $g + r$ only?

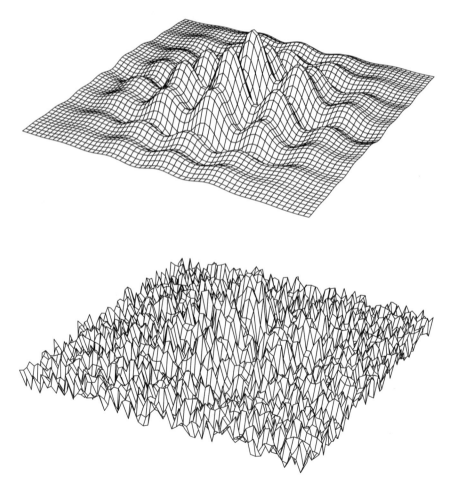

Fig. 7.5 Top: A "true" signal g. Bottom: The "corrupted" signal $g + r$

In Figure 7.5 we illustrate the problem in two dimensions. We might, for example, imagine that the vertical displacement $g(x)$ denotes grayscale level at the point $x = (x_1, x_2)$ in the plane; so g describes a grayscale *image*. Throughout this section we will, in fact, focus on two-dimensional phenomena to fix ideas. We'll also assume relevant functions to be nice enough that their Fourier transforms may be taken, and Fourier inversion applied, when needed. Everything we'll have to say generalizes with little change to the context of similarly reasonable functions on $\mathbb{R}^M$.

The solution to our problem relies on two main ideas. The first is this: The kind of random noise that we see in the bottom part of Figure 7.5, and that corrupts many true signals, is generally *quite unsmooth* compared to those signals. See [42] or Section 6.1 in [10] for more on the mathematics of noise; we'll be content to understand it in a mostly qualitative way.

The second main idea is the local/global principle, which we've encountered several times explicitly and quite a few more times implicitly. Again, that principle states, among other things, that the smoother a function f is, the faster $\widehat{f}$ decays; recall for example Proposition 6.6.2, which readily admits multivariable generalizations. For our present purposes, an illuminating heuristic explanation of this principle is as follows. To say f is relatively smooth is to say it has relatively little *jitter*, or *rapid bouncing around*, or *content at the high frequencies*. So f being relatively smooth indeed implies that $\widehat{f}(s)$ is relatively *small* for $\|s\|$ large.

Let's put these ideas together. Remember that we, as receivers of the corrupted signal, *do not* know g or r explicitly (ignore the top part of Fig. 7.5 for the moment), but we *do* know $g + r$. And again we know, from general considerations, that g is smooth compared to r, so that $\widehat{g}$ will decay faster than $\widehat{r}$. So if we plot or otherwise observe $\widehat{g+r}$—or, more practically, $|\widehat{g+r}|$—over a large enough frequency domain, we should see two things: (i) a large central concentration, where $\widehat{g}$ contributes significantly, and (ii) a low-level "bubbling" away from the center, where $\widehat{g}$ has for the most part died off and essentially only $\widehat{r}$ remains.

Having identified these, we simply mask off, or delete, or *filter* the bubbling, leaving, presumably, most of $\widehat{g}$ and little of $\widehat{r}$. In other words, we're left with an approximation to $\widehat{g}$; using Fourier inversion gives an approximation to g! (The Plancherel theorem says that functions approximate each other, in mean square norm, just as well as their Fourier transforms do.)

Let's recap, a bit more mathematically and with some relevant pictures, the filtering process just described.

1. We observe $|\widehat{g+r}|$. For example, see Figure 7.6.

2. We identify a region, say $\{s\colon \|s\| < \Omega\}$ (for some positive constant Ω), beyond which the central concentration has decayed and is replaced by a low-level bubbling.

3. We discard the information outside this region. To do so, we may multiply $\widehat{g+r}$ by the *filter* $\Phi_\Omega = \chi_{\{s\colon \|s\|<\Omega\}}$ (the characteristic function of the central region in question).

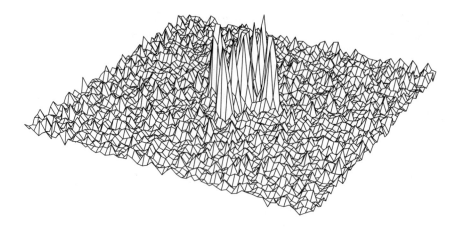

Fig. 7.6 The graph of $|\widehat{g+r}|$, for $g+r$ as in Figure 7.5

Then $(\widehat{g+r})\Phi_\Omega$ should approximate $\widehat{g}$. See Figure 7.7.

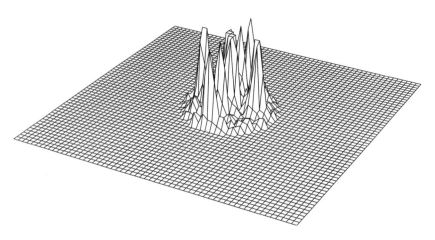

Fig. 7.7 The graph of $|(\widehat{g+r})\Phi_\Omega|$, for $g+r$ as in Figure 7.5 and a choice of Ω motivated by Figure 7.6

4. So, by the Plancherel theorem, $F\big[(\widehat{g+r})(s)\Phi_\Omega(s)\big](x)$ should approximate $F\big[\widehat{g}(s)\big](x)$. The latter is $g(-x)$ by Fourier inversion, so replacing x with $-x$ leads to our final result:

$$g(x) \approx F\big[(\widehat{g+r})(s)\Phi_\Omega(s)\big](-x). \qquad (7.124)$$

See Figure 7.8, and compare it with the top part of Figure 7.5!

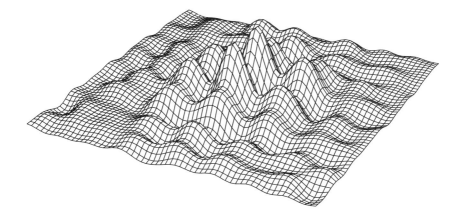

Fig. 7.8 The graph of $F\bigl[(\widehat{g+r})(s)\Phi_\Omega(s)\bigr]^{-}$

The function Φ_Ω is called a *low-pass filter*; it allows only low frequencies (those below a certain fixed threshold) to pass through. There are also *high-pass filters*, *band-pass filters*, and *band-stop filters*, which permit the passage of, respectively, only high frequencies; only those in a given range, say $\Omega_1 < ||s|| < \Omega_2$; and only those outside such a range. (One might call a high-pass filter a *low-stop filter*, and a low-pass filter a *high-stop filter*.) And there are many other kinds of filters.

In fact, the term is used generally, in signal processing, to connote any function Φ that's zero or negligible over certain portions of its domain. (See Chapter 6 in [36]. In other literature, filters may be defined more broadly, and what we just called a filter might be called a *mask filter*. But let's stay with our definition; it's more than broad enough for us.) The idea is that Φ may be considered a function of frequency, and as such may be used to attenuate frequency components of a signal f, much as Φ_Ω was used to diminish the high frequencies in $g + r$ above.

Let's, for example, investigate an implementation of what might be called a *comb filter*. We begin with the line drawing of Figure 7.9, from which we wish to remove some lines. Not all the lines; that would be silly. Just the long vertical ones, say.

We do so by filtering, as follows. We let f be the characteristic function of the drawing; that is, $f(x_1, x_2) = 1$ at any blackened point (x_1, x_2) in the image plane and $f(x_1, x_2) = 0$ elsewhere. We then consider the Fourier transform $\widehat{f}$. See Figure 7.10.

The observed behavior of $\widehat{f}$ may be explained as follows. The vertical lines in our original image are, to a very rough approximation, "constant vertically," meaning they don't change in nature as we scan across the picture from top to bottom. Of course this *is* a very rough approximation—the lines are of finite extent, so they do not really represent vertically constant content. But let's pretend they do; then the only "vertical frequency" they contribute to f is at $s_2 = 0$. They're very roughly P-periodic horizontally for some $P > 0$, so—think of Fourier series!—they contribute "horizontal frequencies" s_1 at integer multiples of $1/P$. The moral is that, because

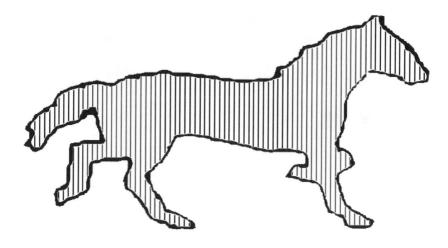

Fig. 7.9 Ceci n'est pas une zebra

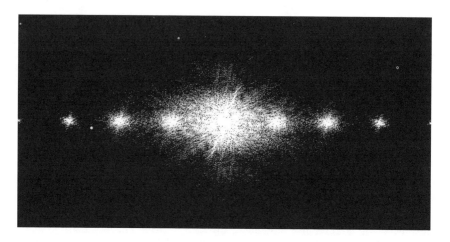

Fig. 7.10 $|\widehat{f}|$, for f the characteristic function of Figure 7.9

of the lines, $\widehat{f}(s_1, s_2)$ will be *large*—that is, it will *have peaks*—at

$$(s_1, s_2) = (n/P, 0) \quad (n \in \mathbb{Z}).$$

This is just what we see in Figure 7.10. (The fact that these peaks decrease in magnitude as s_1 gets larger is explained by the Riemann-Lebesgue lemma, meaning in this case a two-dimensional generalization of Proposition 6.6.1.)

So now we filter: We multiply $\widehat{f}$ by a function Φ_P that's 0 near the noncentral peaks and 1 elsewhere. (We retain the central peak because it represents the "bulk," or "average," of the image in question.) See Figure 7.11.

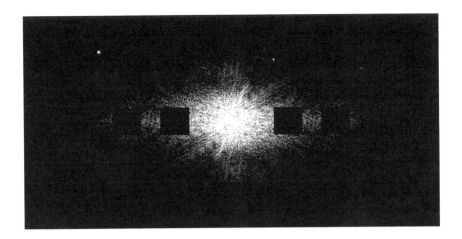

Fig. 7.11 $|\widehat{f}\Phi_P|$, for a suitable filter Φ_P

Finally we transform back, meaning we form the function

$$F\left[\widehat{f}(s)\Phi_P(s)\right]^-, \tag{7.125}$$

which should, in principle, reasonably approximate our drawing minus the vertical lines. And indeed it does (Fig. 7.12)!

Fig. 7.12 It's a horse, of course (the inverse Fourier transform of the function $\widehat{f}\Phi_P$ of Fig. 7.11)

For a somewhat more "real," but similar, application of filtering, to the removal of lines appearing when a photograph is composed from horizontal strips, see Figure 14.13 in [26]. For another related application, to the removal of "nonevent" lines in bubble chamber photographs, see [17]. (In the latter application, the parallel lines to be removed are not necessarily evenly spaced; consequently one must use a "line-stop," instead of a comb, filter.)

Note that filtering as we've described it, by way of our two examples above, is just the process of associating outputs

$$O_\Phi f = F\big[\widehat{f}(s)\Phi(s)\big]^-, \tag{7.126}$$

where Φ is the filter, to inputs f. For instance, f might be a "true" signal g plus random noise r and Φ a low-pass filter Φ_Ω; f might be an image that features some undesired parallel lines and Φ a comb filter Φ_P; and so on. Also note that, if we apply the convolution-inversion theorem (Theorem 6.8.1(d)(i)) to the above formula (7.126) for $O_\Phi f$, then we get (at least for sufficiently reasonable f and Φ) the following fascinating result:

$$O_\Phi f = f * \widehat{\Phi^-}. \tag{7.127}$$

That is, filtering amounts to convolution! So, for example, the recently-witnessed *smoothing* effect of a low-pass filter is essentially the same as the smoothing that results from convolution with elements of an approximate identity. (Recall Section 5.7.)

Exercises

7.7.1 Sometimes, instead of using a characteristic function as a filter, as we did in Figures 7.6 and 7.7 and the surrounding investigations, it's preferable to employ something a bit smoother. We illustrate this as follows: Consider the function $f(t) = 5\,\text{sinc}\,5\pi t = (\pi t)^{-1} \sin 5\pi t$. We'll imagine, for the sake of argument, that f is an audio signal.

Let's attenuate the high-frequency content of f by passing it through the filter $\Phi = \chi_{[-2,2]}$. So the filtered version $O_\Phi f$ is given by $O_\Phi f = F\big[\widehat{f}(s)\Phi(s)\big]^-$, cf. (7.126).

a. Write down an explicit formula for $O_\Phi f$.

b. Evaluate and explain the following statement: "$O_\Phi f$ exhibits spurious 'ringing' artifacts; its sound is not as localized near $t = 0$ as is that of f itself." Hint: Compare the peaks (not including the one at $t = 0$) of f to those of $O_\Phi f$. It might help to plot these functions on the same set of axes and/or to locate the peaks numerically by setting appropriate derivatives equal to zero and solving.

Now let's replace the filter Φ of part a with the function $\Psi = \chi_{[-5/2,5/2]} * \chi_{[-\frac{1}{2},\frac{1}{2}]}$.

c. Sketch the graph of Ψ. How does this graph compare with that of Φ? Hint: See Exercise 7.4.4. Note that our Ψ here is the function h there, with $\Omega = 2$ and $\gamma = \frac{1}{2}$.

d. Compute $O_\Psi f$. Again see Exercise 7.4.4.

e. Evaluate and explain the following statement: "Any spurious ringing of $O_\Psi f$ is less pronounced than that of $O_\Phi f$." Again, some graphs and/or numerical calculations should help.

f. Explain why your observations above are consistent with the local/global principle.

7.8 LINEAR SYSTEMS; DECONVOLUTION

In this section, we consider a notion of supreme importance in mathematics, electrical engineering, acoustics, the geological sciences, chemistry, and elsewhere.

Definition 7.8.1 An *analog, linear, time-invariant, continuous system*, or ALTICS, is a (physical or mathematical) device S with the following properties.

(a) Analogicity: If the input, or *signal*, f is in $L^2(\mathbb{R})$, then the output, or *response*, Sf is in $L^2(\mathbb{R})$.

(b) Linearity: $S(af + bg) = aSf + bSg$ for all $a, b \in \mathbb{C}$ and all inputs f, g.

(c) Translation invariance: For any $t_0 \in \mathbb{R}$ and any input f,

$$S(L_{-t_0} f) = L_{-t_0}(Sf), \tag{7.128}$$

where L_{-t_0} is the "lag operator" by t_0 units:

$$L_{-t_0} f(t) = f(t - t_0) \quad (t_0 \in \mathbb{R}) \tag{7.129}$$

(cf. Lemma 5.3.1).

(d) Continuity: If inputs f_N converge to an input f, then the outputs Sf_N converge to Sf.

We make five notes. First, translation invariance is sometimes called *time invariance*. In practice, inputs to and outputs from ALTICS *are*, often, functions of time. Certainly not always though. This brings us to our second note: The above definition generalizes in natural ways — and is well worth generalizing — to multivariable contexts. See the examples below. For the most part, though, we'll phrase major ideas, arguments, and results in the single variable language, leaving relevant translations to the imagination.

Third, the analogicity condition does not mean all inputs or all outputs need be in $L^2(\mathbb{R})$. Indeed, soon we'll want to allow for complex exponential and even δ "function" signals and responses. Nor do we require that our system accept *all* elements of $L^2(\mathbb{R})$ as inputs or return all such as outputs. We only stipulate that the response be in $L^2(\mathbb{R})$ *if* the signal is. Even this stipulation is not really necessary, but it does help to focus our investigations somewhat. Moreover $||f||^2$ is often interpreted, in applications, as the *energy* of f, and it seems sensible that a finite energy input should yield a finite energy output. (In any case, the *domain* of S should be a vector space and should be closed under delays $t \to t - t_0$, so that linearity and translation invariance make sense).

Fourth, what we call analogicity, some others call continuity! Finally, we confess that *our* definition of continuity is quite imprecise; we've not specified a mode or modes of convergence. This is deliberate. We'll say a bit more about continuity as we proceed. But generally, we'll be content to understand it intuitively, as the requirement that, should two or more signals be "close" to one another, the corresponding responses will be too. And we'll agree to consider an analog, linear, translation invariant system S an ALTICS as long as it satisfies this requirement in some plausible sense.

Here, then, are some examples of ALTICS.

Example 7.8.1 A stereo speaker. The signal f is electromagnetic, and the response Sf is acoustic. In this context, linearity means that an adjustment to the volume control will cause a proportional amplification of what's actually heard and that the output from a combination of inputs will sound as one would expect. (The latter may or may not be a good thing; witness karaoke.) Translation invariance means a CD played today will sound the same as it did yesterday. Continuity means a small scratch on a record (apparently, at one time, there were these things called records) will produce only a small distortion to the sound produced.

Example 7.8.2 Various electrical circuits. See Section 3.5 in [15] or Section 6.8 in [10].

Example 7.8.3 A layer of geological material. Here an acoustic signal f is sent into the material at some point, and an acoustic response Sf is recorded at another. The relation between f and Sf provides useful information concerning the *density* of the material in between those points.

Example 7.8.4 A concert hall. Here the acoustic signal f comes from the stage, and the acoustic response Sf will depend on location within the hall. The goal is to design that hall so that, at any location therein, Sf is as much like f as is possible or is different from f only in aurally pleasant ways.

Example 7.8.5 A linear differential operator L with constant coefficients, under appropriate restrictions on its domain. Here linearity is by definition; translation invariance is because, by the chain rule, $d(f(t - t_0))/dt = f'(t - t_0)$. The "appropriate restrictions" are required to assure analogicity. For example, if L is of second order, then for analogicity it certainly suffices to require that the domain of L be contained in the vector space $\{f \in L^2(\mathbb{R}): f', f'' \in L^2(\mathbb{R})\}$.

The issue of continuity of differential operators is a bit thornier. A sequence f_N can converge to a function f in any number of "standard" ways without f'_N converging in *any* of those ways. For example, $g_N(t) = N^{-1}e^{iNt}$ converges to zero uniformly on $\mathbb{R}$, and converges to zero in the mean and mean square norms on any bounded interval. But $g'_N(t) = ie^{iNt}$ just oscillates more and more rapidly as N increases, and does not converge pointwise, or uniformly, or in mean or mean square norm, on any interval of nonzero length, to anything.

On the other hand, differentiation *is* continuous in the *temperate* sense: If the f_N's and f define tempered distributions and f_N converges temperately to f, then f'_N

converges temperately to f'. See Definition 6.9.3 and Exercise 6.9.2. The vagaries of temperate convergence need not, for the present purposes, concern us excessively, but at any rate this example indicates that we should approach Definition 7.8.1(d) with an open mind.

Example 7.8.6 The earth's atmosphere acts as a three-dimensional ALTICS on electromagnetic radiation from planets, stars, and the like. See Section 95 in [30].

Example 7.8.7 The filtering scheme of (7.126) represents an ALTICS O_Φ in any given number of variables provided the filter Φ is a sufficiently nice function in those variables. This follows from (7.127) and from our next example.

Example 7.8.8 Convolution with a function $h \in L^1(\mathbb{R})$. That is, if we define $Sf = f * h$ for such an h, then S is indeed an ALTICS. Let's check this. First, if $f \in L^2(\mathbb{R})$ then so is Sf, by Proposition 5.6.2. Second, S is linear by the distributive property of convolution (Proposition 5.1.1(c)). Third, translation invariance is by definition of convolution:

$$L_{-t_0}(Sf)(t) = (L_{-t_0}(f*h))(t) = f*h(t-t_0)$$
$$= \int_{-\infty}^{\infty} f(t-t_0-y)h(y)\,dy = \int_{-\infty}^{\infty} (L_{-t_0}f)(t-y)h(y)\,dy$$
$$= (L_{-t_0}f * h)(t) = S(L_{-t_0}f)(t). \tag{7.130}$$

Finally, by linearity and by Proposition 5.6.2 again, we have, for $f_N, f \in L^2(\mathbb{R})$,

$$\|Sf_N - Sf\|_2 = \|S(f_N-f)\|_2 = \|(f_N-f)*h\|_2 \le \|f_N-f\|_2\,\|h\|_1. \tag{7.131}$$

So mean square norm convergence of inputs implies that of outputs, so the conditions of Definition 7.8.1(d) are satisfied in a satisfying way.

The last of these examples is the canonical one. Indeed, it turns out that *any* sufficiently continuous ALTICS S may be described by convolution with a fixed entity h. Moreover, h is just how S responds to the *delta* signal.

Specifically, we have the following amazing principle, which we stop short of calling a proposition, because its statement and proof are a bit vague, though the former can be made precise and the latter rigorous in a number of ways.

Principle 7.8.1 *If S is an ALTICS that's suitably continuous, then for any input f,*

$$Sf = f * h, \tag{7.132}$$

where h, the impulse response *of S, does not depend on f. Indeed,*

$$h = S\delta. \tag{7.133}$$

Proof. We suspend disbelief and write

$$f(t) = \delta * f(t) = \int_{-\infty}^{\infty} \delta(t-y)f(y)\,dy = \int_{-\infty}^{\infty} L_{-y}\delta(t)f(y)\,dy. \qquad (7.134)$$

(See Section 5.7.) We suspend it even further and argue, heuristically, as follows. Since a (sufficiently nice) definite integral is a limit of finite Riemann sums, and since S may, by linearity, be brought inside finite sums and further may, by continuity, be brought inside limits, it may also be brought inside the definite integrals in (7.134). That is,

$$Sf(t) = S\left(\int_{-\infty}^{\infty} L_{-y}\delta(t)f(y)\,dy\right) = \int_{-\infty}^{\infty} S(L_{-y}\delta)(t)f(y)\,dy. \qquad (7.135)$$

(We're thinking of S as operating on functions of t; for any y, $f(y)$ is constant with respect to t and therefore may, by linearity of ALTICS, be brought outside of S.) But by translation invariance of S and the definition (7.133) of the impulse response h,

$$S(L_{-y}\delta)(t) = L_{-y}(S\delta)(t) = L_{-y}h(t) = h(t-y), \qquad (7.136)$$

so (7.135) yields

$$Sf(t) = \int_{-\infty}^{\infty} h(t-y)f(y)\,dy = h * f(t) = f * h(t). \qquad (7.137)$$

We're done. □

To legitimize the above principle and proof, we would need first of all to ascribe definite meanings both to $S\delta$ and to the notion of continuity of S. The former we might define as some kind of limit of $Sg_{[\varepsilon]}$'s, where $\{g_{[\varepsilon]} : \varepsilon > 0\}$ is an approximate indentity. But what kind of limit?

It depends on our definition of continuity! Consider, for example, the identity system $Sf = f$, which certainly satisfies parts (a)–(c) of Definition 7.8.1 and is continuous in many, many plausible senses. (As long as we apply the same interpretation to both instances of the phrase "converge to" in Definition 7.8.1(d), the identity system conforms to that definition.) For this system, of course, $h = S\delta = \delta$ (and (7.132) amounts to the equation $f = f * \delta$, again), so h does *not* make sense as a pointwise or uniform limit, or a limit in mean or mean square norm, of $g_{[\varepsilon]}$'s. But it *does* make sense as a *temperate limit* thereof. See Definition 6.9.3 and Example 6.9.3. Similar remarks apply to the system $Sf = f'$ discussed in Example 7.8.5 above. (There is no actual *function* h such that $f' = f * h$ for all, or even for "most," functions f.)

Should we require that $Sg_{[\varepsilon]}$ converge to a bona fide function h, then we need to impose on S more stringent continuity conditions, such as those of *complete continuity*. See Section 24 of [29] for the definition of the latter. At any rate, we forgo further discussion of exact meanings and proofs of Principle 7.8.1. And we focus on its *gestalt*, which is that the output of an ALTICS S equals the input convolved with the impulse response. In particular, we know S completely if we know what it does to δ. This is true in several variables as it is in one, by arguments nearly identical to those just made.

A fun exemplification of this *gestalt* may be seen in the movie *Jurrasic Park*. A few minutes in, there is a scene in which a shotgun (loaded with blanks, presumably) is fired down into the earth—there's your impulse δ!—and an image of a buried fossil is returned on a monitor—there's your impulse reponse $S\delta$! (Recall Example 7.8.3.)

On the other hand, in real life, feeding δ's, or even approximate ones, into things is not always advisable. It would be a very impractical way of testing the response of a stereo speaker, for example. Fortunately, the *Fourier transform* provides an alternative method.

Principle 7.8.2 *Given a sufficiently reasonable ALTICS S, define the* system function, *also known as the* transfer function, H *as follows:*

$$H(s) = (Se_s)(0), \qquad (7.138)$$

where $e_s(t) = e^{2\pi i s t}$. That is, $H(s)$ is the response, at $t = 0$, of S to the complex exponential e_s.

Then

$$h = \widehat{H^-}, \qquad (7.139)$$

where, as usual, $h = S\delta$ is the impulse response of S.

Proof. By definition (7.138) of the transfer function H and by Principle 7.8.1,

$$H(s) = Se_s(0) = e_s * h(0)$$
$$= \int_{-\infty}^{\infty} e_s(0-y)h(y)\,dy = \int_{-\infty}^{\infty} h(y)e^{-2\pi i s y}\,dy = \widehat{h}(s). \qquad (7.140)$$

Taking inverse Fourier transforms of both sides and applying Fourier inversion yield $\widehat{H^-} = (\widehat{h})^- = h$, as desired. □

So to determine the impulse response of an ALTICS S (and thus to completely determine S itself), one needn't supply it with a δ; in principle, complex exponentials suffice.

The above two principles allow for a suggestive, alternative way of understanding an ALTICS. As follows: We put the recipe $h = \widehat{H^-}$ of Principle 7.8.2 into the formula $Sf = f*h$ of Principle 7.8.1 and apply the composition theorem (Proposition 6.2.1(d)) to the result to get

$$Sf(x) = f * h(x) = f * \widehat{H^-}(x) = \int_{-\infty}^{\infty} \widehat{f}(s)H(s)e^{2\pi i s x}\,ds \qquad (7.141)$$

(for suitable f and S). Thus we obtain the frequency decomposition of a response Sf in terms of that of the signal f. In particular we may, in light of (7.141), think of the transfer function H as telling us "the amounts by which we must weight the individual frequency components of f to get those of Sf."

We conclude by turning all of the above investigations on their heads, sort of. That is, so far we've been discussing ways of understanding systems themselves and of mathematically describing outputs from them in terms of inputs to them. But now let's suppose we know S—that is, we know h or, equivalently, H—and we know a *response* Sf. From these quantities, can we recover the signal f?

In more general terms, what we're asking is: Can we solve $k = f * h$ for f given k and h? Or we might, as one often does, put it this way: Can we *deconvolve*? The answer is: Yes, in many cases. Here's how:

$$k(t) = f * h(t);$$
$$\widehat{k}(s) = \widehat{f * h}(s) = \widehat{f}(s)\widehat{h}(s);$$
$$\widehat{f}(s) = \widehat{k}(s)/\widehat{h}(s);$$
$$f(t) = F[\widehat{f}(s)](-t) = F[\widehat{k}(s)/\widehat{h}(s)](-t). \quad (7.142)$$

Provided the Fourier transforms and divisions and so on behave sufficiently well, we've indeed found our solution f!

Example 7.8.9 Let $h(t) = e^{-2\pi|t|}$. Solve the equation $k = f * h$ for f in terms of k and its derivatives (assume k is as reasonable as is needed). Find f explicitly in the case $k(t) = e^{-\pi t^2}$.

Solution. We have $\widehat{h}(s) = (\pi(1+s^2))^{-1}$ by Example 6.1.2. So by the last line in (7.142),

$$f(t) = \pi F\big[(1+s^2)\widehat{k}(s)\big](-t) = \pi F\big[\widehat{k}(s)\big](-t) + \pi F\big[s^2 \widehat{k}(s)\big](-t)$$
$$= \pi \widehat{\widehat{k}}(-t) + \frac{\pi}{(-2\pi i)^2}\frac{d^2}{dt^2}\widehat{\widehat{k}}(-t) = \pi k(t) - \frac{1}{4\pi}k''(t). \quad (7.143)$$

For the next to the last step we used Proposition 6.2.1(c)(ii), regarding Fourier transforms of derivatives; for the last we used Fourier inversion.

So if $k(t) = G(t) = e^{-\pi t^2}$, then

$$f(t) = \pi e^{-\pi t^2} - \frac{1}{4\pi}(-2\pi + 4\pi^2 t^2)e^{-\pi t^2} = \left(\frac{1}{2} + \pi(1-t^2)\right)e^{-\pi t^2}. \quad (7.144)$$

(See Exercise 7.8.3 for some elaboration on this example.)

Deconvolution, in the above and in analogous multivariable incarnations, is an extremely powerful tool. We already mentioned, in Section 5.7, its utility in the sharpening of images and other signals. Another interesting application is to the interpretation of seismic activity. Here's the idea (cf. Example 7.8.3): A disturbance f is produced at a point A and propagates through the earth, which acts as an ALTICS S and outputs a response Sf, recorded at a point B. As the characteristics of S are at least fairly well understood, the observers at B can deconvolve to deduce the general nature of f. So, for example, it can be determined whether the disturbance corresponds to something dastardly, like violation of a nuclear test ban treaty, occurring at point A or something more "benign," like an earthquake. See [32] and Section 14.5 in [10] for further discussion.

Indeed, there are in fact a great many situations where deconvolution is relevant to the understanding of "true," original signals. This is because the media through which these signals pass, between transmission and reception, very often act like ALTICS—and therefore, by Principle 7.8.1, like convolution operators—on those signals. (The media in question include not only "natural" phenomena intervening between origination and measurement, but also the very instruments with which we measure.) These observations should be compared with (7.127) and the surrounding discussions, where *convolution* is used to get closer to the truth!

Exercises

7.8.1 Show that, for any $t_1 \in \mathbb{R}$, the lag operator L_{-t_1} defined by (7.129) itself is an ALTICS. Be specific about the sense, or at least *a* sense, in which this operator is continuous. Hint: Lemma 5.3.1.

7.8.2 Is the Fourier transform—meaning the "device" S that accepts f as input and returns $Sf = \hat{f}$ as output—an ALTICS? If not, which properies of an ALTICS does it have, and which does it not? Hint for continuity: Use the Plancherel theorem.

7.8.3 Into Example 7.8.9, put $k(t) = e^{At}$, where A is a real constant. For which values of A is the function f of (7.143) an actual solution to the original problem $k = f * h$? For which A is it not? (Note that k is not in $L^1(\mathbb{R})$ or in $L^2(\mathbb{R})$ for any A, so the argument (7.143) is suspect in this case; still, it provides a correct result at least for *some* values of A!)

7.8.4 Verify Example 7.8.9. In other words, show by direct manipulation of the convolution integral that, if f is as in (7.143) and h is as defined in the example, then $f * h = k$. Hint: Integrate by parts. Assume k to be nice enough that this is valid; that "uv" terms coming from the integration by parts go to zero at $\pm\infty$, etc.

7.8.5 Repeat Example 7.8.9 (including the plugging in of $k(t) = G(t) = e^{-\pi t^2}$) with $f(t) = e^{-2\pi|t|}(\cos 2\pi|t| + \sin 2\pi|t|)$ (see Exercise 6.3.6).

7.9 FRAUNHOFER DIFFRACTION AND FOURIER OPTICS

So far, we have detailed only mathematical and computational means of effecting frequency analysis. But there are other means—there are, in fact, various *physical* processes whose outcomes amount to Fourier transformation. Many of these processes hinge on the following beautiful and profound principle: *The Fraunhofer diffraction pattern of an object is essentially the Fourier transform of that object.* Here, we wish to investigate the rudiments of this principle, and of the science of *Fourier optics*, which has this principle at its core.

We need to recall some basic properties of light. First, according to the electromagnetic, or EM, wave theory, light is an electromagnetic field. What this means is that, at any time t and point (r, u, v) in space, light comprises an electric field vector $E = E(r, u, v, t)$ and a magnetic field vector $B = B(r, u, v, t)$. These are vectors in $\mathbb{R}^3$ although, as we'll see, it's sometimes convenient to imagine they have *complex* coordinates.

To simplify our discussions, we impose some restrictions on the nature of our light. First, we assume that it's *planar*. This means E and B depend only on time t and on distance from some fixed plane. By choosing coordinates appropriately, we can, and we do, assume this plane to be the uv plane. Then E and B depend only on r and on t.

From *Maxwell's equations* (cf. [37]), it may be shown that B is similar to and completely determined by E. We will therefore, from now on, concern ourselves only with E.

From these same equations it may further be deduced that, for planar light, the r-component of E is constant. Let's ignore this constant (we may, without suffering excessively, take it to be zero); then we can write

$$E(r, t) = (E_u(r, t), E_v(r, t)). \tag{7.145}$$

See Figure 7.13.

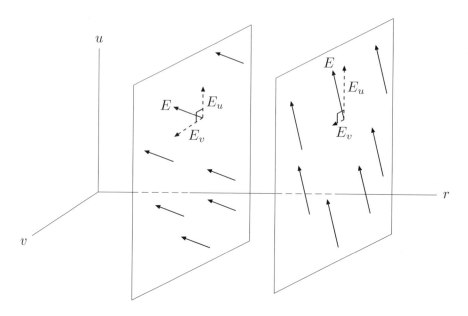

Fig. 7.13 The electric field E for planar light (snapshot at a particular time t)

But now, we further assume that our light is *polarized*, meaning the *direction* of E is constant. We set things up so that this direction is aligned with the u axis; then

E is completely determined by the *scalar* $E_u(r,t)$. For ease of notation, we denote this scalar itself by $E(r,t)$ (that is, we identify $E(r,t)$ with its u component).

Lastly, we take our light to be *monochromatic*, meaning a single frequency (or *color*) s goes into its makeup. Under all these assumptions, then, E may be modeled by an equation of the form

$$E(r,t) = A\cos(2\pi s(r/c - t) + \phi), \qquad (7.146)$$

where c denotes the speed of light, $A > 0$, and $\phi \in (-\pi, \pi]$. In accordance with Section 1.10 above, we call A the *amplitude* and ϕ the *phase* of our planar, polarized, monochromatic light—which we will henceforth refer to as PPM light.

It will be tremendously useful to note the following equivalent form of (7.146):

$$E(r,t) = \mathrm{Re}\left(A\, e^{i(2\pi s(r/c - t) + \phi)}\right), \qquad (7.147)$$

or even better

$$E(r,t) = \mathrm{Re}\left(A_0\, e^{2\pi i s(r/c - t)}\right), \qquad (7.148)$$

where $A_0 = A\, e^{i\phi}$. We call A_0 the *complex amplitude* of E; note that A_0 embodies both the (usual, real) amplitude A *and* the phase ϕ of E! More specifically, we have $A = |A_0|$ and $\phi = \mathrm{Arg}\, A_0$.

Using (7.148) instead of (7.146), we may exploit all the usual advantages of the complex exponential perspective. In particular, (7.148) allows us to *change phase* essentially by multiplying. Indeed, for A_0 as above,

$$A\cos(2\pi s(r/c - t) + (\phi + \Delta\phi)) = \mathrm{Re}\left(A_0'\, e^{2\pi i s(r/c - t)}\right), \qquad (7.149)$$

where $A_0' = A_0\, e^{i\Delta\phi}$.

So from now on, we will in fact model PPM light by the formula

$$E(r,t) = A_0\, e^{2\pi i s(r/c - t)}. \qquad (7.150)$$

This description will be particularly convenient in the analyses that follow; still, we should keep in mind several caveats. First, only the real part of $E(r,t)$ has physical significance. Second, again, $E(r,t)$ represents only the electric "half" of the electromagnetic picture. Third, the human eye, and most other media—screens, films, and so on—commonly used in the capturing and recording of optical infomation, actually "see" not electric or magnetic fields *per se*, but rather *intensity patterns*. These amount mathematically to squares of absolute values of such fields.

In any event, equipped with the above, basic notions, we are now ready to examine the phenomenon of Fraunhofer diffraction itself. We imagine the following situation. Suppose that, in the left half of uvr space—that is, in the subspace $\{(u,v,r): r < 0\}$—we have PPM light as described above, and that we introduce into the uv plane a grayscale, optical *object* k. The latter means that, at each point (u,v) in this plane, a fraction $k(u,v)$ of the incident light is allowed to pass through.

In this context, we call our uv plane the *object plane*. We will assume, for reasons to be made clear as we proceed, that the object is of finite extent, which is to say that k has compact support.

Now let's suppose we also have a *spectrum plane*, meaning a plane parallel to—and, say, r units from—the uv plane (Fig. 7.14). The question we wish to consider is this: What "happens" to the light passing through the object k by the time it reaches this spectrum plane? Or in physical terms, what is the *electric field* $C(x, y)$ at a given point $P(r, x, y)$ in the spectrum plane and a given instant t, assuming this field derives exclusively from the light transmitted through k?

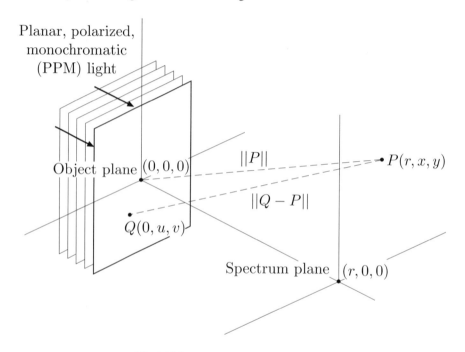

Fig. 7.14 Object and spectrum planes

When r is large compared to the dimensions of the object k, we call $C(x, y)$ the *Fraunhofer* (also sometimes called the *farfield*) *diffraction pattern* of this object. Actually, this pattern will depend on r and t as well; we suppress this dependence from the notation because what we wish to emphasize is the *Fourier transform relationship* between object and spectrum planes. Indeed we wish, and we now proceed, to demonstrate the following amazing result: $C(x, y)$, properly scaled and up to an approximation that improves as r increases, equals the *Fourier transform of* $k(u, v)$!!

To see this, let's first envision an instant τ at which our PPM light hits the object plane. By (7.150) it irradiates this plane with an electric field $E(0, \tau) = A_0 e^{-2\pi i s \tau}$. Emanating *through* this plane, then, at this same instant, we have an electric field

$$D(u, v) = A_0 e^{-2\pi i s \tau} k(u, v). \tag{7.151}$$

We'd like to describe the evolution of this field as it journeys to the spectrum plane.

To do so we invoke *Huygen's principle*, which says we can, at such an instant, treat each point in the object plane as a *point source*, emitting a spherical wave traveling outward (Fig. 7.15). (Huygen's principle itself follows from the EM wave theory, and from laws of three-dimensional wave propagation, cf. Proposition 7.3.1 above.) We find that the electric field element "leaving" a point $Q(0, u, v)$ at some instant τ is augmented by *two factors* before its effects are felt at a point $P(r, x, y)$ and at time t. Those factors are:

1. A factor of $e^{2\pi i s(||Q-P||/c - (t-\tau))}$ ($||Q - P||$ denotes the $\mathbb{R}^3$ distance from Q to P, cf. Section 6.8). Heuristically, this factor arises because the spherical wave from Q "looks flat" to P and thus can be treated, *as far as phase is concerned*, as though it were PPM light traveling in planes perpendicular to the axis determined by Q and P. By (7.150), the phase of such light changes exactly by the factor indicated as we move from τ to t and from Q to P.

2. A factor of $1/||Q - P||$. This factor is dictated by the law of conservation of energy. The total energy in a spherical wave is, essentially, the integral of the *square* of the absolute value of the electric field over the sphere. The surface area of a sphere is proportional to the square of its radius; division of the electric field by this radius therefore assures constant energy as the spherical wave spreads out.

We conclude (using (7.151)) that $D(u, v)$, coming from $Q(0, u, v)$ at some time τ, contributes to the field at $P(r, x, y)$ at the instant t of interest in an amount given by

$$\frac{D(u,v)e^{2\pi is(||Q-P||/c-(t-\tau))}}{||Q-P||} = \frac{A_0 \, k(u,v)e^{2\pi is(||Q-P||/c-t)}}{||Q-P||}. \tag{7.152}$$

The right side is independent of τ. Therefore, to find the *total* electric field $C(x, y)$ at $P(r, x, y)$ and at time t due to the light shining through k, we may *integrate* (7.152) over all (u, v) in the object plane; so

$$C(x,y) = A_0 \, e^{-2\pi ist} \int_{\mathbb{R}^2} \frac{k(u,v)}{||Q-P||} e^{2\pi is||Q-P||/c} \, du \, dv. \tag{7.153}$$

Observe that the variable t appears in the complex exponential factor $e^{2\pi ist}$, but *not* in the integral over the object plane. This integral reflects the *spatial* distribution of our electric field at the spectrum plane. We seek a better understanding of this distribution, so let's peer more closely at this integral.

To this end, we expand out the quantity $||Q - P||$ in the integrand. We have, by the same argument as was used to obtain (3.144),

$$||Q - P|| = \sqrt{||P||^2 - 2\langle Q, P\rangle + ||Q||^2}, \tag{7.154}$$

the inner product here being the one on $\mathbb{R}^3$, cf. (6.162). Now we may as well assume that $||Q||$ is bounded by some positive constant K, for the following reason. Our

408 SPECIAL TOPICS AND APPLICATIONS

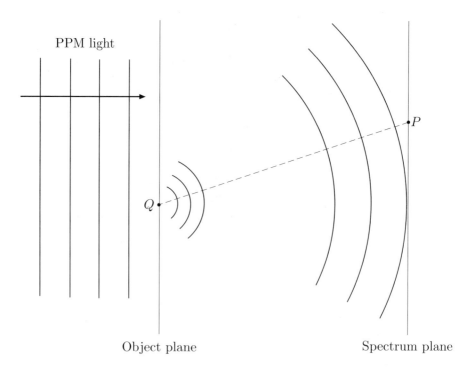

Fig. 7.15 Huygen's principle

interest is in the integral in (7.153), and that integral effectively takes place over only those points $Q(0, u, v)$ such that $k(u, v) \neq 0$. But k has, by assumption, compact support, so $k(u, v) \neq 0$ implies (u, v) is no more than some fixed distance from the origin, which is to say

$$||Q|| = \sqrt{u^2 + v^2} \leq K \qquad (7.155)$$

for some $K > 0$.

From now on, let's assume that the distance r between object and spectrum planes is large compared to this constant K, so that $||P|| = \sqrt{r^2 + x^2 + y^2}$ is too. The $||Q||^2$ term on the right side of (7.154) is then relatively insignificant, so that

$$||Q - P|| \approx \sqrt{||P||^2 - 2\langle Q, P \rangle} = ||P||\sqrt{1 - 2\frac{\langle Q, P \rangle}{||P||^2}} \approx ||P||\left(1 - \frac{\langle Q, P \rangle}{||P||^2}\right). \qquad (7.156)$$

The last step is by the Maclaurin series approximation

$$\sqrt{1 - 2Z} \approx 1 - Z, \qquad (7.157)$$

which is valid for $|Z|$ small. This approximation is applicable to our present case $|Z| = |\langle Q, P \rangle|/||P||^2$ for the following reason. From the Cauchy-Schwarz inequality

(3.15) (which, again, holds in $\mathbb{R}^3$ as it does in $C(\mathbb{T})$) and from (7.155), we have

$$\frac{|\langle Q, P\rangle|}{||P||^2} \leq \frac{||Q||\,||P||}{||P||^2} = \frac{||Q||}{||P||} \leq \frac{K}{||P||}, \tag{7.158}$$

which is indeed small for r as stipulated.

We put the approximation (7.156) into *the complex exponential* in the integrand of (7.153); we also put the simpler, cruder approximation

$$||Q - P|| \approx ||P|| \tag{7.159}$$

(which itself follows from (7.156) and (7.158)) into the *denominator* of that integrand. We get

$$\begin{aligned}
C(x, y) &\approx A_0 \, e^{-2\pi i s t} \int_{\mathbb{R}^2} \frac{k(u, v)}{||P||} e^{2\pi i s (||P|| - \langle Q, P\rangle/||P||)/c} \, du \, dv \\
&= \frac{A_0 \, e^{2\pi i s (||P||/c - t)}}{||P||} \int_{\mathbb{R}^2} k(u, v) e^{-2\pi i s (u x + v y)/(||P|| c)} \, du \, dv \\
&= \frac{A_0 \, e^{2\pi i s (||P||/c - t)}}{||P||} \widehat{k}\left(\frac{s}{||P||c}x, \frac{s}{||P||c}y\right).
\end{aligned} \tag{7.160}$$

So indeed $C(x, y)$ *is*, approximately, a rescaled Fourier transform of the original object k!!

Before proceeding further, we should clarify a certain aspect of the above derivation. Specifically, to obtain the approximation (7.160), we applied to (7.153) two *different* estimates ((7.156) and (7.159)) for $||Q - P||$. Why? Is there a sound motive behind this apparent capriciousness?

Yes there is, as may be understood through a careful error analysis. Let's denote the right side of (7.160) by $C_{\text{appr}}(x, y)$. One then shows, using (for example) the theory of Taylor polynomials, that for some positive number $M > 0$,

$$|C(x, y) - C_{\text{appr}}(x, y)| \leq \frac{M}{r^2} \tag{7.161}$$

for all $r > 0$. (Here M may depend on the dimensions of the object k and on the frequency s, but is independent of r, x, and y.) That is, the error in the approximation (7.160) goes to zero "at least as fast as r^{-2}" as r tends to infinity. We would not get such a nice result were we to apply the rougher estimate (7.159) *throughout* the integral in (7.153).

Now of course very large, or infinite, distances r are impractical or impossible — but there's a straightforward remedy to this problem. Namely, a *lens* of focal length $f > 0$ centered in the plane $\{(f, u, v) : u, v \in \mathbb{R}\}$ will focus the Fraunhofer pattern onto a parallel plane, $2f$ units "downstream" from the original object. So our spectrum plane is effectively brought in to a manageable distance, and optical Fourier transforms become a reality!

An important application of these ideas is to the art/science of *optical filtering*. Here a second object Φ_{diff}, called a *diffraction grating* or — in keeping with our terminology from Section 7.7—a *filter*, is placed in the spectrum plane. According to the principles outlined above, what shines through this latter plane, then, is essentially $\widehat{k}(x,y)\Phi_{\text{diff}}(x,y)$. Next, a second lens is used to focus the resulting Fraunhofer pattern, which itself is in essence the Fourier transform $F\left[\widehat{k}(x,y)\Phi_{\text{diff}}(x,y)\right]$, onto a final plane, called the *image plane*. By appropriate choice of Φ_{diff}, we can thereby alter the frequency content of k in desirable ways, much as in our earlier discussions of filtering. (Because of the nature of Fourier inversion, meaning the fact that, in general, $\widehat{\widehat{f}} = f^-$, the filtered image will be inverted with respect to the original object.)

For some fascinating examples and illustrations of optical filtering, see Chapter 13 in [26].

The phenomenon of Fraunhofer diffraction has implications in many other arenas. Indeed, our above model for light suits other varieties of electromagnetic radiation—radio waves, microwaves, X rays, gamma rays—equally well. (What distinguishes one type of electromagnetic radiation from another is the range of constituent frequencies s, or equivalently *wavelengths* $\lambda = c/s$. Visible light corresponds, roughly, to the range $4000 \leq \lambda \leq 7000$, the units here being *angstroms*, or ten-millionths of a millimeter. Gamma rays, X rays, and ultraviolet light have shorter wavelengths; infrared light, microwaves, and radio waves have longer ones.) And still other physical entities—sound waves, electron beams, and so on—admit, in suitable circumstances, similar models. Wherever such models pertain, Fraunhofer effects are potentially of import. For example, these effects are of critical relevance to the disciplines of *ultrasound imaging*, *crystallography* (recall our discussions at the end of Section 1.10), and *electron microscopy*, to name a few.

Finally, it's interesting to note that Fraunhofer diffraction is not only *neatly explained by* the EM wave theory but is also *largely responsible for the advent of* this theory. Prior to the mid-1600s, the widely held notion was that light was *corpuscular*, or particlelike. But in 1655, Francesco Grimaldi conducted experiments wherein diffraction patterns were plainly manifested. That the corpuscular model could not explain these patterns initiated a gradual turning away from this model.

The emphasis here is on the word "gradual." The corpuscular description continued to hold favor for nearly 150 years after Grimaldi's observations. This was in spite of intervening work by Robert Hooke, Christian Huygens, and others, all of which posited or intimated light's wavelike nature. (The persistence of the corpuscular theory may have been due, in great measure, to the tremendous influence of Sir Isaac Newton, who was a staunch advocate of this theory.) It was only with Thomas Young's famous double-slit experiments, circa 1802, that the wave model really began to gain prominence. (Consult Exercise 7.9.1, below, for more on Young's experiments and their relevance to the Fraunhofer picture.)

Over the course of the nineteenth century, the likes of Jean Foucault, Joseph von Fraunhofer, Augustin Fresnel, Gustav Kirchoff, and James Maxwell infused this model with additional strength and shape. Still, it's not a supermodel: It fails to explain more recently observed, microscopic effects, such as those of absorption and emission. These are, somewhat ironically, more consistent with a corpuscular model.

Hence the modern, quantum theory, according to which light and similar phenomena conform to principles of *wave/particle dualism*.

Exercises

7.9.1 Consider an object plane that is opaque except for two long, narrow, vertical slits.

   a. Explain why the formula
$$k(u, v) = (\delta(u - d) + \delta(u + d))1(v)$$
($1(v)$ is just equal to 1 for all v; we include this factor for the sake of the rest of this exercise) provides an intuitively reasonable model for the double slit. What does d represent here?

   b. Use the facts that $\widehat{\delta} = 1$ and $\widehat{1} = \delta$ (see Examples 6.2.2 and 6.2.4) and the approximation (7.160) for the farfield pattern $C(x, y)$ to deduce that
$$C(x, y) \approx \frac{2A_0 \, e^{2\pi i s (||P||/c - t)}}{||P||} \left(\cos \frac{2\pi ds}{||P||c} x \right) \delta\left(\frac{s}{||P||c} y \right)$$
for the above double slit k.

   c. What one actually *sees* in the spectrum plane, in general, is not the farfield pattern $C(x, y)$, but the corresponding *intensity pattern* $|C(x, y)|^2$. Plot (the approximation to) this intensity pattern as a function of x for $y = 0$. (Of course $\delta(0) = \infty$, but pretend, for the sake of argument, that $\delta(0) = 1$, say. Under this pretense, the above formula gives $C(x, 0) \approx (2A_0 \, e^{2\pi i s(||P||/c - t)}/||P||) \cos(2\pi ds \, x/(||P||c))$.) Note that the absolute value of a complex exponential is 1, and also that $||P|| = \sqrt{r^2 + x^2}$ for $y = 0$. The quality of your graph will depend on your choice of constants; putting $A_0 = r = 1$ and $ds/c = 2$, and plotting over $x \in [-4, 4]$, for example, seem to result in a pretty nice picture.

   d. Using parts b and c above, describe qualitatively what the intensity pattern for a double slit might look like. You might want to compare your analysis to actual pictures, which can be found in just about any undergraduate optics text ([26], for example).

7.9.2 An object (grayscale image of finite extent), call it object A, is illuminated with PPM light. On a distant spectrum plane, evenly spaced dots, all in the same horizontal line (and decreasing in intensity away from the center), are seen. The experiment is repeated with an object B; one sees essentially the same thing, but this time the dots are further apart. Describe objects A and B in general terms, especially insofar as how they compare to each other. (You might want to recall the horse disguised as a zebra in Section 7.7.)

7.9.3 An object C is illuminated with PPM light. On a distant spectrum plane, one sees five dots—one at a point that we'll call the origin, and each of the others at a vertex of a square centered at this origin. (The vertices are dimmer than the center.) Describe the object C in general terms. Hint: Think trig functions. Remember (cf. Example 6.2.5) that the Fourier transform of a complex exponential is a shifted delta.

7.10 FT-NMR SPECTROSCOPY

Consider a sample of some chemical compound situated in an ambient, uniform magnetic field B_0. Suppose a short electromagnetic pulse, of constant magnitude for its duration, is applied to this sample in a direction perpendicular to B_0. An appropriately placed receiver will then detect an output magnetic field, called a *free induction decay*, or *FID*, signal. This signal may, roughly and idealistically (but well enough for our and for many other purposes), be modeled by a *damped harmonic oscillator*

$$g(t) = \chi_{[0,\infty)}(t) \sum_{k=1}^{n} C_k e^{-d_k t} e^{2\pi i s_k t}. \tag{7.162}$$

Here $t \in \mathbb{R}$ is a temporal variable; $n \in \mathbb{Z}^+$; and for $1 \leq k \leq n$, we have $C_k \in \mathbb{C}$, $d_k \in \mathbb{R}^+$, and $s_k \in \mathbb{R}$.

Of particular interest are these s_k's, called the *resonant frequencies of g*: They embody valuable information concerning the various *nuclei* contributing to the molecular makeup of the compound. More specifically, the s_k's are essentially rates of revolution, about certain axes of precession, of the *magnetic moment vectors* associated with these nuclei. (See [28] or [41].) Such information is instrumental to the determination of the natures of not only the nuclei themselves but also the bonding patterns around them.

The above ideas are at the heart of the *pulsed nuclear magnetic resonance* experiment. The theory and practice of such experiments, and the application of discrete Fourier transform methods to analysis of the data they produce, constitute the field of *Fourier transform nuclear magnetic resonance*, or *FT-NMR, spectroscopy*. In the present section we wish to explore this field briefly, by way of some general mathematical discussions and a particular "real-life" example.

We begin with the mathematics, which is rooted in the following problem. While the resonant frequency of a *simple* damped harmonic oscillator g—corresponding to the case $n = 1$ of (7.162)—is readily apparent from a graph of g or even of $\mathrm{Re}\, g$, things are not so evident in the *compound* situation, where $n > 1$. See Figure 7.16.

The solution to this problem, or at any rate a close approximation to a solution, entails Fourier transformation, as follows. Recall that, intuitively speaking, $\widehat{f}(s)$ (for suitable f) is a measure of "the extent to which the complex exponential $e_s(t) = e^{2\pi i s t}$ goes into the makeup of f." Now by its very nature, we would expect the damped harmonic oscillator g of (7.162) to comprise the complex exponentials $e_{s_1}, e_{s_2}, \ldots, e_{s_n}$ in relatively large amounts. So we would expect $|\widehat{g}(s)|$ to be relatively large at $s = s_1, s_2, \ldots, s_n$.

How large? In the case $n = 1$, the answer is "largest." Indeed, we have

$$F\left[\chi_{[0,\infty)}(t) e^{-d_1 t} e^{2\pi i s_1 t}\right](s) = \int_0^{\infty} e^{-t(d_1 + 2\pi i (s - s_1))} \, dt = \frac{1}{d_1 + 2\pi i (s - s_1)}, \tag{7.163}$$

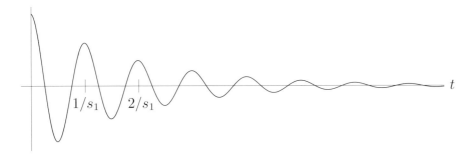

Fig. 7.16 Top: The real part of a simple damped harmonic oscillator. The resonant frequency s_1 is just the reciprocal of the distance between peaks. Bottom: The real part of a compound damped harmonic oscillator. Who knows what its resonant frequencies are?

so that

$$\left|F\left[\chi_{[0,\infty)}(t)e^{-d_1 t}e^{2\pi i s_1 t}\right](s)\right| = \frac{1}{\sqrt{d_1^2 + 4\pi^2(s-s_1)^2}}, \quad (7.164)$$

which is clearly maximized, as a function of s, when $s = s_1$.

Now again, we're more interested in the case $n > 1$ of (7.162). And it's *not quite* true that, in this case, $|\widehat{g}(s)|$ has maxima at $s = s_1, s_2, \ldots, s_n$. Intuitively speaking, this is because each summand on the right side of (7.162) effects something of a "drift" in the frequency content corresponding to each other summand. HOWEVER, if the resonant frequencies of g are far enough apart, we would expect these summands to influence each other only minimally and would therefore expect the peaks of $|\widehat{g}(s)|$ to occur, at any rate, *near* the points $s = s_1, s_2, \ldots, s_n$.

We illustrate with an example—that of the oscillator whose real part is graphed at the bottom of Figure 7.16. We now reveal the formula used to generate that graph:

$$g(t) = \chi_{[0,\infty)}(t)$$
$$\cdot \left[\frac{2}{3}e^{-t}e^{2\pi i(5)t} - \frac{3}{2}e^{-4t/5}e^{2\pi i(12)t} + e^{-3t/4}e^{2\pi i(16)t} - \frac{4}{5}e^{-3t/2}e^{2\pi i(27)t}\right]. \quad (7.165)$$

It's clear from this formula that g has resonant frequencies $s = 5, 12, 16, 27$. And while, again, these frequencies are *not* manifest in the graph of g, they *are* in the graph of $|\hat{g}|$! See Figure 7.17.

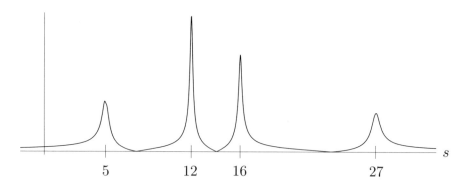

Fig. 7.17 $|\hat{g}|$, for g as in (7.166) (and Re g as depicted in the lower portion of Figure 7.16)

Actually, one checks by numerical means that, for g as in (7.165), $|\hat{g}(s)|$ peaks roughly at $s = 4.92408$, $s = 11.9992$, $s = 16.0036$, and $s = 27.0377$. These numbers are not far off, relatively speaking, from the actual resonant frequencies of g.

One more general note should be made before we turn to a concrete application of the above ideas. In the context of such applications one does not, of course, have precise *a priori* knowledge of the formula (7.162) for the oscillator g under consideration. (If one did, then the resonant frequencies would be transparent, and Fourier analysis would be irrelevant.) What one has, instead, is a finite sequence of *observations* of the phenomenon represented by g. So in practice, the Fourier transforms at issue are *discrete*—or more exactly, *fast*—Fourier transforms. (See Sections 7.5 and 7.6.) Many of these applications would not be feasible without the fast Fourier transform algorithm.

We now turn to our example, from the field of FT-NMR spectroscopy. We begin by considering a sample of ethylbenzene, whose molecular structure is shown in Figure 7.18.

We explain: In accordance with convention, each numbered vertex represents a carbon atom. We'll refer to the carbon atom at vertex k as C-k. Also, each "H" represents a hydrogen atom. Any of the line segments, or the wedges, connecting two atoms represents a bond between those two. The line segments indicate bonds lying in the plane of the paper, while the filled and empty wedges indicate bonds into and out of the plane of the paper, respectively.

The free induction decay of ethylbenzene is shown in Figure 7.19.

Again, the resonant frequencies are not yet evident. But now, we perform a fast Fourier transform of the FID data; the result is depicted in Figure 7.20.

The variable on the frequency (horizontal) axis, in Figure 7.20, is just the coordinate n of the discrete Fourier transform vector. In FT-NMR spectroscopy, frequencies

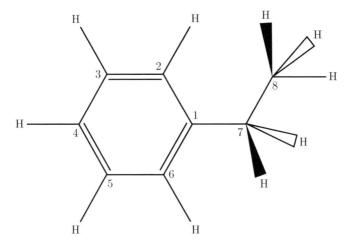

Fig. 7.18 Ethylbenzene

Fig. 7.19 Free induction decay data from a sample of ethylbenzene. Data courtesy of Joseph Hornak, Magnetic Resonance Laboratories, Rochester, New York

are generally calibrated *relative* to that of some known substance included in the sample. In the case of the above FID data, the compound tetramethylsilane is used for calibration; the resonance from this compound is seen in the small, rightmost peak, at about $n = 26000$.

What do the other peaks represent? We scan from right to left. The large peak at about $n = 24000$ represents the three hydrogens bonded to C-8. These hydrogens are *symmetry equivalent* with respect to each other; they occupy symmetric positions within the ethylbenzene molecule. (The bond between C-7 and C-8—see Figure 7.18—is an axis of symmetry between these hydrogens.) Consequently, they give

416 SPECIAL TOPICS AND APPLICATIONS

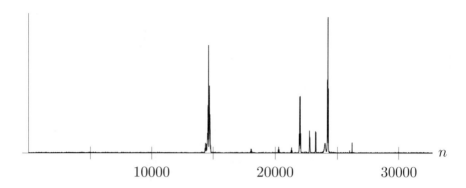

Fig. 7.20 Absolute value of the (fast) Fourier transform of the ethylbenzene FID data

rise to the same resonant frequency. Such considerations indeed suggest the single peak that we see. But wait: In fact there's more to be said, as becomes clear when we zoom in.

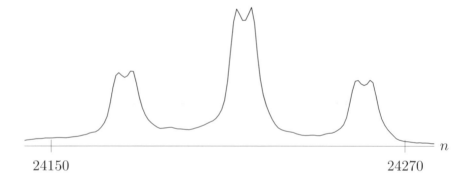

Fig. 7.21 A closer look at the C-8 resonance

We see that the C-8 peak is actually "split" into three smaller ones. What's behind this splitting? The answer lies in the *multiplicity rule*, also called the $N+1$ *rule* (see [18]), which states: Any signal corresponding to a group of symmetry equivalent hydrogens bonded to C-k splits into $N+1$ smaller peaks, where N is the number of hydrogens bonded to carbons *adjacent* to C-k. So, the $3 = N+1$ smaller peaks constituting the large C-8 peak correspond to the $N=2$ hydrogens bonded to the C-7 atom adjacent to C-8.

Some elucidation of the multiplicity rule, as well as an explanation of the 1:2:1 ratio in the heights of the peaks in Figure 7.21, may be found in the following observations. In a given molecule of ethylbenzene, each of the hydrogen protons bonded to C-7 has one of two possible spin orientations: "Spin up," meaning the magnetic moment vector has a component aligned with the ambient field B_0, or "spin down," meaning this vector has a component with the opposite alignment. A C-7 hydrogen in the spin-up state will *increase* the effective magnetic field experienced by the hydrogens at C-8, resulting in an increase in the associated resonant frequency; a C-7 hydrogen in the spin-down state will decrease this field, and consequently this frequency. But, as seen in Figure 7.22, having *both* C-7 hydrogens in the spin-up state is half as likely as having one in each state, which is in turn twice as likely as having both in the spin-down state.

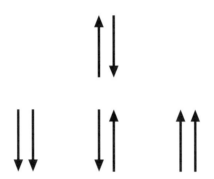

Fig. 7.22 Possible combinations of spin orientations of the two hydrogen protons at C-7

Hence the observed 1:2:1 splitting in the C-8 resonance. (Note how the shape of Fig. 7.21 parallels that of Fig. 7.22.)

Now let's return to Figure 7.20. We skip the two smaller peaks on either side of $n = 23000$—these represent impurities in the sample. The next significant one is roughly at $n = 22000$. This peak results from the two symmetry equivalent hydrogens bonded to C-7, and is in fact composed of a quartet of smaller peaks (see Fig. 7.23), as dictated by the multiplicity rule and the presence of the three hydrogens bonded to C-8.

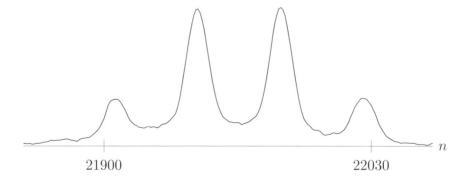

Fig. 7.23 A closer look at the C-7 resonance

The 1:3:3:1 splitting in Figure 7.23 corresponds to Figure 7.24, which depicts possible spin combinations for the C-8 hydrogens.

418   SPECIAL TOPICS AND APPLICATIONS

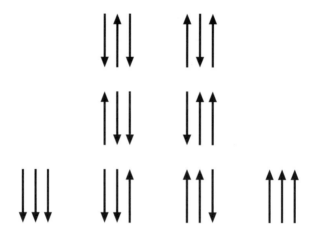

Fig. 7.24   Possible combinations of spin orientations of the three hydrogen protons at C-8

The next, and last, peak of any stature in Figure 7.20 occurs just below $n = 15000$. This peak represents the hydrogens in the "benzene ring," that is, the hydrogens bonded to C-2, C-3, C-4, C-5, and C-6. And as in previous instances, this peak breaks down under closer scrutiny.

Fig. 7.25   A closer look at the benzene ring resonance

The somewhat ragged, complex topography in Figure 7.25 may be attributed to a variety of factors. First, the C-4 hydrogen has its own resonance, as does the symmetry equivalent pair consisting of C-2 and C-6, as does the one consisting of C-3 and C-5. To each of these three resonances corresponds a separate peak—but each of these peaks is further split by effects from adjacent carbons, as prescribed by the multiplicity rule. Resonances even further away also contribute to the overall picture, though to a lesser extent. So do impurities, limitations on the resolution of the experiment, and other sources of noise.

One additional bit of relevant information is reflected in our transformed FID data. Namely, a careful measurement shows that the *areas* under the three largest peaks, in order from right to left, in Figure 7.20 are in the ratio of about 3:2:5. This is in accordance with the fact that the first of these peaks represents the *three* bonds at C-8, the second the *two* bonds at C-7, and the third the *five* benzene ring bonds.

The above study, while perhaps a bit cursory, evinces (we hope) the power of FT-NMR spectroscopy as a tool for determining molecular structure. Nowadays, it is *de rigeur* that any report on the synthesis of new molecules include spectral analyses of FID data. Such analyses, together with analogous examinations of *mass spectra* via FT-ICR methods (ICR stands for *ion cyclotron resistance*) and of *vibrational spectra* via FT-IR methods (IR stands for *infrared*), provide most of the ingredients essential to the understanding of such molecules. (See, for example, [10] for brief discussions of, and [34] or [18] for further details concerning, FT-ICR and FT-IR spectroscopy.)

We make a few closing notes. Until the early 1970's, most nuclear magnetic resonance experiments were of the *continuous wave*, rather than the pulsed, variety. In the former setting, instead of a short, sharp pulse, a *sinusoidal wave* of a definite frequency s is aimed at the sample under study. This results in the excitation of (only) those nuclei whose resonant frequencies are approximately equal to s. That is, as the terminology suggests, such nuclei *resonate* with the wave. (Our discussions here should be compared with those of resonance in Section 2.11.) By iterative application of such waves, each successive one having a different, set frequency s, one can therefore probe the various types of nuclei in the sample *one at a time*.

But this can take a *long* time if a broad range of frequencies needs to be examined at a high resolution. Modern experiments instead use, as noted above, a concentrated, locally constant electromagnetic pulse—in other words, roughly, a constant multiple of a *delta function*!! The advantage of this approach is that, as seen in Example 6.2.2, $\widehat{\delta} = 1$, which is to say δ contains *all* frequencies in equal amounts. Firing a delta pulse at a sample will therefore cause *all* nuclei therein to resonate at once, resulting in the compound damped harmonic oscillation described above.

Such oscillation can be untangled quite efficiently via discrete Fourier transformation and the fast Fourier transform algorithm. The supplantation of continuous wave by pulsed methods, in NMR (and other kinds of) spectroscopy, constitutes a tremendously significant aspect of the fast Fourier transform revolution.

Exercises

7.10.1   a. At what points along the s axis do the maxima of $|\widehat{g}(s)|$ occur for g the compound damped harmonic oscillator

$$g(t) = \chi_{[0,\infty)}(t)e^{-d_1 t}\sin 2\pi s_1 t = \chi_{[0,\infty)}(t)\frac{1}{2i}e^{-d_1 t}(e^{2\pi i s_1 t} - e^{-2\pi i s_1 t})$$

($s_1, d_1 > 0$)? Express your answer in terms of s_1 and d_1. Hint: Maximizing $|\widehat{g}(s)|$ is the same as maximizing $|\widehat{g}(s)|^2$; find an explicit formula for the latter and set the derivative equal to zero.

   b. How must d_1 and s_1 compare in order that there be two distinct peaks? How must they compare in order that there be only one peak?

c. What are the heights of the peaks (in terms of d_1 and s_1) in each of the cases of the previous part of this exercise?

d. How must d_1 and s_1 compare in order that the "drifting effect" described in this section be small? In other words, when can we be sure that the peaks of $|\hat{g}(s)|$ are close to the individual peaks of $|\hat{g_1}(s)|$ and $|\hat{g_2}(s)|$, g_1 and g_2 being the simple oscillators being summed to give g?

e. What does this lead you to believe in general about "relaxation rates" d_k versus resonant frequencies s_k if the latter are going to be determined effectively using the Fourier transform?

7.10.2 **a.** Draw an "arrow diagram," analogous to those of Figures 7.22 and 7.24, indicating possible spin combinations of *four* hydrogens bonded to a single carbon atom.

b. Imagine a molecule in which a group of symmetry equivalent hydrogens is bonded to a carbon atom C-k, and four hydrogens are bonded to a carbon atom *adjacent* to C-k. Draw a rough sketch of what the FID data corresponding to the hydrogens at C-k should look like. (Just try to illustrate general peak structure — in particular, the number of peaks and their relative sizes. Assume no other resonances are affecting the hydrogens at C-k.)

8

Local Frequency Analysis and Wavelets

Because sines and cosines cycle indefinitely, sinusoid components and decompositions are of primary relevance to the understanding of global, or average, oscillatory behavior. But for the same reason, sinusoids are not the ideal atoms for the analysis of *local* oscillations or of *evolution* of cyclical characteristics with time (or whatever the independent variable signifies).

In this chapter, we examine some more suitable atoms—windowed complex exponentials and, better still, wavelets. We pay particular attention to the latter, whose theory and range of applications dramatically augment and continue to expand with modern mathematical and technological landscapes.

8.1 SHORT-TIME, OR WINDOWED, FOURIER TRANSFORMS

Let's begin this section with a reflection on Figure 7.4, which illustrates what might be called a *local frequency analysis* of a certain musical selection. By this we mean a study of the *temporal variation* in the oscillatory features of that selection. We ask: What is the mathematics behind that and similar studies?

Any satisfactory answer to this question will necessarily entail more than just Fourier transformation *per se*. Why? Well, suppose $f \in L^1(\mathbb{R})$ is a function of time, say. (Throughout most of this chapter, we use time as a metaphor for any real variable. It's an apt metaphor: In applications, the phenomena of interest will often be temporal ones.) Then the integral

$$\int_{-\infty}^{\infty} f(t) e^{-2\pi i s t}\, dt = \langle f, e_s \rangle = \widehat{f}(s) \tag{8.1}$$

(again, $e_s(t) = e^{2\pi i s t}$) is a sort of average, over *all* time t, of $f(t)e^{-2\pi i s t}$. The time dependence has been integrated out; $\hat{f}$ reflects global, not local, frequency content of f.

Or to put it another way, the representation

$$f(t) = \int_{-\infty}^{\infty} \hat{f}(s) e^{2\pi i s t} \, dt = \int_{-\infty}^{\infty} \langle f, e_s \rangle \, e_s(t) \, dt \qquad (8.2)$$

(which holds for sufficiently reasonable f; see for example Theorem 6.4.1) decomposes f into the complex exponentials e_s. But while e_s is, for a given s, *perfectly concentrated* in frequency (recall Example 6.2.5 and the surrounding discussions), it is also *uniformly spread out* in time ($|e_s(t)| = 1$ for all t; e_s parcels itself out indefinitely along the t axis, without preference to any point or portion of this axis). So complex exponentials are poorly suited to local frequency analyses of f. What we *really* need for such analyses are functions well localized in both time *and* frequency. Indeed, suppose we have, for any real numbers σ and τ, a function $e_{\sigma,\tau}$ that's concentrated near the point $s = \sigma$ on the frequency axis and near the instant $t = \tau$ on the time axis. That is, suppose $e_{\sigma,\tau}(t)$ is small for t far from τ and $\widehat{e_{\sigma,\tau}}(s)$ is small for s far from σ. In accordance with earlier arguments (see the end of Section 3.7 and the beginning of Section 6.1), $\langle f, e_{\sigma,\tau} \rangle \, e_{\sigma,\tau}$ should, in some sense, represent "that part of f that looks like $e_{\sigma,\tau}$," meaning "that part of f that oscillates roughly like the sinusoid e_σ and lives roughly at the instant $t = \tau$." And such parts of f are what we'd like to understand.

Identification, examination, interpretation, and application of "time-frequency" building blocks $e_{\sigma,\tau}$ constitute the main concerns of this chapter. In this section, we construct some of these building blocks in a conceptually straightforward, though still quite fruitful, way: We *window* the functions e_σ. That is, we *multiply* each of these by a function w_τ concentrated near $t = \tau$, for each $\tau \in \mathbb{R}$. Intuitively, these $e_\sigma w_\tau$'s should enjoy the desired time-frequency localization properties.

To obtain such w_τ's, we may simply *translate* a function w concentrated near $t = 0$. So we are led to:

Definition 8.1.1 (a) A *window function*, or simply *window*, is a $w \in L^2(\mathbb{R})$ of mean square norm 1.

(b) For any window function w and real numbers σ and τ, the *windowed complex exponential* $e_{\sigma,\tau,w}$ is defined by

$$e_{\sigma,\tau,w}(t) = e^{2\pi i \sigma t} w(t - \tau) = e_\sigma(t) w_{(\tau)}(t), \qquad (8.3)$$

where $w_{(\tau)}(t) = w(t - \tau)$ (as before, cf. (6.34)).

(The requirement $||w|| = 1$ is made simply for convenience; as long as $||w|| \neq 0$, the above definition and any results concerning it apply with $w/||w||$ in place of w.)

As just defined, a window function w is "concentrated near $t = 0$" only in the very *mild* sense that, if a function is *too* spread toward the ends of the axis, then it won't have finite mean square area. We'll study the concentration issue more closely by means of some examples at the end of this section and a systematic inquiry in the next one.

Note that the Fourier transform of a window function is another such, by the Plancherel theorem (Proposition 6.5.3(b)). Also observe the following, regarding the $e_{\sigma,\tau,w}$'s. First,

$$|e_{\sigma,\tau,w}| = |w_{(\tau)}|, \tag{8.4}$$

and consequently

$$\|e_{\sigma,\tau,w}\| = \left(\int_{-\infty}^{\infty} |w(t-\tau)|^2\, dt\right)^{1/2} = 1 \tag{8.5}$$

(for the last step we let $t \to t + \tau$). That is, the $e_{\sigma,\tau,w}$'s are window functions too. Second, the Fourier transform of a windowed complex exponential is a windowed complex exponential of a Fourier transform; specifically,

Proposition 8.1.1 *For $e_{\sigma,\tau,w}$ as in (8.3), we have*

$$\widehat{e_{\sigma,\tau,w}} = e^{2\pi i \sigma \tau} e_{-\tau, \sigma, \widehat{w}}. \tag{8.6}$$

Proof. By (6.35) (which was stated for $f \in L^1(\mathbb{R})$, but which generalizes to $L^2(\mathbb{R})$), we have $\widehat{e_\sigma k} = (\widehat{k})_{(\sigma)}$ and $\widehat{k_{(\tau)}} = e_{-\tau} \widehat{k}$ for any $k \in L^2(\mathbb{R})$ and $\sigma, \tau \in \mathbb{R}$. So

$$\widehat{e_{\sigma,\tau,w}} = \widehat{e_\sigma w_{(\tau)}} = (\widehat{w_{(\tau)}})_{(\sigma)} = (e_{-\tau}\widehat{w})_{(\sigma)} = (e_{-\tau})_{(\sigma)}(\widehat{w})_{(\sigma)}$$
$$= (e^{2\pi i \sigma \tau} e_{-\tau})(\widehat{w})_{(\sigma)} = e^{2\pi i \sigma \tau} e_{-\tau, \sigma, \widehat{w}}, \tag{8.7}$$

the next to the last step because

$$(e_{-\tau})_{(\sigma)}(s) = e_{-\tau}(s-\sigma) = e^{-2\pi i \tau(s-\sigma)} = e^{2\pi i \sigma \tau} e_{-\tau}(s). \tag{8.8}$$

We're done. $\square$

From the above proposition and equation (8.4), we deduce that

$$|\widehat{e_{\sigma,\tau,w}}| = |\widehat{w}_{(\sigma)}|. \tag{8.9}$$

We next make:

Definition 8.1.2 For $e_{\sigma,\tau,w}$ as above and $f \in L^2(\mathbb{R})$, the *short-time*, or *windowed, Fourier transform of f* (with respect to w) is the function $\mathrm{ST}_w f \colon \mathbb{R}^2 \to \mathbb{C}$ defined by

$$\mathrm{ST}_w f(\sigma, \tau) = \langle f, e_{\sigma,\tau,w}\rangle = \int_{-\infty}^{\infty} f(t)\overline{e_\sigma(t)w(t-\tau)}\, dt$$
$$= \int_{-\infty}^{\infty} f(t)\overline{w_{(\tau)}(t)} e^{-2\pi i \sigma t}\, dt = \widehat{f\, \overline{w_{(\tau)}}}(\sigma). \tag{8.10}$$

From the stated conditions on w and f, it follows that $\mathrm{ST}_w f(\sigma, \tau)$ is defined for all $(\sigma, \tau) \in \mathbb{R}^2$. One may, in fact, show (among other things) that $\mathrm{ST}_w f$ is continuous on $\mathbb{R}^2$, though we won't.

Note that, by the last equality in (8.10), windowed Fourier transformation of f amounts to Fourier transformation of windowed "glimpses" of f. This is handy—it means, for one thing, that *discrete*, and even *fast*, short-time Fourier transforms of f may readily be obtained, by applying to these glimpses the already available discrete and fast Fourier transform machinery. (One must also discretize judiciously with respect to the translation variable τ.)

It also means that familiar Fourier inversion arguments are relevant to the pursuit of "short-time Fourier inversion" results. Thus we have, for example, the following.

Theorem 8.1.1 *Let $w, f \in \mathrm{PS}(\mathbb{R}) \cap L^2(\mathbb{R})$, with $\|w\| = 1$. Then*

$$\frac{1}{2}(f(t^-) + f(t^+)) = \int_{-\infty}^{\infty} \lim_{b \to \infty} \int_{-b}^{b} \mathrm{ST}_w f(\sigma, \tau) w(t - \tau) e^{2\pi i \sigma t} \, d\sigma \, d\tau$$

$$= \int_{-\infty}^{\infty} \lim_{b \to \infty} \int_{-b}^{b} \langle f, e_{\sigma, \tau, w} \rangle \, e_{\sigma, \tau, w}(t) \, d\sigma \, d\tau \qquad (8.11)$$

for all $t \in \mathbb{R}$.

Proof. The second stated equality follows from the definitions. To get the first we note that, for f and g as stipulated, we have $\widehat{f \, \overline{w_{(\tau)}}} \in \mathrm{PS}(\mathbb{R}) \cap L^1(\mathbb{R})$, so by Fourier inversion on $\mathrm{PS}(\mathbb{R}) \cap L^1(\mathbb{R})$ (Theorem 6.6.2) and by (8.10),

$$\lim_{b \to \infty} \int_{-b}^{b} \mathrm{ST}_w f(\sigma, \tau) e^{2\pi i \sigma t} \, d\sigma = \lim_{b \to \infty} \int_{-b}^{b} \widehat{f \, \overline{w_{(\tau)}}}(\sigma) e^{2\pi i \sigma t} \, d\sigma$$

$$= \frac{1}{2} \left(f(t^-) \overline{w_{(\tau)}(t^-)} + f(t^+) \overline{w_{(\tau)}(t^+)} \right). \qquad (8.12)$$

Consequently

$$\int_{-\infty}^{\infty} \lim_{b \to \infty} \int_{-b}^{b} \mathrm{ST}_w f(\sigma, \tau) \, w(t - \tau) e^{2\pi i \sigma t} \, d\sigma \, d\tau$$

$$= \frac{1}{2} \int_{-\infty}^{\infty} w(t - \tau) \left(f(t^-) \overline{w_{(\tau)}(t^-)} + f(t^+) \overline{w_{(\tau)}(t^+)} \right) d\tau. \qquad (8.13)$$

But w, being piecewise smooth, is continuous almost everywhere, so for fixed t,

$$w_{(\tau)}(t^-) = \lim_{T \to t^-} w_{(\tau)}(T) = \lim_{T \to t^-} w(T - \tau) = w(t - \tau) \qquad (8.14)$$

for almost all τ, and similarly for $w_{(\tau)}(t^+)$. So (8.13) gives

$$\int_{-\infty}^{\infty} \lim_{b \to \infty} \int_{-b}^{b} \mathrm{ST}_w f(\sigma, \tau) w(t - \tau) e^{2\pi i \sigma t} \, d\sigma \, d\tau$$

$$= \frac{1}{2} (f(t^-) + f(t^+)) \int_{-\infty}^{\infty} w(t - \tau) \overline{w(t - \tau)} \, d\tau = \frac{1}{2} (f(t^-) + f(t^+)),$$

$$(8.15)$$

the last step by the substitution $u = t - \tau$ and the fact that $||w||^2 = 1$. □

The conditions $w, f \in \text{PS}(\mathbb{R})$ may be dropped if, instead of pointwise convergence, one is happy with a certain *norm* convergence of truncated integrals of $\langle f, e_{\sigma,\tau,w} \rangle e_{\sigma,\tau,w}$ to f. Conversely, if these conditions are strenghtened somewhat, then "$\lim_{b \to \infty} \int_{-b}^{b}$" simply becomes "$\int_{-\infty}^{\infty}$." But the above theorem as stated is sufficient to communicate the short-time Fourier inversion BIG ideas.

Let's now examine some specific windowed complex exponentials. We begin with particularly simple ones: We pick a positive number b, and define a *rectangular window*

$$r = \frac{\chi_{[-b,b]}}{||\chi_{[-b,b]}||} = (2b)^{-1/2}\chi_{[-b,b]}. \tag{8.16}$$

Certainly the requirements of the above theorem are satisfied when $w = r$, and certainly the building block

$$e_{\sigma,\tau,r}(t) = e_\sigma(t) r_{(\tau)}(t) = \begin{cases} (2b)^{-1/2} e^{2\pi i \sigma t} & \text{if } \tau - b \leq t \leq \tau + b, \\ 0 & \text{if not} \end{cases} \tag{8.17}$$

is well localized in the time domain. In particular, by (8.10), $\text{ST}_r f(\sigma, \tau)$ is, for any τ, just a Fourier transform of a (rescaled) *segment* $f \overline{r_{(\tau)}}$ of f near $t = \tau$. See Figure 8.1.

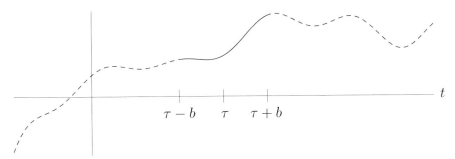

Fig. 8.1 $f \overline{r_{(\tau)}}$ (solid) and $(2b)^{-1/2} f$ (dashed and solid combined), for r as in (8.16)

But how well does this short-time Fourier transform *really* reflect the local frequency content of f? The answer is: Not that well, really! And why not? The problem lies, essentially, with the *discontinuity* of r. Because of this discontinuity $f \overline{r_{(\tau)}}$ will, typically, be less continuous than f itself, and consequently $\text{ST}_r f$ will amount to Fourier transformation of something less continuous than f. But by the local/global principle, less smooth functions have more slowly decaying Fourier transforms, so $\text{ST}_r f$ will, generally, "hang around too long" in the frequency domain to provide a credible reflection of the local frequency content of f.

This can have profound consequences, in practice: It means that signals synthesized from such discontinuous building blocks as the above $e_{\sigma,\tau,r}$'s can exhibit undesired "ringing" effects. Of course Theorem 8.1.1 does say that, in theory, a signal may be *perfectly* synthesized—without ringing or other spurious features—from these building blocks. But in applications one uses a discrete, finite approximation to the integral in that theorem. And finite approximations to infinite processes can be especially crude and inefficient when those infinite processes takes a long time to fade away into negligibility. Witness, for example, Gibbs' phenomenon (cf. Section 1.6).

The difficulty with ST_rf may, equivalently, be attributed to the fact that the windowed complex exponentials $e_{\sigma,\tau,r}$ are *poorly localized* in the frequency domain. Indeed we have, by (8.9) and Example 6.1.1,

$$|\widehat{e_{\sigma,\tau,r}}(s)| = |\widehat{r}(s-\sigma)| = \left|(2b)^{-1/2}\widehat{\chi_{[-b,b]}}(s-\sigma)\right| = \frac{|\sin 2\pi b(s-\sigma)|}{\sqrt{2b}\,\pi|s-\sigma|}. \quad (8.18)$$

The right side peaks at $s = \sigma$, but decays only about like $|s-\sigma|^{-1}$ as $s \to \pm\infty$; such decay is quite slow, in the scheme of things. See the bottom part of Figure 8.2 below.

Such unsatisfactory localization, in one domain or another, may be avoided by further restricting the window function w. We'll delineate the necessary restrictions in the next section; to foreshadow that one and round out this one, let's consider some specific examples that are a bit more satisfying.

Example 8.1.1 A *Bartlett* window, also called a *triangular window*, is an $\ell \in L^2(\mathbb{R})$ obtained by convolving a rectangular window with itself. Specificallly, we define

$$\ell(t) = \sqrt{\frac{3}{2b^3}}\,\chi_{[-b/2,b/2]} * \chi_{[-b/2,b/2]}(t) = \sqrt{\frac{3}{2b^3}}\,(b-|t|)\chi_{[-b,b]}(t) \quad (8.19)$$

(see Exercise 5.2.2), where $b > 0$. Note that

$$\int_{-\infty}^{\infty} |(b-|t|)\chi_{[-b,b]}(t)|^2\,dt = 2\int_0^b (b-t)^2\,dt = 2\int_0^b u^2\,du = \frac{2b^3}{3}, \quad (8.20)$$

so that $||\ell|| = 1$.

Like a rectangular one, a Bartlett window ℓ is compactly supported and therefore well localized in the time domain. By (8.4), then, so is the windowed complex exponential $e_{\sigma,\tau,\ell}$ for any σ and τ.

But unlike a rectangular window, a Bartlett window is *continuous* on $\mathbb{R}$. See the top part of Figure 8.2. This has has a salutary effect on frequency localization: We have

$$\widehat{\ell}(s) = \sqrt{\frac{3}{2b^3}}\left(F[\chi_{[-b/2,b/2]}(t)](s)\right)^2 = \sqrt{\frac{3}{2b^3}}\,\frac{\sin^2 \pi b s}{\pi^2 s^2}, \quad (8.21)$$

cf. Exercise 6.1.4 (or 6.2.4 or 6.3.3). So $\widehat{\ell}(s)$ enjoys a decay about like that of $|s|^{-2}$ as $s \to \pm\infty$, meaning ℓ is better localized in the frequency domain than is the above rectangular window r. (See the bottom part of Figure 8.2.) Consequently, by (8.9), $e_{\sigma,\tau,\ell}$ is better localized in frequency than is $e_{\sigma,\tau,r}$, for any σ and τ.

SHORT-TIME, OR WINDOWED, FOURIER TRANSFORMS

Because of their simplicity and their relatively good localization properties, Bartlett windows *are* employed in various situations involving short-time Fourier transforms. But they are not as prevalent as are *Hann* windows, which we now describe.

Example 8.1.2 A *Hann* window, also called a *Hanning window* or a *cosine bell window*, is a function $h \in L^2(\mathbb{R})$ defined by

$$h(t) = \frac{2}{\sqrt{3b}}\left(\cos^2\frac{\pi t}{2b}\right)\chi_{[-b,b]}(t) = \frac{1}{\sqrt{3b}}\left(1+\cos\frac{\pi t}{b}\right)\chi_{[-b,b]}(t), \quad (8.22)$$

where $b > 0$. (You should check that $\|h\| = 1$; see Exercise 8.1.1(a).) Like rectangular and Bartlett ones, Hann windows are compactly supported; because

$$\cos\frac{\pi(\pm b)}{2b} = \cos\pm\frac{\pi}{2} = 0, \quad (8.23)$$

they're also, like Bartlett windows, continuous on $\mathbb{R}$. But Hann windows take smoothness still one step further; in fact, they're *differentiable* on $\mathbb{R}$. See the top part of Figure 8.2 as well as Exercise 8.1.1(b).

This smoothness translates, again, into concentration with respect to frequency. We compute

$$\begin{aligned}
\widehat{h}(s) &= \frac{1}{\sqrt{3b}}F\left[\left(1+\cos\frac{\pi t}{b}\right)\chi_{[-b,b]}(t)\right](s)\\
&= \frac{1}{\sqrt{3b}}F\left[\left(1+\frac{1}{2}(e^{\pi it/b}+e^{-\pi it/b})\right)\chi_{[-b,b]}(t)\right](s)\\
&= \frac{1}{\sqrt{3b}}\left(\widehat{\chi_{[-b,b]}}(s)+\frac{1}{2}\widehat{\chi_{[-b,b]}}\left(s-\frac{1}{2b}\right)+\frac{1}{2}\widehat{\chi_{[-b,b]}}\left(s+\frac{1}{2b}\right)\right)\\
&= \frac{1}{\sqrt{3b}}\left(\frac{\sin 2\pi bs}{\pi s}+\frac{\sin 2\pi b(s-1/(2b))}{2\pi(s-1/(2b))}+\frac{\sin 2\pi b(s+1/(2b))}{2\pi(s+1/(2b))}\right)\\
&= \frac{\sin 2\pi bs}{\sqrt{3b}\,\pi}\left(\frac{1}{s}-\frac{1}{2s-1/b}-\frac{1}{2s+1/b}\right) = \frac{\sin 2\pi bs}{\sqrt{3b}\,\pi(s-4b^2s^3)}, \quad (8.24)
\end{aligned}$$

the penultimate step because $\sin(\theta\pm\pi) = -\sin\theta$ and the last by obtaining a common denominator (DIY). So $\widehat{h}(s)$ decays about like $|s|^{-3}$ as $s \to \pm\infty$. Hann windows are thus *quite* well localized in the frequency domain—better than are rectangular or Bartlett windows, for example. (See the bottom part of Figure 8.2.) By (8.9), then, the windowed complex exponentials $e_{\sigma,\tau,h}$ are similarly well localized for any σ,τ.

Because of their simplicity, compact support, relative smoothness, and (correspondingly) good frequency concentration, and for related, rather discipline-specific reasons, Hann windows are used extensively in signal processing applications, particularly those involving *audio* signals. For instance, the frames of the animation in Figure 7.4 represent snapshots of a (discretization of a) Hann windowed Fourier transform.

Each of the above windows r, ℓ, and h is of finite duration in the time domain; by reversing the roles of time and the frequency—that is, by replacing each of these windows with its Fourier transform—one obtains windows similarly limited in the frequency domain. One cannot have it both ways; as we saw in Proposition 7.4.3, there exists no nontrivial $w \in L^2(\mathbb{R})$ such that both w and $\widehat{w}$ have compact support. Still, one can have *extremely rapid decay* in both domains; this is the upshot of Example 6.2.3, which we now revisit and reconfigure.

Example 8.1.3 A *Gabor* window, also called a *Gaussian window*, is a function $g \in L^2(\mathbb{R})$ defined by

$$g(t) = \sqrt{\frac{2}{b}} e^{-2\pi t^2/b^2} = \sqrt{\frac{2}{b}} G\left(\frac{\sqrt{2}}{b} t\right), \qquad (8.25)$$

where $b > 0$ and, as always, $G(t) = e^{-\pi t^2}$. That $\|g\| = 1$ follows readily from Lemma 6.2.1.

The rate of decay of g is commensurate with that of $\widehat{g}$ in that, by Example 6.2.3 and properties of the Fourier transform (cf. Proposition 6.2.1(b)(i)),

$$\widehat{g}(s) = \sqrt{\frac{2}{b}} F\left[G\left(\frac{\sqrt{2}}{b} t\right)\right](s) = \sqrt{\frac{2}{b}} \frac{b}{\sqrt{2}} \widehat{G}\left(\frac{b}{\sqrt{2}} s\right) = \sqrt{b}\, G\left(\frac{b}{\sqrt{2}} s\right)$$

$$= \frac{b}{\sqrt{2}} g\left(\frac{b^2}{2} s\right). \qquad (8.26)$$

The time and frequency characteristics of the windows studied so far are compared in Figure 8.2.

Also encountered in applications are the Blackman, Blackman/Harris, Hamming, Kaiser, Lanczos, Parzen, and Welch windows, among others. Some of these are considered in Exercises 8.1.2 and 8.2.2.

We'll see in the next section that Gabor windows have optimal time-frequency localization properties in a certain *theoretical* sense. On the other hand, it turns out that, for most *practical* purposes, Hann and a variety of other windows localize well enough. And they are much more effectively implemented; Gaussians are, mathematically and computationally, rather unwieldy, compared to (truncated) sinusoids, piecewise linear functions, and so on. Because of this, one rarely finds Gabor windows employed in "real-life" local frequency analyses.

Exercises

8.1.1 **a.** Verify that the Hann window h of Example 8.1.2 has norm 1.
   b. Show that h is differentiable on $\mathbb{R}$. Hint: Exercise 1.3.17.

8.1.2 Sketch the graph (for some particular value of b, say $b = 1$, all on the same set of axes) of each of the following functions. Show that each defines a window (that is, has norm 1); find the Fourier transform of each.

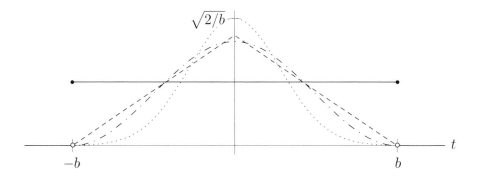

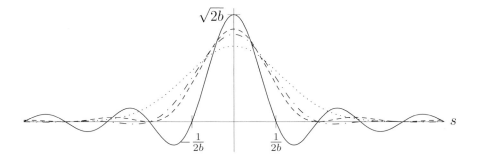

Fig. 8.2 Top: Rectangular (solid), Bartlett (dashed), Hann (dot-dashed), and Gabor (dotted) windows. Bottom: Respective Fourier transforms of these windows

a. The *Blackman* function
$$m(t) = \frac{1}{2\sqrt{1523b}}(42 + 50\cos \pi b^{-1}t + 8\cos 2\pi b^{-1}t)\chi_{[-b,b]}(t) \quad (b > 0).$$

b. The *Lanczos* function $z(t) = b^{-1/2}\operatorname{sinc} \pi b^{-1}t$ $(b > 0)$. (To find $||z||$, use the Plancherel theorem.)

c. The *Welch* function $d(t) = \sqrt{15/16}\, b^{-5/2}(b^2 - t^2)\chi_{[-b,b]}(t)$ $(b > 0)$.

8.1.3 **a.** List the windows of the previous exercise in order of increasing smoothness. Explain. (Use Exercise 1.3.17 if necessary.)

b. List these same windows in order of increasing rate of decay in the frequency domain. Explain. (That is, explain by invoking explicit properties of the relevant Fourier transforms, not by invoking the previous part of this exercise and the local/global principle. Of course, your answer should agree with what you'd deduce by invoking these latter two ideas.)

8.2 FINITE WINDOWS AND THE HEISENBERG UNCERTAINTY PRINCIPLE

To better understand and contrast localization properties of various types of windows, we should better quantify the notion of localization. Here is a time (and frequency) honored, field tested way of doing so.

Definition 8.2.1 Let $f \in L^2(\mathbb{R})$.

(a) We define the *midpoint* $M_f \in \mathbb{R}$ of f by

$$M_f = \frac{1}{||f||^2} \int_{-\infty}^{\infty} t|f(t)|^2\, dt \qquad (8.27)$$

provided the integral on the right exists. If not, we leave the midpoint of f undefined.

(b) Put $f^\times(t) = tf(t)$ (as before, cf. (6.34)). If $f^\times \in L^2(\mathbb{R})$, then we define the *radius* $\Delta_f \in \mathbb{R}^+$ of f by

$$\Delta_f^2 = \frac{1}{||f||^2} \int_{-\infty}^{\infty} (t - M_f)^2 |f(t)|^2\, dt = \frac{1}{||f||^2} \left(\int_{-\infty}^{\infty} t^2 |f(t)|^2\, dt \right.$$

$$\left. - 2M_f \int_{-\infty}^{\infty} t|f(t)|^2\, dt + M_f^2 \int_{-\infty}^{\infty} |f(t)|^2\, dt \right)$$

$$= \frac{1}{||f||^2} \left(\int_{-\infty}^{\infty} t^2 |f(t)|^2\, dt - ||f||^2 M_f^2 \right) = \frac{||f^\times||^2}{||f||^2} - M_f^2. \qquad (8.28)$$

If not, we put $\Delta_f = \infty$.

We will sometimes refer to M_f and Δ_f as the *time domain* midpoint and radius of f and to $M_{\hat{f}}$ and $\Delta_{\hat{f}}$ as the *frequency domain* midpoint and radius of f.

Observe that $t|f(t)|^2 = (t|f(t)|)(|f(t)|)$, so if $f, f^\times \in L^2(\mathbb{R})$ then, by the Cauchy-Schwarz inequality (5.38), the integral on the right side of (8.27) is automatically defined (so that M_f and (8.28) do in fact make sense).

Also note that, for $f, f^\times \in L^2(\mathbb{R})$, we have $f \in L^1(\mathbb{R})$; this too follows from the Cauchy-Schwarz inequality, this time with $F(t) = (1 + |t|)^{-1}$ and $G(t) = (1 + |t|)f(t)$.

Statistically speaking, M_f is a kind of *mean* and Δ_f a *standard deviation* associated with $|f|^2$. *Heuristically* speaking, Δ_f may be seen as a measure of the *spread* of f about its midpoint M_f. Indeed Δ_f^2 is, by the first equality in (8.28), large only if $t - M_f$ and $|f(t)|$ are large *simultaneously*. But $t - M_f$ is large only for t far from M_f, so Δ_f is large only if $|f(t)|$ is large for t far from M_f. Or from the opposite perspective, Δ_f is *small* if f is *concentrated near* its midpoint M_f.

In light of our discussions near the beginning of the previous section, we therefore come to the following decision. For effective local frequency analysis using windowed complex exponentials, we will henceforth demand, first, that $c_{\sigma,\tau,w}$ have midpoint τ and small radius and, second, that $\widehat{e_{\sigma,\tau,w}}$ have midpoint σ and small radius.

Example 8.2.1 The rectangular-windowed complex exponential $e_{\sigma,\tau,r}$ of (8.17) does *not* meet our demands. We have

$$M_{e_{\sigma,\tau,r}} = \int_{-\infty}^{\infty} t |e_\sigma(t) r(t-\tau)|^2 \, dt = \int_{-\infty}^{\infty} (t+\tau)|r(t)|^2 \, dt = \frac{1}{2b} \int_{-b}^{b} (t+\tau) \, dt$$

$$= \tau \tag{8.29}$$

and

$$\Delta^2_{e_{\sigma,\tau,r}} = \int_{-\infty}^{\infty} (t-\tau)^2 |e_\sigma(t) r(t-\tau)|^2 \, dt = \frac{1}{2b} \int_{-b}^{b} t^2 \, dt = \frac{b^2}{3}, \tag{8.30}$$

so everything is copacetic in the time domain. (Whatever we mean by "small," the right side of (8.30) *will* be small for b small enough). But things go badly wrong in the frequency domain:

$$\int_{-\infty}^{\infty} s^2 |\widehat{e_{\sigma,\tau,r}}(s)|^2 \, ds = \frac{1}{2b\,\pi^2} \int_{-\infty}^{\infty} \left(\frac{s \sin \pi b(s-\sigma)}{s-\sigma} \right)^2 ds = \infty, \tag{8.31}$$

since $s/(s-\sigma)$ looks about like 1 for large $|s|$, and the sine function is not in $L^2(\mathbb{R})$. So $\Delta_{\widehat{e_{\sigma,\tau,r}}} = \infty$, which is not small by anyone's standards.

Similarly, the "sinc-windowed" complex exponential $e_{\sigma,\tau,\hat{r}}$, for r as above, does well enough in the frequency domain but has infinite radius in the time domain.

If we're demanding a small radius of both our window and its Fourier transform, then certainly we must exclude from consideration such badly behaved windows as $e_{\sigma,\tau,r}$ and $e_{\sigma,\tau,\hat{r}}$. Let's—for the time being, anyway.

Definition 8.2.2 (a) A *finite function* is a function $w \in L^2(\mathbb{R})$ such that both $w^\times$ and $(\widehat{w})^\times$ belong to $L^2(\mathbb{R})$ (so that, by Definition 8.2.1(b), $\Delta_w, \Delta_{\widehat{w}} < \infty$).

(b) Let w be a finite function. Then

(i) The *center* C_w of w is the point $(M_w, M_{\widehat{w}}) \in \mathbb{R}^2$.

(ii) The *time span* of w is the interval $[M_w - \Delta_w, M_w + \Delta_w] \subset \mathbb{R}$; the *frequency span* of w is the interval $[M_{\widehat{w}} - \Delta_{\widehat{w}}, M_{\widehat{w}} + \Delta_{\widehat{w}}] \subset \mathbb{R}$; the *time-frequency span*, also called simply the *span*, of w is the Cartesian product $[M_w - \Delta_w, M_w - \Delta_w] \times [M_{\widehat{w}} - \Delta_{\widehat{w}}, M_{\widehat{w}} + \Delta_{\widehat{w}}] \subset \mathbb{R}^2$.

(iii) The *width* W_w of w is the length $2\Delta_w$ of its time span; the *height* H_w of w is the length $2\Delta_{\widehat{w}}$ of its frequency span; the *area* A_w of w is the product $4\Delta_w \Delta_{\widehat{w}}$ of its width with its height.

By a *finite window* we mean a finite function w that's also a window function ($\|w\| = 1$).

432 LOCAL FREQUENCY ANALYSIS AND WAVELETS

It's readily checked that, if w is a finite window, then so is the windowed complex exponential $e_{\sigma,\tau,w}$, and moreover

$$M_{e_{\sigma,\tau,w}} = M_w + \tau, \quad \Delta_{e_{\sigma,\tau,w}} = \Delta_w, \quad M_{\widehat{e_{\sigma,\tau,w}}} = M_{\widehat{w}} + \sigma, \quad \Delta_{\widehat{e_{\sigma,\tau,w}}} = \Delta_{\widehat{w}}. \tag{8.32}$$

(See Exercise 8.2.1.) In particular, if w is a finite window with center $(0,0)$, then $e_{\sigma,\tau,w}$ is a finite window with center (τ,σ) and having the *same* dimensions as w. See Figure 8.3.

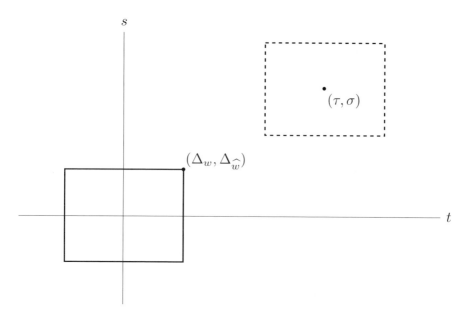

Fig. 8.3 A finite window w centered at $(0,0)$ (solid) and the resulting windowed complex exponential $e_{\sigma,\tau,w}$ (dashed)

The moral is this: To satisfy our above-stated demands, for $e_{\sigma,\tau,w}$'s well localized near $(t,s) = (\tau,\sigma)$ in the time-frequency plane, we need only make sure that w be a finite window, of small dimensions, centered at the origin of this plane. So, for most of the rest of this section, we'll focus on such windows w.

If w is even then so is $\widehat{w}$, by Proposition 6.2.1(b)(ii); if w is odd then so is $\widehat{w}$, by the same result. Consequently, by Definition 8.2.1(a), any even or odd finite function is centered at the origin. In particular, the windows ℓ, h, g considered in the previous section are.

Another nice thing about those particular windows is that, as is readily checked, each of them is in $\text{PSC}(\mathbb{R})$, and has a first derivative in $L^1(\mathbb{R})$. The following result then applies.

Proposition 8.2.1 *Let w be a finite window, with $M_{\widehat{w}} = 0$; suppose also that $w \in \mathrm{PSC}(\mathbb{R})$ and $w' \in L^1(\mathbb{R})$. Then*

$$\Delta_{\widehat{w}} = (2\pi)^{-1}||w'||. \tag{8.33}$$

Proof. Under the stated assumptions, we have $w \in L^1(\mathbb{R})$ as noted above; so $\widehat{w'} = 2\pi i(\widehat{w})^\times$ by Proposition 6.2.1(c)(i). So by the Plancherel theorem,

$$||w'|| = ||\widehat{w'}|| = 2\pi||(\widehat{w})^\times|| = 2\pi\Delta_{\widehat{w}}, \tag{8.34}$$

the last step by the last equality in (8.28). □

Let's now compute the dimensions of $\ell, h,$ and g. (Actually, we'll need Proposition 8.2.1 only for the first two of these.)

Example 8.2.2 For the Bartlett window ℓ of (8.19), we have

$$\Delta_\ell^2 = \int_{-\infty}^{\infty} t^2|\ell(t)|^2\,dt = \int_{-b}^{b} t^2\left(\sqrt{\frac{3}{2b^3}}(b-|t|)\right)^2 dt = \frac{3}{2b^3}\int_{-b}^{b} t^2(b-|t|)^2\,dt$$

$$= \frac{3}{b^3}\int_0^b t^2(b-t)^2\,dt = \frac{b^2}{10}, \tag{8.35}$$

so $W_\ell = 2\Delta_\ell = b\sqrt{\frac{2}{5}}$. We also have, by the above proposition,

$$\Delta_{\widehat{\ell}}^2 = \frac{1}{4\pi^2}||\ell'||^2 = \frac{3}{8\pi^2 b^3}\left(\int_{-b}^{0}(1)^2\,dt + \int_0^b(-1)^2\,dt\right) = \frac{3}{4\pi^2 b^2}, \tag{8.36}$$

so $H_\ell = 2\Delta_{\widehat{\ell}} = b^{-1}\sqrt{3}/\pi$, and

$$A_\ell = W_\ell H_\ell = \frac{1}{\pi}\sqrt{\frac{6}{5}}. \tag{8.37}$$

Next, for the Hann window h of (8.22), we have

$$\Delta_h^2 = \int_{-\infty}^{\infty} t^2|h(t)|^2\,dt = \int_{-b}^{b} t^2\left(\frac{1}{\sqrt{3b}}\left(1+\cos\frac{\pi t}{b}\right)\right)^2 dt$$

$$= \frac{b^2}{3\pi^3}\int_{-\pi}^{\pi} u^2(1+\cos u)^2\,du = \frac{b^2}{3\pi^3}\int_{-\pi}^{\pi} u^2\left(\frac{3}{2} + 2\cos u + \frac{1}{2}\cos 2u\right) du$$

$$= \frac{(2\pi^2 - 15)b^2}{6\pi^2} \tag{8.38}$$

(we used the substitution $u = \pi t/b$, the fact that $\cos^2 u = (1+\cos 2u)/2$, and integration by parts), so $W_h = 2\Delta_h = (b/\pi)\sqrt{(4\pi^2-30)/3}$. Also,

$$\Delta_{\widehat{h}}^2 = \frac{1}{4\pi^2}||h'||^2 = \frac{1}{12\pi^2 b}\int_{-b}^{b}\left(-\frac{\pi}{b}\sin\left(\frac{\pi t}{b}\right)\right)^2 dt = \frac{1}{12\pi b^2}\int_{-\pi}^{\pi}\sin^2 t\,dt$$

$$= \frac{1}{12b^2}, \tag{8.39}$$

so $H_h = 2\Delta_{\widehat{h}} = b^{-1}\sqrt{\frac{1}{3}}$, and

$$A_h = W_h H_h = \frac{1}{\pi}\frac{\sqrt{4\pi^2 - 30}}{3}. \tag{8.40}$$

Finally, for the Gabor window g of (8.25), we make a change of variable, integrate by parts with $u = x$ and $dv = x\,e^{-\pi x^2}\,dx$, and then apply Lemma 6.2.1, as follows:

$$\Delta_g^2 = \int_{-\infty}^{\infty} t^2 g^2(t)\,dt = \frac{2}{b}\int_{-\infty}^{\infty} t^2 e^{-\pi(2t/b)^2}\,dt = \frac{b^2}{4}\int_{-\infty}^{\infty} x^2 e^{-\pi x^2}\,dx$$

$$= \frac{b^2}{4}\left(\left.\frac{xe^{-\pi x^2}}{-2\pi}\right|_{-\infty}^{\infty} + \frac{1}{2\pi}\int_{-\infty}^{\infty} e^{-\pi x^2}\,dx\right) = \frac{b^2}{8\pi}, \tag{8.41}$$

so $W_g = 2\Delta_g = b/\sqrt{2\pi}$. Also, by (8.26),

$$\Delta_{\widehat{g}}^2 = \int_{-\infty}^{\infty} s^2 \left(\frac{b}{\sqrt{2}} g\left(\frac{b^2}{2}s\right)\right)^2 ds = \frac{4}{b^4}\int_{-\infty}^{\infty} u^2 g^2(u)\,du = \frac{4}{b^4}\Delta_g^2 = \frac{1}{2\pi b^2} \tag{8.42}$$

(we put $u = b^2 s/2$), so $H_g = 2\Delta_{\widehat{g}} = b^{-1}\sqrt{2/\pi}$, and

$$A_g = W_g H_g = \frac{1}{\pi}. \tag{8.43}$$

Each of the above windows has width proportional to the parameter b and height proportional to the reciprocal of b. So, while we can improve localization in the time domain by decreasing b, this is at the *expense* of localization in frequency. And vice versa. Here we see the local/global principle at work again.

Since $\frac{6}{5} > \sqrt{4\pi^2 - 30}/3 > 1$, the above example also demonstrates that, by careful choice of the *type* of window function, we can improve *overall* time-frequency localization. But how far? Arbitrarily? The local/global principle suggests not, as does the example itself—in no case did we have a window area less than $1/\pi$.

All of this exemplifies:

Proposition 8.2.2 (the Heisenberg uncertainty principle) *For any window w,*

$$A_w \geq \frac{1}{\pi}, \tag{8.44}$$

with equality holding if and only if w is a Gabor window g.

Proof. This is immediate if w is not finite (since $\Delta_w = \infty$ certainly implies w, and therefore $\widehat{w}$, and therefore $\Delta_{\widehat{w}}$, is nonzero; similarly, $\Delta_{\widehat{w}} = \infty \Rightarrow \Delta_w \neq 0$). So we assume it is.

Now by equations (8.32), we can shift the center of w without affecting the area; so we may also assume that $C_w = (0,0)$.

FINITE WINDOWS AND THE HEISENBERG UNCERTAINTY PRINCIPLE

Finally, we may take $w \in \text{PSC}(\mathbb{R})$ and $w' \in L^1(\mathbb{R})$; the more general statement of our proposition may then be proved using limiting arguments (cf. [15], Section 2.8).

Under these assumptions, we have

$$\begin{aligned} A_w = 4\Delta_w \Delta_{\widehat{w}} &= \frac{2}{\pi}||w^\times||\,||w'|| \geq \frac{2}{\pi}|\langle w^\times, w'\rangle| \\ &= \frac{1}{\pi}\left(|\langle w^\times, w'\rangle| + |\overline{\langle w^\times, w'\rangle}|\right) \geq \frac{1}{\pi}\left(|\langle w^\times, w'\rangle + \overline{\langle w^\times, w'\rangle}|\right) \\ &= \frac{1}{\pi}\left|\int_{-\infty}^{\infty} t\left(w(t)\overline{w'(t)} + w'(t)\overline{w(t)}\right) dt\right| = \frac{1}{\pi}\left|\int_{-\infty}^{\infty} t\frac{d}{dt}\left(w(t)\overline{w(t)}\right)dt\right| \\ &= \frac{1}{\pi}\left|\int_{-\infty}^{\infty} w(t)\overline{w(t)}\,dt\right| = \frac{1}{\pi}. \end{aligned} \qquad (8.45)$$

(For the second step we used the definition of Δ_w and Proposition 8.2.1; for the third we used the Cauchy-Schwarz inequality. The next four steps are straightforward; for the penultimate one, we integrated by parts with $u = t$ and $dv = d(w(t)\overline{w(t)})$; for the last, we used the fact that $||w|| = 1$.) So (8.44) is proved.

To get the remainder of the proposition, we note (cf. Exercise 3.1.2) that the Cauchy-Schwarz inequality is an equality if and only if one of the elements in the inner product there is a constant multiple of the other. But then the first inequality in (8.45) is an equality if and only if

$$w'(t) = \gamma\, tw(t) \qquad (8.46)$$

for some constant γ; the latter happens if and only if $w = Kg$ for g a Gabor window and K a constant. (Recall Exercise 6.2.7.)

Then (8.46) gives $g' = \gamma tg(t)$; from the definition (8.25) of g it follows instantly that $\gamma = 4\pi/b^2$. Also, the assumption $||w|| = 1$ implies $|K| = 1$; so (8.45) gives

$$A_w = \frac{2}{\pi}|\langle w^\times, w'\rangle| = \frac{2}{\pi}|\langle Kg^\times, Kg'\rangle| = \frac{8}{b^2}||g^\times||^2 = \frac{8}{b^2}\Delta_g^2 = \frac{1}{\pi} \qquad (8.47)$$

(the last step by (8.41)), and our proof is complete. □

The Heisenberg uncertainty principle says: If a window is concentrated near (some) $\sigma \in \mathbb{R}$, then its Fourier transform *cannot* be too concentrated near (any) $\tau \in \mathbb{R}$, and vice versa. So, you *can't* get arbitrarily good localization in both time and frequency domains simultaneously.

(In physics, the position of a quantum mechanical particle is represented by the width of a certain window function w, called a *wave function*, associated with the particle, and the momentum of that particle is represented by the height of w. In this context, the Heisenberg uncertainty principle amounts to the assertion that one can *not* simultaneously measure position and momentum to any desired degree of precision; the better a handle one has on the former, the less one can be certain of the latter, and conversely.)

The Heisenberg inequality (8.44) is a fact of life; whatever one's approach to local frequency analysis, one cannot escape the limitations this inequality implies. But certain *other* limitations to our above, windowed complex exponential approach are *unrelated* to this inequality and merit further consideration.

For one thing, might some other approach lead to a *discrete*, or *series*, inversion result, as opposed to the *continuous*, *integral* formula of Theorem 8.1.1? This would be important because infinite series are, generally, more readily approximated than are integrals. And might such a discrete result in fact constitute an *orthonormal decomposition*? If so, then we could bring to bear all of the usual, powerful techniques and ideas associated with such decompositions.

For another thing, we note from (8.32) that the width of a windowed complex exponential $e_{\sigma,\tau,w}$ *does not depend on* the midpoint $M_{\widehat{e_{\sigma,\tau,w}}}$ in the frequency domain. BUT: A low frequency, which is to say a slow oscillation, takes a long time to evolve, so an excessively narrow window cannot properly capture it. Conversely a high frequency, meaning a rapid oscillation, evolves quickly; use of too wide a window to probe it constitutes overkill. So this independence is a liability; for more effective, efficient local frequency analyses, we should employ windows whose widths *vary inversely with* their frequency domain midpoints. Can we construct a theory around windows of this type?

The answers are: Yes it might; yes it might; yes we can. We'll demonstrate this, amply we hope, over the course of the next three sections. And we'll see therein that the windows in question, called *wavelets*, allow for still other nice things (like *multiresolution*, also called *time-scale*, analyses) as well.

Exercises

8.2.1 Verify equations (8.32).

8.2.2 **a.** Determine which of the windows of Exercise 8.1.2 are finite and which are not. Compute the time and frequency midpoints and radii and the areas of the finite ones. Use Proposition 8.2.1 when it's helpful.

b. List the finite windows from the previous part of this exercise, along with those of Example 8.2.2, in order of descending (i) width; (ii) height; (iii) area.

8.3 WAVELETS AND MULTIRESOLUTION ANALYSES: BASICS

At the end of the last section, we expressed a desire for window functions (i) that form an orthonormal basis of $L^2(\mathbb{R})$, and (ii) such that windows concentrated near higher frequencies are better localized in the time domain. In this section, we phrase this desire somewhat more specifically and consider some very satisfying ways of satisfying it.

Definition 8.3.1 Let $\psi \in L^2(\mathbb{R})$; for $j, k \in \mathbb{Z}$ and $t \in \mathbb{R}$, define

$$\psi_{j,k}(t) = 2^{j/2}\psi(2^j(t - k/2^j)) = 2^{j/2}\psi(2^j t - k) = 2^{j/2}(\psi_{(k)})_{\{2^j\}}(t) \quad (8.48)$$

(where $f_{(a)}(t) = f(t-a)$ and $f_{\{b\}}(t) = f(bt)$, as in (6.34)). If $\{\psi_{j,k} : j, k \subset \mathbb{Z}\}$ is an orthonormal basis for $L^2(\mathbb{R})$, then ψ is called a *wavelet*.

(Sometimes, in the literature, wavelets are defined more broadly, and what *we* call a wavelet is termed an *orthonormal wavelet*. But as the theory of orthonormal wavelets is arguably the richest and most applicable, and is less arguably the most elegant, we'll be more than happy to simply call them wavelets.)

Observe that, if ψ is a wavelet, then so is $\psi_{m,n}$ for any given integers m and n; this follows from the fact that, for general $f \in L^2(\mathbb{R})$ and $j, k, m, n \in \mathbb{Z}$,

$$(f_{j,k})_{m,n} = f_{j+m, 2^j n + k}. \tag{8.49}$$

(See Exercise 8.3.1.) It's only a matter of perspective that $\psi = \psi_{0,0}$ has preference over any other $\psi_{j,k}$. To reflect this perspective, one often refers to ψ as the "basic wavelet" (or the "mother wavelet").

An interesting point of comparison between wavelets and windowed complex exponentials is this. The latter are formed by *dilating* the basic complex exponential $e_1(t) = e^{2\pi i t}$ and by *translating* the chosen window w (and by multiplying together the results). Wavelets, on the other hand, are obtained via dilation and translation of the *same* basic function ψ.

An even more interesting, but related, point is that that wavelets do what windowed complex exponentials don't: They conform to the above specification (ii), regarding *location* in frequency versus *localization* in time. The following result demonstrates (not only) this.

Proposition 8.3.1 *Suppose ψ is a finite window, in the sense of Definition 8.2.2(a). Then for any $j, k \in \mathbb{Z}$,*

$$M_{\psi_{j,k}} = 2^{-j}(M_\psi + k), \quad \Delta_{\psi_{j,k}} = 2^{-j}\Delta_\psi, \quad M_{\widehat{\psi_{j,k}}} = 2^j M_{\widehat{\psi}}, \quad \Delta_{\widehat{\psi_{j,k}}} = 2^j \Delta_{\widehat{\psi}} \tag{8.50}$$

(where M and Δ refer to the midpoints and radii prescribed by Definition 8.2.1).

Proof. We have

$$M_{\psi_{j,k}} = 2^j \int_{-\infty}^{\infty} t |\psi(2^j t - k)|^2 \, dt = 2^j \int_{-\infty}^{\infty} 2^{-j}(u+k) |\psi(u)|^2 (2^{-j} du)$$

$$= 2^{-j} \left(\int_{-\infty}^{\infty} u |\psi(u)|^2 \, du + k \int_{-\infty}^{\infty} |\psi(u)|^2 du \right) = 2^{-j}(M_\psi + k), \tag{8.51}$$

by the substitution $u = 2^j t - k$ and the fact that $||\psi|| = 1$. Consequently,

$$\Delta^2_{\psi_{j,k}} = 2^j \int_{-\infty}^{\infty} (t - M_{\psi_{j,k}})^2 |\psi(2^j t - k)|^2 \, dt$$

$$= 2^j \int_{-\infty}^{\infty} (2^{-j}(u+k) - M_{\psi_{j,k}})^2 |\psi(u)|^2 (2^{-j} du)$$

$$= 2^{-2j} \int_{-\infty}^{\infty} (u - M_\psi)^2 |\psi(u)|^2 \, du = 2^{-2j} \Delta^2_\psi, \tag{8.52}$$

as required.

The midpoint and radius of $\widehat{\psi_{j,k}}$ are computed similarly, with the aid of a computation:

$$\widehat{\psi_{j,k}} = 2^{j/2} F[(\psi_{(k)})_{\{2^j\}}(t)] = 2^{j/2} \cdot 2^{-j} F[(\psi_{(k)})(t)]_{\{2^{-j}\}} = 2^{-j/2} \left(e_{-k}\widehat{\psi}\right)_{\{2^{-j}\}}$$
$$= (e_{-k/2^j})(\widehat{\psi})_{-j,0} \tag{8.53}$$

(cf. (6.35)). We leave the remaining details to the reader. □

Suppose ψ is a finite window with $M_{\widehat{\psi}} > 0$. Then by the above proposition, the frequency domain midpoint $M_{\widehat{\psi_{j,k}}}$ of $\psi_{j,k}$ increases with j at the rate of 2^j, while the time domain radius $\Delta_{\psi_{j,k}}$ of $\psi_{j,k}$ decreases with j at the inverse rate. So wavelets do, in the finite case (or in any case where $M_{\widehat{\psi}}$ and Δ_ψ are defined), enjoy time domain narrowing with increasing frequency domain midpoint, as desired.

Actually, this is so even if ψ is not finite, by similar but somewhat less formal arguments. To wit, by (8.53), localization of $\widehat{\psi}$ near $s = \sigma$ means localization of $\widehat{\psi_{j,k}}$ near $2^{-j}s = \sigma$, or $s = 2^j\sigma$. At the same time, by (8.48), $\psi_{j,k}$ is 2^j times as "compressed," in the t direction, as is ψ.

Finite wavelets are generally preferred in applications; still, certain infinite ones are of interest, for the purposes of illustration and comparison if for nothing else. We'll produce an interesting infinite wavelet shortly; to do so, we'll use the following.

Lemma 8.3.1 *Let $\psi \in L^2(\mathbb{R})$; let $\psi_{j,k}$, for $j, k \in \mathbb{Z}$, be as in (8.48). Then ψ is a wavelet if and only if all of the following conditions are met:*

(i) *For $p \leq 0$, $\langle \psi, \psi_{p,q} \rangle = \delta_{0,p}\delta_{0,q}$.*

(ii) *The set $\mathrm{span}(\{\psi_{j,k} : j, k \in \mathbb{Z}\})$ of all finite linear combinations of the $\psi_{j,k}$'s is dense in $L^2(\mathbb{R})$.*

Proof. Suppose ψ is a wavelet. Then $\psi = \psi_{0,0}$ and $\langle \psi_{j,k}, \psi_{m,n} \rangle = \delta_{j,m}\delta_{k,n}$, so the first of the above conditions is met. So is the second, by Corollary 3.7.2.

Conversely, suppose these conditions hold. Then first of all we have, for $m \leq j$,

$$\langle \psi_{j,k}, \psi_{m,n} \rangle = 2^{(j+m)/2} \int_{-\infty}^{\infty} \psi(2^j t - k)\psi(2^m t - n)\, dt$$
$$= 2^{(-j+m)/2} \int_{-\infty}^{\infty} \psi(u)\psi(2^m(2^{-j}(u+k)) - n)\, du$$
$$= 2^{(-j+m)/2} \int_{-\infty}^{\infty} \psi(u)\psi(2^{m-j}u - (n - 2^{m-j}k))\, du$$
$$= \langle \psi, \psi_{m-j, n-2^{m-j}k} \rangle = \delta_{0,m-j}\delta_{0,n-2^{m-j}k} = \delta_{j,m}\delta_{k,n}, \tag{8.54}$$

since $m - j = n - 2^{m-j}k = 0$ if and only if $j = m$ and $k = n$. (We've used the substitution $u = 2^j t - k$ and some algebra.) Because $\langle \psi_{j,k}, \psi_{m,n} \rangle = \overline{\langle \psi_{m,n}, \psi_{j,k} \rangle}$, (8.54) then holds for $m > j$ as well; so the $\psi_{j,k}$'s are orthonormal.

That they constitute an orthonormal basis follows from Corollary 3.7.2, and we're done. □

We now construct some infinite wavelets.

Proposition 8.3.2 *The* Haar function ψ^H, *defined by*

$$\psi^H(t) = \chi_{[0,1/2)}(t) - \chi_{[1/2,1)}(t) = \begin{cases} 1 & \text{if } 0 \le t < \tfrac{1}{2}, \\ -1 & \text{if } \tfrac{1}{2} \le t < 1, \\ 0 & \text{if } t < 0 \text{ or } t \ge 1 \end{cases} \quad (8.55)$$

(see Fig. 8.4) is a wavelet.

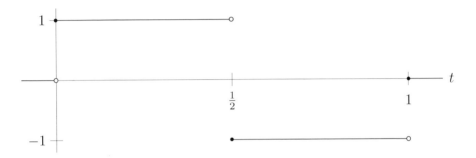

Fig. 8.4 The Haar function ψ^H

Proof. We utilize Lemma 8.3.1. Regarding the requirement (i) there, let's suppose $p \le 0$. We note that $\psi^H(u) \ne 0$ implies $0 \le u < 1$, which implies $0 \le 2^p u < 1$; also, $\psi^H(2^p u - q) \ne 0$ implies $0 \le 2^p u - q < 1$, which implies $q \le 2^p u < q + 1$. SO: $\psi^H(u)\psi^H(2^p u - q)$ is nonzero only if $2^p u \in [0, 1) \cap [q, q + 1)$. If $q \ne 0$, then the intersection on the right is empty, so $\langle \psi^H, \psi^H_{p,q} \rangle = 0$ in this case.

On the other hand,

$$\langle \psi^H, \psi^H_{p,0} \rangle = 2^{-p/2} \int_{-\infty}^{\infty} \psi^H(u)\psi^H(2^p u)\, du. \quad (8.56)$$

If $p = 0$, this inner product equals $\int_0^1 1^2\, dt = 1$. If not then, since $p < 0$, we have $0 < 2^p \le \tfrac{1}{2}$, so $0 \le u < 1$ implies $0 \le 2^p u < \tfrac{1}{2}$, which implies $\psi^H(2^p u) = 1$, so that (8.56) yields

$$\langle \psi^H, \psi^H_{p,0} \rangle = 2^{-p/2} \int_{-\infty}^{\infty} \psi^H(u)\psi^H(2^p u)\, du = 2^{-p/2} \left(\int_0^{1/2} du - \int_{1/2}^1 du \right)$$
$$= 0. \quad (8.57)$$

To summarize, $\langle \psi^H, \psi^H_{p,q} \rangle$ equals 1 if and only if $q = p = 0$; it equals zero if $p < 0$ or $q \ne 0$. So condition (i) *is* met.

It remains to show density in $L^2(\mathbb{R})$ of span($\{\psi_{j,k}^H : j, k \in \mathbb{Z}\}$). We accomplish this in steps. We first observe, by direct computation, that the characteristic function $\chi_{[0,1)}$ is the limit in norm of the k_N's defined by

$$k_N = \sum_{j=-N}^{-1} 2^{j/2} \psi_{j,0}^H. \tag{8.58}$$

(See Fig. 8.5, and Exercise 8.3.2.)

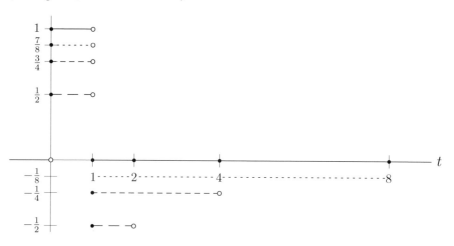

Fig. 8.5 $\chi_{[0,1)}$ (solid), k_1, k_2, and k_3 (shorter dashes correspond to larger subscripts). Here the k_N's are as in (8.58)

From this it may be deduced that span($\{\psi_{j,k}^H : j, k \in \mathbb{Z}\}$) is dense in the vector space of *step functions*, meaning finite linear combinations of characteristic functions of bounded intervals, on $\mathbb{R}$. But it's well known (see, for example, [29]) that the latter vector space is dense in $L^2(\mathbb{R})$. By transitivity of norm density (see Exercise 3.2.6), span($\{\psi_{j,k}^H : j, k \in \mathbb{Z}\}$) is itself dense in $L^2(\mathbb{R})$, and we're done. □

Note that, by Example 6.1.1 and properties of the Fourier transform,

$$\begin{aligned}
\widehat{\psi^H}(s) &= F[\chi_{[0,1/2)}(t)](s) - F[\chi_{[1/2,1)}(t)](s) \\
&= F[\chi_{[-1/4,1/4)}(t-1/4)](s) - F[\chi_{[-1/4,1/4)}(t-3/4)](s) \\
&= (e^{-\pi i s/2} - e^{-3\pi i s/2}) F[\chi_{[-1/4,1/4)}(t)](s) = 2ie^{-\pi i s} \sin\frac{\pi s}{2} \frac{\sin(\pi s/2)}{\pi s} \\
&= ie^{-\pi i s} \frac{1 - \cos \pi s}{\pi s}, \tag{8.59}
\end{aligned}$$

so ψ^H is not a finite wavelet: $\Delta_{\widehat{\psi^H}} = \infty$.

We saw above that $\chi_{[0,1)}$ is an *infinite* linear combination of dyadic dilations (that is, dilations by integer powers of 2) of ψ^H. It's at least as interesting to observe that ψ^H may be expressed as a *finite* linear combination of dilations of tranlations of $\chi_{[0,1)}$. Specifically, we have

$$\psi^H(t) = \chi_{[0,1)}(2t) - \chi_{[0,1)}(2t-1) \tag{8.60}$$

(which is just a rewriting of the definition (8.55) of ψ^H) for all t.
It's also worth noting that $\chi_{[0,1)}$ satisfies the relation

$$\chi_{[0,1)}(t) = \chi_{[0,1)}(2t) + \chi_{[0,1)}(2t-1), \tag{8.61}$$

and that $\chi_{[0,1)}$ and ψ^H are orthogonal:

$$\langle \chi_{[0,1)}, \psi^H \rangle = \int_{-\infty}^{\infty} \chi_{[0,1)}(t)\psi^H(t)\,dt = \int_0^1 \psi^H(t)\,dt = 0. \tag{8.62}$$

The moral is that the Haar wavelet ψ^H is built up from a simpler function $\varphi^H = \chi_{[0,1)}$, which is related to ψ^H in pleasant ways and possesses some nice properties in and of itself. To best discuss these properties, we'll need the following.

Definition 8.3.2 A *multiresolution analysis* $(\mathcal{V}, \varphi)$ is a sequence

$$\mathcal{V} = \ldots, V_{-2}, V_{-1}, V_0, V_1, V_2, \ldots \tag{8.63}$$

of subspaces of $L^2(\mathbb{R})$, together with a function $\varphi \in L^2(\mathbb{R})$, called the *scaling function* for $\mathcal{V}$, such that

(i) (nestedness) $\cdots \subset V_{-2} \subset V_{-1} \subset V_0 \subset V_1 \subset V_2 \subset \cdots$;

(ii) (density) $\cup_{j=-\infty}^{\infty} V_j$ is dense in $L^2(\mathbb{R})$;

(iii) (separation) $\cap_{j=-\infty}^{\infty} V_j = \{0\}$;

(iv) (dilation) $f \in V_0 \Leftrightarrow f_{j,0} \in V_j$ (note $f_{j,0}(t) = 2^{j/2}f(2^jt)$);

(v) (orthonormality) $\{\varphi_{0,k} : k \in \mathbb{Z}\}$ (note $\varphi_{0,k}(t) = \varphi(t-k)$) is an orthonormal basis for V_0.

Suppose the orthonormality condition is met. Then V_0 is certainly closed under integer translations. Now suppose that the dilation condition is also satisfied. Since

$$(f_{j,0})_{0,k} = f_{j,2^jk} = (f_{0,2^jk})_{j,0} \tag{8.64}$$

(cf. (8.49)), each V_j is closed under these translations as well; also, since

$$(\varphi_{0,k})_{j,0} = \varphi_{j,k} \tag{8.65}$$

(again by (8.49)), the set $\{\varphi_{j,k} : k \in \mathbb{Z}\}$ is an orthonormal basis for V_j. We have the following useful consequence: Suppose orthonormality and dilation hold. Then the scaling function φ being a norm convergent linear combination of the $\varphi_{1,k}$'s implies $\varphi \in V_1$, which implies $V_0 \subset V_1$, which implies nestedness.

Let's now consider an example, which we've actually considered already.

Proposition 8.3.3 *Let φ^H denote the characteristic function $\chi_{[0,1)}$ and V_0^H the completion of* $\mathrm{span}(\{\varphi_{0,k}^H : k \in \mathbb{Z}\})$:

$$V_0^H = \left\{ \sum_{k=-\infty}^{\infty} d_k \varphi_{0,k}^H : \sum_{k=-\infty}^{\infty} |d_k|^2 < \infty \right\}. \tag{8.66}$$

(See Exercise 3.10.3.) Also define V_j^H, for $j \in \mathbb{Z}$, by the above dilation condition (iv); that is, $V_j^H = \{f_{j,0} : f \in V_0^H\}$. Then $(\mathcal{V}^H, \varphi^H)$, where $\mathcal{V}^H$ denotes the sequence of the V_j^H's, is a multiresolution analysis.

Proof. That the $\varphi_{0,k}^H$'s form an orthonormal basis for V_0^H follows from Exercise 3.10.3 and Corollary 3.7.2.

Now by definition of the V_j^H's, the dilation condition (iv) is satisfied. So to prove nestedness we need only, as noted above, express φ^H as a linear combination of $\varphi_{1,k}^H$'s. But this is so by (8.61), which says $\varphi^H = 2^{-1/2} \varphi_{1,0}^H + 2^{-1/2} \varphi_{1,1}^H$.

To prove density and separation, we invoke the Haar wavelets $\psi_{j,k}^H$ of Proposition 8.3.2, as follows. First, since $\psi^H = 2^{-1/2} \varphi_{1,0}^H - 2^{-1/2} \varphi_{1,1}^H$ (cf. (8.60)), we have $\psi^H \in V_1^H$, whence $\psi_{0,k}^H \in V_1^H$ for all k, whence in turn $\psi_{j,k}^H \in V_{j+1}^H$ for all j, k. So

$$\mathrm{span}(\{\psi_{j,k}^H : j, k \in \mathbb{Z}\}) \subset \cup_{j=-\infty}^{\infty} V_{j+1}^H; \tag{8.67}$$

the left side is, as previously seen, dense in $L^2(\mathbb{R})$, so the right side is too. Hence, density.

For separation, we note that ψ^H and φ^H are orthonormal (cf. (8.62)) and that $\langle \psi^H, \varphi_{0,k}^H \rangle = 0$ for $k \neq 0$ (since ψ^H is supported on $[0, 1)$ and $\varphi_{0,k}^H$ on $[k, k+1)$); so in fact ψ^H is orthogonal to every element of V_0^H. From this we deduce that $\psi_{j,k}^H$ is, for any j, k, orthogonal to every element of $\cap_{j=-\infty}^{\infty} V_j^H$. From Parseval's equation (Proposition 3.7.1(b)) and the fact that the $\psi_{j,k}^H$'s form an orthonormal basis for $L^2(\mathbb{R})$, it follows that each element in this intersection has zero norm and is therefore zero (in $L^2(\mathbb{R})$, meaning almost everywhere). So the separation condition holds, and we're done. □

In the next section, we'll further detail the association between ψ^H and $(\mathcal{V}^H, \varphi^H)$ and will abstract this association to more general situations. We'll produce scaling functions and wavelets possessing better time-frequency localization properties than φ^H and ψ^H and will develop the notion of a *wavelet decomposition*, wherein scaling functions and wavelets are used to resolve signals at arbitrarily fine (or coarse) levels of resolution.

Exercises

8.3.1 Prove that (8.49) holds for any function f.

8.3.2 Show by direct computation that the k_N's of (8.58) converge in norm to $\chi_{[0,1)}$. Hint: By (3.148) and the orthonormality of the $\psi_{j,0}^H$'s, it suffices to show that $\langle \chi_{[0,1)}, \psi_{j,0}^H \rangle = 2^{j/2}$ for all $j < 0$ and that

$$||\chi_{[0,1)}||^2 = \lim_{N \to \infty} \sum_{j=-N}^{-1} |2^{j/2}|^2.$$

8.3.3 Let $\varphi(t) = \operatorname{sinc} \pi t$; let V_j be the space $L^2(2^{j-1}; \mathbb{R})$ of 2^{j-1}-bandlimited functions in $L^2(\mathbb{R})$ (cf. Section 7.4); let $\mathcal{V}$ be the sequence of V_j's. Use Exercise 7.4.3 to show that $(\mathcal{V}, \varphi)$ is a multiresolution analysis. Hint for density: Given $f \in L^2(\mathbb{R})$, the integral

$$\int_{-2^{j-1}}^{2^{j-1}} \widehat{f}(s) e^{2\pi i s t} \, ds$$

defines a function in V_j. (Why?) Now use Theorem 6.5.2.

8.4 MULTIRESOLUTION ANALYSES AND WAVELETS: A BUILDER'S GUIDE

Here we discuss some general techniques for constructing multiresolution analyses and wavelets, and for using these to produce "wavelet decompositions." We begin with the following, whose proof, being a bit lengthy (though quite illuminating), will be relegated to the next section.

Theorem 8.4.1 *Let $F_1 \in C(\mathbb{T}_1)$ (that is, F_1 is continuous and 1-periodic) satisfy the following conditions:*

$$F_1(0) = 1, \quad |s| \leq \tfrac{1}{4} \Rightarrow F_1(s) \neq 0; \tag{8.68}$$

$$\sum_{L=0}^{1} \left| F_1\left(s + \frac{L}{2}\right) \right|^2 = 1 \quad (s \in \mathbb{R}). \tag{8.69}$$

Also suppose $\Phi \in C(\mathbb{R}) \cap L^2(\mathbb{R})$ is such that

$$\Phi(s) = \Phi\left(\frac{s}{2}\right) F_1\left(\frac{s}{2}\right) \quad (s \in \mathbb{R}), \quad \widehat{\Phi} \in L^1(\mathbb{R}), \quad \Phi(0) = 1. \tag{8.70}$$

Then, putting $\varphi = \widehat{\Phi^-}$, we have the following.

(a) *Let V_0 be the completion of* span$(\{\varphi_{0,k} : k \in \mathbb{Z}\})$:

$$V_0 = \left\{ \sum_{k=-\infty}^{\infty} d_k \varphi_{0,k} : \sum_{k=-\infty}^{\infty} |d_k|^2 < \infty \right\}. \tag{8.71}$$

Also, for $j \in \mathbb{Z}$, define $V_j = \{f_{j,0} : f \in V_0\}$. Then $(\mathcal{V}, \varphi)$, where $\mathcal{V}$ is the sequence of the V_j's, is a multiresolution analysis.

Moreover, we have the "scaling relation"

$$\varphi = \sqrt{2} \sum_{k=-\infty}^{\infty} \overline{c_{-k}(F_1)} \varphi_{1,k}, \tag{8.72}$$

the $c_k(F_1)$'s being the Fourier coefficients of F_1.

(b) *Put*

$$F_2(s) = \overline{e^{2\pi i(s+1/2)} F_1(s + \tfrac{1}{2})}, \tag{8.73}$$

define $\Psi \in L^2(\mathbb{R})$ by

$$\Psi(s) = \Phi\left(\frac{s}{2}\right) F_2\left(\frac{s}{2}\right) = \widehat{\varphi}\left(\frac{s}{2}\right) F_2\left(\frac{s}{2}\right), \tag{8.74}$$

and let $\psi = \widehat{\Psi^-}$. Then

(i) $\{\varphi_{0,k} : k \in \mathbb{Z}\} \cup \{\psi_{0,k} : k \in \mathbb{Z}\}$ *is an orthonormal basis for V_1. Moreover, we have the formula*

$$\psi = \sqrt{2} \sum_{k=-\infty}^{\infty} \overline{c_{-k}(F_2)} \varphi_{1,k} = \sqrt{2} \sum_{k=-\infty}^{\infty} (-1)^k \overline{c_{k-1}(F_1)} \varphi_{1,k}. \tag{8.75}$$

(ii) *ψ is a wavelet.*

(c) *If $K_1, K_2 \in \mathbb{Z}$, $K_1 < K_2$, and $c_k(F_1) = 0$ whenever $k \notin [K_1, K_2]$, then φ is supported on $[-K_2, -K_1]$ and ψ on $[\tfrac{1}{2}(1 + K_1 - K_2), \tfrac{1}{2}(1 + K_2 - K_1)]$.*

In the above situation, we call F_1 the *scale factor for* $(\mathcal{V}, \varphi)$, and ψ the *wavelet associated with* $(\mathcal{V}, \varphi)$.

Example 8.4.1 Let

$$F_1(s) = F_1^H(s) = e^{-\pi i s} \cos \pi s, \quad \Phi(s) = \Phi^H(s) = e^{-\pi i s} \operatorname{sinc} \pi s. \tag{8.76}$$

Then, because

$$\cos \pi(s + \tfrac{1}{2}) = -\sin \pi s \tag{8.77}$$

and because $\cos^2\theta + \sin^2\theta = 1$, the hypotheses of the above theorem (including the hypothesis in part (c), with $K_1 = -1$ and $K_2 = 0$) are satisfied. Moreover, the resulting multiresolution analysis and associated wavelet are, respectively, the "Haar multiresolution analysis" $(\mathcal{V}^H, \varphi^H)$ and the Haar wavelet ψ^H of the previous section. See Exercise 8.4.1.

Before considering other specific examples, we remark on the immensely important application of the above theorem, and of the entities developed therein, to *wavelet decompositions*. The key ideas here are as follows. First, for $j \in \mathbb{Z}$, let W_j be the completion of span($\{\psi_{j,k} : k \in \mathbb{Z}\}$):

$$W_j = \left\{ \sum_{k=-\infty}^{\infty} d_k \psi_{j,k} : \sum_{k=-\infty}^{\infty} |d_k|^2 < \infty \right\}. \tag{8.78}$$

By part (b)(i) of our theorem, each element of W_0 is orthogonal to each element of V_0, and moreover each $f \in V_1$ is a sum of an element of W_0 and an element of V_0. Another way of saying this same thing is to say that V_1 is the *direct sum* of W_0 and V_0, written $V_1 = W_0 \oplus V_0$. Then by dilation arguments,

$$V_j = W_{j-1} \oplus V_{j-1} \tag{8.79}$$

for all j. We note that W_{j-1} has $\{\psi_{j-1,k} : k \in \mathbb{Z}\}$ as a basis for each j.

Now because V_j has the $\varphi_{j,k}$'s ($k \in \mathbb{Z}$) as a basis and because each of these $\varphi_{j,k}$'s has the same radius $\Delta_{\varphi_{j,k}} = 2^{-j}\Delta_{\varphi}$ (recall Proposition 8.3.1), it's fair enough to think of V_j as the subspace of $L^2(\mathbb{R})$ having *jth-level resolution*. Under this interpretation, (8.79) says: The space W_{j-1}, which has the wavelets $\psi_{j-1,k}$ ($k \in \mathbb{Z}$) as a basis, contains the *layer of detail* necessary to pass from the $(j-1)$st level of resolution to the (finer) jth level.

We can extend this analysis by applying (8.79) recursively to itself. We find that, for any $j \in \mathbb{Z}$ and $J \in \mathbb{Z}^+$,

$$V_j = W_{j-1} \oplus W_{j-2} \oplus V_{j-2} = W_{j-1} \oplus W_{j-2} \oplus W_{j-3} \oplus V_{j-3}$$
$$= W_{j-1} \oplus W_{j-2} \oplus W_{j-3} \oplus \cdots \oplus W_{j-J} \oplus V_{j-J} = \left(\bigoplus_{\ell=1}^{J} W_{j-\ell} \right) \oplus V_{j-J}. \tag{8.80}$$

So, If $f \in V_j$, then we can write

$$f = f_{j-1}^+ + f_{j-2}^+ + \cdots + f_{j-J}^+ + f_{j-J}, \tag{8.81}$$

where

$$f_{j-J} = \sum_{k=-\infty}^{\infty} \langle f, \varphi_{j-J,k} \rangle \varphi_{j-J,k} \in V_{j-J} \tag{8.82}$$

is "the resolution of f at the (coarsest) level $j-J$," and for $1 \leq \ell \leq J$,

$$f_{j-\ell}^{+} = \sum_{k=-\infty}^{\infty} \langle f, \psi_{j-\ell,k} \rangle \, \psi_{j-\ell,k} \in W_{j-\ell} \tag{8.83}$$

is the extra layer of detail required to get from resolution level $j - \ell$ to the next finer level.

The expression (8.81) is called a *wavelet decomposition* for $f \in V_j$. For effective implementation of such decompositions, we should have formulas that express the coefficients $\langle f, \varphi_{j-1,k} \rangle$ and $\langle f, \psi_{j-1,k} \rangle$ in terms of the coefficients $\langle f, \varphi_{j,k} \rangle$ of the expansion

$$f = \sum_{k=-\infty}^{\infty} \langle f, \varphi_{j,k} \rangle \, \varphi_{j,k}, \tag{8.84}$$

and vice versa. (Iteration of these formulas will then allow us to construct f, given the $f_{j-\ell}^{+}$'s and the f_{j-J} of (8.81), and vice versa. See, for example, pp. 199–200 in [6] and Exercise 8.4.4 below.) These formulas, called *decomposition* and *reconstruction* formulas, respectively, are spelled out in Exercise 8.4.2.

Now in general, membership in $L^2(\mathbb{R})$ does not assure membership in *any* of the V_j's; so not every element of $L^2(\mathbb{R})$ admits a wavelet decomposition *per se*. BUT: In any given real-life situation, levels of detail beyond a certain threshold level, say the jth, will be negligible. So given $f \in L^2(\mathbb{R})$ we may, for all practical purposes, replace f by its "projection"

$$f_j = \sum_{k=-\infty}^{\infty} \langle f, \varphi_{j,k} \rangle \, \varphi_{j,k} \tag{8.85}$$

onto some V_j and proceed as above. (This projection gives the best approximation in norm to f among all elements of V_j, cf. Proposition 3.7.2).

Wavelet decompositions thus provide us with convenient means of observing, analyzing, and understanding signals at *any resolution level within any desired range* or at many such levels at once. This has profound implications, a particularly dramatic one being to the science of *compression*. To wit, suppose we want, for a given signal f, to discard all details beyond the jth level (they're too fine to be of practical consequence) but to retain all details up to and including the $j-J$ (any coarser resolution is too blurry to be of practical use). Suppose also that we're willing to compromise on the details in between—if these details are *prominent*, we keep them; else, we bid them goodbye. We may accomplish all of this as follows: First, by replacing f with its projection f_j if necessary, we may assume $f \in V_j$. We decompose f as in (8.81); we *subtract*, from each $f_{j-\ell}^{+}$, every "wavelet component"

$$\langle f, \psi_{j-\ell,k} \rangle \, \psi_{j-\ell,k} \tag{8.86}$$

whose coefficient $\langle f, \psi_{j-\ell,k} \rangle$ is small (below some chosen tolerance) in magnitude; we use the new, "deleted" expansion as an approximation to f.

Of course in practice, all of this is done discretely. (The discrete wavelet transform is quite efficient; the number of elementary operations required is roughly proportional to the number of samples, at least if this number is a power of 2 and the wavelet and scaling function are compactly supported. This may be understood as follows. Sampling f at $N = 2^j$ distinct points gives approximations to N of the $\langle f, \varphi_{j,k}\rangle$'s, cf. Exercise 8.4.4. We use these approximations, together with discretizations of the decomposition formulas of Exercise 8.4.2(a), to approximate the $\langle f, \psi_{j-1,\ell}\rangle$'s and the $\langle f, \varphi_{j-1,\ell}\rangle$'s. But because of the 2ℓ's appearing in the indices of the coefficients in these formulas, at most about $N/2 = 2^{j-1}$ of the $\langle f, \psi_{j-1,\ell}\rangle$'s or the $\langle f, \varphi_{j-1,\ell}\rangle$'s will be nonzero. Iterating, we find that at most about $2^{j-1}/2 = 2^{j-2}$ of the $\langle f, \psi_{j-2,\ell}\rangle$'s or the $\langle f, \varphi_{j-2,\ell}\rangle$'s will be nonzero, and so on. We continue until we can no longer subdivide, yielding no more than a constant times $2^{j-1}+2^{j-2}+\cdots+2+1 = 2^j = N$ total operations. For more details, consult Section 7.1 in [6].) Replacing small "discrete wavelet coefficients" by zero results in a reduction of the amount of discrete data. *This* is how wavelets are used in compression. The results can be remarkable. For example, the FBI uses wavelets to achieve roughly 95% compression of fingerprint data. (See [1].) The JPEG2000 and MPEG-4 computer graphics file formats employ wavelets (MPEG-4 is only partly wavelet-based, though), attaining generally from 90% to 99% compression (depending, in essence, on the chosen values of j and J) and improving considerably (under favorable processor conditions) on the capabilities of the older JPEG and MPEG standards. And so on.

We return, now, to the multiresolution analysis/wavelet construction site. We build some structures featuring classic Haar design on the ground floor, but offering more deluxe, smoother accommodations on the upper levels.

Example 8.4.2 For $N \in \mathbb{Z}^+$, let

$$\Theta(s) = (e^{-\pi i s}\operatorname{sinc}\pi s)^N, \quad \Lambda(s) = \left(\sum_{k=-\infty}^{\infty}|\Theta(s+k)|^2\right)^{1/2}. \tag{8.87}$$

We claim that

$$F_1(s) = (e^{-\pi i s}\cos \pi s)^N \frac{\Lambda(s)}{\Lambda(2s)}, \quad \Phi(s) = \frac{\Theta(s)}{\Lambda(s)} \tag{8.88}$$

satisfy the hypotheses of Theorem 8.4.1. Let's demonstrate this.

First, we note that $|\Theta(s)| \leq (\pi s)^{-N}$. From this and the boundedness of the sinc function, it's readily seen that Λ is continuous and that $\Theta \in C(\mathbb{R}) \cap L^2(\mathbb{R})$. Also, Λ is 1-periodic, as may be checked by putting $m = k+1$ into the sum appearing in the definition of $\Lambda(s+1)$. So to show that $F_1 \in C(\mathbb{T}_1)$ and that $\Phi \in C(\mathbb{R}) \cap L^2(\mathbb{R})$, it's enough to prove that $|\Lambda(s)|$ is bounded below by some nonzero constant.

By periodicity, it's sufficient to show that, for some $C > 0$, $|\Lambda(s)| \geq C$ for all $s \in [-\frac{1}{2}, \frac{1}{2}]$. But this is straightforward: Certainly $\Lambda(s)$ is bounded below by $\sqrt{|\Theta(s+0)|^2} = |\operatorname{sinc}\pi s|^N$, and $|\operatorname{sinc}\pi s| \geq |\operatorname{sinc}\pi/2| = 2/\pi$ for $|s| \leq \frac{1}{2}$.

The above and the fact that $e^{-\pi i s}\cos \pi s$ is nonvanishing on $[-\frac{1}{4}, \frac{1}{4}]$ imply the nonvanishing of $F_1(s)$ on this same interval. Since $\Theta(k) = \delta_{k,0}$, we have $\Lambda(0) = 1$, so $F_1(0) = \Phi(0) = 1$. Further, Λ is seen to be infinitely differentiable, so that Φ is too, so that $\hat{\Phi}$ decays more than rapidly enough to be in $L^1(\mathbb{R})$.

Let's now observe that

$$\Theta(2s) = \left(e^{-2\pi i s}\frac{\sin 2\pi s}{2\pi s}\right)^N = \left(e^{-2\pi i s}\frac{\cos \pi s \sin \pi s}{\pi s}\right)^N$$
$$= (e^{-\pi i s}\cos \pi s)^N \left(e^{-\pi i s}\frac{\sin \pi s}{\pi s}\right)^N = (e^{-\pi i s}\cos \pi s)^N \Theta(s). \qquad (8.89)$$

So by the definitions (8.88) of F_1 and Φ,

$$\Phi(s)F_1(s) = \frac{\Theta(s)}{\Lambda(s)}(e^{-\pi i s}\cos \pi s)^N \frac{\Lambda(s)}{\Lambda(2s)} = \frac{\Theta(2s)}{\Lambda(2s)} = \Phi(2s), \qquad (8.90)$$

which is just the first equality in (8.70) (with $2s$ in place of s). It remains only to show that (8.69) holds; we do so as follows. By writing $\mathbb{Z} = \{2\ell: \ell \in \mathbb{Z}\} \cup \{2\ell+1: \ell \in \mathbb{Z}\}$ and noting that $\cos \pi(\theta + \ell) = \pm\cos \pi\theta$ for $\ell \in \mathbb{Z}$, we find that

$$\Lambda^2(2s) = \sum_{k=-\infty}^{\infty} |\Theta(2s+k)|^2 = \sum_{L=0}^{1}\sum_{\ell=-\infty}^{\infty}|\Theta(2s+2\ell+L)|^2$$
$$= \sum_{L=0}^{1}\sum_{\ell=-\infty}^{\infty}\left|\left(\cos\pi\left(s+\ell+\frac{L}{2}\right)\right)^N \Theta\left(s+\ell+\frac{L}{2}\right)\right|^2$$
$$= \sum_{L=0}^{1}\left(\cos\pi\left(s+\frac{L}{2}\right)\right)^{2N}\sum_{\ell=-\infty}^{\infty}\left|\Theta\left(s+\ell+\frac{L}{2}\right)\right|^2$$
$$= \sum_{L=0}^{1}\left(\cos\pi\left(s+\frac{L}{2}\right)\right)^{2N}\Lambda^2\left(s+\frac{L}{2}\right). \qquad (8.91)$$

So, by the definition of F_1 and the 1-periodicity of Λ,

$$\sum_{L=0}^{1}\left|F_1\left(s+\frac{L}{2}\right)\right|^2 = \sum_{L=0}^{1}\left|e^{-\pi i(s+L/2)}\left(\cos\pi\left(s+\frac{L}{2}\right)\right)\right|^{2N}\frac{\Lambda^2(s+L/2)}{\Lambda^2(2s+L)}$$
$$= \frac{1}{\Lambda^2(2s)}\sum_{L=0}^{1}\left(\cos\pi\left(s+\frac{L}{2}\right)\right)^{2N}\Lambda^2\left(s+\frac{L}{2}\right) = 1, \qquad (8.92)$$

as required.

Since

$$\sum_{k=-\infty}^{\infty}\text{sinc}^2\,\pi(s+k) = 1 \qquad (8.93)$$

(see Exercise 3.7.1), the case $N = 1$ of (8.87) gives $\Theta(s) = e^{-\pi i s} \operatorname{sinc} \pi s$ and $\Lambda(s) = 1$. Consequently, the case $N = 1$ of (8.88) gives the F_1 and Φ of (8.76). That is, the above example yields the Haar multiresolution analysis and wavelet, in the case $N = 1$.

For larger N, things look somewhat different. First of all, there are no nice explicit formulas for ϕ and ψ here (although Φ and Ψ can generally be written down quite concretely; see Exercise 8.4.5 for the case $N = 2$). Moreover, while the scaling function and wavelet are, in case $N > 1$, *continuous*—in fact, these functions get smoother with increasing N—and of finite dimensions in the time-frequency plane (again see Exercise 8.4.5), they are *not*, in this case, compactly supported. All of this is in dramatic contrast to the case $N = 1$.

Some suggestive pictures in the case $N = 2$ are depicted in Figure 8.6.

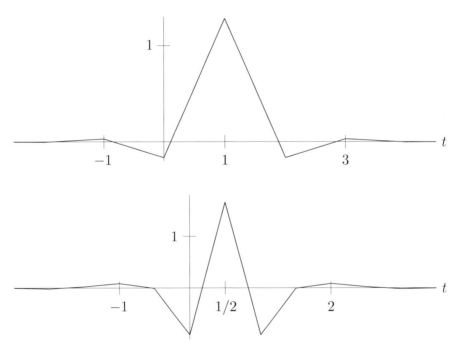

Fig. 8.6 The scaling function φ (top) and wavelet ψ (bottom) of Example 8.4.2 in the case $N = 2$

Around 1988, Ingrid Daubechies [14] made the remarkable discovery of scaling functions and wavelets that are compactly supported, of finite time-frequency dimensions, *and* as smooth as is desired. The *gestalt* behind the Daubechies wavelets is this. Suppose Φ and F_1 satisfy the hypotheses of Theorem 8.4.1, including those of part (c). If the corresponding scaling function φ is not as smooth as we would like it to be, we might try to smooth it out by replacing Φ with a *power* Φ^N for some positive integer $N > 1$. The idea is that, if $\Phi(s)$ decays like $|s|^{-\delta}$ for some $\delta > 0$ as $|s| \to \infty$, then $\Phi^N(s)$ decays at the faster rate of $|s|^{-N\delta}$. And faster decay of a function translates into increased smoothness of its Fourier transform; so the inverse Fourier transform of Φ^N should be smoother than the inverse Fourier transform φ of Φ.

Note that replacing Φ by Φ^N and F_1 by F_1^N preserves the truth value of *most* of the hypotheses of the theorem. Indeed, if F_1 is equal to 1 at 0 and is nonvanishing on $[-\frac{1}{4}, \frac{1}{4}]$, then the same goes for F_1; if $\Phi(s) = \Phi(s/2)F_1(s/2)$ for all s, then $\Phi^N(s) = \Phi^N(s/2)F_1^N(s/2)$ for all s; if the Fourier series for F_1 has only finitely many nonzero terms, then so does F_1^N (although the K_1 and K_2 corresponding to F_1^N may be different from those corresponding to F_1), and so on. So it *seems* as though we might be able to smooth out the Haar scaling function, and consequently the Haar wavelet, while *retaining* compact support, by simply taking Nth powers of the functions F_1^H and Φ^H of (8.76). Can we?

NO! The problem is that the identity (8.69) does *not* hold with $(F_1^H)^N$ in place of F_1^H. Too bad—but all is not lost. Let's just look at things sideways, sort of; that is, instead of hoping that the identity holds for powers, let's *take powers of the identity*! Here's what we mean.

Example 8.4.3 Given $M \in \mathbb{Z}^+$, let's raise both sides of the familiar trig identity $\cos^2 \pi s + \sin^2 \pi s = 1$ (which is just the relation (8.69) in the case $F_1 = F_1^H$) to the $(2M-1)$st power and use the binomial theorem (4.133) as follows:

$$1 = (\cos^2 \pi s + \sin^2 \pi s)^{2M-1}$$
$$= \sum_{m=0}^{2M-1} \left(\frac{(2M-1)!}{m!(2M-1-m)!} \right) \sin^{2m} \pi s \cos^{2(2M-1-m)} \pi s = G_1(s) + G_2(s), \tag{8.94}$$

where G_1 is the first half of the sum, that is,

$$G_1(s) = \sum_{m=0}^{M-1} \left(\frac{(2M-1)!}{m!(2M-1-m)!} \right) \sin^{2m} \pi s \cos^{2(2M-1-m)} \pi s, \tag{8.95}$$

and G_2 the second half. Note that both G_1 and G_2 are 1-periodic, since $\sin^2 \pi(s+1) = (-\sin \pi s)^2 = \sin^2 \pi s$, and similarly for $\cos^2 \pi s$.

Let's now observe that

$$G_2(s) = \sum_{m=M}^{2M-1} \left(\frac{(2M-1)!}{m!(2M-1-m)!} \right) \sin^{2m} \pi s \cos^{2(2M-1-m)} \pi s$$
$$= \sum_{n=0}^{M-1} \left(\frac{(2M-1)!}{(2M-1-n)!n!} \right) \sin^{2(2M-1-n)} \pi s \cos^{2n} \pi s$$
$$= \sum_{n=0}^{M-1} \left(\frac{(2M-1)!}{(2M-1-n)!n!} \right) \cos^{2(2M-1-n)} \pi(s+\tfrac{1}{2}) \sin^{2n} \pi(s+\tfrac{1}{2})$$
$$= G_1(s+\tfrac{1}{2}). \tag{8.96}$$

(We used the substitution $n = 2M - 1 - m$, the trig identity (8.77), and the definition of G_1.) So (8.94) gives

$$1 = G_1(s) + G_1(s + \tfrac{1}{2}). \tag{8.97}$$

Now it follows immediately from the definition of G_1 that it enjoys the following properties: $G_1 \in C(\mathbb{T}_1)$; $G_1(0) = 1$; $G_1(s) \neq 0$ for $|s| \leq \tfrac{1}{4}$; G_1 has only finitely many nonzero Fourier coefficients. So if we can only find a nice enough "square root" of G_1, by which we mean a function F_1 having these same properties and such that

$$|F_1(s)|^2 = G_1(s), \tag{8.98}$$

and then a sufficiently well-behaved function Φ such that $\Phi(s) = \Phi(s/2)F_1(s/2)$, then Theorem 8.4.1 will give us our compactly supported scaling function φ and wavelet ψ. Moreover, if the heuristic discussions preceding this example are to be trusted, φ and ψ should be as smooth as we want provided we choose M large enough.

It turns out that such an F_1, and in fact one with real-valued Fourier coefficients, can always be found. See Theorem 7.17 in [11]. The advantage of the $c_k(F_1)$'s being real numbers is that this assures a real-valued scaling function and wavelet. (That real-valued $c_k(F_1)$'s lead to a real-valued φ may be seen by examining the iterative process that gives Φ in terms of the $\widehat{\varphi_N}$'s in the proof in the next section. That a real-valued ψ ensues may be deduced from (8.75).)

We'll content ourselves with an examination of the case $M = 3$; the procedure here is indicative of what works in general. (See Exercise 8.4.7.) In this case, (8.95) gives

$$\begin{aligned} G_1(s) &= \cos^{10} \pi s + 5 \sin^2 \pi s \cos^8 \pi s + 10 \sin^4 \pi s \cos^6 \pi s \\ &= \cos^6 \pi s (\cos^4 \pi s + 5 \sin^2 \pi s \cos^2 \pi s + 10 \sin^4 \pi s). \end{aligned} \tag{8.99}$$

We'd like to be able to write

$$\cos^4 \pi s + 5 \sin^2 \pi s \cos^2 \pi s + 10 \sin^4 \pi s = |1 + b \sin^2 \pi s + ic \sin \pi s \cos \pi s|^2, \tag{8.100}$$

where $b, c \in \mathbb{R}$. If we can do so, then

$$F_1(s) = e^{\pi i s} \cos^3 \pi s (1 + b \sin^2 \pi s + ic \sin \pi s \cos \pi s) \tag{8.101}$$

will have the desired features, as is readily verified. In particular, the fact that $\sin \pi s$, $\cos \pi s$, and $e^{\pi i s}$ change sign under replacement of s by $s+1$ implies the 1-periodicity of F_1. Also, the identities (1.8) for $\sin \theta$ and $\cos \theta$ in terms of $e^{i\theta}$ and $e^{-i\theta}$ assure that the $c_k(F_1)$'s are real when b and c are, and in fact are zero when $k \notin [-2, 3]$. See Exercise 8.4.6.

Let's, then, seek an expression of the form (8.100). Since $\cos^2 \theta = 1 - \sin^2 \theta$, (8.100) and some algebra give

$$1 + 3\sin^2 \pi s + 6\sin^4 \pi s = 1 + (2b + c^2)\sin^2 \pi s + (b^2 - c^2)\sin^4 \pi s. \tag{8.102}$$

So it will certainly suffice to have $2b + c^2 = 3$ and $b^2 - c^2 = 6$; one set of solutions to these two equations is given by

$$b = -1 - \sqrt{10}, \quad c = -\sqrt{5 + 2\sqrt{10}}. \tag{8.103}$$

So

$$F_1(s) = e^{\pi i s} \cos^3 \pi s \left(1 - (1 + \sqrt{10})\sin^2 \pi s - i\sqrt{5 + 2\sqrt{10}} \sin \pi s \cos \pi s \right) \tag{8.104}$$

does the job!

It remains to find a suitable Φ. Here's a tactic that often works once F_1 is in place: We put

$$\Phi(s) = \prod_{m=1}^{\infty} F_1(2^{-m}s). \tag{8.105}$$

Then, formally at least, we have

$$\Phi(0) = \prod_{m=1}^{\infty} F_1(0) = \prod_{m=1}^{\infty} 1 = 1 \tag{8.106}$$

and

$$\Phi\left(\frac{s}{2}\right) F_1\left(\frac{s}{2}\right) = F_1\left(\frac{s}{2}\right) \prod_{m=1}^{\infty} F_1(2^{-m-1}s) = F_1\left(\frac{s}{2}\right) \prod_{k=2}^{\infty} F_1(2^{-k}s)$$

$$= \prod_{k=1}^{\infty} F_1(2^{-k}s) = \Phi(s). \tag{8.107}$$

Of course these calculations assume suitably nice convergence properties of the product defining Φ; that these properties obtain in the present situation and that the resulting Φ is moreover continuous follow from basic facts concerning infinite products. (See, for example, Section VII.5 in [13].)

We still need to show that $\Phi \in L^2(\mathbb{R})$ and $\widehat{\Phi} \in L^1(\mathbb{R})$. These may be established using perhaps less basic but still not especially exotic results from the theory of infinite products; we refer the reader to Theorem 5.5 in [11].

Our compactly supported scaling function φ and wavelet ψ are now obatained according to the specifications of Theorem 8.4.1, and our construction is complete.

The Φ of the above example does not admit simple formulation, but may be well approximated using the fact that

$$\Phi(s) = \lim_{N \to \infty} \operatorname{sinc} \pi(2^{-N}s) \prod_{m=1}^{N} F_1(2^{-m}s). \tag{8.108}$$

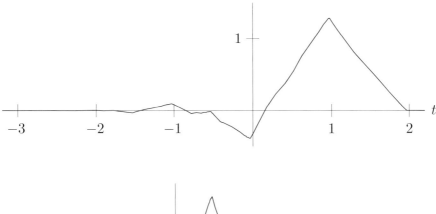

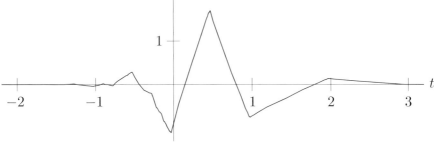

Fig. 8.7 The scaling function φ (top) and wavelet ψ (bottom) of Example 8.4.3 (with $M = 3$)

(That this formula is true pointwise follows immediately from (8.105) and the fact that $\operatorname{sinc} 0 = 1$; in fact, the functions on the right also converge in norm to Φ.) Good pictures of φ and ψ may thereby be obtained. See Figure 8.7.

The pictures reflect the fact (cf. Theorem 8.4.1(c)) that φ is supported on $[-3, 2]$ and ψ on $[-2, 3]$. That φ is *small* on $[-3, -2]$, as is ψ on $[-2, -1]$, is because $c_2(F_1)$ and $c_3(F_1)$ are small compared to the other (nonvanishing) $c_k(F_1)$'s. See Exercise 8.4.6(b).

As the pictures also suggest, φ and ψ are continuous. In fact, they have continuous first, though *not* second, derivatives. Mimicking the above procedure for larger values of M does, as anticipated, yield smoother (but still compactly supported) scaling functions and wavelets. But the improvement is somewhat gradual; M must be about as big as $5J/2$ to guarantee J continuous derivatives of φ and ψ. See [14].

Returning to our case $M = 3$, we see that, because of the factor of $\cos^3 \pi s$ in (8.104), we have $F_1(\frac{1}{2}) = F_1'(\frac{1}{2}) = F_1''(\frac{1}{2}) = 0$. From this and the fact that

$$\Psi(s) = \Phi\left(\frac{s}{2}\right) F_2\left(\frac{s}{2}\right) = \left(\prod_{m=1}^{\infty} F_1(2^{-m-1}s)\right) \overline{e^{\pi i(s+1)/2} F_1\left(\frac{s+1}{2}\right)}, \quad (8.109)$$

we find that $(\widehat{\psi})^{(n)}(0) = \Psi^{(n)}(0) = 0$ for $0 \leq n \leq 2$; this in turn tells us that

$$\int_{-\infty}^{\infty} t^n \psi(t)\, dt = F[t^n \psi(t)](0) = (-2\pi i)^{-n} (\widehat{\psi})^{(n)}(0) = 0 \quad (8.110)$$

for such n. We summarize this by saying ψ has *vanishing first three moments*.

The significance of this is that it makes ψ well suited to *singularity detection*, meaning, in practice, the uncovering and locating of such anomalies as cracks in materials, "scratches" in audio recordings, sudden changes in topography, and so on.

Here's how. Suppose we have a function h that's linear, or even quadratic, except at a single point—say

$$h(t) = \begin{cases} A_1 t^2 + B_1 t + C_1 & \text{if } t < t_0, \\ A_2 t^2 + B_2 t + C_2 & \text{if } t > t_0. \end{cases} \tag{8.111}$$

(Such an h will not be square integrable unless all six of the above coefficients are zero, but no matter. The computations to follow will still make sense, by the compact support and boundedness of ψ. Alternatively, and perhaps more realistically, one can imagine that h is as above but multiplied by a characteristic function.) Then for any given j, k,

$$\langle h, \psi_{j,k} \rangle = 2^{j/2} \int_{-\infty}^{\infty} h(t)\overline{\psi(2^j t - k)}\, dt = 2^{j/2} \int_{2^{-j}(-2+k)}^{2^{-j}(3+k)} h(t)\, \overline{\psi(2^j t - k)}\, dt, \tag{8.112}$$

since ψ being supported on $[-2, 3]$ implies $\psi_{j,k}$ is supported on $[2^{-j}(-2+k), 2^{-j}(3+k)]$. Observe that, if this interval does *not* contain t_0 then, for r equal either to 1 or 2, we have

$$\begin{aligned}\langle h, \psi_{j,k} \rangle &= 2^{j/2} \int_{2^{-j}(-2+k)}^{2^{-j}(3+k)} (A_r t^2 + B_r t + C_r)\overline{\psi(2^j t - k)}\, dt \\ &= 2^{-j/2} \int_{-2}^{3} (A_r(2^{-j}(u+k))^2 + B_r(2^{-j}(u+k)) + C_r)\overline{\psi(u)}\, du = 0, \end{aligned} \tag{8.113}$$

by the vanishing of the first three moments of ψ (and, again, the fact that ψ is supported on $[-2, 3]$).

The moral is this. Let's fix a j, and compute $\langle h, \psi_{j,k} \rangle$ for varying k until we find one for which the inner product is nonzero. Then t_0 *must* belong to the interval $[2^{-j}(-2+k), 2^{-j}(3+k)]$. Since this interval narrows with increasing j, we can get as sharp a bead on t_0 as we would like provided j is large enough.

Of course, if we can see (or otherwise observe) the anomaly directly, then there's no point to the above approach. But a fault in a concrete beam, say, may not be superficially apparent. On the other hand, if the beam is sufficiently ALTICS-like in nature (see Section 7.8), then feeding a signal g *into* the beam gives us an output $g * h$, h being the impulse response of interest. If our input g has the form

$$g(y) = 2^{j/2}\, \overline{\psi(-2^j y)}, \tag{8.114}$$

then the output looks like

$$g * h(y) = 2^{j/2} \int_{-\infty}^{\infty} \overline{g(y-t)} h(t)\, dt = 2^{j/2} \int_{-\infty}^{\infty} h(t)\overline{\psi(2^j t - 2^j y)}\, dt; \tag{8.115}$$

in particular,

$$g * h(2^{-j}k) = 2^{j/2} \int_{-\infty}^{\infty} h(t)\overline{\psi(2^j t - k)}\, dt = \langle h, \psi_{j,k} \rangle. \quad (8.116)$$

So, by recording this output, we *can* get a handle on the wavelet coefficients $\langle h, \psi_{j,k} \rangle$, and therefore, by the above discussions, on the location of the fault! (If the transfer function is better modeled, away from the fault, by a higher degree polynomial, then one should use a wavelet with a higher number of vanishing moments. See Exercise 8.4.7(b).)

These and related features of wavelets have important implications for faults occuring in other contexts, such as that of San Andreas. Indeed, wavelets were first introduced to the real world, and vice versa, by seismologists.

Exercises

In the following exercises, all quantities are as in Theorem 8.4.1.

8.4.1 Let $F_1 = F_1^H$ and $\Phi = \Phi^H$ be as in (8.76).
  a. Compute the Fourier series for F_1.
  b. Show that F_1 and Φ satisfy the hypotheses of Theorem 8.4.1.
  c. Compute F_2 (cf. (8.73)) and its Fourier series; also compute Ψ (cf. (8.74)).
  d. Show that the prescriptions $\varphi = \widehat{\Phi^-}$ and $\psi = \widehat{\Psi^-}$ of the theorem give the Haar scaling function $\varphi = \varphi^H$ and Haar wavelet $\psi = \psi^H$, respectively.
  e. Verify that the identities (8.72) and (8.75) are satisfied. (For the latter, check both equalities given there.)

8.4.2 Note by (8.79) that, for $f \in V_j$,

$$f = \sum_{k=-\infty}^{\infty} \langle f, \varphi_{j,k} \rangle \varphi_{j,k} = \sum_{k=-\infty}^{\infty} \langle f, \psi_{j-1,k} \rangle \psi_{j-1,k} + \sum_{k=-\infty}^{\infty} \langle f, \varphi_{j-1,k} \rangle \varphi_{j-1,k}. \quad (*)$$

This exercise shows how to pass back and forth between these two representations of f.

  a. Derive the *decomposition formulas* for $f \in V_j$:

$$\langle f, \psi_{j-1,\ell} \rangle = \sqrt{2} \sum_{m=-\infty}^{\infty} (-1)^m \overline{c_{m-2\ell-1}(F_1)} \langle f, \varphi_{j,m} \rangle,$$

$$\langle f, \varphi_{j-1,\ell} \rangle = \sqrt{2} \sum_{m=-\infty}^{\infty} \overline{c_{2\ell-m}(F_1)} \langle f, \varphi_{j,m} \rangle.$$

Hint: For the second formula, use (8.49) and the scaling relation (8.72) to write $\varphi_{j-1,\ell}$ as a linear combination of $\varphi_{j,2\ell+k}$'s. Make a substitution in the index of summation, then take inner products with f throughout, to finish. Similarly, for the first formula, use (8.49) and (8.75).

b. Put $f = \varphi_{j,k}$ into the above decomposition formulas to deduce that

$$\langle \varphi_{j,k}, \psi_{j-1,\ell} \rangle = \sqrt{2}(-1)^k \overline{c_{k-2\ell-1}(F_1)}, \quad \langle \varphi_{j,k}, \varphi_{j-1,\ell} \rangle = \sqrt{2}\,\overline{c_{2\ell-k}(F_1)}.$$

c. Derive the *reconstruction formula*

$$\langle f, \varphi_{j,\ell} \rangle = \sqrt{2} \sum_{k=-\infty}^{\infty} \left((-1)^\ell \overline{c_{\ell-2k-1}(F_1)} \langle f, \psi_{j-1,k} \rangle + \overline{c_{2k-\ell}(F_1)} \langle f, \varphi_{j-1,k} \rangle \right).$$

Hint: Take the inner product with $\varphi_{j,\ell}$ of both sides of the second equality in (*), directly above. Use the formulas from part b of this exercise (with the roles of k and ℓ reversed) to finish.

8.4.3 Let $f = \chi_{[-1/2,-1/4]} + 2\chi_{[-1/4,0]} - \chi_{[0,1/2]}$. Also let $(\mathcal{V}^H, \varphi^H)$ and ψ^H denote the Haar multiresolution analysis and wavelet, as usual.

a. Sketch the graph of f.
b. Show that $f \in V_2^H$.
c. Use the decomposition formulas of the previous exercise to obtain explicitly the wavelet decomposition $f = f_1^+ + f_0^+ + f_0$, where $f_1^+ \in W_1^H$; $f_0^+ \in W_0^H$; $f_0 \in V_0^H$. (The W_j^H's are defined in terms of the V_j^H's in the usual way—that is, $V_j^H = W_{j-1}^H \oplus V_{j-1}^H$.) Sketch the graphs of f_1^+, f_0^+, and f_0.
d. Build f back up from f_1^+, f_0^+, and f_0 using the reconstruction formula of the previous exercise. (Make sure you really do get back f!)

8.4.4 Let $f \in L^2(\mathbb{R})$ be bounded and continuous. For a given integer $m \in [0, 2^j - 1]$, define $S_{j,m}(f) = 2^{-j/2} f(m/2^j)$. Show that $S_{j,m}(f) \approx \langle f, \varphi_{j,m} \rangle$, the approximation improving as j increases. Hint: Arguing much as in (8.114)–(8.116), show that

$$2^{j/2} \langle f, \varphi_{j,m} \rangle = f * \left(\overline{\varphi^-} \right)_{[2^{-j}]} (m/2^j),$$

where as usual $g_{[\varepsilon]}(y)$ denotes $\varepsilon^{-1} g(\varepsilon^{-1} y)$. Now use (8.70) to show that $\overline{\varphi^-}$ satisfies the hypotheses of Proposition 5.7.1; apply part (a) of that proposition.

8.4.5 Write down, explicitly, the F_1, Φ, and Ψ of Example 8.4.2 in the case $N = 2$. You'll need to evaluate the series

$$\Lambda(s) = \left(\sum_{k=-\infty}^{\infty} |(e^{-\pi i(s+k)} \operatorname{sinc}^2 \pi(s+k))|^2 \right)^{1/2};$$

for this, see Exercise 3.7.2.

8.4.6 a. Show that, if $b, c \in \mathbb{R}$, then the function F_1 of (8.101) has real Fourier coefficients $c_k(F_1)$ and that these coefficients vanish for $k \notin [-2, 3]$.
b. Find these $c_k(F_1)$'s explicitly in the specific case of Daubechies' scale factor (8.104).

8.4.7 a. Beginning with the *third* power of the identity $\cos^2 \pi s + \sin^2 \pi s = 1$, meaning the case $M = 2$ of the situation described in Example 8.4.3, use the ideas of that example to find a scale factor F_1, a scaling function φ, and an associated wavelet ψ such that the latter is real-valued and compactly supported and has vanishing first *two* moments. (Write down F_1 and its Fourier coefficients explicitly; describe φ and ψ in terms of F_1.)

b. Repeat the above in the case $M = 4$ (so that, this time, ψ will have vanishing first *four* moments). Warning: This gets a bit messy.

8.5 PROOF OF THEOREM 8.4.1

We begin with:

Lemma 8.5.1 *Let $f \in L^2(\mathbb{R})$. Then*

(a) $\{f_{0,k} : k \in \mathbb{Z}\}$ *is an orthonormal set if and only if*

$$1 = \sum_{m=-\infty}^{\infty} |\widehat{f}(s+m)|^2 = \sum_{L=0}^{1} \sum_{\ell=-\infty}^{\infty} |\widehat{f}(s+2\ell+L)|^2 \tag{8.117}$$

for almost all $s \in \mathbb{R}$.

(b) *Suppose also that $g \in L^2(\mathbb{R})$. Then each $f_{0,k}$ is orthogonal to each $g_{0,n}$ if and only if*

$$0 = \sum_{m=-\infty}^{\infty} \widehat{f}(s+m)\overline{\widehat{g}(s+m)} = \sum_{L=0}^{1} \sum_{\ell=-\infty}^{\infty} \widehat{f}(s+2\ell+L)\overline{\widehat{g}(s+2\ell+L)} \tag{8.118}$$

for almost all $s \in \mathbb{R}$.

Proof. The second equality in either part follows from the representation $\mathbb{Z} = \{2\ell : \ell \in \mathbb{Z}\} \cup \{2\ell + 1 : \ell \in \mathbb{Z}\}$.

We next consider the first equality in part (b). The $f_{0,k}$'s are orthogonal to the $g_{0,n}$'s if and only if, for all $k, n \in \mathbb{Z}$, $\langle f_{0,k}, g_{0,n}\rangle = 0$. But

$$\langle f_{0,k}, g_{0,n}\rangle = \langle \widehat{f_{0,k}}, \widehat{g_{0,n}}\rangle = \langle e_{-k}\widehat{f}, e_{-n}\widehat{g}\rangle$$

$$= \int_{-\infty}^{\infty} e_{n-k}(v)\widehat{f}(v)\overline{\widehat{g}(v)}\, dv = \sum_{m=-\infty}^{\infty} \int_{m}^{m+1} e_{n-k}(v)\widehat{f}(v)\overline{\widehat{g}(v)}\, dv$$

$$= \sum_{m=-\infty}^{\infty} \int_{0}^{1} e_{n-k}(s+m)\widehat{f}(s+m)\overline{\widehat{g}(s+m)}\, ds$$

$$= \int_{0}^{1} e^{-2\pi i(k-n)s} \sum_{m=-\infty}^{\infty} \widehat{f}(s+m)\overline{\widehat{g}(s+m)}\, ds. \tag{8.119}$$

We used the Plancherel theorem for the first equality, (8.53) for the second, the definition of the inner product for the third, a breaking up of $\mathbb{R}$ into disjoint intervals of length one for the fourth, the substitution $v = s + m$ for the fifth, and an exchange of a sum with an integral for the last. This exchange is valid because $f, g \in L^2(\mathbb{R})$ (cf. the corollary on p. 307 of [29]).

Now (8.119) says that the series in the rightmost integrand has $(k-n)$th Fourier coefficient equal to $\langle f_{0,k}, g_{0,n} \rangle$. So the $f_{0,k}$'s are orthogonal to the $g_{0,n}$'s if and only if this series has $(k-n)$th Fourier coefficient equal to zero. By uniqueness of Fourier expansions (that is, by the $L^1(\mathbb{T}_1)$ analog of Corollary 3.6.1), this is true if and only if the series equals zero almost everywhere.

So part (b) is proved. For the remaining equality in part (a) we proceed similarly, replacing $g_{0,n}$ with $f_{0,n}$ throughout (8.119), noting that the $f_{0,k}$'s are orthonormal if and only if $\langle f_{0,k}, f_{0,n} \rangle = \delta_{k,n} = \delta_{k-n,0}$ for all k, n, and observing that the only element of $L^1(\mathbb{R})$ with $(k-n)$th Fourier coefficient equal to $\delta_{k-n,0}$ is the constant element 1. $\square$

We now prove Theorem 8.4.1. We consider part (a) first. To demonstrate that the $\varphi_{0,k}$'s form an orthonormal basis for V_0, it suffices, by the definition (8.71) of V_0, to show that they form an orthonormal *set*. (Recall Exercise 3.10.3 and Corollary 3.7.2.)

To do this, we construct auxiliary functions φ_N ($N \in \mathbb{N}$) that converge in norm to φ as $N \to \infty$ and such that $\{(\varphi_N)_{0,k} : k \in \mathbb{Z}\}$ is an orthonormal set for each N. The $(\varphi_N)_{0,k}$'s will then converge in norm to $\varphi_{0,k}$ for each k. And because, for $k, n \in \mathbb{Z}$,

$$|\langle \varphi_{0,k}, \varphi_{0,n} \rangle - \langle (\varphi_N)_{0,k}, (\varphi_N)_{0,n} \rangle| \qquad (8.120)$$

$$\leq A \|\varphi_{0,n} - (\varphi_N)_{0,n}\| + B \|\varphi_{0,k} - (\varphi_N)_{0,k}\| \qquad (8.121)$$

for some $A, B > 0$ (recall the arguments leading to (3.70)), the desired orthonormality of the $\varphi_{0,k}$'s will follow.

We define φ_N by way of its Fourier transform. We first put

$$\widehat{\varphi_0}(s) = e_a(s) \operatorname{sinc} \pi s, \qquad (8.122)$$

where a is, for the moment, arbitrary. (The choice of a will concern us in our proof of part (c).) Then for any given s,

$$\sum_{m=-\infty}^{\infty} |\widehat{\varphi_0}(s+m)|^2 = 1, \qquad (8.123)$$

as we saw in (8.93). So the $(\varphi_0)_{0,k}$'s are orthonormal by Lemma 8.5.1(a).

For $N \in \mathbb{N}$, we recursively set

$$\widehat{\varphi_{N+1}}(s) = \widehat{\varphi_N}\left(\frac{s}{2}\right) F_1\left(\frac{s}{2}\right) \qquad (8.124)$$

and let $S(N)$ be the statement that $\{(\varphi_N)_{0,k} : k \in \mathbb{Z}\}$ is an orthonormal set. We've just proved $S(0)$. So let's assume $S(N)$; if we can deduce $S(N+1)$ from this, we'll have the desired demonstration of $S(N)$ for all N, by mathematical induction.

PROOF OF THEOREM 8.4.1

We first note that each φ_N is indeed in $L^2(\mathbb{R})$, by (8.124) and the continuity (whence boundedness) of F_1. Now

$$\sum_{L=0}^{1}\sum_{\ell=-\infty}^{\infty}\left|\widehat{\varphi_{N+1}}(s+2\ell+L)\right|^2 = \sum_{L=0}^{1}\sum_{\ell=-\infty}^{\infty}\left|\widehat{\varphi_N}\left(\frac{s+L}{2}+\ell\right)F_1\left(\frac{s+L}{2}+\ell\right)\right|^2$$

$$= \sum_{L=0}^{1}\left|F_1\left(\frac{s+L}{2}\right)\right|^2 \sum_{\ell=-\infty}^{\infty}\left|\widehat{\varphi_N}\left(\frac{s+L}{2}+\ell\right)\right|^2. \tag{8.125}$$

(We've applied (8.124) and used the periodicity of F_1.) If both sums over ℓ on the right equal 1 then, by (8.69), so does the sum on the left. Or in other words, by Lemma 8.5.1(a), $S(N) \Rightarrow S(N+1)$, as required.

We still need to show that $\lim_{N\to\infty}\|\varphi_N - \varphi\| = 0$; by the Plancherel theorem, it suffices to show that $\lim_{N\to\infty}\|\widehat{\varphi_N} - \Phi\| = 0$. To this end we note that, for $N \in \mathbb{N}$,

$$\Phi(s) = \Phi(2^{-N}s)\prod_{m=1}^{N} F_1(2^{-m}s) \tag{8.126}$$

by the first identity in (8.70), while

$$\widehat{\varphi_N}(s) = \widehat{\varphi_0}(2^{-N}s)\prod_{m=1}^{N} F_1(2^{-m}s) \tag{8.127}$$

by (8.124). Combining these equations gives

$$\widehat{\varphi_N}(s) = \frac{\widehat{\varphi_0}(2^{-N}s)\Phi(s)}{\Phi(2^{-N}s)} = \frac{\Phi(s)}{\Phi(2^{-N}s)} e_a(2^{-N}s)\operatorname{sinc}\pi(2^{-N}s); \tag{8.128}$$

the assumption $\Phi(0) = 1$ and the fact that $e_a(0)\operatorname{sinc} 0 = 1$ then imply that the $\widehat{\varphi_N}$'s converge *pointwise* to Φ. To get norm convergence we need only, by the Lebesgue dominated convergence theorem, produce a function $h \in L^2(\mathbb{R})$ such that $|\widehat{\varphi_N}(s) - \Phi(s)| \le |h(s)|$ for all s and N.

By (8.126) and the assumption that F_1 is continuous and nonzero on the interval $[-\frac{1}{4}, \frac{1}{4}]$, we find that Φ is continuous and nonzero on $[-\frac{1}{2}, \frac{1}{2}]$. (Were Φ equal to zero at some point s_0 on that interval, then by (8.126) it would be zero at $2^{-N}s_0$ for any N, contradicting the continuity of Φ and the assumption $\Phi(0) = 1$.) So $|\Phi|$ attains a minimum $C > 0$ on this interval; consequently $|\Phi(2^{-N})s| \ge C$ for $-2^{N-1} \le s \le 2^{N-1}$. So by (8.128), $|\widehat{\varphi_N}(s)| \le C^{-1}|\Phi(s)|$, so that $|\widehat{\varphi_N}(s) - \Phi(s)| \le (1 + C^{-1})|\Phi(s)|$ for all s and N, as needed.

So our present φ and V_0 do indeed satisfy the orthonormality condition (v) of Definition 8.3.2. That the dilation condition is met follows from our definition of the V_j's. To show nestedness, it suffices, as noted earlier, to show that φ is a norm convergent linear combination of $\varphi_{1,k}$'s, which we accomplish as follows. Writing out F_1 in a Fourier series

$$F_1 = \sum_{k=-\infty}^{\infty} c_k(F_1)e_k, \qquad (8.129)$$

we find from (8.70) that

$$\Phi = \sqrt{2} \sum_{k=-\infty}^{\infty} c_k(F_1)e_{k/2}\Phi_{-1,0}. \qquad (8.130)$$

Now some Fourier transform calculus involving (6.35) and (8.53) shows that, for general $f \in L^2(\mathbb{R})$,

$$F[e_{k/2}(s)f_{-1,0}(s)]^- = (\widehat{f^-})_{1,-k} \qquad (8.131)$$

(see Exercise 8.5.1), so taking the inverse Fourier transform of each side of (8.130) gives

$$\varphi = \widehat{\Phi^-} = \sqrt{2}\sum_{k=-\infty}^{\infty} c_k(F_1)(\widehat{\Phi^-})_{1,-k} = \sqrt{2} \sum_{k=-\infty}^{\infty} c_{-k}(F_1)\varphi_{1,k}. \qquad (8.132)$$

(The norm convergence of the series on the right follows from the fact that F_1, being continuous on $\mathbb{T}_1$, is square integrable there; so its sequence of Fourier coefficients is square summable.) In particular, the scaling equation (8.72) holds.

To complete our proof of part (a), it remains to demonstrate separation and density, under the stated assumptions. For the first of these it suffices to show that, for any $f \in L^2(\mathbb{R})$,

$$\lim_{j \to -\infty} \sum_{k=-\infty}^{\infty} |\langle f, \varphi_{j,k}\rangle|^2 = 0. \qquad (8.133)$$

Why does this suffice? Because if $f \in V_j$ then, by Parseval's equation and the fact that the $\varphi_{j,k}$'s ($k \in \mathbb{Z}$) form an orthonormal basis for V_j, the sum in (8.133) equals the square of the norm of f. So if (8.133) is true, then

$$f \in \bigcap_{j=-\infty}^{\infty} V_j \Rightarrow \lim_{j\to-\infty} ||f||^2 = 0 \Rightarrow ||f||^2 = 0 \Rightarrow f = 0 \qquad (8.134)$$

(almost everywhere), so that separation holds.

Now the space of step functions is dense in $L^2(\mathbb{R})$ (see the proof of Proposition 8.3.3), so in fact, it will be enough to demonstrate (8.134) for f equal to a characteristic function $\chi_{(a,b)}$ whenever $-\infty < a < b < \infty$. Let's do this. We have

$$|\langle \chi_{(a,b)}, \varphi_{j,k}\rangle|^2 = |\langle \chi_{(a,b)}, \chi_{(a,b)}\varphi_{j,k}\rangle|^2 \le ||\chi_{(a,b)}||^2 \, ||\chi_{(a,b)}\varphi_{j,k}||^2$$

$$= (b-a)\int_{-\infty}^{\infty} |2^{j/2}\chi_{(a,b)}(t)\varphi(2^j t - k)|^2 \, dt$$

$$= (b-a)\int_{-\infty}^{\infty} \chi_{(2^j a - k, 2^j b - k)}(u)|\varphi(u)|^2 \, du. \qquad (8.135)$$

PROOF OF THEOREM 8.4.1

(We used the fact that $\chi_{(a,b)}(2^{-j}(u+k)) = \chi_{[2^j a-k, 2^j b-k]}(u)$.) Now we are concerned with the limit as $j \to -\infty$, so we may as well assume j to be a negative number of sufficiently large absolute value that the intervals $[2^j a - k, 2^j b - k]$ ($k \in \mathbb{Z}$) are disjoint. (This will certainly be the case if $2^j(b-a) < 1$.) Then (8.135) gives

$$\sum_{k=-\infty}^{\infty} |\langle \chi_{(a,b)}, \varphi_{j,k} \rangle|^2 \le (b-a) \int_{-\infty}^{\infty} \chi_{S_j(a,b)}(u) |\varphi(u)|^2 \, du, \qquad (8.136)$$

where

$$S_j(a,b) = \bigcup_{k=-\infty}^{\infty} (2^j a - k, 2^j b - k). \qquad (8.137)$$

Now the intersection, over all j, of the $S_j(a,b)$'s is $\cup_{k=-\infty}^{\infty}(-k,-k)$, which is empty, so the pointwise limit as $j \to -\infty$ of the integrand in (8.136) is zero. This integrand is bounded by $|\varphi(u)|^2$; since $\varphi \in L^2(\mathbb{R})$, Proposition 3.5.1 tells us that the limit as $j \to -\infty$ of the right side of (8.136) is zero. So we have separation.

Next we consider density. Let $f \in L^2(\mathbb{R})$; it will be enough to show that

$$\sum_{k=-\infty}^{\infty} \langle f, \varphi_{j,k} \rangle \varphi_{j,k}, \qquad (8.138)$$

which belongs to V_j, converges in norm to f as $j \to \infty$. By orthonormality of the $\varphi_{j,k}$'s (for a fixed j) and by (3.148), this is equivalent to showing that

$$||f||^2 = \lim_{j \to \infty} \lim_{N \to \infty} \sum_{k=-N}^{N} |\langle f, \varphi_{j,k} \rangle|^2. \qquad (8.139)$$

Moreover, as the space of step functions is, again, dense in $L^2(\mathbb{R})$, so is the space of functions whose Fourier transforms are step functions (by the Plancherel theorem); so it will in turn suffice to demonstrate (8.139) for $\hat{f}$ equal to a characteristic function $\chi_{(a,b)}$, whenever $-\infty < a < b < \infty$.

But for such f we have, by the Plancherel theorem and (8.53),

$$\langle f, \varphi_{j,k} \rangle = \langle \chi_{(a,b)}, \widehat{\varphi_{j,k}} \rangle = \langle \chi_{(a,b)}, e_{-k/2^j}(\hat{\varphi})_{-j,0} \rangle = \langle \chi_{(a,b)}, e_{-k/2^j} \Phi_{-j,0} \rangle$$

$$= 2^{-j/2} \int_a^b \overline{\Phi(2^{-j}u)} e_{k/2^j}(u) \, du; \qquad (8.140)$$

so

$$\sum_{k=-N}^{N} |\langle f, \varphi_{j,k} \rangle|^2$$

$$= 2^{-j} \sum_{k=-N}^{N} \int_a^b \overline{\Phi(2^{-j}u)} e_{k/2^j}(u) \int_a^b \Phi(2^{-j}v) e_{-k/2^j}(v) \, dv \, du$$

$$= \int_a^b \overline{\Phi(2^{-j}u)} \left(\sum_{k=-N}^{N} e_{k/2^j}(u) \frac{1}{2^j} \int_a^b \Phi(2^{-j}v) e_{k/2^j}(-v) \, dv \right) du. \qquad (8.141)$$

(For the first step, we used (8.140) and Fubini's theorem to write $|\langle f, \varphi_{j,k}\rangle|^2$ as a double integral; this is readily justified under the given assumptions on φ.)

Now we're concerned with the limit as $j \to \infty$, so we may as well take j large enough that $-2^{j-1} < a < b < 2^{j-1}$. But then the quantity in parentheses in (8.141) is just the Nth partial sum $S_N^P(u)$ of the Fourier series for the function $P_j \in L^2(-2^{j-1}, 2^{j-1})$ defined by

$$P_j(v) = \Phi(2^{-j}v)\chi_{(a,b)}(v). \tag{8.142}$$

These partial sums converge in norm to P_j as $N \to \infty$; from the Cauchy-Schwarz inequality, it follows that $\langle S_N^{P_j}, P_j\rangle$ converges to $\langle P_j, P_j\rangle$ as $N \to \infty$. So (8.141) implies

$$\begin{aligned}\lim_{j\to\infty}\lim_{N\to\infty}\sum_{k=-N}^{N}|\langle f, \varphi_{j,k}\rangle|^2 &= \lim_{j\to\infty}\lim_{N\to\infty}\int_a^b \overline{\Phi(2^{-j}u)}S_N^{P_j}(u)\,du \\ &= \lim_{j\to\infty}\lim_{N\to\infty}\langle S_N^{P_j}, P_j\rangle = \lim_{j\to\infty}\langle P_j, P_j\rangle \\ &= \lim_{j\to\infty}\int_a^b |\Phi(2^{-j}v)|^2\,dv = \int_a^b |\Phi(0)|^2\,dv \\ &= (b-a)|\Phi(0)|^2 = b-a = ||\chi_{[b-a]}||^2 = ||f||^2.\end{aligned} \tag{8.143}$$

(We used the Lebesgue dominated convergence theorem, readily justified in the present situation; the fact that $\Phi(0) = 1$; and the Plancherel theorem.) So density holds and we are, *finally*, done with our proof of part (a).

On to part (b). It will be convenient to write

$$\Omega_1 = \Phi, \quad \omega_1 = \widehat{\Omega_1^-} = \varphi, \quad \Omega_2 = \Psi, \quad \omega_2 = \widehat{\Omega_2^-} = \psi, \tag{8.144}$$

so that (8.70) and (8.74) read

$$\Omega_q(s) = \Omega_1\left(\frac{s}{2}\right)F_q\left(\frac{s}{2}\right) = \widehat{\omega_1}\left(\frac{s}{2}\right)F_q\left(\frac{s}{2}\right) \quad (q \in \{1,2\}). \tag{8.145}$$

We then have

$$\begin{aligned}&\sum_{L=0}^{1}\sum_{\ell=-\infty}^{\infty}\widehat{\omega_2}(s+2\ell+L)\overline{\widehat{\omega_q}(s+2\ell+L)} \\ &= \sum_{L=0}^{1}\sum_{\ell=-\infty}^{\infty}\Omega_2(s+2\ell+L)\overline{\Omega_q(s+2\ell+L)} \\ &= \sum_{L=0}^{1}\sum_{\ell=-\infty}^{\infty}\left|\Omega_1\left(\frac{s+L}{2}+\ell\right)\right|^2 F_2\left(\frac{s+L}{2}+\ell\right)\overline{F_q\left(\frac{s+L}{2}+\ell\right)}.\end{aligned} \tag{8.146}$$

Now it follows readily from the relation (8.69) concerning F_1 and from the definition (8.73) of F_2 that

$$\sum_{L=0}^{1} F_p\left(s+\frac{L}{2}\right)\overline{F_q\left(s+\frac{L}{2}\right)} = \delta_{p,q} \quad (p,q \in \{0,1\}, s \in \mathbb{R}). \tag{8.147}$$

This, (8.146), the periodicity of F_1 and F_2, the already established orthonormality of the functions $\varphi_{0,k} = (\omega_1)_{0,k}$, and Lemma 8.5.1(a) then give

$$\sum_{L=0}^{1}\sum_{\ell=-\infty}^{\infty} \widehat{\omega_2}(s+2\ell+L)\overline{\widehat{\omega_q}(s+2\ell+L)}$$

$$= \sum_{L=0}^{1} F_2\left(\frac{s+L}{2}\right)\overline{F_q\left(\frac{s+L}{2}\right)} \sum_{\ell=-\infty}^{\infty}\left|\Omega_1\left(\frac{s+L}{2}+\ell\right)\right|^2$$

$$= \sum_{L=0}^{1} F_2\left(\frac{s+L}{2}\right)\overline{F_q\left(\frac{s+L}{2}\right)} = \delta_{2,q} \tag{8.148}$$

almost everywhere. The case $q = 2$ of this and Lemma 8.5.1(a) tell us that the functions $(\omega_2)_{0,k} = \psi_{0,k}$ are orthonormal; the case $q = 1$ and part (b) of that lemma tell us that every $(\omega_2)_{0,k} = \psi_{0,k}$ is orthogonal to every $(\omega_1)_{0,n} = \varphi_{0,n}$.

We still need to demonstrate that $\{(\omega_q)_{0,n} : q \in \{1,2\}, n \in \mathbb{Z}\}$ is an orthonormal basis for V_1. To do this, it will be sufficient, by the converse to Parseval's equation (that is, by the implication (ii)$\Rightarrow$(i) of Corollary 3.7.2), to show that

$$\sum_{q=1}^{2}\sum_{n=-\infty}^{\infty} |\langle f, (\omega_q)_{0,n}\rangle|^2 = \|f\|^2 \tag{8.149}$$

for all $f \in V$. In fact, since the $\varphi_{1,m}$'s do constitute an orthonormal basis for V_1, it will be enough to show that (8.149) holds for each $f = \varphi_{1,m}$.

Let's choose an m and write $m = 2k + p$, where p equals 1 if m is odd and 2 if m is even. By (8.49),

$$\varphi_{1,m} = (\varphi_{1,p})_{0,k}, \tag{8.150}$$

so by (8.119) with $f = \varphi_{1,p}$ and $g = \omega_q$,

$$\langle \varphi_{1,m}, (\omega_q)_{0,n}\rangle = \langle(\varphi_{1,p})_{0,k}, (\omega_q)_{0,n}\rangle$$

$$= \sum_{L=0}^{1}\int_0^1 e^{-2\pi i(k-n)s} \sum_{\ell=-\infty}^{\infty} \widehat{\varphi_{1,p}}(s+2\ell+L)\overline{\widehat{\omega_q}(s+2\ell+L)}\,ds$$

$$= \sum_{L=0}^{1}\int_L^{L+1} e^{-2\pi i(k-n)u} \sum_{\ell=-\infty}^{\infty} \widehat{\varphi_{1,p}}(u+2\ell)\overline{\widehat{\omega_q}(u+2\ell)}\,du$$

$$= \int_0^2 e^{-2\pi i(k-n)u} \sum_{\ell=-\infty}^{\infty} \widehat{\varphi_{1,p}}(u+2\ell)\overline{\widehat{\omega_q}(u+2\ell)}\,du \tag{8.151}$$

(for the second step, we put $u = s + L$ and used the fact that $e^{2\pi i(n-k)L} = 1$). But

$$\sum_{\ell=-\infty}^{\infty} \widehat{\varphi_{1,p}}(u+2\ell) \overline{\widehat{\omega_q}(u+2\ell)} = \sum_{\ell=-\infty}^{\infty} e_{-p/2}(u+2\ell)(\widehat{\varphi})_{-1,0}(u+2\ell) \overline{\widehat{\omega_q}(u+2\ell)}$$

$$= e_{-p/2}(u) \sum_{\ell=-\infty}^{\infty} (\Omega_1)_{-1,0}(u+2\ell) \overline{\Omega_q(u+2\ell)}$$

$$= \frac{1}{\sqrt{2}} e_{-p/2}(u) \sum_{\ell=-\infty}^{\infty} \Omega_1\left(\frac{u}{2}+\ell\right) \overline{\Omega_q(u+2\ell)}$$

$$= \frac{1}{\sqrt{2}} e_{-p/2}(u) \overline{F_q\left(\frac{u}{2}\right)} \sum_{\ell=-\infty}^{\infty} \left|\Omega_1\left(\frac{u}{2}+\ell\right)\right|^2$$

$$= \frac{1}{\sqrt{2}} e_{-p/2}(u) \overline{F_q\left(\frac{u}{2}\right)}, \qquad (8.152)$$

the first equality by (8.53); the second by definition of Ω_1 and Ω_2; the third by definition of $f_{-1,0}$; the fourth by the scaling equation (8.145) for Ω_q, together with the periodicity of F_q; and the last by the orthonormality of the functions $(\omega_1)_{0,k} = \varphi_{0,k}$, combined with Lemma 8.5.1(a). So (8.151) gives

$$\langle \varphi_{1,m}, (\omega_q)_{0,n} \rangle = \frac{1}{\sqrt{2}} \int_0^2 e^{-2\pi i(k-n)u} e^{-\pi i p u} \overline{F_q\left(\frac{u}{2}\right)} du$$

$$= \sqrt{2} \int_0^1 e^{-2\pi i(m-2n)x} \overline{F_q(x)} \, dx = \sqrt{2}\, c_{m-2n}\left(\overline{F_q}\right) \qquad (8.153)$$

(we substituted $x = u/2$ and used the fact that $2k + p = m$).

We observe that, by (8.73),

$$c_j\left(\overline{F_2}\right) = \int_0^1 \overline{F_1(s+\tfrac{1}{2}) e^{2\pi i(s+1/2)}} e^{-2\pi i j s} \, ds$$

$$= (-1)^j \int_{1/2}^{3/2} \overline{F_1(u)} e^{-2\pi i(j-1)u} \, du = (-1)^j c_{j-1}(\overline{F_1})$$

$$= (-1)^j \overline{c_{1-j}(F_1)} \qquad (8.154)$$

(we substituted $u = s + \tfrac{1}{2}$ and noted that $e^{\pi i j} = (-1)^j$), so by (8.153) and Parseval's equation,

$$\sum_{q=1}^{2} \sum_{n=-\infty}^{\infty} |\langle \varphi_{1,m}, (\omega_q)_{0,n} \rangle|^2 = 2 \sum_{n=-\infty}^{\infty} \left(|c_{m-2n}(\overline{F_1})|^2 + |c_{1-m+2n}(\overline{F_1})|^2 \right)$$

$$= 2 \|F_1\|^2. \qquad (8.155)$$

(For m fixed and even, $m - 2n$ ranges over all even integers and $1 - m + 2n$ over all odd integers as n ranges over $\mathbb{Z}$. Similarly, for m fixed and odd, $m - 2n$ gives all odd integers and $1 - m + 2n$ all even ones.) But by a simple change of variable and the periodicity of F_1,

$$||F_1||^2 = \int_0^1 |F_1(s)|^2 \, ds = \int_0^1 |F_1(u + \tfrac{1}{2})|^2 \, du, \tag{8.156}$$

so that

$$2||F_1||^2 = \int_0^1 \left(|F_1(u)|^2 + |F_1(u + \tfrac{1}{2})|^2\right) du = \int_0^1 du = 1, \tag{8.157}$$

so (8.155) gives

$$\sum_{q=1}^{2} \sum_{n=-\infty}^{\infty} |\langle \varphi_{1,m}, (\omega_q)_{0,n} \rangle|^2 = 1 = ||\varphi_{1,m}||^2, \tag{8.158}$$

which *is* (8.149) in the case $f = \varphi_{1,m}$.

To complete our proof of part (b)(i), we must show that the formulas (8.75) hold. But the first formula there is derived almost exactly as was the scaling equation (8.61); the second one follows from the first one and (8.154).

For part (b)(ii), we show first that $\{\psi_{j,k} : j, k\}$ is an orthonormal set. We've already seen the $\psi_{0,k}$'s to be orthonormal to each other; so it will be sufficient, by dilation and translation arguments, to show that $\psi_{-1,0}$ is orthonormal to any $\psi_{0,n}$.

But ψ is in V_1, so $\psi_{-1,0}$ is in V_0. Since the $\varphi_{0,k}$'s form an orthonormal basis for V_0 and since each $\varphi_{0,k}$ is, as already seen, orthogonal to each $\psi_{0,n}$, so is $\psi_{-1,0}$, as required.

We next show that the $\psi_{j,k}$'s form an orthonormal *basis* for $L^2(\mathbb{R})$, by showing that

$$f_\psi = f - \sum_{j,k \in \mathbb{Z}} \langle f, \psi_{j,k} \rangle \psi_{j,k} = 0 \tag{8.159}$$

for all $f \in L^2(\mathbb{R})$. It will suffice, by density and dilation, to show that this is true for any $f \in V_0$.

We observe that, by orthogonality of the $\psi_{j,k}$'s,

$$\langle f_\psi, \psi_{m,n} \rangle = \langle f, \psi_{m,n} \rangle - \sum_{j,k \in \mathbb{Z}} \langle f, \psi_{j,k} \rangle \langle \psi_{j,k}, \psi_{m,n} \rangle = \langle f, \psi_{m,n} \rangle - \langle f, \psi_{m,n} \rangle$$

$$= 0 \tag{8.160}$$

for any m, n; so f_ψ is orthogonal to all $\psi_{m,n}$'s. So f_ψ is orthogonal to $W_{-\ell}$ for all ℓ. But by (8.80),

$$V_0 = \left(\bigoplus_{\ell=1}^{J} W_{-\ell}\right) \oplus V_{-J} \tag{8.161}$$

for any $J \in \mathbb{Z}$. So $f \in V_{-J}$ for all J. The intersection of the V_{-J}'s is zero by separation, and we're done with part (b).

For part (c), suppose K_1, K_2, and the $c_k(F_1)$'s are as hypothesized there. To see that $\varphi(t) = 0$ for almost all $t \notin [-K_2, -K_1]$, we first take inverse Fourier transforms on either side of the recurrence formula (8.124) for the $\widehat{\varphi_N}$'s. Proceeding much as we did in deriving (8.132), we get

$$\varphi_{N+1}(t) = \sum_{k=-K_2}^{-K_1} c_{-k}(F_1)(\varphi_N)_{1,k}(t) = \sqrt{2} \sum_{k=-K_2}^{-K_1} c_{-k}(F_1)\varphi_N(2t - k). \tag{8.162}$$

We claim that (8.162) has the following implication: If $\varphi_N(t)$ vanishes outside of $[-K_2, -K_1]$, then so does $\varphi_{N+1}(t)$. If we can demonstrate this claim, we'll be done for the following reason. Since φ_0 may be chosen to vanish outside of $[-K_2, -K_1]$ (recall that $\varphi_0 = F[e_a(s) \operatorname{sinc} \pi s] = \chi_{[a-1/2, a+1/2]}$, where a may be any real number; so we need only choose $a = -K_2 + \frac{1}{2}$), the claim will imply that φ_1, and therefore $\varphi_2, \ldots$, and therefore φ_N vanishes outside of $[-K_2, -K_1]$, for *any* N. But φ is the norm limit of the φ_N's; a norm limit of functions all vanishing outside the same interval itself vanishes almost everywhere outside that interval. Hence the desired result for φ; the corresponding one for ψ will then follow from (8.75) in a similar fashion.

Let's prove our claim. If $t < -K_2$ or $t > -K_1$ and $K_1 \le k \le K_2$, then $2t - k < -K_2$ or $2t - k > -K_1$. So, if $\varphi_N(t) = 0$ whenever $t < -K_2$ or $t > -K_1$, then $\varphi_N(2t - k) = 0$ for $t < -K_2$ or $t > -K_1$, for *all* $k \in [K_1, K_2]$. But the latter means the right, and therefore the left, side of (8.162) equals zero for $t < -K_2$ or $t > -K_1$, as claimed.

So part (c) of our theorem is proved, and we're done. □

Exercises

8.5.1 Prove (8.131) for $f \in L^2(\mathbb{R})$. (Use (6.35) and (8.53).)

8.5.2 "Prove" the *Poisson summation formula*

$$\sum_{k=-\infty}^{\infty} f(k) = \sum_{k=-\infty}^{\infty} \widehat{f}(k)$$

for sufficiently reasonable functions f on $\mathbb{R}$. Hint: Put $n = 0$ and $g = \delta$ into (8.119); sum both sides over k. Express each summand on the left side of your result as a convolution, and use the "fact" that $f * \delta = f$; then, using the "fact" that $\widehat{\delta} = 1$, realize the right side of your result as a Fourier series for an appropriate function.

8.5.3 The Poisson summation formula of the previous exercise is seen to hold if f is continuous and, for some $C > 0$ and $q > 1$, $|f(x)|, |\widehat{f}(x)| < C|x|^{-q}$ for all x. Use this fact to evaluate the following.
    a. $\sum_{k=1}^{\infty} 1/(1 + k^2)$. (Compare with Example 1.5.2.) Example 6.4.2 and the geometric series formula $\sum_{k=1}^{\infty} r^k = r/(1 - r)$ $(r < 1)$ should help.

b. $\lim_{t\to 0^+} t^{1/2} \sum_{k=1}^{\infty} e^{-\pi k^2 t}$. (This limit is relevant to the study of the "circle problem," which counts the number of points with integer coordinates inside a sphere of radius $R > 0$ in $\mathbb{R}^m$. See, for example, p. 40 in [44].) Hint: If $k \neq 0$, then $e^{-\pi k^2/t}$ converges rapidly to zero as t approaches zero from the right—rapidly enough that the *sum*, over $k \neq 0$, of $e^{-\pi k^2/t}$ does the same.

Appendix
Complex Numbers

It's nice, as noted in our Introduction, that the second derivative of a sinusoid is a constant times that sinusoid. But sometimes we want to differentiate just once. So an even nicer thing would be a sinusoid whose *first* derivative is proportional to that sinusoid.

Suppose we have such a sinusoid, let's call it e_s, of frequency $s > 0$. Then by assumption, $e'_s = \lambda e_s$ for some constant λ. And consequently, $e''_s = (e'_s)' = (\lambda e_s)' = \lambda e'_s = \lambda^2 e_s$. On the other hand, $e''_s = -(2\pi s)^2 e_s$, since any sinusoid f_s of positive frequency s satisfies the differential equation (I.3). So $\lambda^2 e_s = -(2\pi s)^2 e_s$, which, assuming e_s is not identically zero, implies $\lambda^2 = -(2\pi s)^2$.

Since we're assuming $s > 0$, $-(2\pi s)^2$ is strictly negative, whence so is λ^2. The moral is this: If we seek a nonzero sinusoid e_s of positive frequency s that's proportional to its first derivative, then we must allow for *square roots of negative numbers*, meaning "numbers" λ whose squares are negative.

The proper framework for such numbers is the complex plane $\mathbb{C}$, which we now describe briefly.

Definition A.0.1 The *complex plane* $\mathbb{C}$ is the set of ordered pairs (x, y) of real numbers, endowed with the following operations of addition ("+") and multiplication ("·"):

$$(x_1, y_1) + (x_2, y_2) = (x_1 + x_2, y_1 + y_2), \tag{A.1}$$

$$(x_1, y_1) \cdot (x_2, y_2) = (x_1 x_2 - y_1 y_2, x_1 y_2 + x_2 y_1). \tag{A.2}$$

Elements of $\mathbb{C}$ are called *complex numbers*.

What does all this have to do with square roots of negative numbers, and what's the logic or intuition behind our definition of multiplication anyway? To answer, it's helpful to describe how the real line $\mathbb{R}$ is considered a subset of $\mathbb{C}$. A good choice is to think of $\mathbb{R}$ as being the horizontal axis in $\mathbb{C}$; that is, $\mathbb{R}$ is the set of ordered pairs $\{(x, 0): x \in \mathbb{R}\}$. Since addition and multiplication of these particular ordered pairs, according to (A.1), (A.2), cause their x coordinates to add and multiply in the usual way, without affecting their y coordinates:

$$(x_1, 0) + (x_2, 0) = (x_1 + x_2, 0 + 0) = (x_1 + x_2, 0), \tag{A.3}$$
$$(x_1, 0)(x_2, 0) = (x_1 x_2 - 0 \cdot 0, x_1 \cdot 0 + x_2 \cdot 0) = (x_1 x_2, 0), \tag{A.4}$$

there will be no confusion if, for real numbers considered as points in the complex plane, we simply write x instead of $(x, 0)$.

Now let i denote the complex number $(0, 1)$. Then

$$i^2 = (0, 1) \cdot (0, 1) = (0 \cdot 0 - 1 \cdot 1, 0 \cdot 1 + 0 \cdot 1) = (-1, 0) = -1, \tag{A.5}$$

so i is a square root of -1. Of course $-i$, by which we mean $-1 \cdot i$, is also such a root, as the reader may check.

Next, we have the following particularly convenient way to write and think of complex numbers:

Proposition A.0.1 *The complex number (x, y) is equal to $x + iy$.*

Proof.

$$\begin{aligned} x + iy &= (x, 0) + (0, 1)(y, 0) = (x, 0) + (0 \cdot y - 1 \cdot 0, 0 \cdot 0 + 1 \cdot y) \\ &= (x, 0) + (0, y) = (x, y). \quad \square \end{aligned} \tag{A.6}$$

We call $x + iy$ the *standard form* for the complex number (x, y). It should be noted that the standard form is sometimes tweaked for aesthetic reasons, particularly when x and y have definite numerical values. For example, we would likely write $3 + 5i$ rather than $3 + i5$. Further, we'd write $3 - 5i$ rather than $3 + i(-5)$, or even $3 + (-5)i$. The latter two are somewhat awkward.

By virtue of the above proposition, the definitions (A.1), (A.2) may be rewritten

$$(x_1 + iy_1) + (x_2 + iy_2) = (x_1 + x_2) + i(y_1 + y_2), \tag{A.7}$$
$$(x_1 + iy_1)(x_2 + iy_2) = (x_1 x_2 - y_1 y_2) + i(x_1 y_2 + x_2 y_1). \tag{A.8}$$

In other words, to add or multiply complex numbers you just do what comes naturally, and remember at the end (in the case of multiplication) that $i^2 = -1$.

We observe that we may take, in $\mathbb{C}$, the square root—two of them, actually—of any negative number, as follows. If x, y, and a are real then, by (A.8), $(x + iy)^2$ equals $a + i0$ if and only if $x^2 - y^2 = a$ and $2xy = 0$. But if $a < 0$, then these latter two equations obtain if and only if $x = 0$ and $y = \pm\sqrt{-a}$. That is, $z^2 = a$, for $a < 0$, if and only if $z = \pm i\sqrt{-a}$. (More generally, *any* nonzero complex number has exactly two square roots; see Exercises A.0.4 and A.0.13.)

What about subtraction and division in $\mathbb{C}$? The first is easy: We define it simply by replacing, in (A.1), each plus sign with a minus sign. Division requires a bit more work: To define z_1/z_2 for z_1 and z_2 complex, it's convenient to first define the reciprocal of z_2. By this we mean the complex number z_2^{-1} satisfying $z_2 z_2^{-1} = 1$. Of course, we'd expect z_2^{-1} to make sense only for $z_2 \neq 0$. We have:

Proposition A.0.2 *If $z = x + iy \in \mathbb{C}$ is not equal to zero, then the reciprocal z^{-1}, also denoted $1/z$, is given by*

$$z^{-1} = \frac{x}{x^2 + y^2} - i\frac{y}{x^2 + y^2}. \tag{A.9}$$

Proof. For such z, we have

$$zz^{-1} = (x + iy)\left(\frac{x}{x^2 + y^2} - i\frac{y}{x^2 + y^2}\right) = \frac{x \cdot x - y \cdot (-y)}{x^2 + y^2} + i\frac{x \cdot (-y) + xy}{x^2 + y^2}$$

$$= \frac{x^2 + y^2}{x^2 + y^2} + i\frac{0}{x^2 + y^2} = 1 + i0 = 1. \quad \square \tag{A.10}$$

Division in $\mathbb{C}$ is now defined by $z_1/z_2 = z_1 z_2^{-1}$, for $z_1, z_2 \in \mathbb{C}$ and $z_2 \neq 0$.

It is not hard to show, and it is well-known, that the set of complex numbers with the above operations is a *field*. This means all the nice algebraic things happen: Order of addition or multiplication doesn't matter; multiplication distributes over addition; a product of nonzero elements is nonzero; and so on. See [21].

We now define some useful quantities associated with a complex number z, and explore some properties of and relations between these quantities.

Definition A.0.2 Let $z = x + iy$, with $x, y \in \mathbb{R}$.

(a) The *real part* $\operatorname{Re} z$ of z is defined by $\operatorname{Re} z = x$.

(b) The *imaginary part* $\operatorname{Im} z$ of z is defined by $\operatorname{Im} z = y$.

(c) The *complex conjugate* $\bar{z}$ of z is defined by $\bar{z} = x - iy$.

(d) The *modulus*, or *absolute value*, $|z|$ of z is defined by $|z| = \sqrt{x^2 + y^2}$.

(e) The *argument* $\operatorname{Arg} z$ of $z \neq 0$ is defined by

$$\operatorname{Arg} z = \operatorname{sgn} y \arccos\left(\frac{x}{\sqrt{x^2 + y^2}}\right), \tag{A.11}$$

where

$$\operatorname{sgn} y = \begin{cases} -1 & \text{if } y < 0, \\ 1 & \text{if not}. \end{cases} \tag{A.12}$$

We also define $\operatorname{Arg} 0 = 0$.

All five of these quantities have nice geometric interpretations: $\operatorname{Re} z$ and $\operatorname{Im} z$ are, respectively, the horizontal and vertical displacements of z from the origin; $|z|$ is the *distance* from z to the origin (and more generally, $|z_1 - z_2|$ is the distance between z_1 and z_2, for $z_1, z_2 \in \mathbb{C}$); $\overline{z}$ is the reflection of z about the horizontal axis; and $\operatorname{Arg} z$, at least for $z \neq 0$, is the angle of inclination of z. (When we say *the* angle of inclination, we're referring to the one in $(-\pi, \pi]$; the range of the arc cosine function is $[0, \pi]$, so the range of $\operatorname{Arg} z$, as defined above, is indeed $(-\pi, \pi]$.) See Figure A.1.

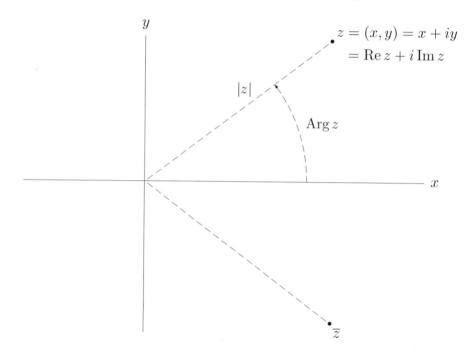

Fig. A.1 Real and imaginary parts, complex conjugate, modulus, and argument

In particular, z has *polar coordinates* $(|z|, \operatorname{Arg} z)$.

If $\operatorname{Im} z = 0$, then z is (identified with) a real number; if $\operatorname{Re} z = 0$, we say z is *imaginary*.

The quantities introduced in the above definition have a variety of other notable properties. A few of these are summarized by the following.

Proposition A.0.3 (a) *For any complex number z, we have:*

(i) $\overline{(\overline{z})} = z$, (ii) $|\overline{z}| = |z|$, (iii) $|\operatorname{Re} z|, |\operatorname{Im} z| \leq |z|$, (iv) $z\overline{z} = |z|^2$,

(v) $\operatorname{Re} z = \dfrac{z + \overline{z}}{2}$ and $\operatorname{Im} z = \dfrac{z - \overline{z}}{2i}$.

(b) *For complex numbers z_1 and z_2, we have:*

(i) $|z_1 z_2| = |z_1||z_2|$, (ii) $\overline{z_1 + z_2} = \overline{z_1} + \overline{z_2}$, (iii) $\overline{z_1 z_2} = \overline{z_1}\,\overline{z_2}$,
(iv) $\overline{1/z_2} = 1/\overline{z_2}$, (v) $\overline{z_1/z_2} = \overline{z_1}/\overline{z_2}$.
(*For the last two parts we assume, of course, that* $z_2 \neq 0$.)

(c) **(the triangle inequality in $\mathbb{C}$)** *If* $z, w \in \mathbb{C}$, *then*

$$|z + w| \leq |z| + |w|. \tag{A.13}$$

Proof. We prove selected parts. For part (a)(iv) we note that, by Proposition A.0.2 and Definition A.0.2(c,d), $z^{-1} = \bar{z}|z|^{-2}$. Multiplying both sides by $z|z|^2$ gives the desired result. (This works only for $z \neq 0$, but (a)(iv) is clearly true if $z = 0$.)

For part (b)(ii), we write $z_1 = x_1 + iy_1$ and $z_2 = x_2 + iy_2$, whence

$$\begin{aligned}\overline{z_1 + z_2} &= \overline{x_1 + x_2 + i(y_1 + y_2)} = x_1 + x_2 - i(y_1 + y_2) \\ &= (x_1 - iy_1) + (x_2 - iy_2) = \overline{z_1} + \overline{z_2}.\end{aligned} \tag{A.14}$$

The proofs of the remainders of parts (a) and (b) are left to the exercises below. Part (c) goes like this:

$$\begin{aligned}|z+w|^2 &= (z+w)(\overline{z+w}) = (z+w)(\bar{z}+\bar{w}) = z\bar{z} + w\bar{w} + z\bar{w} + w\bar{z} \\ &= |z|^2 + |w|^2 + 2\operatorname{Re} z\bar{w} \leq |z|^2 + |w|^2 + 2|z\bar{w}| \\ &= |z|^2 + |w|^2 + 2|z|\,|w| = (|z|+|w|)^2.\end{aligned} \tag{A.15}$$

(We used part (a)(iv) for the first step, (b)(ii) for the second, algebra for the third, (a)(iv) and (a)(v) for the fourth, (a)(iii) for the fifth, (a)(ii) and (b)(i) for the sixth, and algebra for the last.) Taking square roots of both sides gives the desired result. □

Let's return to the discussions commencing this Appendix. We noted that, should a sinusoid

$$e_s(x) = A \cos 2\pi s x + B \sin 2\pi s x, \tag{A.16}$$

not identically zero, satisfy $e'_s = \lambda e_s$ for some $\lambda \in \mathbb{C}$, then necessarily $\lambda^2 = -(2\pi s)^2$. The solutions to the latter equation are, as we've seen, given by $\lambda = \pm 2\pi i s$. To be definite, we now choose the "+" sign; that is, we consider the differential equation

$$e'_s = 2\pi i\, s e_s. \tag{A.17}$$

How might we solve this equation?

One way would be to substitute (A.16) into it and see what this tells us about A and B. Certainly this approach would work in a fairly straightforward manner. But it's more illuminating to approach (A.17) by relating it to a situation with which we're already quite familiar, that of *exponentiation of real numbers*.

Specifically, since

$$\frac{d}{dx} e^{2\pi a s x} = 2\pi a s\, e^{2\pi a s x} \tag{A.18}$$

for any real constant a, we'd be eminently justified in speculating that, for some reasonable notion of exponentiation of *imaginary* numbers, we have

$$\frac{d}{dx}e^{2\pi i s x} = 2\pi i s\, e^{2\pi i s x}. \qquad (A.19)$$

If our speculation is on target, then $e_s(x) = e^{2\pi i s x}$ provides the desired solution to (A.17).

The following definition supplies the requisite notion, confirms our speculation, and more.

Definition A.0.3 For $z \in \mathbb{C}$, the *natural exponential* $e^z \in \mathbb{C}$ is defined by

$$e^z = \sum_{n=0}^{\infty} \frac{z^n}{n!}. \qquad (A.20)$$

When z is real, the series on the right gives the usual, familiar power series representation of the real exponential function e^z. So the above definition amounts to an extension of the domain of definition of that function.

Of course we're glossing over some important details regarding convergence of infinite series of complex numbers. We we take it on faith (it's true!) that the series for e^z converges "very nicely," meaning we can rearrange or manipulate it, when we need to, essentially as if it were a finite sum.

Such manipulations allow us to reveal some things not immediately apparent, namely, the relation of e^z to sinusoids and to the differential equation (A.17). To see these relations, let's put an *imaginary* $z = i\theta$ into the definition (A.20) of e^z. We get

$$\begin{aligned}
e^{i\theta} &= \sum_{n=0}^{\infty} \frac{(i\theta)^n}{n!} = 1 + \frac{i\theta}{1!} + \frac{i^2\theta^2}{2!} + \frac{i^3\theta^3}{3!} + \frac{i^4\theta^4}{4!} + \frac{i^5\theta^5}{5!} + \frac{i^6\theta^6}{6!} + \cdots \\
&= \left(1 + (-1)\frac{\theta^2}{2!} + (-1)^2\frac{\theta^4}{4!} + (-1)^3\frac{\theta^6}{6!} + \cdots\right) \\
&\quad + i\left(\frac{\theta}{1!} + (-1)\frac{\theta^3}{3!} + (-1)^2\frac{\theta^5}{5!} + \cdots\right) \\
&= \sum_{k=0}^{\infty} \frac{(-1)^k \theta^{2k}}{(2k)!} + i\sum_{k=0}^{\infty} \frac{(-1)^k \theta^{2k+1}}{(2k+1)!} = \cos\theta + i\sin\theta \qquad (A.21)
\end{aligned}$$

by the usual Maclaurin series for the cosine and sine functions. That is, the natural exponential of an imaginary variable $i\theta$ is a sinusoid in θ.

In particular,

$$e^{2\pi i s x} = \cos 2\pi s x + i\sin 2\pi s x \qquad (A.22)$$

is a sinusoid of frequency s if $s \geq 0$. Of course $e^{2\pi i s x}$ makes sense for negative s as well and has, for such s, frequency $-s$ (since, for $\theta < 0$, $\cos\theta = \cos(-\theta)$ and $\sin\theta = -\sin(-\theta)$). Or to put it another way: Given $s \geq 0$, both $e^{2\pi i s x}$ and $e^{-2\pi i s x}$ are sinusoids of frequency s.

COMPLEX NUMBERS

We observe that $e^{2\pi i s x}$ does what we want it to from the point of view of differential equations. That is,

Proposition A.0.4 *The function* $e_s(x) = e^{2\pi i s x}$ *satisfies the differential equation* (A.17).

Proof.

$$e'_s(x) = \frac{d}{dx}\cos 2\pi s x + i\frac{d}{dx}\sin 2\pi s x = 2\pi s(-\sin 2\pi s x + i\cos 2\pi s x)$$
$$= 2\pi i s(i\sin 2\pi s x + \cos 2\pi s x) = 2\pi i s\, e_s(x) \tag{A.23}$$

(since $-1 = i^2$). □

Of course the given function e_s is *complex-valued*, meaning $e_s(x)$ is, in general, a complex number. Implicit in the above proof is the fact that *differentiation* of complex-valued functions is defined in the expected, natural way. That is, we define $(u + iv)' = u' + iv'$ if u and v are real-valued functions (and both derivatives on the right exist). Certainly this definition is not surprising, and certainly any other would be.

While we're at it, we remark that calculus involving complex numbers may be and is, in general, defined in terms of the real and imaginary parts of these numbers. And with such definitions in place, the "expected" things happen. For example, we define

$$\lim_{x\to x_0} f(x) = \lim_{x\to x_0}\operatorname{Re} f(x) + i\lim_{x\to x_0}\operatorname{Im} f(x) \tag{A.24}$$

for f complex-valued (when the limits on the right exist), and find that the limit of a sum is the sum of the limits, that

$$\lim_{x\to x_0} f(x) = L \quad \text{if and only if} \quad \lim_{x\to x_0} |f(x) - L| = 0, \tag{A.25}$$

and so on. Also, we put

$$\int_a^b f(x)\,dx = \int_a^b \operatorname{Re} f(x)\,dx + i\int_a^b \operatorname{Im} f(x)\,dx \tag{A.26}$$

(when the integrals on the right exist) and find that things behave much as in the real-valued case; in particular, we may evaluate suitably nice definite integrals by antidifferentiating.

But back to our program. We now catalog some other salient properties of the natural exponential function.

Proposition A.0.5 (a) $e^{z_1+z_2} = e^{z_1}e^{z_2}$ *for* $z_1, z_2 \in \mathbb{C}$.

(b) $(e^z)^k = e^{kz}$ *for* $z \in \mathbb{C}$ *and* $k \in \mathbb{Z}$.

(c) $|e^{i\theta}| = 1$ *for* $\theta \in \mathbb{R}$.

476 COMPLEX NUMBERS

(d) $\overline{e^{i\theta}} = e^{-i\theta}$ for $\theta \in \mathbb{R}$.

(e) $|e^z| = e^{\operatorname{Re} z}$ for $z \in \mathbb{C}$.

Proof. For part (a) we note that, by Definition A.0.3,

$$e^{z_1}e^{z_2} = \sum_{n=0}^{\infty}\sum_{m=0}^{\infty}\frac{z_1^n z_2^m}{n!\,m!} = \sum_{n=0}^{\infty}\sum_{k=n}^{\infty}\frac{z_1^n z_2^{k-n}}{n!\,(k-n)!}, \tag{A.27}$$

the last equality resulting from the substitution $k = m + n$ in the sum on m. Now the double sum on the right ranges over the set S of all pairs (n, k) of nonnegative integers with $k \geq n$. Note that S may, equivalently, be described as the set of all pairs (n, k) of nonnegative integers with $n \leq k$. So (A.27) gives

$$e^{z_1}e^{z_2} = \sum_{k=0}^{\infty}\left[\sum_{n=0}^{k}\frac{z_1^n z_2^{k-n}}{n!\,(k-n)!}\right] = \sum_{k=0}^{\infty}\frac{(z_1+z_2)^k}{k!} = e^{z_1+z_2}; \tag{A.28}$$

the second equality is by the binomial theorem (4.133), and the last by Definition A.0.3. (Here again we've performed some infinite series manipulations without explicit justification.)

Part (b) of our proposition, in the case $k = 0$, is a matter of definition and of the fact that $e^0 = 1$. In the case $k > 0$, it follows from repeated application of part (a) to itself—or more formally from mathematical induction applied to that part. In the case $k < 0$, it follows from mathematical induction and the case $k = -1$, which reads $1/e^z = e^{-z}$. This latter identity is, in turn, just part (a) with $z_2 = -z_1$. See Exercise A.0.12.

Part (c) amounts to a direct computation:

$$|e^{i\theta}| = |\cos\theta + i\sin\theta| = \sqrt{\cos^2\theta + \sin^2\theta} = \sqrt{1} = 1. \tag{A.29}$$

Since $|z|^2 = z\bar{z}$, this last result implies $e^{i\theta}\overline{e^{i\theta}} = 1$, so that

$$\overline{e^{i\theta}} = \frac{1}{e^{i\theta}} = e^{-i\theta}, \tag{A.30}$$

the last step by part (b) with $z = i\theta$ and $k = -1$. Hence part (d).

Finally,

$$|e^z| = |e^{\operatorname{Re} z}e^{i\operatorname{Im} z}| = |e^{\operatorname{Re} z}||e^{i\operatorname{Im} z}| = |e^{\operatorname{Re} z}| = e^{\operatorname{Re} z}, \tag{A.31}$$

the first step by part (a), the third by part (c), and the last because $e^x > 0$ for $x \in \mathbb{R}$. □

It's worth noting that part (b) does *not* hold for arbitrary $k \in \mathbb{C}$. Indeed, let $z = 2\pi i$ and $k = (2\pi)^{-1}$. Then

$$(e^z)^k = (e^{2\pi i})^{(2\pi)^{-1}} = (\cos 2\pi + i\sin 2\pi)^{(2\pi)^{-1}} = 1^{(2\pi)^{-1}} = 1, \tag{A.32}$$

but
$$e^{kz} = e^{(2\pi)^{-1} \cdot 2\pi i} = e^i = e^{i \cdot 1} = \cos 1 + i \sin 1 \approx 0.540302 + i(0.841471). \tag{A.33}$$

We conclude by applying the natural exponential function toward another convenient way of expressing complex numbers. Namely, we observe from Definition A.0.2(d,e) (or Figure A.1) that, for $z \in \mathbb{C}$, we have $\operatorname{Re} z = |z| \cos \operatorname{Arg} z$ and $\operatorname{Im} z = |z| \sin \operatorname{Arg} z$ (note that this holds even for $z = 0$). We therefore can write

$$\begin{aligned} z &= |z| \cos \operatorname{Arg} z + i|z| \sin \operatorname{Arg} z \\ &= |z|(\cos \operatorname{Arg} z + i \sin \operatorname{Arg} z) = |z| e^{i \operatorname{Arg} z}. \end{aligned} \tag{A.34}$$

The right side is called the *complex exponential form* for z; this form is to polar coordinates what the standard form $\operatorname{Re} z + i \operatorname{Im} z$ is to Cartesian coordinates.

Exercises

A.0.1 Let $z_1 = 2 + 3i$ and $z_2 = -1 - 2i$. Find (and write in standard form):
  a. $z_1 - z_2$;
  b. $1/z_2$;
  c. $z_1 + 3z_2$;
  d. z_1/z_2;
  e. z_1^4;
  f. $z_1 z_2^2$.

A.0.2 Prove the remainder of Proposition A.0.3(a)(b).

A.0.3 By squaring, show that the imaginary number ib ($b \in \mathbb{R}, b \neq 0$) has the two square roots $\pm(1 + i \operatorname{sgn} b)\sqrt{|b|/2}$, where $\operatorname{sgn} b$ denotes the sign of b, cf. (A.12).

A.0.4 Find and express in standard form the two square roots of the arbitrary nonzero complex number $z = u + iv$. Hint: Solve $(x + iy)^2 = u + iv$ for x and y.

A.0.5 *By direct computation* (that is, by actually evaluating the relevant sines and cosines and then cubing), show that the complex number $z = \cos(2\pi/3) + i \sin(2\pi/3)$ is a cube root of unity; that is, show that $z^3 = 1$. Show the same for $z = \cos(4\pi/3) + i \sin(4\pi/3)$ and $z = \cos(6\pi/3) + i \sin(6\pi/3)$.

A.0.6 Similarly, show that $z = \cos(\pi/4) + i \sin(\pi/4)$ and $z = \cos(7\pi/4) + i \sin(7\pi/4)$ are eighth roots of unity.

A.0.7 Write the following numbers in complex exponential form:
  a. $1 + i$;
  b. $1 - i$;
  c. $-2\sqrt{3} + 2i$;
  d. i;

e. $7 - 3i$;

f. $e^{3\ln 2 + i\pi/4}$. Also write this last number in standard form.

A.0.8 Show that, for $z \in \mathbb{C}$, $\bar{z} = |z|e^{-i\,\mathrm{Arg}\,z}$.

A.0.9 For which $z \in \mathbb{C}$ is it true that $\mathrm{Arg}\,\bar{z} = -\mathrm{Arg}\,z$? For which z is it false? In the latter case, what *is* $\mathrm{Arg}\,\bar{z}$ in terms of $\mathrm{Arg}\,z$? Hint: Think negative.

A.0.10 Use the definition (A.20) to show that, for x a *real* number, $e^x = 1$ if and only if $x = 0$. Hint: Consider first positive x; then use Proposition A.0.5(b), with $k = -1$, for negative x.

A.0.11 Show that, for z *complex*, $e^z = 1$ if and only if z is an integer multiple of $2\pi i$. (Use the previous exercise. You may take it on faith that, for y *real*, $\cos y = 1$ if and only if y is an integer multiple of 2π and $\sin y = 0$ if and only if y is an integer multiple of π.)

A.0.12 Use mathematical induction to prove Proposition A.0.5(b).

A.0.13 Find and express in complex exponential form the two square roots of the nonzero complex number $z = re^{i\theta}$.

A.0.14 Repeat Exercises A.0.5 and A.0.6, this time using complex exponentials (and Proposition A.0.5(b)).

A.0.15 Put $z_1 = i\theta_1$ and $z_2 = i\theta_2$ into Proposition A.0.5(a) and equate real and imaginary parts of the result to deduce the identities

$$\cos(\theta_1 + \theta_2) = \cos\theta_1 \cos\theta_2 - \sin\theta_1 \sin\theta_2,$$
$$\sin(\theta_1 + \theta_2) = \cos\theta_1 \sin\theta_2 + \sin\theta_1 \cos\theta_2.$$

A.0.16 Use the results of the previous exercise to derive the identity

$$\cos\theta_1 \cos\theta_2 = \frac{1}{2}(\cos(\theta_1 - \theta_2) + \cos(\theta_1 + \theta_2))$$

and similar identities for $\cos\theta_1 \sin\theta_2$ and $\sin\theta_1 \sin\theta_2$.

References

1. E. Aboufadel and S. Schlicker, *Discovering Wavelets*, John Wiley and Sons, New York, 1999.

2. R. Apéry, Irrationalité de $\zeta(2)$ et $\zeta(3)$, *Astérisque* **61**, 11–13 (1979).

3. T. Apostol, *An Introduction to Analytic Number Theory*, Springer-Verlag, New York, 1976.

4. L. Baggett and W. Fulks, *Fourier Analysis*, Anjou Press, Boulder, CO, 1979.

5. R. G. Bartle, *A Modern Theory of Integration*, American Mathematical Society Graduate Studies in Mathematics **32**, Amercian Mathematical Society, Providence, RI, 2001.

6. A. Boggess and F. J. Narcowich, *A First Course in Wavelets with Fourier Analysis*, Prentice-Hall, Upper Saddle River, NJ, 2001.

7. J. W. Brown and R. V. Churchill, *Fourier Series and Boundary Value Problems* (6th ed.), McGraw-Hill, Boston, 2001.

8. E. B. Burger and R. Tubbs, *Making Transcendence Transparent: An Intuitive Approach to Classical Transcendental Number Theory*, Springer-Verlag, New York, 2004.

9. L. Carleson, On the convergence and growth of partial sums of Fourier series, *Acta Mathematica* **116**, 135–157 (1966).

10. M. Cartwright, *Fourier Methods for Mathematicians, Scientists, and Engineers*, Ellis Horwood, New York, 1990.

11. C. K. Chui, *An Introduction to Wavelets*, Academic Press, San Diego, 1992.

12. J. W. Cooley and J. W. Tukey, An algorithm for the machine computation of complex Fourier series, *Mathematics of Computation* **19**, 297–301 (1965).

13. J. B. Conway, *Functions of One Complex Variable* (2nd. ed.), Springer-Verlag, New York, 1978.

14. I. Daubechies, The wavelet transform, time-frequency localization and frequency analysis, *IEEE Transactions on Information Theory* **36**, 961–1005 (1990).

15. H. Dym and H. P. McKean, *Fourier Series and Integrals*, Academic Press, New York, 1972.

16. C. H Edwards and D. E. Penney, *Calculus* (6th ed.), Prentice-Hall, Upper Saddle River, NJ, 2002.

17. D. G. Falconer, Optical processing of bubble chamber photographs, *Applied Optics* **5**, no. 9, 1365–1369 (1966).

18. R. J. Fessenden and J. S. Fessenden, *Organic Chemistry* (2nd ed.), Willard Grant Press, Boston, 1982.

19. G. B. Folland, *Fourier Analysis and its Applications*, Wadsworth and Brooks/Cole, Pacific Grove, CA, 1992.

20. J. Fourier, *The Analytical Theory of Heat* (translated by A. Freeman), Dover, New York, 1955.

21. J. B. Fraleigh, *A First Course in Abstract Algebra* (6th ed.), Addison-Wesley Longman, 1999.

22. J. W. Gibbs, *Nature* **59**, 606 (1899).

23. J. W. Gibbs, *Elementary Principles in Statistical Mechanics*, Yale University Press, New Haven, 1902.

24. R. A. Gordon, *The Integrals of Lebesgue, Perron, Denjoy, and Henstock*, Graduate Studies in Mathematics **4**, American Mathematical Society, Providence, RI, 1994.

25. I. S. Gradshteyn and I. M. Rhyzik, *Tables of Integrals, Series, and Products* (corrected and enlarged edition), edited by A. Jeffreys, Academic Press, New York, 1980.

26. Hecht, *Optics*, Academic Press, New York, 1972.

27. M. T. Heideman, D. H. Johnson, and C. S. Burrus, Gauss and the history of fast Fourier transform, *Archive for the History of the Exact Sciences* **21**, 129–160 (1979).

28. R. W. King and K. R. Williams, The Fourier transform in chemistry, part 1. Nuclear magnetic resonance: Introduction, *Journal of Chemical Education* **66**, no. 9, A213–A219 (1989).

29. A. N. Kolmogorov and S. V. Fomin, *Introductory Real Analysis* (translated by R. A. Silverman), Dover, New York, 1970.

30. T. W. Körner, *Fourier Analysis*, Cambridge University Press, Cambridge, 1988.

31. P. D. Lax, *Functional Analysis*, John Wiley & Sons, New York, 2002.

32. J. Leggett, The noise of the Bomb, *New Scientist* **108**, no. 1482, 39–42 (1985).

33. J. E. Marsden and M. J. Hoffman, *Elementary Classical Analysis* (2nd ed.), W. H. Freeman and Company, New York, 1993.

34. A. G. Marshall and F. R. Verdun, *Fourier Transforms in NMR, Optical, and Mass Spectrometry: A User's Handbook*, Elsevier, New York, 1990.

35. E. J. McShane, *Integration*, Princeton University Press, Princeton, 1944.

36. A. Papoulis, *Signal Analysis*, McGraw-Hill, London, 1977.

37. E. M. Purcell, *Electricity and Magnetism* (2nd ed.), McGraw-Hill, New York, 1985.

38. P. Reina and A. Cho, Spans sway underfoot in Europe, *Engineering News Record* **245**, no. 2, 14–15 (2000).

39. B. Riemann, *The Collected Works of B. Riemann*, edited by H. Weber, Dover, New York, 1953.

40. T. Rivoal, La fonction zêta de Riemann prend une infinité de valeurs irrationnelles aux entiers impairs, *C. R. Acad. Sci. Paris Sér. I Math.* **331**, no. 4, 267–270 (2000).

41. L. J. Schwartz, A step-by-step picture of pulsed (time-domain) NMR, *Journal of Chemical Education* **65**, no. 11, 959–963 (1988).

42. R. H. Shumway, *Applied Statistical Time Series Analysis*, Prentice-Hall, Englewood Cliffs, NJ, 1988.

43. D. W. Stroock, *A Concise Introduction to the Theory of Integration* (3rd ed.), Birkhauser, Boston, 1999.

44. A. Terras, *Harmonic Analysis on Symmetric Spaces and Applications I*, Springer-Verlag, New York, 1985.

45. A. Terras, *Fourier Analysis on Finite Groups and Applications*, Cambridge University Press, Cambridge, 1999.

46. E. C. Titchmarsh, *Eigenfunction Expansions Associated with Second Order Differential Equations*, vol. I, Oxford University Press, Oxford, 1946.

47. G. P. Tolstov, *Fourier Series* (translated by R. A. Silverman), Dover, New York, 1976.

48. J. S. Walker, *Fast Fourier Transforms*, CRC Press, Boca Raton, FL, 1991.

49. G. N. Watson, *A Treatise on the Theory of Bessel Functions* (2nd ed.), Cambridge University Press, Cambridge, 1944.

50. A. I. Zayed, *Advances in Shannon's Sampling Theory*, CRC Press, Boca Raton, FL, 1993.

51. W. Zudilin, Irrationality of values of the Riemann zeta function, *Izvestiya: Mathematics* **66**, no. 3, 489–542 (2002).

52. A. Zygmund, *Trigonometric Series* (3rd ed.), Cambridge University Press, Cambridge, 2002.

Index

2Ω-sampling, 375

A

Aliasing, 376, 384
Almost everywhere, 167
ALTICS, 399
Amplitude, 60, 407
    complex, 407
Area, 433
Average value, 11

B

Base pitch, 63, 127
Basis
    orthogonal, 190, 222
    orthonormal, 190, 222
Bay City Rollers, 372
Benzene ring, 420
Bernoulli, Daniel, xxi
Bessel functions
    and circular drums, 150, 244
    and Fourier-Bessel series, 239
    and Fourier transforms of radial functions, 342
    as Fourier coefficients, 54
    of half-integer order, 239
    of integer order, 234
    of order $\nu > 0$, 289
    versus sinusoids, 237
Bessel's inequality, 43, 198

Best approximation theorem, 199
Binomial theorem, 251
Boundary conditions, 80
    continuity of, 85, 220
    Dirichlet, 81
    homogeneous, 88
    Neumann, 81
    periodic, 100, 220
    self-adjoint, 221
    separated, 220
Boundary value problem, 79
    Dirichlet, 92
    existence of solutions, 123
    Neumann, 92
    uniqueness of solutions, 85, 123, 126
    weak solutions, 126
Bounded interval, 8

C

Carleson's theorem, 216
Cauchy-Schwarz inequality, 162
Cauchy sequence, 165
Center, 433
Complex exponential, 4, xxiii
    windowed, 423–424
Complex number, 471
    argument of, 473
    complex conjugate of, 473
    complex exponential form for, 479

imaginary part of, 473
modulus (absolute value) of, 473
real part of, 473
standard form for, 472
Complex vector space, 10
Composition theorem, 305, 337, 339
Convolution, 263
   and ALTICS, 401
   and boundedness, 273
   and compact support, 267
   and continuity, 273
   and differentiability, 277, 280
   and filtering, 398
   and integrability, 285
   and smoothing, 294
   and square integrability, 287
   approximate identity for, 292
   (lack of an) identity function for, 289
   of a tempered distribution with a Schwartz function, 349
   semigroup, 288, 323
Convolution-inversion theorem, 322, 326, 341
Convolution theorem, 306, 339
Cooley, J. W., 390
Cup holders, 130
Cylindrical coordinates, 105

D

Damped harmonic oscillator, 414
   compound, 414
   simple, 414
Decibel (dB), 63
Deconvolution, 294, 404
   and seismic disturbances, 404
   and sharpening, 294
Density with respect to the norm, 170
   transitivity of, 176
Density, 443
Diffraction
   farfield, 408
   Fraunhofer, 65, 405, 408
   grating, 411
   X-ray, 65
Dilation, 305, 439, 443
   dyadic, 442
Direct sum, 447
Dirichlet, P. G. L., 24
Dirichlet kernel, 27
Dirichlet problem, 92
   for a disk, 107
   for a half-plane, 362
   for a half-space, 373
   for a semiannulus, 230
   for a semidisk, 363
   for a solid ball, 259
   for a solid half-ball, 262

Discrete wavelet transform, 448
Distribution, xxiii, 126
   tempered, 126, 296, 347, 401
   associated with a function f, 348
   delta, 349
DIY, 19
Double slit experiment, 412–413

E

Eigenfunction, 92
Eigenspace, 223
Eigenvalue, 92
   problem, 92
Eigenvector, 92
Electromagnetic (EM) wave theory, 406
Electrostatic potential in a crystal, 3, 153
Elementary operation, 387, 448
Energy, 400, 409
   conservation of, 409
Equivalence class, 180
Equivalence relation, 179
Ethylbenzene, 416
Euler, Leonard, xxi
Evaluation of infinite series, 31, 52, 54, 198, 203, 380, 468
Evaluation operator, 90
Even extension, 336

F

Filter, 393, 395, 412
   band-pass, 395
   band-stop, 395
   comb, 395
   high-pass, 395
   line-stop, 398
   low-pass, 395
   mask, 395
Filtering, 378, 392
   and convolution, 398
   optical, 411
For almost all x, 167
Fourier, Joseph, xix–xx, 79, 390
Fourier coefficients, 11, 56, 67
Fourier cosine and sine transforms, 335
   inversion of, 336–337
Fourier inversion
   of bandlimited functions, 374
   of discrete data, 385
   on $FL^1(\mathbb{R})$, 318–319
   on $FL^1(\mathbb{R}^M)$, 341
   on $L^2(\mathbb{R})$, 326, 328
   on $L^2(\mathbb{R}^M)$, 341
   on $PS(\mathbb{R}) \cap L^1(\mathbb{R})$, 332
   on $TD(\mathbb{R})$, 352
   short-time, 426
Fourier method, 85, 88

INDEX **485**

Fourier optics, 406
Fourier series, 11
  amplitude-phase form, 60
  antidifferentiation of, 49
  associated to f, 11
  complex exponential form, 11
  convergence of
    absolute, 45
    norm, 182
    pointwise, 27
    uniform, 44
  cosine series, 72
  cosine-sine form, 11
  differentiation of, 41, 43, 47
  for functions of three variables, 151
  for functions of two variables, 148, 209
  halfsine series, 75
  Nth partial sum of, 25, 55
  sine series, 72
  standard, 67
  uniqueness of, 194
Fourier transform
  discrete, 382
  fast (FFT algorithm), 387
  in several variables, 338
  of characteristic function, 301, 315
  of complex exponential, 312
  of constant function, 311, 351
  of Dirac delta function, 308, 351
  of Gaussian, 310
  of piecewise polynomial function, 316
  of piecewise smooth function, 313
  of radial function, 342
  of sinc function, 327
  of windowed complex exponential, 425
  on $L^1(\mathbb{R})$, 301
  on $L^1(\mathbb{R}^M)$, 338
  on $L^2(\mathbb{R})$, 325
  on $L^2(\mathbb{R}^M)$, 341
  on $S(\mathbb{R})$, 350
  on $\mathrm{TD}(\mathbb{R})$, 351
  properties summarized, 308
  short-time (windowed), 425
Free induction decay (FID) signal, 414
Frequency, xix
  analysis, xx
  component, 5, xx, 60, 73, 299
  decomposition, 3–4, xx, 73, 299
  domain, 3, xx, 73, 299
  fundamental, 63, 127
  local
    analysis, 423
    content, 390
  Nyquist, 375
  resonant, 139, 414
  span, 433

Fubini's theorem, xxiii, 188
Function
  antiperiodic, 19, 22
  aperiodic, 263
  associated Legendre, 260
  averaged, 29
  bandlimited, 374
  beta, 281
  characteristic, 188
  compactly supported, 184
  complex-valued, 477
  Dirac delta, 263, 295, 345
  Dirichlet's, 1, 7, 175
  even, 16
  finite, 433
  gamma, 281
  generalized, 345
  harmonic, 111
    maximum modulus theorem for, 111
    mean value theorem for, 111
  Hermite, 355
  Kronecker delta, 5
  (Lebesgue) integrable, 178, 185
  (Lebesgue) square integrable, 175, 185
  measurable, 187
  odd, 16
  periodic, 1
  piecewise continuous, 8
  piecewise smooth, 29
  real-valued, 5, 59
  reasonable, 8, xix
  sawtooth, 36
  sign, 23
  sinc, 302
  step, 442
  window, 424
Functional, 346
Fundamental cube, 153
Fundamental period, 1

G

Gauss, Carl Friedrich, 390
Gaussian, 310
Gibbs, J. W., 36
Gibbs' phenomenon, 36
Glass, Philip, 3
Grimaldi, Francesco, 412

H

Hankel transform, 343
Harmonic, 63
  first, 127
  higher, 127
  oscillator, 358–359
Heat
  conduction, xxi

equation, 94
    infinite, 361
    inhomogeneous, 136, 145, 150, 366
    steady state, 80
Height, 433
Heisenberg uncertainty principle, 436
Hendrix, Jimi, 140
Hermite equation, 355
Hermite polynomials, 360
Hilbert space, xxiii, 213
Hologram, 64
Homogeneous equation, 88
Huygen's principle, 409

I

Image plane, 412
Impulse response, 402
Infinity norm, 273
    See also Supremum
Initial conditions, 95
Inner product
    on $C(\mathbb{T})$, 162
    on $L^2(\mathbb{T})$, 177
    on $L^2_w(a,b)$, 222
    on $\mathbb{R}^2$, 164
    on $\mathbb{R}^M$, 338
Intensity pattern, 407
Inverse Fourier transform, 320
Isometry, 212
Isomorphism between inner product spaces, 214
Isoperimetric problem, 210

J

Jurrasic Park, 403

K

Karaoke, 400
Kilohertz (kHz), 374

L

Lag operator, 399
Laguerre polynomials, 359
Laplace equation, 80
Laplacian (Laplace operator), 90, 92
    in cylindrical coordinates, 107
    in polar coordinates, 107
    in spherical coordinates, 113
Lebesgue
    dominated convergence theorem, xxiii, 186
    integral, xxiii, 178
Lebesgue, Henri, 178
Legendre polynomial, 252
Legendre's equation, 252
Lens, 411
    focal length, 411
Light, 406
    monochromatic, 407
    planar, 406
    polarized, 407
    PPM, 407
Limit
    ε-N and ε-δ definitions, xxiii, 184
    left-hand, 8
    right-hand, 8
Linear
    combination, 11
        closure under, 11
        infinite, 84
    differential equation, 88
    operator, 88
    second-order differential operator, 89
Linearity
    of Fourier coefficients, 13
    of Fourier transform, 301
    of tempered distributions, 347
Linear system, *See* ALTICS
Local/global principle, 22, 49, 54, 303, 331, 379, 393
Lord Kelvin, 36

M

Maclaurin series, 21
    for $\cos\theta$, 476
    for e^z, 476
    for $\sin\theta$, 476
    for $z/(1-z)$, 20
Maxwell's equations, 406
Mean norm, 178
Mean square convergence, 158
Mean square distance, 158
Mean square norm
    on $\mathbb{C}^M$, 158
    on $C(\mathbb{T})$, 158
    on $L^2_w(a,b)$, 222
    on $\mathbb{R}^M$, 158, 341
Metric, 165
    space, 165
        complete, 165
        completion of, 212
Midpoint, 432
    frequency domain, 432
    time domain, 432
Monotone sequence property, 26
Multiplicity rule, 418
Multiresolution analysis, 438, 443, 445
    Haar, 444, 446

N

$N+1$ rule, *See* Multiplicity rule
Natural exponential function, 476
Nestedness, 443
Newton's law of cooling, 102, 229

INDEX **487**

Nobel Prize
    Hauptmann and Karle, 65
    Michelson, A. A., 36
Noise, 392
Nuclear magnetic resonance
    continuous wave, 421
    pulsed, 414

O

Object plane, 408
Observable energy levels, 358–359
Odd extension, 337
Orthogonality, 7, 190
Orthogonal polynomials, 249
Orthonormality, 190, 443
Oversampling, 378
Overshoot, 38
Overtones, 127

P

Parseval's equation, 198
Periodic extension, 66, 73
    even, 71
    odd, 71
Periodicity, xxi
Phase, 60, 407
    deafness, 184
    problem of, 63
Plancherel theorem, 322, 326, 337, 342
Planck's constant, 358
Poisson summation formula, 468
Polar coordinates, 105
Prime number theorem, 35
Product domain, 186
    orthonormal basis for, 205
Product theorem, 322, 326, 342
Pseudometric, 165
    space, 165
Pythagorean theorem, 162

R

Radius, 432
    frequency domain, 432
    time domain, 432
Real analysis, xxii
Realization, 168
Resonance, 139
Reverse 3.5 somersault pike, 116
Riemann
    integral, xxiii, 35, 178
    sum, xxiii, 35, 402
        approximation to Fourier coefficients, 381
    zeta function, 34
Riemann, Bernhard, 34
Riemann-Lebesgue lemma
    for $L^1(\mathbb{R})$, 330

    for $L^1(\mathbb{R}^M)$, 341
    for $PC(\mathbb{T})$, 25
    for $PC(0, \pi)$, 84
Rodrigues' formula, 255

S

Scale factor, 446
    Daubechies, 454
Scaling function
    Daubechies, 451
    noncompactly supported, 449
Scaling relation, 446
Schrödinger equation, 355, 358
Schwartz convergence, 347
Schwartz function, 346
Self-similarity, 443
Separation of variables, xxi, 81
Separation, 443
Set
    ε-small, 167
    Cantor, 168
    countable, 168
    countably infinite, 168
    negligible, 167
    of measure zero, 168
    orthogonal, 7, 190, 222
    orthonormal, 190, 222
Shannon sampling theorem, 374
Sinusoid, xix
    P-periodic, 3
Sinusoid decomposition, *See* Frequency
    decomposition
Space, *See* Complex vector space
Span, 201
Spatial symmetry, 92
Spectroscopy
    FT-ICR, 421
    FT-IR, 421
    FT-NMR, 414
Spectrum, 62, 360
Spectrum plane, 408
Spherical coordinates, 111
Spin orientation, 418
Splat, 265
Square summable sequences, 214
Squeeze law II, 181, 184
Sturm-Liouville
    differential equation, 220
    problem, 219
        regular, 223
        singular, 223
Superposition, xix
Superposition principle, 81, 90, 132
Supremum, 41
Surface conductance, 102
Symmetry equivalence, 417

System function, 403

T

Tacoma Narrows Bridge, 140
Tchebyschev polynomial, 250
Tchebyschev's equation, 250
Tchebyschev's inequality, 169
Temperate convergence, 296, 350, 401
Tension constant, 116
Thermal conductivity, 102
Thermal diffusivity, 94
Time-frequency
   building blocks, 424
   localization, 424
   span, 433
Time-scale analysis, 438
Time span, 433
Torus, 9, 55
Transfer function, 403
Triangle inequality, 162, 475
Tukey, J. W., 390

U

UltraCauchy sequence, 168

V

Variation of parameters, 137
Vector space, *See* Complex vector space

W

Wave equation, xxi, 116
   d'Alembert's form of the solution, 128
   for a circular drum, 148, 244, 246
   for a rectangular drum, 146
   Fourier's form of the solution, 127
   infinite
     in one dimension, 373
     in three dimensions, 368
     in two dimensions, 370
   inhomogeneous, 132, 136, 246

Wavelength, 1, 412
Wavelet component, 448
Wavelet decomposition, 447
   compression
     FBI, 449
     JPEG2000, 449
     MPEG-4, 449
   decomposition formulas, 448
   layer of detail, 447
   level of resolution, 447
   reconstruction formula, 448
Wavelet, 439
   decomposition, 444
   orthonormal, 439
Wavelets, 390, 423, 445
   and singularity detection, 456
   associated with multiresolution analyses, 446
   basic (mother), 439
   Daubechies, 451
   finite, 440
   Haar, 441, 446
   noncompactly supported, 449
   vanishing moments, 456
Wave/particle dualism, 413
Weierstrass' function, 46
Weight function, 220
Width, 433
Window, 424
   Bartlett (triangular), 428
   Blackman, 431
   finite, 433
   Gabor (Gaussian), 430
   Hann (Hanning; cosine bell), 429
   Lanczos, 431
   rectangular, 427
   Welch, 431
Wirtinger's inequality, 211
Wronskian, 221

Z

Zebra, 395

PURE AND APPLIED MATHEMATICS

A Wiley-Interscience Series of Texts, Monographs, and Tracts

Founded by RICHARD COURANT
Editors Emeriti: MYRON B. ALLEN III, DAVID A. COX, PETER HILTON, HARRY HOCHSTADT, PETER LAX, JOHN TOLAND

ADÁMEK, HERRLICH, and STRECKER—Abstract and Concrete Catetories
ADAMOWICZ and ZBIERSKI—Logic of Mathematics
AINSWORTH and ODEN—A Posteriori Error Estimation in Finite Element Analysis
AKIVIS and GOLDBERG—Conformal Differential Geometry and Its Generalizations
ALLEN and ISAACSON—Numerical Analysis for Applied Science
*ARTIN—Geometric Algebra
AUBIN—Applied Functional Analysis, Second Edition
AZIZOV and IOKHVIDOV—Linear Operators in Spaces with an Indefinite Metric
BERG—The Fourier-Analytic Proof of Quadratic Reciprocity
BERMAN, NEUMANN, and STERN—Nonnegative Matrices in Dynamic Systems
BERKOVITZ—Convexity and Optimization in $\mathbb{R}^n$
BOYARINTSEV—Methods of Solving Singular Systems of Ordinary Differential Equations
BURK—Lebesgue Measure and Integration: An Introduction
*CARTER—Finite Groups of Lie Type
CASTILLO, COBO, JUBETE, and PRUNEDA—Orthogonal Sets and Polar Methods in Linear Algebra: Applications to Matrix Calculations, Systems of Equations, Inequalities, and Linear Programming
CASTILLO, CONEJO, PEDREGAL, GARCIÁ, and ALGUACIL—Building and Solving Mathematical Programming Models in Engineering and Science
CHATELIN—Eigenvalues of Matrices
CLARK—Mathematical Bioeconomics: The Optimal Management of Renewable Resources, Second Edition
COX—Galois Theory
†COX—Primes of the Form $x^2 + ny^2$: Fermat, Class Field Theory, and Complex Multiplication
*CURTIS and REINER—Representation Theory of Finite Groups and Associative Algebras
*CURTIS and REINER—Methods of Representation Theory: With Applications to Finite Groups and Orders, Volume I
CURTIS and REINER—Methods of Representation Theory: With Applications to Finite Groups and Orders, Volume II
DINCULEANU—Vector Integration and Stochastic Integration in Banach Spaces
*DUNFORD and SCHWARTZ—Linear Operators
    Part 1—General Theory
    Part 2—Spectral Theory, Self Adjoint Operators in Hilbert Space
    Part 3—Spectral Operators
FARINA and RINALDI—Positive Linear Systems: Theory and Applications
FOLLAND—Real Analysis: Modern Techniques and Their Applications
FRÖLICHER and KRIEGL—Linear Spaces and Differentiation Theory
GARDINER—Teichmüller Theory and Quadratic Differentials

*Now available in a lower priced paperback edition in the Wiley Classics Library.
†Now available in paperback.

GILBERT and NICHOLSON—Modern Algebra with Applications, Second Edition
*GRIFFITHS and HARRIS—Principles of Algebraic Geometry
GRILLET—Algebra
GROVE—Groups and Characters
GUSTAFSSON, KREISS and OLIGER—Time Dependent Problems and Difference Methods
HANNA and ROWLAND—Fourier Series, Transforms, and Boundary Value Problems, Second Edition
*HENRICI—Applied and Computational Complex Analysis
    Volume 1, Power Series—Integration—Conformal Mapping—Location of Zeros
    Volume 2, Special Functions—Integral Transforms—Asymptotics— Continued Fractions
    Volume 3, Discrete Fourier Analysis, Cauchy Integrals, Construction of Conformal Maps, Univalent Functions
*HILTON and WU—A Course in Modern Algebra
*HOCHSTADT—Integral Equations
JOST—Two-Dimensional Geometric Variational Procedures
KHAMSI and KIRK—An Introduction to Metric Spaces and Fixed Point Theory
*KOBAYASHI and NOMIZU—Foundations of Differential Geometry, Volume I
*KOBAYASHI and NOMIZU—Foundations of Differential Geometry, Volume II
KOSHY—Fibonacci and Lucas Numbers with Applications
LAX—Functional Analysis
LAX—Linear Algebra
LOGAN—An Introduction to Nonlinear Partial Differential Equations
MARKLEY—Principles of Differential Equations
MORRISON—Functional Analysis: An Introduction to Banach Space Theory
NAYFEH—Perturbation Methods
NAYFEH and MOOK—Nonlinear Oscillations
PANDEY—The Hilbert Transform of Schwartz Distributions and Applications
PETKOV—Geometry of Reflecting Rays and Inverse Spectral Problems
*PRENTER—Splines and Variational Methods
RAO—Measure Theory and Integration
RASSIAS and SIMSA—Finite Sums Decompositions in Mathematical Analysis
RENELT—Elliptic Systems and Quasiconformal Mappings
RIVLIN—Chebyshev Polynomials: From Approximation Theory to Algebra and Number Theory, Second Edition
ROCKAFELLAR—Network Flows and Monotropic Optimization
ROITMAN—Introduction to Modern Set Theory
*RUDIN—Fourier Analysis on Groups
SENDOV—The Averaged Moduli of Smoothness: Applications in Numerical Methods and Approximations
SENDOV and POPOV—The Averaged Moduli of Smoothness
*SIEGEL—Topics in Complex Function Theory
    Volume 1—Elliptic Functions and Uniformization Theory
    Volume 2—Automorphic Functions and Abelian Integrals
    Volume 3—Abelian Functions and Modular Functions of Several Variables
SMITH and ROMANOWSKA—Post-Modern Algebra
STADE—Fourier Analysis
STAKGOLD—Green's Functions and Boundary Value Problems, Second Editon
STAHL—Introduction to Topology and Geometry
STANOYEVITCH—Introduction to Numerical Ordinary and Partial Differential Equations Using MATLAB®

*Now available in a lower priced paperback edition in the Wiley Classics Library.
†Now available in paperback.

*STOKER—Differential Geometry
*STOKER—Nonlinear Vibrations in Mechanical and Electrical Systems
*STOKER—Water Waves: The Mathematical Theory with Applications
 WATKINS—Fundamentals of Matrix Computations, Second Edition
 WESSELING—An Introduction to Multigrid Methods
†WHITHAM—Linear and Nonlinear Waves
†ZAUDERER—Partial Differential Equations of Applied Mathematics, Second Edition

*Now available in a lower priced paperback edition in the Wiley Classics Library.
†Now available in paperback.